Developing Technologies in E-Services, Self-Services, and Mobile Communication:

New Concepts

Ada Scupola
Roskilde University, Denmark

Information Science
REFERENCE

Senior Editorial Director:	Kristin Klinger
Director of Book Publications:	Julia Mosemann
Editorial Director:	Lindsay Johnston
Acquisitions Editor:	Erika Carter
Development Editor:	Hannah Abelbeck
Production Editor:	Sean Woznicki
Typesetters:	Mike Brehm, Jennifer Romanchak, Natalie Pronio, Milan Vracarich, Jr., Deanna Zombro
Print Coordinator:	Jamie Snavely
Cover Design:	Nick Newcomer

Published in the United States of America by
Information Science Reference (an imprint of IGI Global)
701 E. Chocolate Avenue
Hershey PA 17033
Tel: 717-533-8845
Fax: 717-533-8661
E-mail: cust@igi-global.com
Web site: http://www.igi-global.com

Library of Congress Cataloging-in-Publication Data

Developing Technologies in E-Services, Self-Services and Mobile Communication: New Concepts / Ada Scupola, editor.
p. cm.
Includes bibliographical references and index.
Summary: "This book presents perspectives on the management, delivery and consumption, and creation of new e-services technologies and tools, reviewing specific e-services such as e-banking and e-health initiatives and offering new perspectives useful to industry professionals and academicians interested in the creation, application, and evaluation of a number of different mobile and electronic services"--Provided by publisher.
ISBN 978-1-60960-607-7 (hardcover) -- ISBN 978-1-60960-608-4 (ebook) 1. Electronic commerce. I. Scupola, Ada.
HF5548.32 .D484
658.8'72--dc23

2011016797

British Cataloguing in Publication Data
A Cataloguing in Publication record for this book is available from the British Library.

Table of Contents

Section 2
E-Services: Adoption, Perceptions and User Values

Section 3
Mobile Services: Acceptance and Consumer Values

Detailed Table of Contents

Organizations increasingly cooperate in organizational networks to remain competitive. Electronic intermediaries can provide all kinds of e-services to support the creation and management of such networks. While there has been substantial discussion on intermediaries matching demand and supply, there has been little analysis in relation to the management and orchestration of organizational networks. In this chapter, we analyze an intermediary that uses e-services for orchestrating a network in the consumer electronics industry. The empirical results show that the coordination and management of networks requires specific expertise and skills which could be provided by specialized intermediaries. The primary value creation activity of the intermediary is leveraging the products, activities, and knowledge of the specialized companies and providing e-services for orchestrating the organizational network to create short lead times, improve customer responsiveness, and ensure adaptability.

The concept of e-services has gained increasing use during the past decade. There is, however, no general agreement as to the precise meaning and scope of the term. The research purpose of the present chapter is, therefore, to discuss the e-service concept, its strengths and scope, and thereby contribute to the general understanding and definition of the term. More specifically, the chapter aims at examining one of the primary conditions for the development of e-services, namely the codification of knowledge in connection with knowledge intensive services. The e-service concept is used in relation to private as well as public e-services. The present chapter examines the e-service concept at a generic level and does not differentiate between private and public e-services.

This chapter explores the key challenges faced by IT service providers offering after-sales support services to enterprise customers via the Web. The chapter is based on insights provided by representatives of six multi-national IT service organizations. It considers how IT service providers use knowledge-based approaches to offer such services. The chapter highlights the trend towards the employment of networks of business partners to assist with the processes of support delivery. By exploring the perceived needs of the different stakeholders involved in IT support provision, and the transfer of knowledge across these organizations and individuals, the key challenges in IT support provision are identified. The chapter proposes a range of solutions to address the key challenges.

This research investigated the website use of Australian accounting practices in metropolitan Melbourne, Australia in the period 2006-2010. This occurred through an investigation of the websites of 100 practices and interviews involving selected practices at the start of the period and a further investigation of the websites at the end of the period. The analysis revealed a heavy emphasis towards website features that identify the businesses and the services that they offer, which is consistent with other website studies involving professional services. More than other businesses, accounting practices rely on keeping their clients up to date and thus the provision of information such as taxation news updates, changes to superannuation and regular newsletters to clients through their websites was identified as being an important tool for many practices. None of the accounting practices offered online order or payment features over the period. Again this is consistent with other studies of similar professional services businesses. Although some of the practices had log-in features in 2006, only two contained features that could perhaps be classed as 'e-services'. However, the 2010 review found an increase in the number of accounting practices with a log-in feature on their website. This suggests that the issues of trust in the technology are being addressed and that they could be moving towards a situation where e-services may become more commonplace in the future.

The introduction of self-service applications by online service providers provided an effective solution to the lack of direct personal contact over the Internet, and offered a quick way to expand business op-

and the effects of the factors that precede consumers' acceptance of mobile advertisement. The results suggest that attractive content and trust in advertisers are key predictors of mobile device users' behavioral intention towards mobile advertising. The findings lead to a modification and extension of Fishbein's behavioral intention model and its application in the research on advertising effectiveness.

M-banking has been highlighted as one of the most promising e-commerce services. However, the adoption of financial mobile services is still far behind market expectations. This study seeks to provide academics and bank managers alike with a better understanding of resistance to m-banking. Based on Ram and Sheth's (1989) Theory of Innovation Resistance, a SEM model was built to describe resistance barriers. A total number of 3,852 observations, of which 2,344 were effective for SEM analysis, were obtained from an Internet survey at a Portuguese bank. The "functional barriers" revealed more severe barriers than the "psychological" ones. Latent scores were used to compare consumers' perceptions and behaviour. Results indicate a significantly higher resistance among non-users, and demographic and behavioural profiles were established, promoting wider knowledge and possible enhancement of m-banking adoption. Conclusions and managerial implications are provided.

Context aware services have the ability to utilize information about the user's context and adapt services to a user's current situation and needs. In this article the authors consider users' perceptions of the added value of location awareness and presence information in mobile services. The authors use an experimental design, where stimuli comprising specific bundles of mobile services were presented to groups of respondents. The stimuli showed increasing, manipulated, levels of context-awareness, including location of the user and location and availability of buddies as distinct levels. Their results indicate that simply adding context aware features to mobile services does not necessarily provide added value to users, rather the contrary. The potential added value of insight in buddies' location and availability is offset by people's reluctance to share location information with others. Although the average perceived value overall is rather low there exists a substantial minority that does appreciate the added context aware features. High scores on constructs like product involvement, social influence and self-expressiveness characterize this group. The results also show that context aware service bundles with utilitarian elements have a higher perceived value than bundles with hedonic elements. On the basis of the different results some guidelines for designing context aware mobile services are formulated.

tion. We posit that telehealth assimilation can only be understood by taking a multilevel approach to the phenomenon, whereby assimilation starts at the individual level, permeates through other organizational levels such as groups, and finally ends at the organizational and inter-organizational level. Further, assimilation of technological innovations must be considered within their institutional context. Derived from our conceptual model, we make several propositions and hope that our work will significantly guide future research and managerial actions geared toward integrating healthcare in the workplace.

As one of the top issues for executives nowadays the recruiting, retaining, and skilling of scarce talent has received a lot of attention both in research and practice. While companies are in dire need of new strategies and integrated approaches in human resources, research has predominately observed the skill profile and general attitude of professionals to their work, their individual incentives, and consequentially, their turnover intention. We aim to relieve these needs in practice by introducing a first step towards a holistic system architecture of recruiting, retention, and development processes. For this purpose we use a design science approach to develop a talent management system architecture containing all respective processes and their related subsystems based on the next-generation holistic e-recruiting system invented by Lee (2007). This architecture could increase knowledge transfer and thereby improve the adjustment and performance of the entire HR process. Furthermore the increased amount of data gained through new implemented performance measurement subsystems offers management a variety of new strategic options for continuing to battle in a global "War for Talent."

Today's executives are challenged to deliver value to their shareholders in a global marketplace. To compete, organizations are setting business strategies for the entire organization. This management change presents a significant change for Information Technology (IT) organizations that have historically dealt with individual business units or functional domains. Now IT must support the enterprise's business strategy and the entirety of the enterprise. Technology organizations must deliver interoperation of processes, people, and technology to the entire enterprise. To design, deliver, and manage IT services to meet an agreed level of quality, organizations are implementing IT Service Management (ITSM) and creating quality standards, which use the best practices of IT Infrastructure Library (ITIL). ITIL tells IT management what needs to be done and how it will get done from the process perspective. When undertaking an ITSM a project to implement ITIL, ITIL does not provide a method for measuring quality or identifying and completing process improvement projects. By integrating Six Sigma, a quality methodology, IT management will have the methodology and tools for measuring quality and improving processes. Adopting Six Sigma principles also will help IT focus on their customers and the business strategy, manage proactively based on facts, and reinforce collaboration across the enterprise. The framework in our exploratory research has been built upon a deductive study which has been developed through a literature review and synthesis and an exploratory inductive research which has been developed using a qualitative case study. It makes the case for leveraging ITIL and Six Sigma with ITSM in practice and opportunities for future research.

Introductory Chapter

UNDERSTANDING E-SERVICES, SELF SERVICES AND MOBILE COMMUNICATIONS: PAST RESEARCH AND FUTURE DIRECTIONS

As the introductory chapter to the first book of "*Advances in E-Services and Mobile Applications series*", this chapter provides a short overview of the e-service concept and discusses similarities and differences of the e-service concept in relation to other well established concepts such as services, goods, and self services. In addition the chapter provides a broad overview of the type of research conducted within the field of mobile applications and communications. Finally the chapter concludes with some directions for future research in these fields.

Introduction

With the advent of information and communication technologies, many of the traditional face to face services have been transformed first into self services, then with the advent of the World Wide Web into e-services and more recently into m-services with the spreading of mobile networks. For example the payment for fuel at a gas station was earlier done by going to the cashier and paying there before or after the fuel was filled. Nowadays this has become a self service where the customer pays by using the credit card at a machine and often fills the fuel by himself/herself. Another classic example of self services is ATM machines. More recently other types of services such as filling of tax forms, searching for products information, comparing product types that also requested a face to face interaction between the service provider and the service customer have been automatized with the advent of the Worldwide Web. This type of e-services has gone even further by sometimes involving communities of users or customers to develop online services themselves as for example in the cases of online marketplaces or wikis or to even contribute to further development of goods or services produced by companies as for example in the case of ski equipment, the design of new products as for example in the case of t-shirts or simply contributing to service development as in the case of library services (e.g. Nicolajsen et al. 2009). As the editor of the book series entitled "*Advances in E-Services and Mobile Applications series,*" I am planning in this introductory chapter to the first book of the series to mainly give an overview of the e-service and mobile application research with some recommendations for future research.

As pointed out in the literature (e.g. Scupola et al., 2009; Stafford, 2003) there is still much disagreement as to what e-services are. There have been a number of empirical studies focussing on e-services spanning a range of fields and applications varying from government to business e-services or consumer-to-consumer e-services. This is also reflected in the range of articles' subjects that have been submitted and accepted for publication in the *Journal of E-Services and Mobile Applications* (IJESMA), for which I am editor in chief. Over the last couple of years in fact the subjects of the accepted articles for IJESMA have been varying from e-health, to online auctions as e-services, to e-recruiting systems, Web-based IT support services, e-banking, as well as e-services used by accounting companies. Also by looking at the geographical regions where the articles submitted or accepted are coming from, it can be concluded that e-services is a phenomenon or concept that is increasingly gaining attention and investigation in all the continents and countries, including USA, Europe, Canada, Australia, Japan, Russia, or Jamaica to name a few. In this introductory chapter to the series, I will summarize what it is intended in the literature with e-services and how this concept differs from services, goods, and self services concepts. In addition I will touch upon the types of research conducted within mobile applications and will show some trends within these fields both from a theoretical and empirical perspective.

The chapter is structured as follows. In this introduction the background, the purpose, and the motivation of the chapter is given. The following section presents a short overview and definitions of e-services and the e-service concept. The next section discusses the self service concept and its similarities and differences with the e-service concept, and it is followed by a section presenting major trends within the mobile applications and communication field. Finally, the last section will provide some concluding remarks and suggestions for future research.

Understanding E-Services

E-Services Definitions

As already pointed out by many authors (e.g. Singh, 2002; de Ruyter et al., 2000) there is no clear and unique definition of e-services, and the definition often depends on the context in which e-services are studied, such as, for example, in the context of e-government field or business to consumer field. Based on an extensive literature review, Scupola et al. (2009) have categorized the definitions of e-services into two main groups as shown in Table 1: the first group focuses mainly on the delivery and infrastructure of e-services, while the other group focuses on the production, delivery, and outcome of e-services. For example, within the delivery and infrastructure view, e-services are seen as the "provision of services over electronic networks" (Rust, 2001; Rust and Kannan, 2003). Within the production, delivery, and outcome view instead, e-services are defined as "services that are produced, provided, and/or consumed through the use of ICT-networks such as Internet-based systems and mobile solutions" (Scupola, 2008).

In addition, Scupola et al. (2009) have distinguished 4 main categories of e-services summarized in Table 2: business to business (B-to-B), business to consumer (B-to-C), government to business and to consumer (G-to-B, G-to-C), and Customer to Customer (C-to-C). As Table 2 shows, B-to-B e-services are usually used in building collaboration and relationships between companies, while business to consumer e-services are mainly used to sell to the customer or to retain the customer through e-enabled customer support, provision of information about the products/services, implement user driven innovation, et cetera. Government-to-business and government-to-consumer are mainly targeted to reduce the digital divide among city and rural areas, empowerment of the citizens in their use of governmental services, or e-democracy. Finally the fourth category, consumer-to consumer e-services (C-to-C), are developed by citizens and are directed to citizens and include for example online communities, wikis such as Wikipidia (http://en.wikipedia.org/), or simply online markets as for example online markets for second hand objects and things.

Table 1. A summary of e-services definitions

Delivery and infrastructure view	Production, delivery and outcome view
Those services that can be delivered electronically (Javalgi, Martin and Todd, 2004)	Any asset that is made available via the Internet to drive revenue streams or create new efficiencies (Piccinelli and Stammers, 2001)
Provision of services over electronic networks (Rust, 2001; Rust and Kannan, 2003)	An act or performance that creates and provides benefits for customers through a process that is stored as an algorithm and typically implemented by networked software (Hofacker et al., 2007)
Interactive services that are delivered on the Internet using advanced telecommunications, information, and multimedia technologies (Boyer, Hallowell and Roth, 2002)	E-services as Internet-based applications that fulfill service needs by seamlessly bringing together distributed, specialized resources to enable complex (often real-time) transactions (Tiwana and Balasubramaniam, 2001)
E-service is deeds, efforts or performance whose delivery is mediated by information technology (including Web, information kiosks and mobile devices). Such e-service includes the service element of e-tailing, customer support and service, and service delivery (Rowley, 2006)	E-services are defined as services that are produced, provided and/or consumed through the use of ICT-networks such as Internet-based systems and mobile solutions (Scupola, 2008)

E-Services, Services and Goods

As already said, a number of studies have tried to define and conceptualize the e-service concept (e.g. *Singh*, 2002; de Ruyter et al., 2000; Hofacker et al., 2007) and tried to position it in relation to established concepts such as goods, services, and self services. For example, Scupola et al. (2009) conducted a thorough literature review of e-services in an attempt to identify the strengths and the limitations of the e-service concept in relation to other concepts such as services and goods. They concluded that the e-service concept has features, relationships, and developments that differentiate it from and add something to other and already well-established concepts such as services or goods. Scupola et al. (2009) argue that the e-service concept in particular adds value when it is seen at the convergence of goods and services. They observe that services, which normally depend on face-to-face encounters, can increasingly be digitized and delivered on communication networks whereby the connection between the immediate service producer and the service provided is changed, turning the service into a good-like product. Conversely, some goods such as software or printing, which normally were sold to the customers as goods (for example in the form of a CD ROM) are increasingly delivered as service to customers for example as print on demand in the case of printing services.

Scupola et al. (2009) summarize the major characteristics of goods, e-services, and services as shown in Table 3 below. For example, goods are tangible and can be inventoried and patented. Their use usu-

Table 2. E-Services Categories (Scupola et al., 2009)

Types of e-services	B-to-B	B-to-C	G-to-B and G-to-C	C-to-C
Characteristics/focus	Collaboration and relationship building	Selling to and retaining the customer	User/citizen empowerment, e-democracy, city/rural areas divide	Peer-to-peer value creation
Examples	Supply chain management in outsourced printing services and facilities, SaaS	E-retailing, e-customer relationship management, e-banking, e-newspapers, Web portals	Online tax returns, e-voting, e-libraries, telemedicine, remote social security services	Online auctions, consumer driven e-marketplaces, online gaming, online communities (newsgroups). wikis

Table 3. Distinguishing features of goods, e-services and services

Goods	E-services	Services
Tangible	Intangible, but need tangible media	Intangible
Can be inventoried	Can be inventoried	Cannot be inventoried
Separable consumption	Separable consumption	Inseparable consumption
Can be patented	Can be copyrighted, patented	Cannot be patented
Homogeneous	Homogeneous	Heterogeneous
Easy to price	Hard to price	Hard to price
Can't be copied	Can be copied	Can't be copied
Can be shared	Can be shared	Can't be shared
Use equals consumption	Use does not equal consumption	Use equals consumption
Based on atoms	Based on bits	Based on atoms

Source: Scupola et al. (2009): adopted from Charles F. Hofacker et al. (2007): E-services: A synthesis and research agenda, Journal of Value Chain Management 1(1/2).

ally implies consumption and their consumption can be shared among different consumers. In addition they are homogenous and are easy to price. On the contrary, services are intangible and it is impossible to inventory them, to copy or patent them; they are heterogeneous and hard to price. E-services on the other hand are intangible, but need tangible media such as a computer, can be inventoried and patented, and they can be shared and copied. Their use does not necessarily imply consumption.

Therefore, as can be seen in Table 3, e-services share some of the characteristics of services, such as intangibility, or difficulty of pricing, while simultaneously sharing some of the characteristics of goods, such as the possibility to be inventoried and the possibility to be copyrighted. However, e-services have also some characteristics that are unique and not common either to goods or services: e-services are based on bits, and the use of an e-service does not equal consumption of the e-service.

Furthermore, Scupola et al. (2009) pointed out that when the e-service concept is defined as the convergence of some types of services as for example knowledge services (e.g. customer service) and goods, codification of the production or the content of the service becomes essential. Such codification can be obtained either by codifying the service content or by developing ICT-based tools through which knowledge, or at least some type of knowledge, can be stored and thus provided to the customer.

However, Scupola et al. (2009) also raise the question to what degree the knowledge embedded in the service provision can be codified, entered on ICT, and sold on markets as informational goods. Based on the theories of knowledge creation, storage, and interpretation (e.g. Polanyi, 1958; Nonaka, 1994; Davenport and Pruzak, 1998) they conclude that even for services with a great amount of codified knowledge, it is not certain that this service can be fully sold digitally or as a stand-alone e-service. They state that all the knowledge that can be codified and put on a piece of paper can be entered on digital media. However, this does not mean that it can be sold as an all-digitized service or e-service. For example legal services or real estate services are based on laws and other information that can be easily put in digital forms, but need the right competences for interpretation. For example in the case of real estate, many legal documents and other kinds of information related to the buying and selling of real estate can be codified and digitized. However, real estate agents still exist as intermediaries of these processes, mainly due to the fact that a lot of specialized knowledge is required in handling real estate, knowledge that not all customers feel confident to possess. The implication is that the production and marketing of knowledge services is limited to contexts in which there is a match of interpretation. Therefore many core knowledge-intensive services might not easily lend themselves to become e-services (Scupola et al., 2009). This applies, for example, to services that are not easily or fully codifiable because the knowledge is too complex, too situational, or dependent on experiences or trust.

Conceptualization of Self-Services vs. E-Services

Having conceptualized e-services in relation to goods and services, the question remains: what are self services and how do they differentiate themselves from our conceptualization of e-services? The concept "self service" has, similarly to the concept of e-services, a lot of connotations and definitions and a lot of confusion exists in the literature regarding the meaning of the concept. Like e-services concept, the self service concept primarily has its roots in the service marketing literature. For example Meuter et al. (2000) states that the strategic importance of self service technologies (SSTs) is due to the radical changes in service delivery that it has brought to many industries such as airline, banking, travel, hotel, financial, and retailing. Meuter et al. (2000) define self service technologies as technological interfaces that allow customers to produce services independent of direct service employee involvement. The use of self service technologies (SSTs) implies primarily the elimination (or at least a significant reduction in involvement) of a service representative and the empowerment of the customer in the service delivery. As for e-services, self service technologies (SSTs) are a cost-effective way of providing a greater level of customer service for corporations. In fact, convenience, ubiquitous availability, and time and money savings, are some of the benefits provided by the use of self services for consumers. Cunningham et al. (2009) state that Parasuraman regards this radical progression from traditional interpersonal service encounters to self service technologies, replacing human-to-human contact with human-machine interaction, as a fundamental shift in the nature of services. In their study comparing consumer views of traditional services and self-service technologies, Cunningham et al. (2009) concentrate on the following services that can be provided with the use of self service technologies: online banking, distance education, airline reservations, tax software, retail self-scanning, online auctions, pay at the pump, ATMs, online brokerage, interactive phone, Internet search, and online car buying. According to e-services literature, many of the above are services that can be transformed into e-services or mobile services by using internet and mobile communications.

In addition, Fitzsimmons (2003) in a short article published under "The guru's view" of the journal *Managing Service Quality* states that services have migrated from human interaction to substitution of machines for service employees or, where feasible, to anywhere-anytime electronic service. According to Fitzsimmons (2003),

This trajectory is similar to the past experience in the agricultural and manufacturing sectors of the economy where human labor has been driven out of the production process relentlessly ... The initial targets of SST were service transactions that did not add value or have revenue-enhancing opportunity because the substitution of technology for employee labor achieved cost savings. For example, the introduction of ATMs by banks a quarter of a century ago saved teller costs but also provided customers with place-and-time convenience. However, services that can be digitized and delivered via the Internet such as entertainment, information, and training represent new opportunities for SST (Fitzsimmons, 2003, p. 444).

He then summarizes what he calls the evolution of self- services in some key service industry, demonstrating that there is an evolution from services to machine assisted services such as vending machine, check in kiosks, et cetera to electronic services (e-services) according to the definition and understanding of this chapter, which is online banking, online shopping, or online ordering.

In addition, Fitzsimmons (2003) states that high touch services such as healthcare, fire fighting, dentistry, et cetera should be not so impacted by self services, even though this might not be entirely true for healthcare. In fact, nowadays e-health is becoming a more and more important field for e-services as it is showed by the flourishing literature on e-health and telemedicine within the e-services literature (e.g. Gogia, 2008).

So one may wonder: what is the difference between e-services and self services? Are they the same thing or are there differences between them? The self service technology concept has been initially introduced when the advancement in technology, and especially Information Technology replaced conventional face-to-face services with innovative self-service technology (SST) such as ATM. Later, as the technology has evolved, in the marketing literature, the term has been extended to include into self service technologies, and also technologies such as Internet banking (e.g. Hensmans *et al.,* 2001). In reality, as I see it, Internet banking has followed self service technologies such as telephone banking and ATMs. From our brief discussion here about self services and what has been discussed in the previous session about e-services, it can be concluded that both e-services and self services try to substitute human interaction with machines or computers, but self services do not necessarily need to be Web based as e-services (as in the case of an ATM machine or kiosk ticketing.

Therefore, self services can be seen as the first era of substitution of humans with machines in service provisions, while e-services and mobile services are the most advanced arena therein, offering something more than self services (interactivity between the service provider and the customer) as e-services can use chats. This interactivity aspect of e-services has been also emphasized by Rust and Lemon (2001), who stated that in the service provisions, three main changes have been possible due to the Internet: true interactivity with the consumer, customer specific, situational personalization, and the opportunity for real-time adjustments to a firm's offering to customers. Self services instead have the main aim of substituting humans with machines and give little possibility of transferring knowledge. This is possible with e-services as in the case of news as e-services (Ihlstrom et al., 2008) or software transmitted online. It can therefore be concluded that e-services can be conceptualized as a natural, more complex form and evolution of self services technologies. However, establishing the right relationship between self services and e-services might require an in-depth and thorough literature review of the e-service and self service research. Therefore this might be subject for future research as the purpose of this chapter is mainly to focus on e-services and mobile applications and shortly describe its relation to goods, services, and self services.

Mobile Applications and Communication

Lately, a new generation of e-services is taking place due to the wireless revolution and the explosion of mobile commerce, implying that mobile devices, including mobile phones, are becoming a critical component of the new digital economy. The transactions over electronic networks are therefore transformed from fixed locations to anytime, anywhere, and anyone transactions. New forms of mobile technologies are rapidly transforming the marketplace and the e-services landscape as mobile applications and wireless networks are enabling and transforming many of the standard e-services into m-services.

Such applications include mobile financial services, user and location specific mobile advertising, and mobile inventory management. Ngai and Gunasekaran (2007) conduct a thorough literature review of mobile commerce research and applications published from 2000 through 2003. On the base of research conducted previously by Varshney and Vetter (2002) and Mennecke and Strader (2003) they develop a classification framework to classify the articles dealing with m-commerce research. Their classification framework categorizes m-commerce research into five levels that they describe as follows:

1. M-commerce theory and research. This is the lowest level of the framework. The articles included here deal with issues or subjects such as the development of m-commerce applications and guidelines, behavioural issues such as consumer behaviour, the acceptance of technology, and the diffusion of m-commerce applications and services, as well as m-commerce economics, strategy, and business models. Legal and ethical issues such as privacy, regulations, and the legal environment when using m-commerce are also included in this level of the framework by Ngai and Gunasekaran (2007). Articles dealing with a general introduction to m-commerce, foundational concepts of m-commerce, and so forth were also included into this level.

The model then presents 3 pillars that include research articles addressing more technical characteristics of the wireless infrastructure that enables mobile commerce, and subsequently, m-services. These 3 pillars are: wireless network infrastructure, mobile middleware, and wireless user infrastructure, and they are described below, mainly based on Ngai and Gunasekaran (2007).

2. Wireless network infrastructure is the first of the three pillar technologies of m-commerce that supports the development of m-commerce applications, including m-services. According to Ngai and Gunasekaran (2007),

Wireless network infrastructure plays an important role in m-commerce as this is the core part of m-commerce technology. It provides wireless networks and network standards such as the Global System for Mobile Communication (GSM), Bluetooth, the wireless local area network (WLAN), radio frequency identification (RFID), the Third- generation (3G) network, et cetera. In order to ensure the reliability and efficiency of the m-commerce applications and services running in a mobile environment, it is necessary for various networking requirements to be implemented in the wireless and mobile networks (p. 6).

Articles on the wireless infrastructure requirements of m-commerce, such as location management, multicast support, network dependability, quality-of-service, and roaming across multiple networks are also included under this pillar.

3. Mobile middleware refers to the software layer between the wireless networks and the operating systems of the mobile devices to connect the m-commerce applications. Based on previous literature, Ngai and Gunasekaran (2007) identified four research issues for mobile middleware. The first issue is called "agent technologies" and includes publications dealing with the use of software agents or mobile agents to support m-commerce activities as for example searching for products. The second is called "database management" and includes articles dealing with mobile database management. The third deals with "security issues" and includes articles that discuss the security issues in m-commerce. An example is the design of a secure wireless network infrastructure for

m-commerce applications using public key infrastructure or other techniques. The fourth is "wireless and mobile communication systems" that deal with techniques and methods to connect and manage m-commerce applications.

4. Wireless user infrastructure consists of two parts, i.e., software and hardware. Ngai and Gunasekaran (2007) state that software refers to the operating systems and their interfaces, while hardware deals with the mobile devices needed to communicate with the m-commerce applications, such as PDAs and mobile phones. An important research sub-theme identified in this category deals with mobile interfaces since these are very important in a mobile environment due to the fact that mobile applications normally run on small and portable mobile hand-held devices.
5. Mobile commerce applications and cases include literature that focuses on specific applications or cases. In categorizing such a literature Ngai and Gunasekaran (2007) used classes of mobile applications that had been earlier developed by Varshney and Vetter (2002). They include mobile financial applications, mobile advertising, mobile inventory management, locating and shopping for products, proactive service management, wireless re-engineering, mobile auctions or reverse auctions, mobile entertainment services and games, mobile offices, mobile distance education, and wireless data centres.

The *International Journal of E-Services and Mobile Applications* (IJESMA), for which I am editor in chief, has published and is in the process of publishing several articles and special issues addressing mainly category 1 and 2 of the research framework proposed by Ngai and Gunasekaran (2007): M-commerce theory and research and Mobile commerce applications and cases. For example, Kushchu (2010) edited a special issue on mobile government, and in his editorial, he states that a mobile life "emerges from communications via mobile technologies in our personal, family, and business (public or private) interactions, and influences, in a positive or negative way, all aspects of life on earth" (p. i). In addition he mentions a number of positive and negative influences that mobility may have on society as a whole. For example based on a study that the Mobile Government Consortium International (mGCI – www.mgovernment.org) has made about the contributions of mobile phones to our lives, Kushchu (2010) states that the positive contributions can be seen at least at four different levels:

1. Contributions to personal and primary relations
2. Contributions to the society in general,
3. Contributions to public and private sector organisation, and
4. Contributions to the economies (especially developing economies).

Among the negative influence of mobile phone use the Mobile Government Consortium International Study found that:

1. Members of the society do not have a strict love and hate relationship with mobile phones. Rather, there are ranges of negative feelings and the negative feelings may change from one circumstance to another and from one person to another.
2. Members of the society feel negative and unpleased under various situations, but the following top three circumstances of mobile use seem to be critical:
    - When mobile use may lead to a danger.
    - When mobile phones are intrusive or imposing on others.

- When mobile phones start to change our lives in the way we do not wish to and when we become so dependent on them.

3. It has been also observed that being in a particular age group does not necessarily lead to a dislike of mobile phones but older members of society seems to be more sensitive to bad manners when mobile phones are being used.

Another special issue of the *International Journal of E-Services and Mobile Applications* (IJESMA) focused on mobile services. In this special issue, Salo (2009) collects four papers focusing on different aspects of mobile services and on different types of mobile service applications. Westerlund et al. (2009) focus on the mobile advertising field and suggest that attractive content and trust in advertisers are key predictors of mobile deviser users′ behavioural intention towards mobile advertising. Cruz et al. (2009) explore resistance to mobile banking in Portugal with extensive empirical data. Their results indicate that functional barriers are more severe than traditional psychological barriers. Their research theoretically extends the multidisciplinary approach to explaining resistance of technology in different contexts, while practically, the study emphasizes the importance of decreasing functional barriers relating to mobile banking as services provided should be easily usable and provide high value-for-money. Finally, the studies by de Vos et al. (2009) and Almasri et al. (2009) focus both on a specific application: Location Based Mobile Services. De Vos et al. (2009) examine consumer perceived value of mobile services that are context aware and location sensitive. The study shows that mobile services that are context aware, do not provide added value to customers, and that utilitarian elements have higher perceived value than hedonic elements. Almasri et al. (2009)'s study investigates micro-scale navigation issues relating to GPS systems by especially focussing on pedestrian navigation. The study provides insights about how to improve both wireless networks and devices to improve services provided.

Conclusion and Recommendations for Further Research

As the introductory chapter to the first book of "*Advances in E-Services and Mobile Applications series,*" this chapter provides a short overview of the e-service concept and discusses similarities and differences of the e-service concept in relation to other well established concepts such as services, goods, and self services. In addition, the chapter provides an overview of the research focusing on mobile applications and highlights how important research in this field as well as in the e-services field is published in the International Journal of E-Services and Mobile Applications. This chapter deals with all these concepts in a theoretical way by summarizing and analyzing important literature in these fields. However, the issues dealt with in this chapter are only the most general issues and are provided and discussed here to understand at a general level the most recent trends and the concepts within e-services and mobile communication field.

There are however a number of issues that are important within this field and that can be subject for further research. Two aspects worth mentioning here are: e-service quality and technologies for e-services and self services. In the service field, there has been a lot of literature addressing service quality and trying to find dimensions to measure the quality of a service. One of the most used and valid instruments that has been developed to measure the service quality has been the SERVQUAL instrument developed by Parasuraman et al. (2008). Similarly, in the last few years some studies have also tried to understand and study e-services quality. According to *Tsikriktsis* (2002) three major studies within the service marketing field have tried to establish some dimensions of e-service quality: the study conducted by Zeithaml et al. (2000), the one conducted by Voss (2000), and the one conducted by Loiacono et al. (2000). It could be interesting to conduct an in depth literature review of the e-service quality research in order to understand what we mean by e-service quality and the dimensions that can be used to implement it. Another important field for future research is the investigation of the technologies for e-services, self services, and mobile communication. Some studies exist about this, as pointed out by Singh (2002) and Ngai and Gunasekaran (2007), however it could be interesting to conduct an updated literature review. In addition more cases are needed to understand and explain these phenomena. Finally, further research could also focus on how e-services, self services, and mobile communication can be used to drive innovation and company strategy as already done by some companies, such as Starbucks, that make uses of social media in order to involve users in the innovation process and to make company's and innovation strategies.

Ada Scupola
Roskilde University, Denmark

REFERENCES

Almasri, S., Alnabhan, M., Hunaiti, Z., & Sedoyeka, E. (2009). Location-based services (LBS) in micro-scale navigation: Shortcomings and recommendations. *International Journal of E-Services and Mobile Applications, 1*(4). doi:10.4018/jesma.2009070804

Boyer, K., Hallowell, R., & Roth, A. V. (2002). E-services: Operating strategy—A case study and a method for analyzing operational benefits. *Journal of Operations Management, 20*(2). doi:10.1016/S0272-6963(01)00093-6

Cruz, P., Laukkanen, T., & Muñoz, P. (2009). Exploring the factors behind resistance to mobile banking in Portugal. *International Journal of E-Services and Mobile Applications, 1*(4). doi:10.4018/jesma.2009070802

Cunningham, L. F., Young, C. E., & Gerlach, J. (2009). A comparison of consumer views of traditional services and self-service technologies. *Journal of Services Marketing, 23*(1), 11. doi:10.1108/08876040910933057

de Ruyter, K., Wetzels, M., & Kleijnen, M. (2000). Customer adoption of e-service: An experimental study. *International Journal of Service Industry Management, 12*(2), 184–207. doi:10.1108/09564230110387542

de Vos, H., Haaker, T., Teerling, M., & Kleijnen, M. (2009). Consumer value of context aware and location based mobile services. *International Journal of E-Services and Mobile Applications, 1*(4). doi:10.4018/jesma.2009070803

Fitzsimmons, J. A. (2003). Is self-service the future of services? *Managing Service Quality, 13*(6), 443–445. doi:10.1108/09604520310506496

Gogia, S. B. (2008). Providing tele-mental health services after disasters – A case based on the post tsunami experience . In Scupola, A. (Ed.), *Cases on managing e-services*. Hershey, PA: Idea Group Inc. doi:10.4018/9781605660646.ch016

Hensmans, M., Van den Bosch, F. A. J., & Volberda, H. W. (2001). Clicks vs bricks in the emerging online financial services industry. *Long Range Planning, 34*, 231–247. doi:10.1016/S0024-6301(01)00030-9

Hofacker, (2007). E-services: A synthesis and research agenda. *Journal of Value Chain Management, 1*(1/2).

Ihlström Eriksson, C., Kalling, T., Åkesson, M., & Fredberg, T. (2008). Business models for m-service–Exploring the e-newspaper case from a consumer view. *Journal of Electronic Commerce in Organizations, 6*(2).

Javalgi, R. G., Martin, C. L., & Todd, P. R. (2004). The export of e-services in the age of technology transformation: Challenges and implications for international service providers. *Journal of Services Marketing, 18*(7), 560–573. doi:10.1108/08876040410561884

Kushchu, I. (2010). Editorial for the theme issue of IJESMA on mobile government. *International Journal of E-Services and Mobile Applications, 2*(2), i.

Loiacono, E. T., Watson, R. T., & Goodhue, D. L. (2000). *WebQual: A website quality instrument.* (Working Paper 2000-126-0), University of Georgia.

Mennecke, B. E., & Strader, T. J. (2003). *Mobile commerce: Technology, theory, and applications.* Hershey, PA: Idea Group Publishing.

Meuter, M. L., Ostrom, A. L., Roundtree, R. I., & Bitner, M. J. (2000). Self-service technologies: Understanding customer satisfaction with technology-based service encounters. *Journal of Marketing, 64*, 50–64. doi:10.1509/jmkg.64.3.50.18024

Ngai, E. W. T., & Gunasekaran, A. (2007). A review for mobile commerce research and applications. *Decision Support Systems, 43*, 3–15. doi:10.1016/j.dss.2005.05.003

Nonaka, I. (1994). A dynamic theory of organizational knowledge creation. *Organization Science, 5*(1), 14–37. doi:10.1287/orsc.5.1.14

Parasuraman, A., Zeithaml, V. A., & Berry, L. L. (1988). Servqual: A multiple-item scale for measuring consumer perceptions of service quality. *Journal of Retailing, 64*(1), 12–41.

Piccinelli, G., & Stammers, E. (2001). *From e-processes to e-networks: An e-service-oriented approach.* Retrieved from http://www.research.ibm.com /people/b/bth/OOWS2001 /piccinelli.pdf

Polanyi, M. (1958). *The tacit dimension.* New York, NY: Doubleday.

Rowley, J. (2006). An analysis of the e-service literature: Towards a research agenda. *Internet Research, 16*(3), 339–359. doi:10.1108/10662240610673736

Rust, R. (2001). The rise of e-services [Editorial]. *Journal of Service Research, 3*(1).

Rust, R., & Kannan, P. K. (2003). E-service: A new paradigm for business in the electronic environment. *Communications of the ACM, 46*(6). doi:10.1145/777313.777336

Rust, R., & Lemon, K. N. (2001). E-service and the consumer. *International Journal of Electronic Commerce, 5*(3), 83–99.

Salo, J. (2009). Mobile services special issue editorial. *International Journal of E-Services and Mobile Applications, 1*(4).

Scupola, A., Henten, A., & Nicolajsen, H. (2009). E-services: Characteristics, scope and conceptual strengths. *International Journal of E-Services and Mobile Applications*, *1*(3). doi:10.4018/jesma.2009070101

Singh, M. (2002). E-services and their role in B2C e-commerce. *Managing Service Quality*, *12*(6), 434–445. doi:10.1108/09604520210451911

Stafford, T. (2003). E-services. *Communications of the ACM*, *46*(6), 26–28. doi:10.1145/777313.777333

Tiwana, A., & Balasubramaniam, R. (2001). *E-services, problems, opportunities, and digital platforms*. 34th Hawaii International Conference on System Sciences. March 2001.

Tsikriktsis, N. (2002). Does culture influence web site quality expectations? An empirical study. *Journal of Service Research*, *5*(2), 101–113. doi:10.1177/109467002237490

Varshney, U., & Vetter, R. (2002). Mobile commerce: Framework, applications and networking support. *Mobile Networks and Applications*, *7*(3), 185–198. doi:10.1023/A:1014570512129

Voss, C. (2000). Developing an e-service strategy. *Business Strategy Review*, *2*(1), 21–33. doi:10.1111/1467-8616.00126

Westerlund, M., Rajala, R., Tuunainen, T., & Salo, J. (2009). The influence of content and trust on consumers' intention to accept mobile advertisements. *International Journal of E-Services and Mobile Applications*, *1*(4). doi:10.4018/jesma.2009070801

Zeithaml, V. A., Parasuraman, A., & Malhotra, A. (2000). *A conceptual framework for understanding e-service quality: Implications for future research and managerial practice*. Working Paper No. 00115, Marketing Science Institute, Cambridge, MA.

Section 1

E-Service and Self Services:

Concepts, Challenges and Applications

Chapter 1
E-Services for Managing and Orchestrating Organizational Networks

Marijn Janssen
Delft University of Technology, The Netherlands

ABSTRACT

Organizations increasingly cooperate in organizational networks to remain competitive. Electronic intermediaries can provide all kinds of e-services to support the creation and management of such networks. While there has been substantial discussion on intermediaries matching demand and supply, there has been little analysis in relation to the management and orchestration of organizational networks. In this chapter, we analyze an intermediary that uses e-services for orchestrating a network in the consumer electronics industry. The empirical results show that the coordination and management of networks requires specific expertise and skills which could be provided by specialized intermediaries. The primary value creation activity of the intermediary is leveraging the products, activities, and knowledge of the specialized companies and providing e-services for orchestrating the organizational network to create short lead times, improve customer responsiveness, and ensure adaptability.

INTRODUCTION

Organizations increasingly find that they must rely on effective supply chains formed in networks, to successfully compete (Hagel-III, Durchslag, & Brown, 2002). During the past decades, globalization, outsourcing and all kinds of information and communication technology have enabled many organizations to successfully operate solid collaborative networks in which each specialized business partner focuses on only a few key strategic activities. In such a network complimentary resources exist in a number of cooperating and sometimes competing companies. One of the main advantages for companies cooperating in a network is the ability to deliver products by selecting the resources and appropriate companies who are able to deliver these service elements (Provan & Milward, 1995). The aim of networks

DOI: 10.4018/978-1-60960-607-7.ch001

is to provide a range of products against low costs, which single organizations are not able to offer.

Such a network needs to be coordinated to make it work in concert. This can be done by using of all kinds of e-services. E-services are services that are produced, provided and/or consumed through the use of ICT-networks and also includes e-services for supporting the purchasing and delivery of physical goods (Scupola, 2008). Such e-services are provided and used by various organizations in a network to coordinate their interdependent activities. The larger the number of companies the more complex the management of organizational networks becomes. Furthermore, the coordination overhead increases due to the the friction costs. To satisfy and respond quickly to customer demand, effective management is necessary to strengthen the competitiveness of the network.

It is often predicted that by using the Internet the number of traditional intermediaries will be reduced (Gellman, 1996; Malone, Yates, & Benjamin, 1987) and new types of electronic intermediaries will come into existence (Del Aguila-Obra, Padilla-Melendex, & Serarols-Tarres, 2007; Janssen & Verbraeck, 2005; Sarkar, Butler, & Steinfield, 1995). Recently it is suggested that intermediaries have evolved from being merely matchmakers to providers of a set of e-services (Bhargava & Choudhary, 2004; Wise & Morrison, 2000). Bhargava and Choudhary investigated information intermediaries and Wise and Morrison analyzed intermediaries in business exchanges. In this research we focus on intermediaries providing e-services to manage organizational networks.

Although there is much research in the field of intermediaries operating electronic markets for matching supply and demand there is little work analyzing the range of e-services provided by intermediaries in the management and orchestration of organizational networks. Therefore, we opted for case study research to analyze how and why e-services are used. Case study research is well-suited for investigating a contemporary phenomenon in its natural setting, especially when boundaries between phenomenon and context are not clearly evident (Yin, 1989) and answering 'how and 'why' questions (Checkland, 1981). This chapter explores the concept of orchestrating by intermediaries and demonstrates the potential value of intermediaries by analyzing a case study in the consumer electronics industry. In the following section an overview of the relevant, state-of-the-art theoretical frameworks around intermediaries is provided. Based on these theories a case study is analyzed and discussed in section. Next, the findings are discussed in the light of these theories. Finally, conclusions and recommendations are drawn.

THEORETICAL BACKGROUND

Intermediaries

The organizational network setting can be described using three main types of players, which can be labeled as intermediary, service provider and service requester. *Service providers* can be all kind specialized organization involved in the delivery of a physical product or service. This implies that also manufacturing and assembling companies are viewed as service providers, as they provide a set of services that result in the production and delivery of a product. *Service requesters* are the consumers who buy the product or services. An intermediary can be defined as an autonomous organization aimed at bringing together demand and supply (Bailey & Bakos, 1997). This definition primarily emphasizes the market role of intermediaries. Another role of intermediaries can be the creating and coordinating of networks of competing and cooperating service providers to fulfill the demand of service requesters. Malone and Crowston (1994) define *coordination* as managing the dependencies between activities. Based on coordination theory, we follow Janssen and Verbraeck (2005) and define intermediaries as organizations supporting the coordination of the

Figure 1. Levels of intermediation

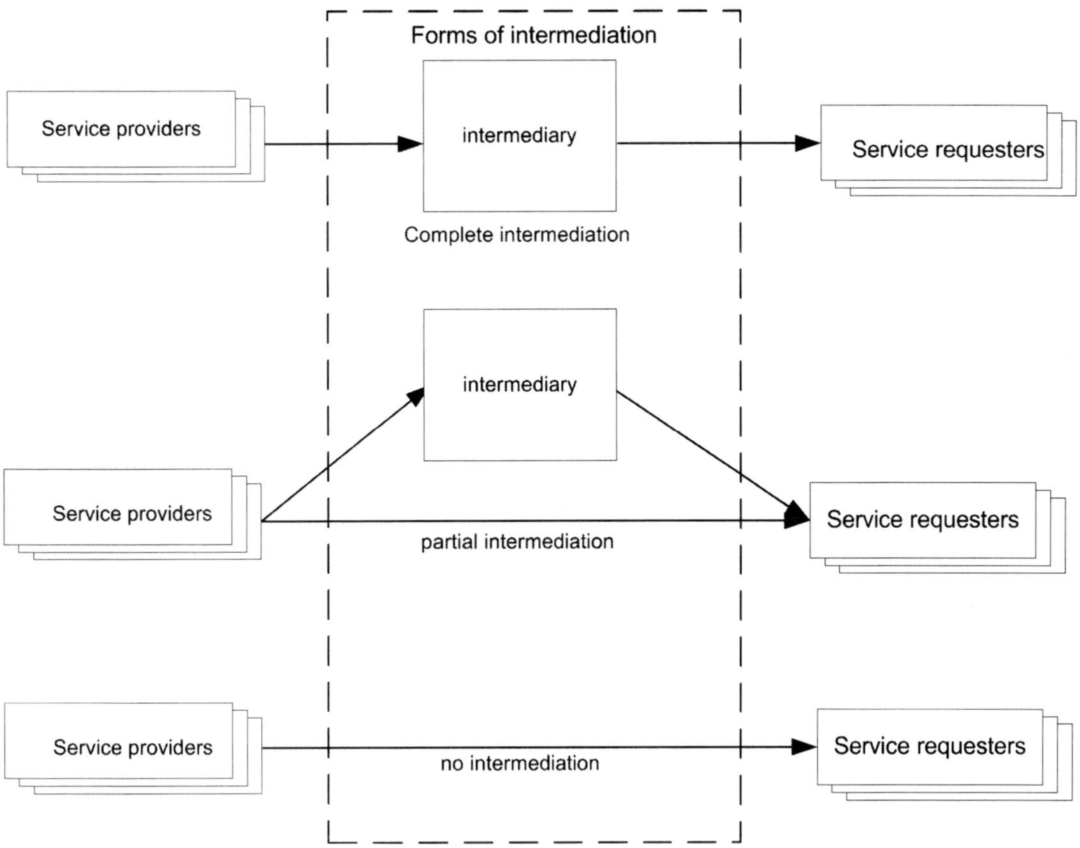

activities of companies cooperating in a networks to connect to consumers.

In a general sense, there are two possible ways to coordinate between service provider and requesters: bilateral or intermediated contact. This is too simplistic, as in reality an almost infinite number of structures are possible. Figure 1 shows three possibilities. Intermediaries might only facilitate a market transaction and not being involved in the actual transaction or transfer of (physical) products (e.g. Garbade, 1982). Furthermore, even more than one intermediary might be involved in the process, as each intermediary might provide different added value services (e.g. Del Aguila-Obra, et al., 2007).

The literature on electronic intermediaries is related to the *electronic brokerage effect* as mentioned in electronic market theory (Bakos, 1991; Malone, et al., 1987). The *electronic brokerage effect* is the prediction that the use of electronic markets would result in the bypassing of intermediaries who previously matched buyers and sellers. This theory has triggered researchers to responses and new theories have been developed which have resulted in a shift in focus in thinking about electronic intermediaries.

Institutional theories, especially transaction cost theory, have been used to make predictions about the need for intermediaries (Malone, et al., 1987; Sarkar, et al., 1995). They are often criticized for not taken into account aspects like, trust, relationships, complexity of product, the need for information collecting and processing resources and specialized knowledge. Aspects that buyers or sellers want to or can manage directly. Moreover, intermediary mechanisms depend on the situational characteristics and circumstances (Janssen & Verbraeck, 2005). As such, there is a need for more explorative and in depth research which takes the situation and circumstances into account.

A continuing discussion is going on about the added value of intermediaries and the re- and disintermediation of these intermediaries (Chircu & Kauffman, 1999; Janssen & Sol, 2000; Janssen & Verbraeck, 2005; Malone, et al., 1987). Cybermediaries (Sarkar, et al., 1995) or electronic intermediaries (Janssen & Verbraeck, 2005) can introduce new complexities to the supply chain. The current status of this discussion is that new roles for electronic intermediaries are coming into existence (e.g. Sarkar, et al., 1995). Yet, the functions executed and e-services provided by intermediaries in the future are currently only roughly known, as these kinds of e-services are still in their infancy, and the first type of intermediaries, that provide these services, is now arising (Österle, Fleisch, & Alt, 2000). Therefore, we wanted to investigate intermediaries providing e-services to manage and orchestrate organizational networks.

Holland and Lockett (1997) found that the roles of electronic intermediaries are dependent on organization strategies. Furthermore, intermediation, disintermediation and reintermediation (IDR) are dependent on the life cycle of an industry, and the strategies and the assets of players (Chircu & Kauffman, 1999). It is important to analyze at value chain levels instead of market level (Chircu & Kauffman, 1999). No single variable-based theory, even if intuitively appealing, will successfully predict all events in a phenomenon as complex as the structure of markets (Hess & Kemerer, 1994). Therefore case study research is necessary to investigate one intermediary in detail including the provided e-services.

E-Services

In the preceding parts we argued the rise of intermediaries using e-services to completely or partly intermediate between service requesters and providers. E-services are produced, provided and/or consumed through the use of ICT-networks (Scupola, 2008). An intermediary can play a relatively inactive role by merely providing services, such as secure connections, and leaving the management to others. On the other hand they can play an active role by not only providing facilitating services, but also actively managing the interactions.

E-service definitions emphasize a simultaneous or near-simultaneous exchange of production and consumption, transformation in the experience and value that customers receive from engagement with providers, (Rai & Sambamurthy, 2006). Grönroos (2001) identified three basic characteristics of services 1) services are processed consisting of activities or a series of activities rather than things, 2) services are at least to some extent produced and consumed simultaneously, and 3) the customer participants in the service delivery process. E-services are customer specific and this challenges the standardization of delivery processes and the economies of scale. Additionally, services cannot be inventoried, because the co-production of services requires a high degree of interaction between customers and many service providers.

Spulber (1996) found that intermediaries can provide a whole range of services, including price setting, clearing, matching, searching, inventory management, liquidity, immediacy, quality guarantees and monitoring. Kalakota and Whiston (1996) found that an intermediary can perform the following tasks: (1) processing and retrieving information, (2) searching, filtering, and summarizing large volumes of data, (3) translating or passing on information requests (4) maintaining information directories and (5) monitoring usage patterns and information changes. Bailey and Bakos (1997) analyzed a number of case studies and identified four roles of intermediaries including information aggregating, providing trust, facilitating and matching. Janssen (2005) found that an intermediary or broker can decouple and aggregate supply and demand. The intermediary seeks to balance between optimizing the whole supply chain and taking the self-interest of actors into account. More recently Del Aguila-Obra, Padilla-Melendex and Serarols-Tarres (2007) investigated infomediaries in the digital, online news industry and found that new Internet intermediaries have appeared aggregating demand and supply in this industry. All these researches take the demand and supply aspect as central to the role of intermediaries. There has been little analysis in relation to the use of e-services for the management and orchestration of organizational networks.

The activities of an intermediary managing and orchestrating organizational networks can systematically be derived by taking a life-cycle view on the service delivery process. In general, an intermediary orchestrating a network performs the following main process phases, in which each phase e-services can be provided.

1. *Service Selection*: the appropriate services provided by service providers need to be determined to fulfill a customer need. Often a service catalogue will be needed. E-services in this phase include searching and browsing products, selecting products based on multiple criteria and product ordering;
2. *Network Configuration*: the number and location of service providers like suppliers, production facilities, distribution centers, warehouses and customers need to be selected. In this phase e-services include asking for availability of products and production capacity, negotiating prices, lead-times and other attributes;
3. *Service delivery strategy*: this concerns how the service will be delivered to the consumer. In this phase e-services for finding and ordering transport capacity are used.
4. *Tracking and tracing*: Integrate systems and processes through the supply chain to monitor the status and take appropriate measure if necessary. E-services are necessary for sharing valuable information, including demand signals, forecasts, monitoring, inventory and transportation and prioritizing of orders.
5. *Settlement*. The financial settlement by making use of billing and payment e-services.
6. *After-sales service*. The checking and evaluation of the performance of the network in order to continuously improve and adapt to changing circumstances. This includes e-services for surveying customer satisfaction and the handling of complaints.

There has been little analysis of intermediaries in relation to the management and orchestration of organizational networks and the type of e-services offered by intermediaries are changing. Due to the complexity and dependency among many factors, case study research was chosen to investigate the changing roles of intermediaries.

RESEARCH APPROACH

Different types of theory exist in information systems research that all can potentially be valuable (Gregor, 2006). Gregor (2006) makes a distinction between theories for 1) analysis, 2) explanation, 3) predication, 4) explanation and prediction and 5) design and action. Our aim is to analyze for design and action purposes which is labeled as design science (e.g. Hevner, March, Park, & Ram, 2004). The selection of research instruments depends on the amount of existing theory available, the type and complexity of the problem under study and on the research objective or questions. Although some theory available, the theory concerning e-services and intermediaries orchestrating organizational networks is scarce and there is a need for in-depth investigation of an orchestration intermediary from multiple views. Dunn (1981) provides an overview of the characteristics of well-structured, moderately structured, and ill-structured problems by considering variation in their common elements. Our domain under study is an ill-structured problem as we have to deal with (1) many decision-makers, (2) they might have opposing perspectives, (3) there are many alternatives and (4) the effectiveness and/or the efficiency of the courses of actions cannot be evaluated using only numerical data. According to Yin (1989) case study research is most appropriate in scenarios where the research is exploratory in nature and focuses on contemporary events that occur beyond control of the investigator. Therefore, a qualitative approach based on a case study research was adopted for this research.

Case study research can be characterized as qualitative and observatory, using predefined research questions (Yin, 1989). The case study research methodology is particularly well-suited to information systems (IS) research, since the object of the discipline is the study of IS in organizations (Benbasat, Goldstein, & Mead, 1987). The case study analysis was based on semi-structured interviews with nine representatives representing service requesters, providers and intermediaries. Two interviewees represented the intermediaries and both line managers as administrative staff of the organizations interacting with the intermediaries were interviewed. All the interviews lasted between one and two hours. These interviews were geographically limited to one country, whereas the network is operated on a world scale. In addition five phone interviews were held with interviewees outside the border lasting between 20 and 40 minutes.

Single case study research is often criticized for its limited generalizability, the possibility for researcher bias and its tendency to be open to different interpretations by different researchers (King & Applegate, 1997). The use of multiple research instruments is often encouraged to overcome these limitations (Yin, 1989). Multiple instruments enable a researcher to compare and contrast the analyses derived from using different research instruments. Using a variety of research instruments can cancel the pitfalls and problems encountered with individual instruments and a richer and more objective picture of the research object can be created. Therefore documents relating to the history, the set up and operation of the intermediary were gathered and examined in order to acquire a good understanding of the operational, technical, management and organizational aspects as well as issues and problems faced during the operation. The researcher was requested to keep the name and products of the company confidential; therefore, the intermediary under study will be denoted as Consumer Electronics Company (CEC).

CASE STUDY: CONSUMER ELECTRONICS COMPANY

Background

CEC is an international company divided into a large number the business units and sells over a thousand different products. This company was founded as a manufacturing company. In the eighties and nineties the production was shifted to countries with low wages and over time more and more of the production units were sold and some were outsourced. In the late nineties it was decided that marketing, research and development and managing the supply chain would become their core business and many of the manufacturing business units were sold.

We investigated only one product line, in which the company solely performed the role of intermediary by orchestrating the demand and supply of products. In this segment the demand for innovation, flexibility, and a shorter time-to-market for new products and the desire to create new revenue sources has led to rethinking the company's activities. CEC played only an intermediary role, as they were not directly involved in the physical production and distribution of the product. CEC's main goal in this product line is defined as the effective management and coordination of the production network by making use of e-services.

Intermediary Position

The intermediary aims at fulfilling customer demand by selecting and evaluating appropriate business partners and creating a customer-driven chain of activities by orchestrating the activities in a particular sequence. This is not simple, as each organization in the network might have different expertise, budget, resources and capabilities. Moreover, most of the organizations are part of more than one chain and other companies claim and make use of their scarce resources. In periods with economic growth, there is a competition for acquiring access to the scarce resources of each company, whereas in economic decline there is a surplus of resources.

The end-to-end supply chain involves complex interactions in which a number of e-services are used. The supply chain consists of manufacturing companies who provide dear parts to the assembling company, which use these parts to create products. The products are then transported to a warehouse for distribution to dealers. Dealers sell their products to the customers and place their orders at the intermediary. The intermediary manages the product and information flow among the supply chain partners. The intermediary is involved in demand forecasting and discussions about new product releases. Dealers often have one or more shops and an Internet presence. Some of the dealers only have an internet presence. The dealer orders products at the intermediary who annually creates a product catalogue and can ask wholesalers to deliver products from their stock to the dealers.

CEC also takes care of selecting appropriate manufacturers and assemblers. Periodically the contracts are renegotiated based on the evaluation of their past performance. Several manufacturers produce parts of the products and there are two or three assemblers. A conscious strategy is to have at least 2 organizations in the network, who are able to perform the same type of tasks. In this way these organizations compete with each other to attract a larger amount of orders and are forced to continuously improve. Typically the best performing organization is allocated more than half of the orders and the worst performing organization is allocated a smaller amount. This allocation is dependent on the performance differential, the capacity available at the respective organizations and other circumstances. For example, the intermediary takes care that the worst performing organizations will not get bankrupt due to an abrupt decline in the number of orders, as this would ultimately weaken the position of CEC.

A number of transportation companies are involved as many of the products are manufactured overseas and need to be distributed to many geographically dispersed dealers. For overseas transportation long term contracts are used, whereas inland transportation company have shorter term contracts and are changed more frequently. In the terms introduced in the background part of this chapter the service requesters are the dealers, CEC is the intermediary and the service providers are the manufacturing, assembling and transporting companies. The physical flow is going directly from one organization to another organization and there is no intermediation. Although a number of information flows concerning the operational process flow directly from one-organization to another organization most of the information flows are mediated by CEC. This enabled the intermediary to create an overview of the whole network, take appropriate actions if necessary, continuous evaluate the business partner and create new supply chains. For this purpose a pallet of e-service has been developed and is being used.

E-Services

CEC plays a number of roles and fulfills each role by providing e-services. Unified Modeling

Figure 2. Interaction diagram for an e-order

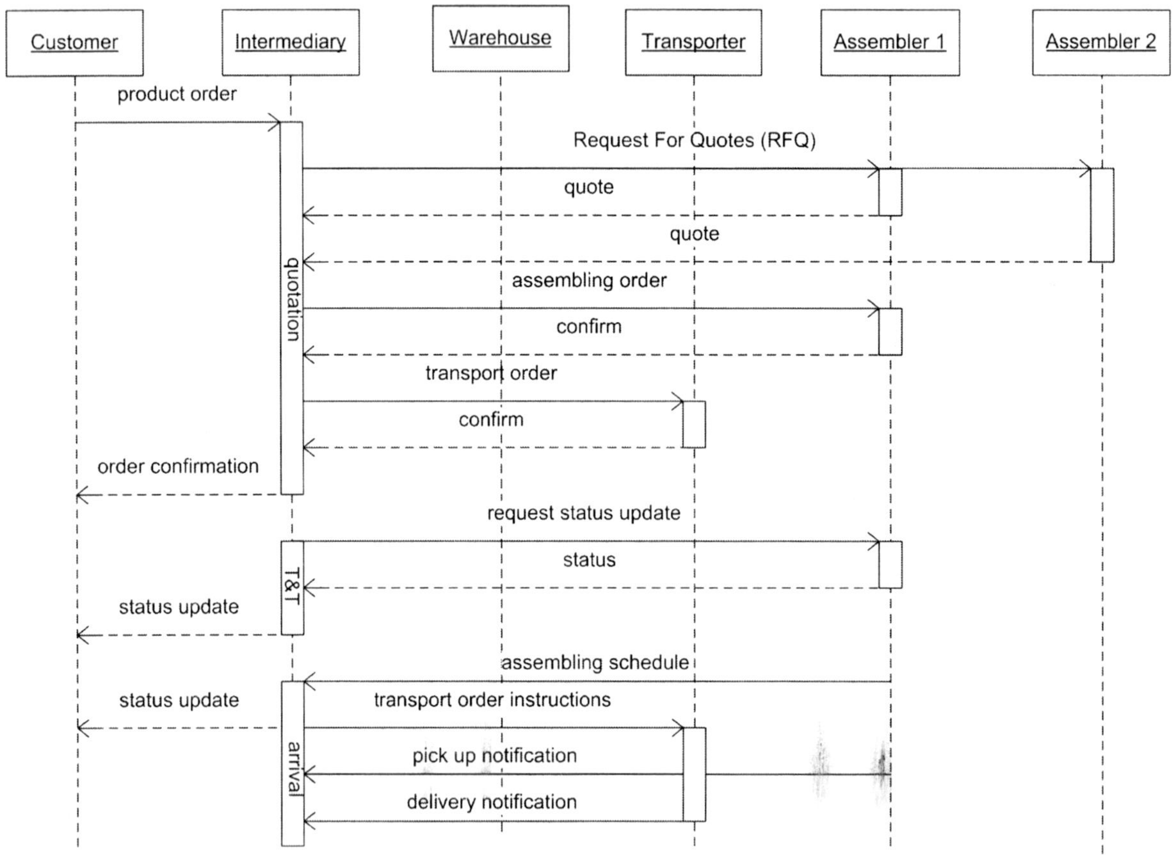

Language (UML) is a defacto standard for modeling software systems. UML Sequence Diagrams display the time sequence of the objects participating in the interaction (Fowler, 2000). In Figure 2 the ordering e-service is described in which actors are taken as the objects. This figure shows that a product order results in a quotations process which results in the selection of an assembling party and the ordering of transport. Next the status is tracked. Finally the assembler alerts the intermediary that the goods will be available at a certain date and time. The intermediary sends the detailed transport order instructions to arrange the transport from the assembler to the customer and get informed of status changes. All the interactions and the resulting e-services are well-defined and loosely coupled.

This example clearly shows the complexity of an e-service as many alternative sequences of activities are possible. Typical examples of variations are direct ordering for assembling without a quotation process, the involvement of manufacturers and the use of a quotation process for selecting a transportation company. Moreover, the boundaries of an e-service are not always clear and are dependent on the stakeholder under consideration. As the customer might consider the product ordering process as an e-service, whereas the intermediaries might view this as a combination of several e-services, including quotation, assembling order and transport order services. The customer might consider the ordering e-service ended after the ordering confirmation or after the delivery of the physical products. For the intermediary the delivery notification of the

Table 1. Types of e-services

Level	Type of e-services
Strategic	• Strategic network optimization, including the number of service providers, location, knowledge, quality and reliability • Network evaluation and reporting • Product and service innovation • Development of IT facility and infrastructure, to support the organizational network • Select new network participants, access quality and risks • Align overall network strategy with the strategies of the individual participants • Forecasting of demand and need for new products
Tactical	• Negotiation of contracts, service levels and other decisions • Marketing and advertising • Comparison of prices and quality information • Product and network improvement • Workflow coordination (including tasks, quantity, location, quality, and frequency) • Mid-term forecasting • Evaluating of performance of trading partners and feedback • Billing and payment • Financial reporting • Adapting to changing circumstances
Operational	• Maintain catalogues containing products and their attributes • Maintain catalogues of service providers and their past performance • Daily orchestration of network activities • Sourcing and planning • Tracking and tracing of status • Progress reporting • Taking measure to ensure meeting of service levels required by customers

transporter informs the intermediary that the product has been delivered at the customer and the financial settlement can start which requires billing and payment services. In summary, a high-granular e-service might consist of small granular services and what is viewed as an e-service is at least dependent o the stakeholder position.

Apart from the operational e-services necessary to create the orders for products, raw material and transportation, the intermediary provides a number of e-services at the strategic, tactical and operational level. Table 1 depicts a systematic overview of the e-services as found in our case study. The e-services are aimed at supporting the life-cycle phases. In the following paragraphs, we will briefly describe the roles using the facilitation, trusted, information aggregation and matching roles of Bailey and Bakos (1997).

The intermediary performs the *facilitating* role creating a platform supporting the interactions with all their business partners. The shared infrastructure makes it possible to connect to each other and the intermediary provides all kinds of e-services residing on top of this infrastructure to enable effective collaboration. Using this infrastructure the intermediaries supports the close integration with the systems of the business partners, and also provides web-based forms for enabling loosely coupled interactions.

The infrastructure is of vital importance for fulfilling the *trusted* role. The intermediary takes care of securing communication channels and ensures that data will not be lost. Moreover, privacy of companies should be ensured as competing organizations are often involved in the network and these might benefit from gaining information on the other companies. One of the biggest risks is that information will become available to the wrong party. Moreover, the companies in the network are very afraid that this information might be used to lower prices.

The perceived independent or neutral position of the intermediary is crucial for fulfilling the trusted role. If their business partners get the feeling that the intermediary cannot be trusted or provide information to their competitors they are not likely to continue collaboration. A quote of one of the interviewees underlines this observation "*as soon as we got the feeling that they are biased and more orders go to our competitors without informing us and having clear arguments we will stop our partnership and seek for others*". As such the e-services creating a reliable and secure platform are essential.

A large number of e-services are available for collecting, processing and disseminating information, thus for fulfilling the *information aggregation role*. The intermediary provides e-services for ordering, tracking and tracing of orders, scheduling, and also collects additional data to evaluate the quality and performance of the supplies. This includes client information such as the number of sold products, returned products and the number of repairs necessary. The intermediary uses this system to collect, aggregate, process and disseminate information. The intermediary takes the responsible for overseeing the execution of orders from start to completion and for handling exceptional requests and complaints.

The *matching role* is fulfilled by selecting business partners and creating a unique chain of activities for each order. First, the partners are selected by looking at their past performance and schedule. E-services provide insight into the available resources and the utilization of the manufacturing and assembling capacity of the business partners. Second, the intermediary has the flexibility of invoking various sequences of loosely coupled web services and in this way can creates a unique chain of activities for each product.

DISCUSSION AND EVALUATION

The Rise of Orchestration Intermediaries

Disintermediation of physical distribution elements in the network has brought about the advent of a new form of electronic intermediary, acting as a manager and orchestrator of the physical activities performed by other organizations using e-services through which delivery and reception of goods is realized. As a former manufacturing organization, the consumer electronic company had the necessary starting point for becoming the orchestrating intermediary of the network. One interviewee stated very strongly "*The reason why we became the spider in the web was that we had the expertise, the funding, knew the business partners and competitors and could even have produced it ourselves*". Later this high-ranked interviewee formulated it in relationship to the competitive environment "*luckily most of the business partners viewed us not as competitors .... they knew that it would be too expensive for us to do it ourselves.... they merely viewed us as a customer or as an additional sales channel*". Although on some aspects they were competing, especially in the field of research and development,

A traditional intermediary is an organization who is in contact with many potential buyers and suppliers and helps to match them (Malone, et al., 1987). Such an intermediary substantially reduces the need for buyers and suppliers to contact a large number of alternative partners individually. The electronic brokerage effect predicts that by electronically connecting many different buyers and suppliers an intermediary would not be necessarily anymore. The Internet allows a buyer to search and select suppliers by comparing offerings quickly and inexpensively (Malone, et al., 1987). In our case study of the consumer market we did not find this effect at all. The intermediary has expertise in selecting the best service providers and has all kind of tactical knowledge considering their past performance and quality that was used to select the service providers and orchestrate the supply chain. Another reason for not observing this effect might be that the types of products under consideration are not commodity goods but the products are rather difficult to understand and orchestration requires extensive product knowledge. This type of knowledge can also not be easily stored in a system or transferred. Although the electronic brokerage effect might be true for commodity markets, it does not apply to markets for complex, technological products. Hence, we expect that the electronic brokerage effect will not hold for non-commodity markets and we expect the rise of new types of intermediaries aimed at orchestrating and managing networks for producing and delivering non-commodities.

E-Services for Network Orchestration

The basis of the management and orchestration of the chain of activities in the organizational network are the e-services which we briefly described using the facilitation, trusted, information aggregation and matching roles and can be at the strategic, tactical and operational level. We extended the roles of Bailey and Bakos (1997) and found them generic enough to be used to classify a whole range of e-services provided by the intermediary. We did not decide to introduce a new role for describing the management and orchestration aspects, as most elements related to the management and orchestration are already covered by the four roles.

The strength of the four roles, the generality which makes them applicable for various situations, is also their weakness, as inevitably they remain at an abstract level. The extension of the roles by looking at the strategic, tactical and operational level helps to acquire a better overview by capturing long, mid and short term aspects. Although the roles are suitable for describing all types of intermediaries, in contrast to market intermediaries other e-services were found when using these roles to orchestrating intermediaries. Thus, at a high abstraction level these roles might seem to be similar for intermediaries, whereas at a lower level of abstraction these roles can be completely different fulfilled. This pleas for more in-depth case studies within a certain domain and also for comparative case studies of different types of intermediaries, including market intermediaries, cybermediaries, infomediaries and orchestration intermediaries.

The process of network decomposition and integration is realized through the use of e-services aimed at integrating the efforts of the single business partners. The intermediary in our case study concentrates on performing mere business process management and orchestration activities and has given up performing traditional tasks. The intermediary composes and orchestrates different service providers which are specialized companies in their respective domain. The whole interaction is primarily based on loosely coupled and well-standardized e-services. New business partners can make use of these loosely coupled e-services to connect to the network. In some exceptional cases the systems are tightly integrated. The intermediary has created its own platform and e-services for dealing with these kinds of interaction fitting their own orchestration needs. This platform matches the complex product characteristics and helps to decompose the product into activities which can be performed by the service providers. The case study demonstrates the important role played by componentization and by service orientation in interorganizational networks. Componentization enables a business to operate in a network of partnerships with customers and suppliers supported by real-time information flows and information technology systems. All kinds and types of e-services are necessary to achieve seamless integration of business components.

Many of the activities are related to the orchestration of the organizational network, however, there are also many standalone e-services, for example financial reporting. As a result a differentiation between orchestration services and additional service can be made. The intermediary has a number of basic and specialized e-services. The basic e-services might be provided by other type of intermediaries, whereas, the specialized e-services might be typical for orchestration intermediaries.

Although intuitively it is clear what an e-services is, the formal description of an e-services is much more difficult. E-services are produced and consumed during the interactions among the participants and a pre-defined process flow cannot describe all possible combinations. The boundaries of an e-service are not always clear and are dependent on the stakeholder under consideration and the outcomes of the interactions. The sequence diagram shown in Figure 2 demonstrates that what is viewed as an e-service is at least dependent on the stakeholder position. Furthermore, e-services might be created by deploying other e-services; a high-granular e-service is constructed of small granular e-services.

The Added Value of the Orchestrating Intermediary

The creation of flexible, temporary arrangements results in the creation of business processes that are no longer self-contained within a single organization. The reliability of business processes depends more and more on the performance of external partners. Each organization has different strategic objectives, capabilities and resources available. Moreover, potential partners may have different processes and levels of IT sophistication that need to be synchronized before services can be shared. The dynamic nature of organizational networks, for example, the changing number and/or types of partners, and the involvement in several networks increases the difficulty and complexity to understand the dependencies in the supply chain. One of the intermediary's main challenges is to understand the relation between demand and the challenges of coordinating the network. One interviewee stated that "*the basic of our added value is not only understanding the need of our customers and innovating products, but consist of understanding what is going on at our business partners, who in fact are also our customers*". Knowledge of the specialties of their business partners and their strengths and weakness are key to selection and creation of a new chain for producing a product. The intermediary's ability to integrate, build, and reconfigure the external resources to address changing needs is their key to success. The e-services help a firm to spot changes and adjust its resource mix and thereby maintaining the sustainability of the network. The intermediary needs knowledge of the companies to create new products and orchestrate the organizational network to create short lead time, improve customer responsiveness and ensure adaptability to changing circumstance. In this way the network profits from specialization (e.g. Clemons & Row, 1992). Each network member specializes in performing certain types of roles and activities. In turn, specialization in a network creates dependencies among participants that need to be orchestrated.

The use of orchestration intermediaries can improve the quality of the network as only the most suitable partners are selected for each service and duplication of efforts can be avoided. The intermediary creates an overview of the total network and ensures that activities are only performed once, at the best suitable service provider. In summary the main added value of the use of intermediaries as mentioned by the interviewees are

- Creating an overview of the network and reacting to events and changing circumstances
- Knowledge of the business partners
- Continuous improvement of the network
- Reduction of friction costs by making use of specialized e-services
- Avoiding duplication of activities

- Ensuring information quality and availability
- Market and product knowledge

An intrinsic benefit of the intermediary is that it is an intermediary and is not tied to one of the organizations participating in the network. The perceived independent or neutral position of the intermediary is viewed as a condition and not as a benefit by the interviewees. CEC profits also from economies of scale and scope (e.g. Lee & Clark, 1997). Once they are developed CEC faces relatively small incremental costs if new participants make use of their e-services. Furthermore the technological and organizational resources and expertise acquired during the development and operation of in this product line is transferred to other products resulting in economies of scope.

CONCLUSION AND RECOMMENDATIONS

As the process of disintegration and reintegration of organizational networks continues, emerging intermediary players can play a crucial role in managing and orchestrating the network. Our case study demonstrates how a traditional company performing all kinds of physical functions is reborn as an intermediary primarily performing supply chain management and orchestratration activities. For these purposes the intermediary uses a large number of well-defined and loosely coupled e-services. Although transaction cost theory can be used to predict the disappearance of intermediaries, our case study shows that this prediction overlooks the reality that the orchestration of organizational networks requires the solving of complicated issues which needs all kinds of dedicated services. The coordination and management requires specialized expertise and skills which result in the rise of intermediaries orchestrating these networks.

Our in-depth case study shows that the main types of e-services can be described by extending the four roles of Baily and Bakos (1997). The four roles are generic enough to support the analysis of e-services. The extension of the roles by looking at the strategic, tactical and operational level helps to acquire a better overview by capturing long, mid and short term aspects. Although the roles are suitable for describing intermediaries in general, e-services differ per type of intermediary. More in-depth studies into orchestrating intermediaries and comparative case studies among types of intermediaries are recommendations for further research.

In the case study the primary value creation activities of the intermediary are the leveraging of activities and knowledge of the specialized companies and orchestrating the organizational network to create short lead time, improve customer responsiveness and react to changing circumstances. The type of e-services might be dependent on market characteristics and conditions, product complexity and circumstances. More research is necessary in the type of e-services and the added value provided by intermediaries.

Since we have only looked at one case, albeit extensively, the findings can only be analytically generalized based on the distinguished characteristics (Yin, 1989). Our findings call for additional research in the field of e-services, to further analyze orchestrating intermediaries within different settings and to further statistically generalize the findings.

REFERENCES

Bailey, J. P., & Bakos, J. Y. (1997). An exploratory study of the emerging role of electronic intermediaries. *International Journal of Electronic Commerce*, *1*(3), 7–20.

Bakos, J. Y. (1991). A strategic analysis of electronic marketplaces. *Management Information Systems Quarterly, 15*(3), 294–310. doi:10.2307/249641

Benbasat, I., Goldstein, D. K., & Mead, M. (1987). The case research strategy in studies of Information Systems. *Management Information Systems Quarterly, 11*(3), 369–386. doi:10.2307/248684

Bhargava, H. K., & Choudhary, V. (2004). Economics of an information intermediary with aggregation benefits. *Information Systems Research, 15*(1), 22–36. doi:10.1287/isre.1040.0014

Checkland, P. (1981). *Systems thinking, systems practice*. Chichester, UK: Wiley.

Chircu, A. M., & Kauffman, R. J. (1999). Strategies of Internet middlemen in the intermediation/ disintermedation/ reintermediation cycle. *Electronic Markets, 9*(2), 109–117. doi:10.1080/101967899359337

Clemons, E. K., & Row, M. C. (1992). Information Technology and industrial cooperation: The changing economics of coordination and ownership. *Journal of Management Information Systems, 9*(2), 9–28.

Del Aguila-Obra, A. R., Padilla-Melendex, A., & Serarols-Tarres, C. (2007). Value creation and new intermediaries on Internet. An exploratory analysis of the online news industry and the Web content aggregators. *International Journal of Information Management, 27*, 187–199. doi:10.1016/j.ijinfomgt.2006.12.003

Dunn, W. N. (1981). *Public policy analysis*. Englewood Cliffs, NJ: Prentice-Hall.

Fowler, M. (2000). *Distilled: A brief guide to the standard object modeling language*. Addison-Wesley Professional Series.

Garbade, K. (1982). *Securities markets*. New York, NY: McGraw-Hill.

Gellman, R. (1996). Disintermediation and the Internet. *Government Information Quarterly, 13*(1), 1–8. doi:10.1016/S0740-624X(96)90002-7

Gregor, S. (2006). The nature of theory in Information Systems. *Management Information Systems Quarterly, 30*(3), 611–642.

Grönroos, C. (2001). *Service management and marketing: A customer relationship management approach*. Chichester, UK: Wiley.

Hagel-III. J., Durchslag, S., & Brown, J. S. (2002). *Orchestrating loosely coupled business processes: The secret to successful business collaboration*. Unpublished manuscript.

Hess, C. M., & Kemerer, C. F. (1994). Computerized loan originating systems: An industry case study of the electronic markets hypothesis. *Management Information Systems Quarterly, 18*(3), 251–274. doi:10.2307/249618

Hevner, A., March, S., Park, J., & Ram, S. (2004). Design scienc in Information Systems research. *Management Information Systems Quarterly, 28*(1), 75–105.

Holland, C. P., & Lockett, A. G. (1997). Mixed mode network structures: The strategic use of electronic communication by organizations. *Organization Science, 8*(5), 475–488. doi:10.1287/orsc.8.5.475

Janssen, M., & Sol, H. G. (2000). Evaluating the role of intermediaries in the electronic value chain. *Internet Research: Electronic Networking Applications and Policy, 19*(5), 406–417.

Janssen, M., & Verbraeck, A. (2005). Evaluating the information architecture of an electronic intermediary. *Journal of Organizational Computing and Electronic Commerce, 15*(1), 35–60. doi:10.1207/s15327744joce1501_3

Kalakota, R., & Whinston, A. B. (1996). *Frontiers of electronic commerce*. Reading, MA: Addison-Wesley.

King, J., & Applegate, L. (1997). The crisis in the case of research crisis. In DeGross, J., Lee, A., & Lieberman, J. (Eds.), *Information Systems and qualitative research*. London, UK: Chapman and Hill.

Lee, H. G., & Clark, T. H. (1997). Market process reengineering through electronic market systems: Opportunities and challenges. *Journal of Management Information Systems, 13*(3), 113–136.

Malone, T. W., & Crowston, K. (1994). The interdisciplinary study of coordination. *ACM Computing Surveys, 26*(2), 87–119. doi:10.1145/174666.174668

Malone, T. W., Yates, J., & Benjamin, R. I. (1987). Electronic markets and electronic hierarchies. *Communications of the ACM, 30*(6), 484–497. doi:10.1145/214762.214766

Österle, H., Fleisch, E., & Alt, R. (2000). *Business networking. Shaping enterprise relationships on the Internet*. Springer Verlag.

Provan, K. G., & Milward, H. B. (1995). A preliminary theory of network effectiveness: A comparative study of four community mental health systems. *Administrative Science Quarterly, 40*(1), 1–33. doi:10.2307/2393698

Rai, A., & Sambamurthy, V. (2006). Growth and interest in service management. *Information Systems Research, 17*(4), 327–331. doi:10.1287/isre.1060.0108

Sarkar, M. B., Butler, B., & Steinfield, C. (1995). Intermediaries and cybermediaries: A continuing role for mediating players in the electronic marketplace. *Journal of Computer-Mediated Communication, 1*(3).

Scupola, A. (2008). Special issue on e-services. *Journal of Electronic Commerce in Organizations (JECO)*.

Spulber, D. F. (1996). Market microstructure and intermediation. *The Journal of Economic Perspectives, 10*(3), 135–152.

Wise, R., & Morrison, D. (2000). Beyond the exchange: The future of B2B. *Harvard Business Review, 78*(6), 86–89.

Yin, R. K. (1989). *Case study research: Design and methods*. Newbury Park, CA: Sage Publications.

Chapter 2
E-Services and the Codification of Informational Services

Ada Scupola
Roskilde University, Denmark

Anders Hente
Aalborg University – Copenhagen, Denmark

Hanne Westh Nicolajsen
Aalborg University – Copenhagen, Denmark

ABSTRACT

The concept of e-services has gained increasing use during the past decade. There is, however, no general agreement as to the precise meaning and scope of the term. The research purpose of the present chapter is, therefore, to discuss the e-service concept, its strengths and scope, and thereby contribute to the general understanding and definition of the term. More specifically, the chapter aims at examining one of the primary conditions for the development of e-services, namely the codification of knowledge in connection with knowledge intensive services. The e-service concept is used in relation to private as well as public e-services. The present chapter examines the e-service concept at a generic level and does not differentiate between private and public e-services.

INTRODUCTION

During especially the past two to three decades, services have increasingly come to the fore in social sciences. Consequently, discussions on the concept of services have erupted time and again - from Hill (1977) to Chesbrough and Spohrer (2006). More recently, after the advent of e-commerce and e-business (e.g. Barua et al., 2001; Kalakota & Whinston, 1996), researchers and academics alike started discussing the concept of e-services (Evanschitzky & Iyer, 2007). However, a clear definition and full understanding of e-services and the e-service concept and paradigm is lacking (e.g. Scupola, 2008). This chapter, which is conceptual in nature, has the major purpose of contributing to the literature on e-services by characterizing the e-service concept - its strengths and scope - and by examining one of the primary conditions for the development of

DOI: 10.4018/978-1-60960-607-7.ch002

e-services, namely the codification of knowledge in connection with knowledge intensive services. The main research question is: How can we characterize the e-service concept and what are the conditions for its further development? The goal of the chapter is achieved by shedding light on the e-service concept as convergence of services and goods and, in addition, by discussing the conditions relating to the codification of knowledge for the development of e-services. The reason why the knowledge codification issue is important for the development of e-services is that knowledge codification is a requirement for some services, such as knowledge services, to be easily produced and distributed over online networks.

The chapter is structured as follows. The first section is the introduction, which mainly presents the purpose of the chapter. The second section provides an overview of the distinguishing features of goods, e-services and services, thus laying the foundation for the understanding of the e-service concept. A literature review of e-service definitions, characteristics and e-service research approaches are then provided in the third section. The following section then discusses knowledge codification and its importance for e-service development. Finally, the last section provides concluding remarks and challenges for further research.

E-SERVICES AS CONVERGENCE OF SERVICES AND GOODS

In contrast to goods, which can be separated from the immediate producers and sold on an anonymous market, services are delivered by their immediate producers and are not anonymous (Henten, 1994). Consumers will know who the immediate producers are (or will at least have the possibility to do so). It is often said that services are produced and consumed simultaneously and, therefore, require face-to-face contact between the producers and the consumers in the production/ consumption phase (Rust, 2001; Hill, 1977). This may not always hold entirely true, but the consumption will at least start right after the end of production – as in the case of repair work. The basic definition of services has nothing inherently to do with whether it is material or immaterial. The repair work of a plumber, for instance, is material in nature, but is a service. Indeed, information and communication technologies (ICTs) affect all kinds of goods and services with respect to their transaction on the market (e-business) (e.g. Penttinen et al., 2008). However, in the case of data, information and knowledge services (informational services), it is the product itself which is affected (e.g. Gullkvist, 2008). With ICTs, it is possible to enter data, information and knowledge (to the extent it can be codified) on digital media and use communication networks for transportation (e.g. Ahonen et al., 2008). This means that data, information or knowledge services increasingly can be separated from the immediate producers and sold on anonymous markets. They become goods in a sense. Formerly, paper was the primary physical media for turning informational services into goods. Presently, electronic media increasingly dominate (e.g. Ihlström Eriksson et al., 2008).

At the same time, for some categories of goods, producers seek to customize their products to meet the individual demands of customers (Sundbo, 1997). One of the most heralded examples is Dell and their use of Internet to receive information from customers regarding their specific computer configuration choices (Dell, 2000). Furthermore, a wide variety of services surrounding the goods, e.g. after sales services and information services, are developed in order to provide a better customer experience and a more personalized/customized environment (e.g. Pine and Gilmore, 1999). Nevertheless, most goods will remain mass-produced and maintain their anonymous character.

Software is increasingly sold as a service via electronic networks (Tiwana and Balasubramaniam, 2001). This applies to custom-made software solutions for business enterprises and it

Table 1. Distinguishing features of goods, e-services and services

Goods	E-services	Services
Tangible	Intangible, but need tangible media	Intangible
Can be inventoried	Can be inventoried	Cannot be inventoried
Separable consumption	Separable consumption	Inseparable consumption
Can be patented	Can be copyrighted, patented	Cannot be patented
Homogeneous	Homogeneous	Heterogeneous
Easy to price	Hard to price	Hard to price
Can't be copied	Can be copied	Can't be copied
Can be shared	Can be shared	Can't be shared
Use equals consumption	Use does not equal consumption	Use equals consumption
Based on atoms	Based on bits	Based on atoms

Source: Charles F. Hofacker et al. (2007): E-services: A synthesis and research agenda, Journal of Value Chain Management 1(1/2).

applies to mass-produced software transferred on networks. In the case of custom-made software, we are already dealing with a service according to the above definition of services. In the case of mass-produced software, we are concerned with a good. In both cases, however, the new feature is that the ownership rights of the software are not transferred to the user and the operation of the software will be likely to be executed on the computers of the organisation delivering the software service. Therefore software, which has hitherto been sold as goods to end-users, is increasingly sold as services. To the extent that this develops, it could be claimed that there is a degree of convergence of goods and services enabled by the use of digital communications. As always, there is no sharp and unequivocal dividing line between goods and services. The most important thing in the context of the present chapter, though, is the trend regarding informational services to acquire elements of the basic characteristics of goods concerning separation from the immediate producers and anonymity. It is this convergence of goods and services when transmitted over online networks that we call e-services as the e-service concept is helpful in conceptualizing these developments (see the following section for definitions and characteristics of e-services).

In a seminal paper by Hofacker et al. (2007) a similar understanding of the e-service term is developed. E-services are situated between services and goods (Table 1). E-services have characteristics in common with services as well as with goods. The authors of the present chapter do not necessarily agree with all specific characterizations of all three product areas illustrated in Table 1. However, the authors agree that the main idea and thrust in the characterizations illustrate very well the e-service concept.

For a concept to gain and maintain scientific value, it must refer to features, relationships or developments in a manner, which differentiates it from and adds something to other and already well-established concepts. As discussed above and based on previous work, especially the work by Hofacker et al. (2007), we think the e-service concept denotes and connotes such features and developments. Its primary scientific value relates to the ongoing convergence of services and goods facilitated by digital communications. Services, which hitherto depended on face-to-face encounters, can increasingly be digitized and delivered on communication networks whereby the connection between the immediate producer and the service is changed turning the service into a good-like product. Conversely, some goods such as software (Tiwana and Balasubramaniam, 2001) and printing services (e.g. Penttinen et al. 2008; WhatIs.com) which formerly were sold to the customers as goods are increasingly delivered as service to customers. In both cases we talk about e-services. The core strength of the e-service concept lies therefore in the conceptualization of this convergence of informational services and goods. Defined this way codification becomes essential to e-services development and diffusion.

E-SERVICES: DEFINITIONS, CHARACTERISTICS AND THEORETICAL APPROACHES

Having discussed the primary strengths of the e-service concept as lying in e-services being conceptualized as the convergence of informational services and goods, we now turn to the more formal definitional exercise and to the characteristics and the provenance of the term in some of the many contributions that have been made during the past ten years. Prior to this, research in the area was mainly conducted under the heading of technology-based self-service (e.g. Dabholkar, 1991). Many of the same themes are taken up in the e-service as in the self-service literature. The tendency, however, is that the self-service literature is more demand-oriented, while the e-service literature is more supply-oriented.

E-services have been defined in the literature as a new phenomenon at the border between two main domains of studies: marketing and e-commerce/e-business (Hoffman, 2003; Hofacker et al. 2007). As such this new concept attracts the attention of researchers and business people alike and creates a number of research challenges especially in terms of definitions, characteristics and theoretical foundations. This section aims to address some of these challenges.

E-Service Definitions

There are many definitions of e-services. Some focus on the delivery and delivery infrastructure (digital networks) while others emphasize both the delivery process and the benefits or outcome of the service (Hofacker et al. 2007). However, what is in common to all these definitions is that e-services are characterized by the electronic delivery of the service.

One of the early definitions of e-services was provided by Tiwana and Balasubramaniam in 2001. In this chapter, Tiwana and Balasubramaniam define e-services in the following manner:

We view e-services as Internet-based applications that fulfil service needs by seamlessly bringing together distributed, specialized resources to enable complex, (often real-time) transactions. Examples of e-services include supply chain management, customer relationship management, accounting, order processing, resource management, and other services that are electronically delivered through the Internet.

Their paper is about software and the trends towards selling software as a service (SaaS). Service in the SaaS concept means that software vendors develop software applications and host and operate the applications for the users. Users do not pay for owning the software, but for using it. The 2001 definition of Tiwana and Balasubramaniam thus emphasizes one of the two main strengths of the e-service concept, namely the trend towards delivering software functions as a service.

However, the general use of the term e-service is closer to the e-service concept developed by Hewlett-Packard. As in the case of the e-business concept, which was first coined by IBM, the term e-service was also originally developed by an IT-company, Hewlett-Packard. According to WhatIs.com:

E-services, a business concept developed by Hewlett Packard (HP), is the idea that the World Wide Web is moving beyond e-business and e-commerce (that is, completing sales on the Web) into a new phase where many business services can be provided for a business or consumer using the Web. Some e-services, such as remote bulk printing, may be done at a Web site; other e-services, such as news updates to subscribers, may be sent to your computer. Other e-services will be done in the background without the customer's immediate knowledge. (WhatIs.com)

This conception of e-services is wider than SaaS. It encompasses informational services offered to the end-users (e.g. news updates) in addition to the services 'done in the background without the customer's immediate knowledge', i.e. back-office operations (see e.g. Janssen, 2008). It includes all services delivered digitally and it applies whether the services are delivered to end-users (Business-to-Consumer (B2C)) or to business users (Business-to-Business (B2B)). In a paper written by Piccinelli & Stammers (2001) working for Hewlett-Packard, an e-service is defined as

... any asset that is made available via the Internet to drive revenue streams or create new efficiencies (Piccinelli & Stammers, 2001).

This definition is sufficiently broad and, at the same time, narrow, in the sense that it excludes services that cannot be delivered via the Internet. This means that the services, which are marketed and sold on the net but cannot be delivered (i.e. transported) on the net, are excluded from the e-service concept. It must be services that can be entered on digital media and, therefore, transported on digital networks in order to be included in the e-service concept.

This does not exclude services delivered as support to services which cannot themselves be digitally delivered. This applies, for instance, to travel agencies, which sell travel services but do not deliver the travel or accommodation service themselves. This kind of service is, for example, used as examples of e-services by Iqbal, Verma, and Baran (2003).

While most definitions of e-services mainly focus on the delivery part (via digital networks), other researchers also stress the production of the e-service in addition to the delivery part of the service. For example Scupola (2008) states that:

E-services are defined as services that are produced, provided and/or consumed through the use of ICT-networks such as Internet-based systems and mobile solutions (Scupola, 2008).

Furthermore, some definitions also emphasize the outcome and the benefits of e-services in addition to the delivery mode and delivery infrastructure of e-services. This has been noted, for example, by Hofacker et al. (2007). By following this line of thought, we develop a classification of e-services according to the delivery and infrastructure view and the production, delivery and outcome view in Table 2.

Characteristics of E-Services

Having presented definitions of e-services in the previous session, we will discuss e-service characteristics as summarized in Table 3 in the following. For this purpose, we use a classification developed by Hofacker et al. (2007), who identify three prototypes of e-service:

1. E-services as complements to existing offline services and goods. Examples can be online seat reservations offered by airlines and travel agencies (e.g. Hofacker et al., 2007).
2. E-services as substitutes for existing offline services. Examples can be e-newspapers (e.g. Ihlstrom Eriksson et al., 2008) or online auctions such as e-bay (e.g. Wang et al., 2008).

Table 2. A summary of e-services definitions

Delivery and infrastructure view	Production, delivery and outcome view
Those services that can be delivered electronically (Javalgi, Martin and Todd, 2004)	Any asset that is made available via the Internet to drive revenue streams or create new efficiencies (Piccinelli and Stammers, 2001)
Provision of services over electronic networks (Rust, 2001; Rust and Kannan, 2003)	An act or performance that creates and provides benefits for customers through a process that is stored as an algorithm and typically implemented by networked software (Hofacker et al., 2007)
Interactive services that are delivered on the Internet using advanced telecommunications, information, and multimedia technologies (Boyer, Hallowell and Roth, 2002)	E-services as Internet-based applications that fulfill service needs by seamlessly bringing together distributed, specialized resources to enable complex (often real-time) transactions (Tiwana and Balasubramaniam, 2001)
E-service is deeds, efforts or performance whose delivery is mediated by information technology (including Web, information kiosks and mobile devices). Such e-service includes the service element of e-tailing, customer support and service, and service delivery (Rowley, 2006)	E-services are defined as services that are produced, provided and/or consumed through the use of ICT-networks such as Internet-based systems and mobile solutions (Scupola, 2008)

Table 3. A summary of e-service characteristics

Types of e-services	B-to-B	B-to-C	G-to-B and G-to-C	C-to-C
Characteristics/focus	Collaboration and relationship building	Selling to and retaining the customer	User/citizen empowerment, e-democracy, city/rural areas divide	Peer-to-peer value creation
Examples	Supply chain management in outsourced printing services and facilities, SaaS	E-retailing, e-customer relationship management, e-banking, e-newspapers, Web portals	Online tax returns, e-voting, e-libraries, telemedicine, remote social security services	Online auctions, consumer driven e-marketplaces, online gaming, online communities (newsgroups). wikis

3. Uniquely new core e-services. Examples can be online computer games, or search engines as for example Google (e.g. Vragov, 2009).

In addition, according to Scupola (2008), there are three main characteristics of e-services:

- The service is accessible via the Internet or other electronic networks.
- The service is consumed either directly or indirectly via the Internet or other electronic networks.
- There might be a fee that the consumer pays the provider for using the e-service (e.g. a surcharge for buying a movie ticket online), but that might not always be the case as is exemplified by some e-services offered by the government.

Much of the theoretical literature on e-services is found within the service and service marketing field and focuses mainly on e-services that are substitutes or complements to offline goods or services. This stream of literature takes mainly a commercial approach to investigating e-services (e.g. Hofacker et al., 2007). However, there are many studies investigating e-services that do not have an immediate commercial return, but are offered for example by government agencies or are e-services provided and co-produced by users such as wikis (e.g. Gogia, 2008). Therefore, if we take a broader approach to e-services than just service marketing and e-commerce, we can distinguish the following 4 main groups of e-services:

- Business-to-business (B-to-B): these play an important role in the trend towards supply chain integration and coordination (Iyer et al., 2004) as for example in the case of outsourced printing services and facilities (e.g. Penttinen et al., 2008). B2B suppliers often create more powerful electronic fulfilment mechanisms than do B2C retailers. In addition, they might function also in the back stage (e.g. Janssen, 2009).

- Business-to-consumer (B-to-C): these are mostly commercial e-services as for example e-banking (e.g. Toufaily et al. 2009), e-newspapers (e.g. Ihlstrøm et al., 2008) and some types of portals (e.g. Pejic-Bach and Pejic-Bach, 2008).
- Government to business (G-to-B) or to consumer (G-to-C): these e-services are not commercial in nature and examples can be social security services provided online to remote areas (e.g. Rubeck and Miller, 2008), telemedicine (e.g. Gogia, 2008) or e-libraries (e.g. Scupola and Nicolajsen, 2009).
- Consumer-to-consumer (C-to-C): this includes most of the literature on virtual worlds and online communities as for example wikis and online dating (e.g. Krishnan et al., 2003).

Theoretical Approaches to E-Services

While the emerging e-service concept has its roots in the service marketing and e-commerce literature (e.g. Hofacker et al., 2007; Hoffmann, 2003), e-services are being studied from different empirical and theoretical perspectives among which innovation and innovation diffusion literature, management and strategy literature as well as e-government literature. For example, at the border between e-services as defined in this chapter and innovation and diffusion theory, a number of studies have addressed adoption and diffusion of business to consumer e-services (e.g. Toufaily et al., 2009), e-services business models (e.g. Ihlstrom et al. 2008), user-driven innovation in e-services (Scupola and Nicolajsen, 2009), business-to-business e-services innovations and transformations (e.g. Legner, 2008; Cooper et al., 2009; Janssen, 2009), and e-government (e.g. Lawson-Body, 2008).

Furthermore, Gurao (2008) divides the marketing literature focusing on e-services into two major streams: one investigating consumer readiness and one addressing e-service quality. According to Gurao (2008), consumer readiness in e-services focuses on three main elements: role clarity, motivation and ability. Regarding e-service quality, different studies have found different variables which include ease of use, aesthetic design, security, intangibility, reliability, availability. Table 4 presents selected studies of e-services.

E-SERVICES CHALLENGES

As our literature review reveals there are many definitions and descriptions of e-services. Each definition and description has implications for theorizing and empirically investigating e-services. However, when we define the e-service concept as the convergence of some types of services (e.g. knowledge services) and goods, codification of the production or the content of the service becomes essential. This can be reached either by codifying the service content or by developing ICT-based tools through which knowledge is embedded and thus provided to the customer. This section addresses the second part of the research question and attempts at understanding the degree to which the knowledge embedded in the service provision can be codified, entered on ICT, and sold on markets as informational goods or e-services. This is important because there are different types of services and e-services. Some can be characterized as homogenous, inventoried and with separable consumption. Others, which are more personalised, require personal involvement at each end. Here the service provided can

Table 4. Theoretical approaches to investigating e-services- selected readings

Innovation and innovation diffusion	Management and strategy	Marketing	
		Consumer readiness	*E-service quality*
Black et al. (2001)	Boyer et al. (2002)	Schneider and Bowen (1995)	Parasuraman et al (2005)
Toufaily et al. (2009)	Pejic-Back and Pejic-Bach (2008)	Meuter et al. (2005)	Gournais and Dimitriadis (2003)
Scupola and Nicolajsen (2009)	Gullkvist (2008)	Bobbitt and Dabhholkar (2001)	Bauer (2006)
Walker et al. (2002)			Collier and Bienstock (2006)
Gerrard and Cunningham (2003)			Jun and Cai (2001)
			Zeithaml et al. (2003)

neither be inventoried, nor is it homogenous, and it cannot be copied as it involves social relations.

Codification of Knowledge

Codification is not always an easy task especially if complex knowledge, building on experience, is involved (Nonaka, 1994). In the second session of this chapter, the concept of 'data, information and knowledge' was used. However, there is an important difference between data, information and knowledge. While the data concept denotes the simple elements, information denotes a more complex structure of data elements put into context, whereas knowledge constitutes an even more complex phenomenon, namely an understanding of relationships between different data elements, i.e. a conceptual framework including norms (Davenport and Pruzak, 1998). A persuasive conceptualization of 'data, information and knowledge' is that knowledge is a tool used to turn data into information, meaning that knowledge is used as a framework for putting data into context transforming it into information (Valentin and Hansen, 2004). This discussion resembles the distinction between knowledge as product and knowledge as process (e.g. see Newell et al., 2002 or Cook and Brown, 1999). Cook and Brown (1999) argue that knowledge as a product needs to be recreated (reprocessed) in order to become valuable in a new context (knowledge as process). This recreation process can be more or less complicated and feasible depending on 1) the knowledge to be exchanged as well as 2) the actors involved and their capabilities in communicating and understanding the knowledge. There is thus a process of codifying knowledge as well as one of decoding or reinterpretation.

Following the above understanding, the question is to what extent it is possible to codify knowledge and use it in the production of informational services for digital distribution or e-services? The question raised is the degree to which the knowledge embedded in the service provision can be codified, entered on ICT, and sold on markets as informational goods.

In discussions on the codifiability of knowledge, the issue of tacitness is often invoked (Polanyi, 1958; Cowan, David and Foray, 2000, Nonaka, 1994). The question of codifiability concerns the degree to which tacit knowledge can be turned into explicit or codified knowledge. Limitations on the codification of knowledge are often related to lack of language and implicit and complex relations without clear cause and effect. This is further explained by Blackler (1995) as he lists five different images or sources of knowledge. He argues that codified knowledge is a special and incomplete type of one of the four other types of knowledge: embodied, embedded, embrained or encultured knowledge. The knowledge we use when providing a service is most often personal as well as social. Our bodily, cognitive and emotional personal competences and the greater systems, norms and structures that we are part of are involved.

According to Newell et al. (2002) a number of reasons can be given as to why it may not be worthwhile or possible to codify knowledge. First of all, it can be too *difficult to explain*, it can be *imprecise* or difficult to *verify*, or it can be too *time consuming* because of changes in conditions and the knowledge needed. Newell et al. (2002) also argue that there may be *political issues* involved, making codification sensitive. In the service provision this could be related for example to revealing inherent norms or differences in the treatment of customers. There can also be too big a need for variations, like customizing the service to the receiver (reducing the possibility of standardization).

Difficulties in codification are often related to differences in competences, knowledge and understanding between the sender and receiver which in our case equal the service provider and the service customer. Markus (2001) has examined such challenges arguing that the knowledge base and especially the differences in knowledge base are highly important and affect the possibility of knowledge transfer and also how much is needed to be codified to make sense. When two communicating persons have the same background and experiences it may be easier for them to communicate and exchange knowledge (Markus, 2001). However, if an expert communicates with a less experienced receiver it is necessary to consider the knowledge base of the newcomer to judge what he needs to know to get a valuable service. This includes considerations on how much detail and explanations is needed to make the receiver understand what is conveyed.

Even in services with a great amount of codified knowledge, it is not certain that this service can be fully sold digitally or as a stand-alone e-service. All the knowledge that can be codified and put on a piece of paper, can be entered on digital media. However, this does not mean that it can be sold as an all-digitized service or e-service. Legal services, for instance, are based on laws and other legal documents that can easily be put in digital form, but need the right competences for interpretation. Another example is the buying and selling of real estate. All legal documents and other kinds of information related to the buying and selling of real estate can be codified and digitized. However, real estate agents still make money on the intermediation of these processes. A major reason is that buyers and sellers of real estate do not have sufficient confidence in their own handling of these affairs. They want an intermediary with experience regarding decoding/interpretation in these matters (Hagedorn-Rasmussen, 2006).

This issue is closely related to the process of interpretation. Customers need an interpretation of the facts and documents. Such interpretations can, of course, be written down. However, an overview of the different kinds of interpretations requires experience, which is what service providers offer. This is even more apparent if dealing with people from different countries or cultures (knowledge as encultured). The implication is that the production and marketing of knowledge and knowledge services is limited to contexts in which there is a match of interpretation. International engineering consultancy services in the house building market provide a good example: local offices of engineering consultancy companies are often essential, as the requirements and environments are different from country to country (Henten, 2005).

We can then conclude that many core knowledge-intensive services might not easily lend themselves to become e-services. This applies for example to services that are not easily codified because the knowledge is too complex, too situational or dependent on experiences or trust, issues that delimit the codification needed in what we understand as the core of e-services in this chapter.

In addition to these core e-services, there are a number of services, which are delivered via digital networks, but which are not totally codified. This is illustrated in Figure 1, which

Figure 1. E-services

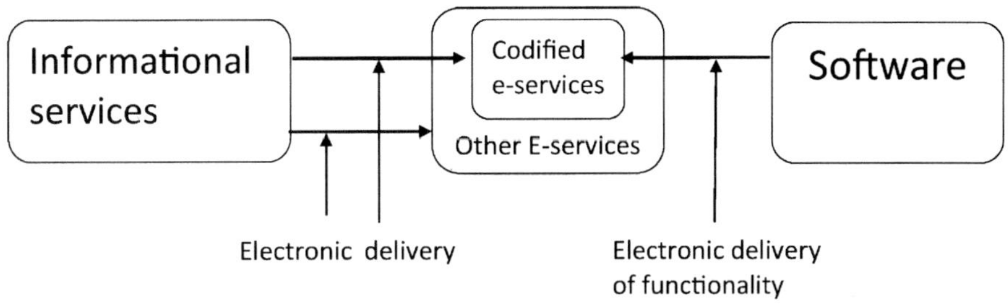

differentiates between 'codified e-services' and 'other e-services'. Codified e-services include informational services, which are codified and delivered on digital networks, as well as software functionalities delivered as services. Other e-services are the non- or not entirely codifiable services delivered on digital networks.

CONCLUSION

The purpose of this chapter has been to add to the literature on e-services by discussing the e-service concept - its strengths and scope - and importance of knowledge codification for further development of e-services. As already said, for a concept to gain and maintain scientific value, it must refer to features, relationships or developments in a manner, which differentiates it from and adds something to other already well-established concepts. In this chapter we have discussed different interpretations of the e-service concept and we have related e-services to services and goods. The conclusion is that the e-service concept denotes and connotes such features and developments.

Specifically, its primary scientific value relates to the ongoing convergence of services and goods facilitated by digital communications. Services, which hitherto depended on face-to-face encounters, can increasingly be digitized and delivered on communication networks whereby the connection between the immediate producer and the service is changed turning the service into good-like product. Conversely, some other products that formerly were sold to the customers as goods are increasingly delivered as services to customers. The e-service concept thus encompasses all informational services (data, information and knowledge) and software delivered on digital networks.

This understanding of the e-service concept gives emphasis to standardized and modularized services. However, there are also e-services, which are less standardized. This applies, for instance, to consultation services, which are delivered via communication networks, but consist of non-standardized interpersonal communications between service providers and service users (e.g. Gogia, 2008; Rubeck and Miller, 2008).

With respect to the conditions for the development and growth of e-services, the chapter points at the codifiability and interpretation of the knowledge embedded in the service provision as crucial aspects, thus emphasizing the delivery of standardized services, including software as a service, via digital networks.

We suggest that future work on e-services must centre on the conditions for turning traditional informational services into e-services including the drivers and barriers for the codification and decodification of knowledge. Also, the drivers and barriers for software functionalities to be sold as a service need to be further examined. For software functionalities to be sold as e-services, the major challenges are related to changing business models and to the market acceptance of new ways of acquiring the utilities of software.

The chapter suggests a number of themes related to the e-service concept which are necessary to take into consideration in scientific analyses of e-service developments. These themes need further elaboration. It could, for example, be interesting to deepen the discussion on service and goods characteristics in order to better explain and understand the concept of e-services. Nevertheless, we believe that the chapter has made a contribution to understanding the e-service concept, which might be of interest to researchers and managers alike.

REFERENCES

Barua, A., Konana, P., Whinston, A. B., & Yin, F. (2001). Driving e-business excellence. *Sloan Management Review*, *43*(1), 36–44.

Black, N. J., Lockett, A., Wiklofer, H., & Ennew, C. (2001). The adoption of Internet financial services: A qualitative study. *International Journal of Retail & Distribution Management*, *29*(8), 390–398. doi:10.1108/09590550110397033

Bobbitt, L. M., & Dabholkar, P. A. (2001). Integrating attitudinal theories to understand and predict use of technology-based self-service: The Internet as an illustration. *International Journal of Service Industry Management*, *12*(5), 423–450. doi:10.1108/EUM0000000006092

Boyer, K., Hallowell, R., & Roth, A. V. (2002). E-services: Operating strategy—A case study and a method for analyzing operational benefits. *Journal of Operations Management*, *20*, 175–188. doi:10.1016/S0272-6963(01)00093-6

Chesbrough, H., & Spohrer, J. (2006). A research manifesto for services science. *Communications of the ACM*, *49*(7), 35–40. doi:10.1145/1139922.1139945

Collier, J. E., & Bienstock, C. C. (2006). Measuring service quality in e-retailing. *Journal of Service Research*, *8*(3), 260–275. doi:10.1177/1094670505278867

Cooper, V., Lichtenstein, S., & Smith, R. (2009). Successful Web-based IT support services: Service provider perceptions of stakeholder-oriented challenges. *International Journal of E-Services and Mobile Applications*, *1*(1), 1–20. doi:10.4018/jesma.2009092201

Cowan, R., David, P. A., & Foray, D. (2000). The explicit economics of knowledge codification and tacitness. *Industrial and Corporate Change*, *9*(2), 211–253. doi:10.1093/icc/9.2.211

Dabholkar, P. A. (1991). Using technology-based self-service options to improve perceived service quality. In M. C. Gilly, et al. (Eds.), *AMA Summer Educator's Conference Proceedings* (pp. 534-535). Chicago, IL: American Marketing Association.

Dabholkar, P. A. (1996). Consumer evaluations of new technology-based self-service options: An investigation of alternative models of service quality. *International Journal of Research in Marketing*, *13*(1), 29–51. doi:10.1016/0167-8116(95)00027-5

Dell, M. (2000). *Direct from Dell: Strategies that revolutionized an industry*. New York, NY: Collins.

Evanschitzky, E., & Iyer, G. (Eds.). (2007). *E-services – Opportunities and threats. Journal of Value Chain Management, 1(1/2) (Special Issue)*. Wiesbaden, Germany: Deutscher Universitäts-Verlag.

Gerrad, P., & Cunningham, J. B. (2003). The diffusion of Internet banking among Singapore consumers. *International Journal of Bank Marketing*, *21*(1), 16–28. doi:10.1108/02652320310457776

Gogia, S. B. (2008). Providing tele-mental health services after disasters – A case based on the post tsunami experience. In Scupola, A. (Ed.), *Cases on managing e-services* (pp. 238–252). Hershey, PA: Idea Group Inc. doi:10.4018/9781605660646.ch016

Gournaris, S., & Dimitriadis, S. (2003). Assessing service quality on the Web: Evidence from business-to-consumer portals. *Journal of Services Marketing, 17*(5), 529–548. doi:10.1108/08876040310486302

Gullkvist, B. (2008). Emerging e-services in accounting: A longitudinal case study. In Scupola, A. (Ed.), *Cases on managing e-services*. Hershey, PA: Idea Group Inc. doi:10.4018/9781605660646.ch007

Hagedorn-Rasmussen, P. (2006). *Making sense of 'e-knowation' – Exploring the relationship between emerging strategy, innovation and entrepreneurial nets of critical capabilities and resources*. Working paper nr. 8. Lyngby: E-service project.

Hansen, M. T., Nohria, N., & Tierney, T. (1999). What's your strategy for managing knowledge? *Harvard Business Review, 77*(2), 106–116.

Henten, A. (1994). *Impacts of information and communication technologies on trade in services*. Lyngby, Denmark: Technical University of Denmark.

Henten, A. (2005). *Internationalisation of knowledge services: The case of engineering consultancy services – Ramboll*. Presentation at Nordic Conference on Innovation and Value Creation in the Service Economy. Oslo, 25-26 October 2005.

Hill, T. P. (1977). On goods and services. *Review of Income and Wealth, 23*(4), 314–339. doi:10.1111/j.1475-4991.1977.tb00021.x

Hofacker, C. F., Goldsmith, R. E., Bridges, E., & Swilley, E. (2007). E-services: A synthesis and research agenda. *Journal of Value Chain Management, 1*(1/2), 13–44.

Hoffman, K. D. (2003). Marketing + MIS = E-service. *Communications of the ACM, 46*(6), 53–55. doi:10.1145/777313.777340

Ihlström Eriksson, C., Kalling, T., Åkesson, M., & Fredberg, T. (2008). Business models for m-service– Exploring the e-newspaper case from a consumer view. *Journal of Electronic Commerce in Organizations, 6*(2), 29–57. doi:10.4018/jeco.2008040103

Iqbal, Z., Verma, R., & Baran, R. (2003). Understanding consumer choices and preferences in transaction-based e-services. *Journal of Service Research, 6*(1), 51–65. doi:10.1177/1094670503254276

Iyer, K. N. S., Germain, R., & Frankwick, G. L. (2004). Supply chain B2B e-commerce and time-based delivery performance. *International Journal of Physical Distribution & Logistics Management, 34*(8), 645–661. doi:10.1108/09600030410557776

Janssen, M. (2009). The rebirth of intermediaries: New intermediaries for managing and orchestrating organizational networks using e-services. *International Journal of E-Services and Mobile Applications, 1*(1). doi:10.4018/jesma.2009010104

Javalgi, R. G., Martin, C. L., & Todd, P. R. (2004). The export of e-services in the age of technology transformation: Challenges and implications for international service providers. *Journal of Services Marketing, 18*(7), 560–573. doi:10.1108/08876040410561884

Jun, M., & Cai, S. (2001). The key determinants of Internet banking service quality: A content analysis. *International Journal of Bank Marketing, 19*(7), 276–291. doi:10.1108/02652320110409825

Kalakota, R., & Whinston, A. B. (1996). *Frontiers of electronic commerce*. Reading, MA: Addison-Wesley Publishing Company, Inc.

Krishnan, R., Smith, M. D., & Telang, R. (2003). The economics of peer-to-peer networks. *Journal of Information Technology Theory*, *5*(3), 31–44.

Newell, S., Robertson, M., Scarbrough, H., & Swan, J. (2002). *Managing knowledge work*. New York, NY: Palgrave.

Nonaka, I. (1994). A dynamic theory of organizational knowledge creation. *Organization Science*, *5*(1), 14–37. doi:10.1287/orsc.5.1.14

Pejic-Bach, M., & Pejic-Bach, M. (2008). Profightstore.com – Developing an online store for the niche market. In Scupola, A. (Ed.), *Cases on managing e-services* (pp. 76–88). Hershey, PA: Idea Group Inc. doi:10.4018/9781605660646.ch006

Penttinen, E., Saarinen, T., & Sinervo, P. (2008). The role of e-services in the transition from the product focus to the service focus in the printing business. In Scupola, A. (Ed.), *Cases on managing e-services* (pp. 156–165). Hershey, PA: Idea Group Inc.doi:10.4018/9781605660646.ch011

Piccinelli, G., & Stammers, E. (2001). *From e-processes to e-networks: An e-service-oriented approach*. Retrieved from http://www.research.ibm.com/people/b/bth/OOWS2001/piccinelli.pdf

Pine, B. J., & Gilmore, J. H. (1999). *The experience economy*. Boston, MA: Harvard Business School Press.

Polanyi, M. (1958). *The tacit dimension*. New York, NY: Doubleday.

Rowley, J. (2006). An analysis of the e-service literature: Towards a research agenda. *Internet Research*, *16*(3), 339–359. doi:10.1108/10662240610673736

Rubeck, R., & Miller, G. A. (2008). vGOV! Remote video access to government services. In A. Scupola (Ed.), *Cases on managing e-services* (pp. 253-268). Hershey, PA: Idea Group Inc.

Rust, R. T. (2001). Editorial: The rise of e-services. *Journal of Service Research*, *3*(1), 283–284. doi:10.1177/109467050134001

Rust, R. T., & Kannan, P. K. (2003). E-service: A new paradigm for business in the electronic environment. *Communications of the ACM*, *46*(6), 37–42. doi:10.1145/777313.777336

Scupola, A. (2008a). Conceptualizing competences in e-services adoption and assimilation in SMEs. *Journal of Electronic Commerce in Organizations*, *6*(2). doi:10.4018/jeco.2008040105

Scupola, A. (2008b). E-services: Definition, characteristics and taxonomy: Guest editorial preface, Special Issue on E-Services. *Journal of Electronic Commerce in Organizations*, *6*(2), i–iii.

Scupola, A. (Ed.). (2008c). *Cases on managing e-services*. Hershey, PA: Idea Group Inc.

Scupola, A., & Nicolajsen, H. W. (2009). The role of customer involvement in library e-services. In D'Atri, A., & Saccà, D. (Eds.), *Information Systems: People, organizations, institutions, and technologies*. Springer. doi:10.1007/978-3-7908-2148-2_3

Sundbo, J. (1997). Management of innovation in services. *The Service Industries Journal*, *17*(3), 432–455. doi:10.1080/02642069700000028

Sweeney, J. C., & Lapp, W. (2004). Critical service quality encounters on the Web: An exploratory study. *Journal of Services Marketing*, *18*(4), 276–289. doi:10.1108/08876040410542272

Tiwana, A., & Balasubramaniam, R. (2001). *E-services, problems, opportunities, and digital platforms*. 34th Hawaii International Conference on System Sciences. March 2001.

Toufaily, E., Daghfous, N., & Toffoli, R. (2009). The adoption of e-banking by Lebanese banks: Success and critical factors. *International Journal of E-Services and Mobile Applications, 1*(1). doi:10.4018/jesma.2009010105

Valentin, F., & Hansen, P. V. (2004). *Udvikling af videnservice (Development of knowledge service)*. Copenhagen, Denmark: Nyt Teknisk Forlag.

Vragov, R. (2009). Sponsored search as a strategic e-service. *International Journal of E-Services and Mobile Applications, 1*(1). doi:10.4018/jesma.2009010102

Walker, R. H., Craig-Lees, M., Hecker, R., & Francis, H. (2002). Technology-enabled service delivery: An investigation of reasons affecting customer adoption and rejection. *International Journal of Service Industry Management, 13*(1), 91–106. doi:10.1108/09564230210421173

Wang, J., Yao, J., Yan, R., & Hsu, J. (2008). E-bay: An e-titan success story. In Scupola, A. (Ed.), *Cases on managing e-services* (pp. 126–140). Hershey, PA: IGI Global. doi:10.4018/9781605660646.ch009

WhatIs.com. (n.d.). *E-services*. Retrieved from http://searchcio.techtarget.com/sDefinition/0,sid19_gci535452,00.html

Zeithaml, V., Parasuraman, A., & Malhotra, A. (2003). Service quality delivery through Web sites: A critical review of extant knowledge. *Journal of the Academy of Marketing Science, 30*(4), 362–375. doi:10.1177/009207002236911

Chapter 3
Key Challenges for Delivering Web-Based IT Support Services:
The Provider Perspective

Vanessa Cooper
RMIT University, Australia

Sharman Lichtenstein
Deakin University, Australia

Ross Smith
RMIT University, Australia

ABSTRACT

This chapter explores the key challenges faced by IT service providers offering after-sales support services to enterprise customers via the Web. The chapter is based on insights provided by representatives of six multi-national IT service organizations. It considers how IT service providers use knowledge-based approaches to offer such services. The chapter highlights the trend towards the employment of networks of business partners to assist with the processes of support delivery. By exploring the perceived needs of the different stakeholders involved in IT support provision, and the transfer of knowledge across these organizations and individuals, the key challenges in IT support provision are identified. The chapter proposes a range of solutions to address the key challenges.

INTRODUCTION

Experts forecast that, post the recent economic downturn, a revolution in information technology (IT) services is on the way with the advent of cloud computing and software as a service (McCarthy & Matzke, 2010). They note that the IT services market brought in $450 billion in 2010. Other reports suggest that "managed IT services" are booming not only for large client businesses but, increasingly, small client businesses (Shukla, 2010). While lower costs for client firms are predicted as part of the coming IT services revolution, providers will also have very good opportunities for profit making (McCarthy & Matzke, 2010) with increasing leverage of IT and the Web be-

DOI: 10.4018/978-1-60960-607-7.ch003

ing employed to deliver value-added customer services (Ragsdale, 2010).

The delivery of electronic services ("e-services") to client firms via the World Wide Web ("Web") has grown in the past decade as a novel way to realise customer service value (Hall, 2008). Web-enabled managed IT support, where IT support services are offered to enterprises via the Web, is a popular e-service and is the subject of this chapter. So far IT service providers have received little insight or guidance from scholars with regard to facilitating IT support delivery success via such approaches. For Web-enabled managed IT support, complex systems involving multiple stakeholders such as business partners and contact centres, are often used. Such systems have been termed "Web-based self-service systems" (WSSs) as customer self-service is an important feature of such systems (Cooper, 2007). WSSs reach out to many regions, each with its own laws and cultures. They leverage important evolving concepts such as knowledge-based IT solutions. As the complexities of offering IT support to enterprises via WSSs are considerable, IT service providers would benefit from having access to published guidelines and models that can assist them with their efforts.

As a first step, IT service providers would benefit from an understanding of the key influences and challenges for successful IT support delivery when WSSs are involved. There are four major rationales as to why the key influences on successful WSSs for managed IT support should be explored. First, traditional customer service influences such as quality of service have mainly been identified in the context of serving consumers rather than enterprises. Second, prior research has focused on exploring key influences for services delivered by traditional channels such as face-to-face or the telephone rather than the Web channel, or multiple channels with the Web a key channel. Of those that do investigate traditional influences on customer service in the Web-based context, these tend to focus on end-user consumers rather than enterprises (e.g. Udo et al., 2010). Third, with multiple stakeholders involved in IT support service delivery, issues surrounding networks and relationships may be more important than in the simpler contexts of yesteryear. Fourth, critical success factors (CSFs) for customer service delivery have mainly been researched from the customer perspective. Only now is research appearing that takes a provider perspective (e.g. Cho & Menor, 2010). Finally, industry-consortium based guidelines such as knowledge-centred support or "KCS" (Consortium Service Innovation, 2002) are available, however the key enablers and challenges for such a methodology are not available, nor are best practice implementations of this approach.

This chapter will identify the CSFs and key challenges for the use of WSSs for managed IT support. Success is conceptualized as the transfer of relevant knowledge along the chain of IT support service delivery via WSSs with a view to ultimately increasing customer satisfaction while decreasing the cost of support provision. As foreshadowed above, the chapter takes the service provider perspective. To focus this chapter, the provision of operational IT support services relating to (1) assembling and operating the core IT environment, and (2) providing key value-adding after-sales services such as the Service (Help) Desk, is examined (Peppard, 2003).

Following this brief introduction, background literature is presented. We then describe the design of the research study. Next, the findings are presented, highlighting the key issues for Web-based enterprise IT support provision, as identified by participants from six multi-national IT service organizations. Here, the discussion centres on outlining the CSFs and key stakeholder-oriented challenges. Finally the key points from the chapter are discussed and new directions for IT service providers in the challenging environment of WSS-enabled managed IT support are proposed.

BACKGROUND

Customer Relationship Management

There are benefits for increasingly commoditised industries such as IT services for businesses to draw closer to their customers (Reimann et al., 2010). The development of a valuable customer-provider relationship can be addressed by a strategy of customer relationship management ("CRM"). This strategy requires the aggregation of data, information and knowledge about the customer via specialised software that performs electronic CRM (eCRM). eCRM applications may employ a Web interface, supported by database and data mining tools that record past customer transactions and analyse data to identify customer segments, match products to customer profiles, and better understand target demographics and psychographic characteristics (Brohman et al., 2003). Customer service agents can also utilise this "customer intelligence" to potentially up-sell or cross-sell products and services (Brohman et al. 2003). A broader definition of eCRM has evolved as "a kind of CRM that is channel independent in the sense that it uses one company-wide set of data and one set of business rules to manage customer interaction via any channel at any time and can increase customer loyalty" (Simons et al., 2009, p591). Building customer loyalty through eCRM requires understanding what customers regard as important at each service stage. During post-sales support for their problems, customers are more loyal when the service provider "takes them seriously" and solves the problem for them (Simons et al., 2009).

WSS and After-Sales IT Support for Enterprise Customers

A WSS is an important type of operational eCRM application (Geib et al., 2006; Khalifa & Shen, 2005) and is discussed below in the context of managed IT support services. A WSS is an information system that enables organizations to move from labour-intensive manual processes towards low-cost automated Web-based self-service (Pujari 2004). WSSs can facilitate the offering of customer support services for pre-sales, sales, and after-sales activities. They are underpinned by complex information systems, complemented by a customer contact centre, and integrated with a multi-channel service strategy (Negash *et al.* 2003).

In managed IT service environments, WSSs offer informational, transactional, and proactive support services. Informational support includes "break-fix" support which provides customers experiencing technical problems with resolutions to their problems. This includes (1) unassisted support such as answers to Frequently Asked Questions (FAQs) and the provision of downloadable software patches and (2) assisted support such as peer-to-peer online forums, e-mail and online chat. Other informational support includes the provision of information and knowledge to assist with enquiries and enable customers to access best-practice – for example, by the publication of White Papers. Transactional support includes case tracking, whereby the customer initially documents their IT problem scenario, requests assistance from the service provider, and is subsequently able to monitor the service provider's progress in resolving this problem. Proactive support includes the embedding of problem detection support software on customer end-user computers, and personalised messages directing the customer to potential product or service purchases. This chapter is primarily concerned with the use of WSSs to provide informational and transactional after-sales IT support services.

WSSs offer important advantages to service providers and customers (Geib et al., 2006; Pujari 2004). These include electronically leveraging the Web interface, customer/service-provider (and customer/customer) interactions, knowledge management (KM) and self-service principles, in order to capture and provide information and knowledge useful for pre-sales, sales and after-sales support. WSSs can also reduce customer support expenses by empowering employees (Support Agents). These employees provide support in the form of knowledge such as solutions to customers' IT problems. WSSs boost Support Agent productivity by providing access to a comprehensive knowledge base that includes many solutions, delivering convenient, consistent and higher quality support and thus increasing customer loyalty and retention (Consortium Service Innovation, 2002) – a key goal of CRM. Non-IT businesses outsource IT support services to service providers because IT is not their core competency and outsourcing is perceived to offer opportunities to improve service levels and reduce the total cost of ownership (Barthelemy, 2008). Furthermore, client firms are often receptive to Web-based support delivery (SSPA 2004). Thus the Web is considered by customer firms as an important channel for IT support delivery.

A study of Internet Service Providers (ISPs) in Skaanning (2005) illustrates many of these concepts. Here the use of Web-based self-service was able to deflect 50% of support calls that traditionally make up 30-60% of business costs (Potaroo in Skaanning, 2005). This deflection was achieved through transferring support-oriented knowledge to a sophisticated knowledge-base and making this available to customers directly via the Web. Arguably customers are increasingly familiar with the Web and are better able to exploit information services and would prefer to help themselves than spend time waiting for and accessing phone-based services. Importantly, where customers perceived the self-service option offered equal or better service than that provided by the ISP's Service Desk there was a positive impact on customer satisfaction and retention (Skaanning, 2005) – a significant benefit to an industry which traditionally experiences a 25% annual customer "churn rate" (ISP-Planet in Skaanning, 2005) and where the cost of gaining a new customer is four times the cost of retaining an existing customer (Skaanning, 2005).

Despite the potential benefits of WSSs to service providers and customers, there are many reports of dissatisfaction with Web-based self-service (Ragsdale, 2010). One study found that successful Web-based self-service interactions with customers declined from 48% in 2003 to 40% in 2007 (Subramaniam, 2009). These reports suggest that the CSFs and challenges involved for such systems should be identified and addressed.

This chapter identifies a set of CSFs and important challenges, as perceived by IT service providers, in the provision of Web-based IT support (Cooper, 2007). As a stakeholder-based approach can be useful for such a complex inter-organizational setting, a stakeholder-oriented framework is presented next.

Stakeholder-Oriented Relational Web-Based Enterprise Customer Service

Stakeholder needs, roles, responsibilities, relationships and other interactions are especially important to the success of business-to-business (B2B) commerce (Chung et al., 2009; Pan, 2005). In multi-stakeholder networks such as B2B environments issues tend to be complex and it is important to business success that a firm resolve conflicting stakeholder interests (Roloff, 2008). One useful model of stakeholder collaboration in B2B highlights three behavioural processes: knowledge sharing, participative decision-making and conflict governance (Chi & Holsapple, 2005). Other reasons for understanding stakeholder issues in B2B commerce include the need to manage stakeholder expectations (Singh & Byrne, 2005).

Stakeholder relationships in services are associated with emotions and expectations (Gutek et al., 1999). Consequently relationships and

Figure 1. A Stakeholder-oriented relational framework for Web-based enterprise customer service (Cooper et. al., 2006a)

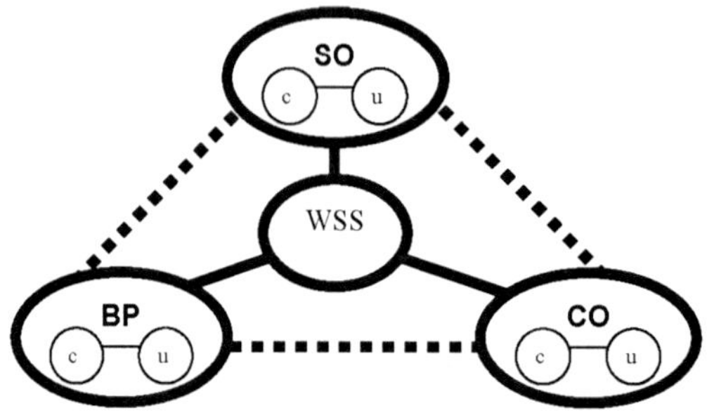

LEGEND
SO = Support Organization
BP = Business Partner Organization
CO = Customer Organization
c = Corporate Entity
u = End-User Entity
WSS = Web-based Self-service System
___ = Relationship via WSS
--- = Direct Relationship

associated knowledge flows are important in managed IT services environments (Dahlberg & Nyrhinen, 2006) as the quality of knowledge exchange influences the quality of outsourcing relationships (Gong et al., 2007). Of interest to this chapter, is how the adoption of WSSs might strengthen stakeholder relationships (Bhappu & Schultze, 2006).

A framework that conceptualises key stakeholders and relationships in a B2B service context is shown in Figure 1 (Cooper et. al., 2006a). The framework depicts key relationships and knowledge flows between the main stakeholder types and a WSS. It shows three key types of stakeholder organizations that may be involved in support provision – a service provider (hereon termed "support organization"), business partner and customer organization – and their interaction with one another directly and indirectly via a WSS. At each organization there are corporate entity representatives (for example, managers) interacting with end-users. The framework clearly highlights the interdependencies found in the multi-stakeholder managed IT support environment. In the next section we explore these interdependencies from a knowledge transfer perspective.

Knowledge Transfer for IT Support of Enterprise Customers using WSS

Below we describe the knowledge transfer process in the managed IT support context, as understood from a review of literature on knowledge transfer and IT support coupled with insights from participants of six multi-national IT Support Organizations (see Cooper et. al., 2006, Cooper, 2007).

IT Support Organizations respond to enterprise customers' after-sales enquiries, incidents and problems regarding core IT products and services by providing and transferring support-oriented knowledge (Figure 2). The knowledge is complex and based on solutions to the enquiries, incidents and problems reported.

The knowledge transfer process is activated when end-users at a customer organization experience IT incidents – or an IT Support Organization recognizes a pattern of incidents at the customer organization and identifies a related problem (Figure 2). An end-user at the customer firm may elect to telephone the IT Support Organization's Service Desk to access assisted support via support staff. Alternatively, he or she may directly access the IT Support Organization's WSS seeking a resolution. Where an end-user does not find a solution accessible as explicit knowledge,

Figure 2. Knowledge transfer in managed after-sales Web-based IT support (Cooper et. al., 2006a)

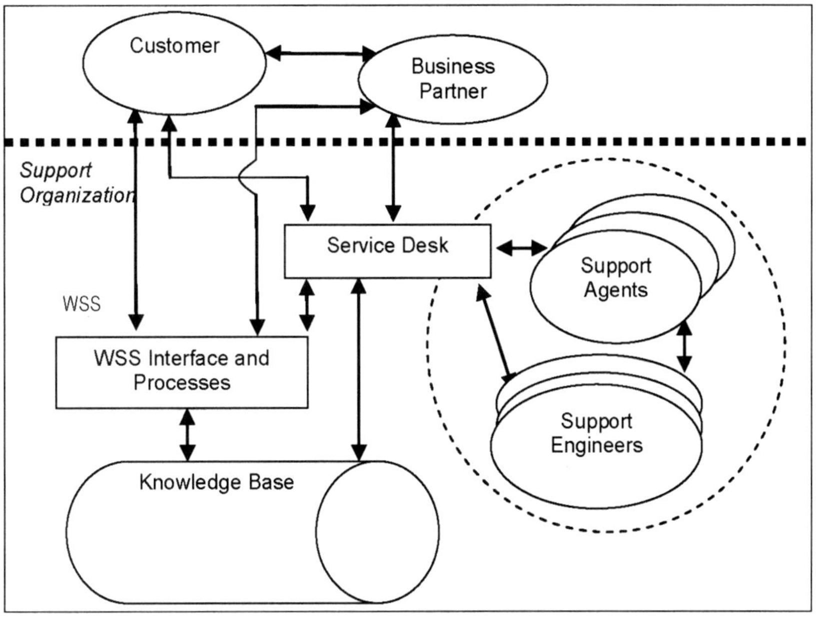

first tier Support Agents at the IT Support Organization identify potentially successful solutions from their tacit knowledge of the subject matter or by searching explicit solutions from the knowledge base. Complex problems are escalated to experienced second and third tier Support Agents. Downstream are several tiers of Support Engineers – technology specialists who ultimately resolve the most difficult problems by drawing on expert tacit knowledge. New and evolving solutions are captured in the knowledge base as explicit knowledge and organised for reuse. Where IT support is provided by a business partner, the customer's IT professionals may communicate directly with that partner's IT professionals. Business partners can access the IT Support Organization's knowledge base via the WSS.

RESEARCH DESIGN

Initially a comprehensive literature review was undertaken. As the environment in which a WSS is planned, developed, implemented and used is a human activity system, we adopted an interpretive research approach (Walsham, 1995) that accounts for the subjectivity and social construction of perceptions. To this end, we conducted an in-depth interpretive case study at the Australian headquarters of a large best-in-class multinational IT services organisation, "ServIT" (a pseudonym). ServIT provides remote IT support to enterprise customers globally and was selected for this study because of its progressive approach to IT support, award-winning customer support Web site, extended enterprise status, and mature, successful WSS strategy integrated with multi-channel, multi-vendor support. The choice of such an organization also enabled issues of high service volume, complexity, dispersion and electronic business to surface. Case data was collected from semi-structured interviews with twelve senior customer service managers, IT support managers and support staff, observations of WSS use, document collection, and a focusing workshop (Rockart, 1979) with participation by five senior customer service managers and IT support managers. Case study data were analysed using qualitative con-

*Table 1. Summary of represented IT support organizations (*pseudonyms)*

Organization*	Description
DataCorp	A large software company, specializing in database, middleware and enterprise management software. The organization's focus is on providing information management technologies and services to its customers. The primary customer base includes small, medium and large organizations.
DataHouse	Provides plan, build and support services for IT projects. The three major areas of business include: systems integration; computer training and IT certification; and IT distribution. Its customers are primarily medium and large enterprises.
DistSystems	A supplier of IT hardware, software and consulting services to consumers, small, medium and large enterprises. The consulting services offered to enterprise customers include: building IT infrastructures; CRM solutions; operations outsourcing; security services; change management; and supply chain operations.
OpSys	An international software company, which focuses on platform products and developing business and entertainment software. The organization also offers consulting services through a global network of business partners. Its customer base includes enterprise customers and consumers.
I-Systems	A supplier of IT hardware, software and consulting services to consumers, small, medium and large enterprises. The consulting services offered to enterprise customers include: building IT infrastructures; CRM solutions; operations outsourcing; security services; change management; and supply chain operations.
ServIT	A supplier of IT hardware, software and managed services to consumers, small, medium and large enterprises. The managed services offered to enterprise customers centre on building, implementing and managing IT infrastructures.

tent analysis techniques (Mayring 2000) by two researchers working independently to identify the key challenges relating to stakeholders. A set of CSFs for the transfer of after-sales IT solution-oriented knowledge to enterprise customers when a WSS is used, and a set of key challenges have been identified.

A focus group of Australian senior customer service and IT support managers from an additional five large multinational IT service provider organisations, all with successful WSSs, was conducted, seeking to confirm and extend the findings from the ServIT case study. A summary of the organizations represented in the study is shown in Table 1. A three hour session took place and feedback was collected by electronic mail. From the focus group discussion it became clear that, while the WSS of each company was successful, various challenges remained. Two researchers working independently confirmed and extended the set of CSFs and key challenges working inductively from the transcript.

The quality of the findings was assured by appropriate techniques. Achieving validity in qualitative research requires a fair, honest and balanced account from the viewpoint of someone who lives in the social situation (Neuman, 2006). Construct validity was assured in the study by use of multiple sources of evidence. Participants were also given copies of interview summaries and the outcomes of each phase of the research were provided to participants for confirmation. Reliability was increased via a case study protocol (Yin 2003) to document all procedures and problems. Triangulation was achieved by collecting and analyzing data from a variety of sources at ServIT and establishing consistency of results. The results of the cross-organizational focus group added another level of triangulation.

FINDINGS

Having introduced the context of enterprise Web-based after-sales IT support provision, the remainder of this chapter provides a discussion of provider perceptions of the key issues for B2B after-sales IT support provision via WSS. As will be shown, many of these issues were found to concern stakeholder-based issues and these are the focus of this chapter.

Critical Success Factors for B2B After-Sale IT Support via WSS

In offering Web-based self-service, participants from each organization identified that the key objective of so doing, is to increase customer satisfaction while reducing the cost of IT support provision. Thus success in this context was defined as meeting this objective and twenty-seven factors were identified as critical in this pursuit. While these CSFs are not the focus of this chapter they are summarised for reference in the Appendix (further details on the CSFs can be found in Cooper et. al., 2006b and Cooper, 2007). The CSFs were also classified non-orthogonally into six categories which were believed by the IT service provider participants to apply to *all* three stakeholder organizations (that is, those shown in Figure 1):

- *Organizational Commitment and Readiness*: The organization must manage the policies, processes and cultural issues which will affect its ability and willingness to embrace Web-based Self-service.
- *Manage for Strategic and Operational Benefits*: The WSS strategy must assist the organization in attaining its strategic and operational objectives.
- *KM Capabilities and Processes*: The organization must practice the principles of knowledge management and implement associated knowledge management processes, to maximise the benefits received from the WSS strategy.
- *IT Infrastructure Capability*: The organization must have an adequate IT infrastructure in place, to enable it to participate in Web-based Self-service.
- *Experience Management*: The WSS should manage the stakeholder's experience, both at the corporate and end-user level. The stakeholder experience will directly affect satisfaction levels and therefore ongoing use of the WSS.
- *Content*: The WSS must contain useful, accurate and up-to-date content in order to resolve the end-user's support issue or knowledge requirement.

Competing Stakeholder Requirements

Participants from IT service provider firms believe that all stakeholder viewpoints should be considered by an IT support organization when planning, implementing and managing a WSS for after-sales IT support provision. This finding is consistent with recommendations for electronic business settings more generally (Kandampully, 2003; Singh & Byrne, 2005). However IT support organizations emphasize that in a B2B WSS context there is the potential for different stakeholder types to hold unique perceptions. For example, while a support organization and some end-user customers with specialized job roles such as Database Administrators may find the transfer of IT solutions to a customer firm highly desirable, it was questioned by IT service provider participants whether all end-users at a customer enterprise would feel the same way:

I think the provider [support organization] is interested in transferring knowledge so [that] they don't have a problem any more and they can manage their costs and help the customers. [However, the customer firm's end-user is thinking] I am interested, not in receiving knowledge... [but] I am interested in my problem being fixed. Don't give me all this stuff [the details of the problem and solution]. Tell me what the problem is so I can fix it. I don't want the transferring of any knowledge. (Senior IT Architect, I-Systems)

If an end-user at a customer firm does not wish to learn from a provided IT solution, how can a support organization ensure such learning? This is an important issue deserving of further research as it may impact on the number of "repeat calls" received by the IT support organization which ultimately has implications for cost effectiveness of IT support provision.

Cost-Benefit Analysis of WSS Using a Multi-Stakeholder Perspective

The IT support organization's motivation for knowledge transfer highlights the proposed cost and efficiency benefits to a support organization in offering Web-based self-service. However there is a lack of evidence as to whether WSS has the same benefits for the customer or whether Web-based self-service may even have negative implications for customer satisfaction. While senior management at ServIT could readily cite the results of studies reporting the cost savings to the company in providing Web-based self-service as opposed to telephone support, detailed studies investigating the costs of this move to customers and the associated implications for customer satisfaction were not as readily available, for example:

One would assume that if [customers] are not having to pick up the telephone and call us that they have sort of got what they want [particularly] if satisfaction levels are still chugging along... I am pretty sure the sales staff conduct a questionnaire [on satisfaction with the WSS]. (Customer Technology Division Manager, ServIT)

Thus further studies of the cost and benefits of Web-based self-service from the customer perspective are needed. An additional issue which highlights the need for further consideration of the costs and benefits of knowledge transfer via WSS, is the potential of enterprise customers to become increasingly independent of the support organization. If such independence negatively impacts on the profitability of the support organization there may be a level of knowledge that a support organization may wish to retain and not transfer to customer firms. For example:

The more you allow a customer to self-help, self-heal, you can see a very logical link, which is the more products you will sell. But you can also see a logical link to a reduction in service and revenue. I guess the strategy is [based] around 'where do you put your investment dollar?' It's like the old light globes. They can manufacture light globes that will last forever, but they will never sell another light globe – that sort of thing. (IT Solutions Manager, ServIT)

Need to Address CSFs for WSS Using a Stakeholder-Based Perspective

While the CSFs were identified as relevant to all stakeholder organizations, participants from IT support organizations questioned whether the different stakeholder types might interpret the requirements of the CSFs differently. For example, in terms of *CSF-9 Ease of use,* participants mentioned that regular end-users of the WSS interface would prefer an efficient interface, whereas novice users would prefer easy-to-use interfaces. As another example, in terms of *CSF-1 Cost-effectiveness* some participants mentioned that cost-effectiveness when a WSS is used to a support organization, is not the same as cost-effectiveness from the perspective of an end-user at the customer firm. End-users are concerned with efficiency gains and usefulness of the knowledge gained from the WSS to perform their jobs (Cenfetelli et al., 2005), rather than the financial costs involved.

Cost effectiveness, from both the end-user of the service and for ServIT, are equally important, but I think you will find some subtleties in [how they are] both explained as cost-effectiveness. For an end-user it's 'Can I actually get the results quickly from my perspective?' From ServIT's perspective, it is 'Can we actually reduce the cost of this service for our customers so that we are actually making a profit? (Consulting Services Division Manager, ServIT)

Performance Management and Measurement of CSFs for WSS in B2B Context

The six CSF categories were found to apply to *all three* stakeholder organizations. Consequently, capabilities and interdependencies within an industry will have implications for WSS success, particularly around performance management and measurement.

First, industries have been found to have different levels of capability and motivation when it comes to Internet-based trading and adoption of inter-organizational systems (Kurnia & Johnston, 2002). Thus if customer organizations operate within industries with low-levels of adoption of Internet-based inter-organizational systems, this may influence the adoption and use of Web-based self-service for after-sales IT support by particular customer segments. Second, as electronic commerce environments are increasingly interdependent (Kumar & van Dissel, 1996; Kurnia & Johnston 2002; Cameron 2006), if corporate customers and business partners in a managed IT support situation do not manage the CSFs, the support organization's WSS may not be successful. This is consistent with previous research investigating self-service technologies which also argues that customers are often partly responsible for a service provider's success (Schultze & Bhappu, 2005). For example, regardless of whether the support organization has in place a state-of-the-art IT infrastructure to support the WSS, if customer or partner organizations do not ensure that they have an adequate IT infrastructure capability, the WSS will not be successful. However, a support organization will have little control over whether a customer organization and business partner organization do, in fact, address the six CSF categories. Thus a support organization may benefit from providing further assistance to business partners and customer organizations to better manage the key factors.

To address these issues, one possible strategy may be for support organizations to conduct education and training programs with partner and customer organizations to increase their awareness and understanding of CSFs. Such education and training could be provided by online support, training, newsletters and so on. These programs in addition to providing knowledge to increase customer and business partner awareness of WSSs, may also offer support organizations opportunities to develop relationships with customers and business partners and thus ultimately assist in meeting key objectives for CRM. This finding supports current literature highlighting the importance of developing relationships with enterprise customers when providing support services (Peppers & Rogers, 2001; Pujari, 2004).

A second strategy would be to ensure when developing external service contracts and associated service level agreements (SLAs), that customer and business partner organizations are obliged to meet minimum standards relating to the CSFs. Any failure to do so would acknowledge that this would affect the ability of a support organization to service them effectively. Of course, it would be important to carefully consider both customer and business partner perspectives of such a strategy, as it may be incompatible with the development of improved relationships between stakeholders.

Implications for B2B Relationships in WSS Context

There are interesting implications for relationships in the B2B WSS context. One perspective questions whether relationships between the end-user and the support organization are possible in a B2B WSS context, particularly when the end-user uses the WSS anonymously. At most, this type of interaction would be perceived as one-way only. An alternative perspective considers that two-way relationships between the support organization and customer and business partner organizations are possible via WSSs. From this perspective, if an end-user trusts the Web-site and the organization behind it, this trust forms part of the relationship, with the WSS simply providing the connection:

Although it is a piece of software, it is the front end of a company. For instance, if I feel I have a positive relationship with my Netbank Web-site because I trust it, then that is part of the relationship. (IT Customer Consultant, DistSystems)

Further, stakeholder relationships may also be more complex than Figure 1 allows as relationships with stakeholder types may vary. For example, when supporting some customer firms a business partner is involved, while when supporting other customers there are no business partners involved:

We can have the relationship with either both the customer and the business partner or we can have it directly with the customer or we could just have it with the business partner…I mean there are just so many different combinations that that relationship can actually take. (IT Customer Consultant, DistSystems)

Given that support organizations are highly interested in meeting objectives for CRM while simultaneously reducing costs via Web-based Self-service strategies, all of these issues are worthy of further investigation.

Implications for Intellectual Capital in B2B WSS Context

Intellectual property (IP), security and privacy issues can be problematic in complex multi-stakeholder contexts as firms in collaborative relationships must protect strategic knowledge (Solitander, 2006). IT support organizations, business partners and customers frequently operate across international borders thus support organizations must consider issues of security, privacy and IP within the context of off-shore environments (Rai & Sambamurthy, 2006).

IT service providers have expressed uncertainty about the ability of legal systems to deal with such complex issues. Consequently, sensitive information, knowledge and IP may indeed be compromised (Rai & Sambamurthy, 2006). For example, customer end-users may be unaware of the location where their personal information is stored and retrieved. Further, while customer organizations may be aware that, in some instances, third parties such as a support organization's business partners have access to their stored personal information, they may not be aware of the eventual *use* of this information or its accessibility. A perceived potential privacy violation may lead to unforeseen problems – for example, where legal systems across countries vary and where business partners of the support organization compete with a customer enterprise:

A really interesting one that is going through the legal battles now and it comes back to privacy. As a main [support] organization, you have information that could be of a private nature. If you then make the information available to another organization that does not have the same privacy policy as you, and then [if they] were to use that information in such a way that it violates somebody's privacy, who is ultimately responsible? That is a huge legal question that is being tested now. Does it come back to the original person or the person who let the information go? (Emerging Technology Consultant, OpSys)

In the after-sales WSS context, some support organizations are also concerned about the IP that is leveraged in user forums. In such forums, end-users may provide solutions to reported problems however it can be unclear who owns the IP that is the solution. An additional issue faced by support organizations stems from the knowledge posted in online forums not being captured more formally into the WSS. This represents a possible loss of useful knowledge. Greater consideration of the costs and benefits associated with such capture is needed.

Complexities in Alliances in B2B WSS Context

Business partner alliances are complex and there are varying perceptions held about the distinction between a business partner relationship and a customer/supplier relationship. For example, in the cross-organizational focus group, one participant claimed that his organization did not have business partners, while other participants claimed that their organizations were, in fact, partners of his organization. It was subsequently advocated that the distinction between a business partner and customer/supplier relationship, surrounds whether the relationship is a one-off transaction, or whether it is an ongoing relationship. In an enterprise IT services context, however, there is an apparent blurring of relationships. In some projects, organizations may be considered business partners, while in others they would be considered competitors:

DistSystems, DataCorp, OpSys and I-Systems can say, we have been working together as business partners and as competitors and sometimes even on the same bid... on the one hand you compete, in the next 30 seconds, I might be talking to OpSys about something we are competitors about, and in another thing we are working together on, you switch hats, you switch alliances, it is just the way things are these days... (Senior Systems Consultant, I-Systems)

The complexity of relationships also increases as stakeholder organizations increase in size, as multiple relationships are developed at the individual, departmental and corporate levels. This is significant as while it has been identified that developing relationships with partners is important, there is less understanding around partner-unit-based relationships (Vlachopoulou & Manthou, 2003).

Some of the complexities of business partner alliances surround intellectual property, security and privacy, discussed earlier. Lei (1997) argues that, with respect to strategic alliances, regardless of the various types of legal structures which may be put in place, over time companies will absorb and internalise skills, regardless of the amount of formal, legal ownership that is demarcated by the alliance structure. Further research on illegal inter-organizational knowledge transfer is also needed (Ferdinand & Simm, 2007).

Customer Co-Contribution to Service

Managing customer contributions to service is an important challenge in the provision of Web-based self-service. Customer co-production ("co-production"), is reported to result in productivity benefits but it also gives rise to a range of managerial challenges when customers become "partial employees" (Benapudi & Leone, 2003; Nambisan & Baron, 2010). The advantages of co-production include decreased cost of service provision and greater control and autonomy for the customer, while disadvantages include the difficulty of controlling service quality when customers are actively involved in the production process (Schultze & Bhappu, 2005).

In the provision of after-sales IT support, IT service providers have raised concerns relating to the contribution of knowledge by customers to their own service fulfillment, optimisation and improvement. The more common concern, however, surrounds the contribution of customers to the service provision of other customers. For example, when customers interact and share resolution-knowledge in online forums, this may result in customer dissatisfaction if the offered advice results in additional support issues:

ServIT has ventured into the hosting of forums. Solutions can be provided by non-ServIT people and that's a potential conflict between the reliability of our knowledge and the fact that we are opening up [knowledge] to end-users, which could have a good result, but there is a danger that we are facilitating the [incorrect] solution. (IT Manager, ServIT)

To address this concern support organizations typically invest significant resources in their legal teams and explicitly state in the terms and conditions of using their WSSs and in support contracts that they will not be responsible for any degradation of a customer's systems, if a customer decides to act upon information derived from the online user forums. Further, while support organizations may monitor forum content to some extent, often they rely more heavily on a "merit" based systems whereby forum users are allocated merit points by posters (of questions) when they provide valuable responses. This system of points aggregation is designed to motivate users to share valuable knowledge and thus contribute to service provision as well as to provide potential users an indication of how reliable the knowledge may be.

Other potentially negative impacts of customer co-contribution include potential defamation in the forums. While the terms and conditions of posting to online forums state that users should not post defamatory statements, such defamation sometimes eventuates. Scholars have noted that in customer co-production of service, as illustrated by this situation, new models of digital governance and customer-based innovation are needed (Rai & Sambamurthy, 2006).

DISCUSSION

The provision of customer service via Web- based channels offers organizations an opportunity to draw closer to their customers and thus potentially a key differentiator in the marketplace. The provision of integrated, customer-centric, self-service solutions, are increasingly important for competitive success (Brohman et al., 2009). This chapter has highlighted that in offering such solutions in an enterprise IT support context there is a need for industry sector attention to stakeholder-based issues.

The inter-dependencies in this B2B context emphasize the need for IT support organizations to learn at the industry level. This is consistent with wider studies that suggest the importance of business learning in a network context (Knight & Pye, 2007). Learning at the industry level would require new collaborative electronic business projects focused on IT support provision involving IT service providers, business partners and customer organizations. There are four important influences to consider for successful electronic business collaboration: motivation, capability, communication and coordination (Cameron, 2005). First, organizations must be motivated to participate in collaborative projects. Second, the desire to increase organizational capability (skills and knowledge) can be a powerful motivator. Third, it is important to communicate the value of such collaborative projects within organizations. Fourth, industry groups can assist in coordinating such projects for successful conclusions.

As many of the key challenges in the provision of after-sales IT support via WSS centre on the potential for stakeholder conflict, careful attention is needed to devise suitable governance structures, mechanisms and methods for collaborative electronic business projects (Cameron 2006). In this pursuit, the development of individual IT service contracts (including SLAs) with customer firms and business partners would be helpful. This process should entail greater clarification and negotiation of stakeholder needs. The objectives of each organization should be articulated during joint planning, which is an important aspect of business collaboration and decision making in supply chains (Holsapple & Jin, 2007). The stakeholder-based issues described in this chapter suggest a need for focused studies seeking to identify key problem points in decision-making where stakeholder conflicts are influential so that the overall process can be improved to avoid or manage such conflicts. The model for stakeholder collaboration in B2B provided by Chi and Holsapple (2005) may prove useful in this pursuit. This model highlights the need for attention to behavioural processes such as knowledge sharing, participative decision-making and governance.

The issues raised in this chapter also suggest a need for supporting infrastructure in the provision of after-sales IT support via WSS. Electronic marketplaces offer a recognised structure which may be useful to support both collaborative IT support projects and specific supply chains (Markus & Christiaanse, 2003; Wang & Archer, 2007). However there are also challenges in using electronic marketplaces successfully. In a recent case study of an electronic marketplace, the key influences affecting its success were the loss of social capital, nature of communication channels used, time taken to reach critical mass, and power imbalances among participants (Driedonks et al., 2005). It is possible that some of these challenges will be relevant to the Web-based IT support provision, given the issues highlighted in this chapter.

The risk of reduced service quality was identified in the practice of customer co-contribution to service. This concern was noted more generally for online service provision by Schultze and Bhappu (2005). There may be ways for IT service firms to better monitor customer contributions via user forums and capture the high quality solutions (which typically are not captured formally by support organizations). Related questions of knowledge ownership must be resolved. However regulatory issues at different geographic locations cloud the resolution of such questions.

Indeed there were several areas identified where IT service providers suffer from inadequate regulatory support. The security, privacy and IP issues experienced in the often offshore managed IT support environment highlight the need for clarification and awareness of relevant regulations and laws at industry, national and international levels. IT support companies also seek to understand how to share knowledge with business competitors within strategic alliances while maintaining a competitive advantage. New theories are sought to underpin such knowledge sharing in an increasingly collaborative global business environment.

It was uniformly noted by participants in this study that the key objective for use of a WSS for after-sales IT support was to increase customer satisfaction while reducing support costs. This objective was expected to be achieved by relational as well as transactional methods. However as the discussion in this chapter has shown, the successful accomplishment of both goals can be problematic (also found by Bunduchi, 2005). Research is beginning to appear on the enabling of relationships in service provision when the Internet is used. In a business to consumer (B2C) context, Sigala (2007) explored online travel service provision and found that the communication aspect of the online service played a key role in relationship development. Where provider-consumer communication was enhanced by use of relevant communication tools, relationships and client satisfaction were improved. Thus IT support organizations may find that the path to satisfying the relational aspect of WSS success is by better leveraging the internet's communication tools and customising communication-oriented content.

Some of the challenges identified by IT support organizations relate to the process of knowledge transfer. For example, it was noted that a customer firm end-user may not be interested in institutionalizing transferred IT solutions. If a customer firm does not prioritise the institutionalization of IT solutions throughout the firm, its employees will not make the effort to learn the solutions transferred from the service provider and will simply apply them to resolve the initial problem, potentially resulting in repeat enquiries if the issue is experienced again in the future. This suggests a need for IT service providers to educate their customers about the importance of institutionalizing IT solutions to their organization. In other studies investigating inter-organizational knowledge transfer, the importance of a receiving organization being capable of absorbing shared knowledge has been identified (Nieminen, 2005), as has the need for a fit between the knowledge received and the receiving organization's objectives (Cranefield & Yoong, 2005). Support organizations should remain mindful that such absorptive capacity and fit might be missing from the customer firms that are receiving IT solutions.

CONCLUSION

This chapter has shown that the success of WSS in the provision of enterprise after-sales IT support is dependent on a series of complex inter-dependent stakeholder relationships.

Theoretically this chapter provides numerous insights into the key challenges faced by IT service providers in the provision of B2B after-sales IT support services via WSSs. Importantly, it highlights that issues related to eCRM and in particular, relationship development via WSS, is poorly understood and further investigation is needed to develop new understandings. Further work is also needed to better understand the cost-benefit of WSS from the perspective of business partner and customer perspectives since these stakeholders may have needs which compete with those of the IT support organization.

This chapter also highlights that in practice, IT service providers should consider all stakeholder viewpoints when planning, implementing and managing WSSs in the managed after-sales IT support context. Addressing diverse stakeholder needs may be particularly challenging in some areas such as security, privacy and IP. Such emerging challenges highlight the complexities where customers co-contribute to service provision and in working with business partners which are also considered competitors. Greater collaboration is needed with better supporting infrastructure and regulation.

While the findings from this chapter are limited by the context (managed after-sales IT support) and scope (only the IT support organization perspective was represented), they are indicative of possible concerns that other types of service providers may have in offering supplementary E-services using WSSs. Thus the findings provide a foundation for exploration in other settings. Investigating the customer and business partner perspectives would also provide valuable balance to the views expressed and analysed in this chapter.

REFERENCES

Barthelemy, J. (2008). The determinants of total IT outsourcing: An empirical investigation of French and German firms. *The Journal of Computer Information Systems*, April. Retrieved on November 29, 2010, from http://www.allbusiness.com/human-resources/workforce-management-hiring-consulting/1036323-1.html

Benapudi, N., & Leone, R. P. (2003). Psychological implications of customer participation in co-production. *Journal of Marketing*, *67*(1), 14–28. doi:10.1509/jmkg.67.1.14.18592

Bhappu, A. D., & Schultze, U. (2006). The role of relational and operational performance in business-to-business customers' adoption of self-service technology. *Journal of Service Research*, *8*(3), 372–385. doi:10.1177/1094670506286571

Brohman, M. K., Piccoli, G., & Martin, P. (2009). A design theory approach to building strategic network-based customer service systems. *Decision Sciences*, *30*(3), 403–430. doi:10.1111/j.1540-5915.2009.00242.x

Brohman, M. K., Watson, R. T., Piccoli, G., & Parasuraman, A. (2003). Data completeness: A key to effective net-based customer service systems. *Communications of the ACM, 46*(6), 47–51. doi:10.1145/777313.777339

Bunduchi, R. (2005). Business relationships in Internet-based electronic markets: The role of goodwill trust and transaction costs. *Information Systems Journal, 15*(4), 321–341. doi:10.1111/j.1365-2575.2005.00199.x

Cameron, J. (2005). Ten concepts for an e-business collaborative project management framework. In *Proceedings of eIntegration in Action: 18th Bled eConference*, June 6-8 2005, Bled, Slovenia.

Cameron, J. (2006). Governance structure, mechanisms and methods for managing collaborative e-business projects. *Journal of Chain and Network Sciences, 6*(2), 155–174. doi:10.3920/JCNS2006.x072

Cenfetelli, R., Benbasat, I., & Al-Natour, S. (2005). Information Technology mediated customer service: A functional perspective. In *Proceedings of International Conference of Information Systems*, Las Vegas, (pp. 725-738).

Chi, L., & Holsapple, C. (2005). Understanding computer-mediated interorganizational collaboration: A model and framework. *Journal of Knowledge Management, 9*(1), 53–75. doi:10.1108/13673270510582965

Cho, Y. K., & Menor, L. J. (2010). Toward a provider-based view on the design and delivery of quality e-service encounters. *Journal of Service Research, 13*(1), 83–95. doi:10.1177/1094670509350490

Chung, W., Chen, H., & Reid, E. (2009). Business stakeholder analyzer: An experiment of classifying stakeholders on the Web. *Journal of the American Society for Information Science and Technology, 60*(1), 59–74. doi:10.1002/asi.20948

Consortium Service Innovation. (2002). *Getting started with KCS.* Official Consortium Service Innovation Site. Retrieved on November 30, 2010, from http://www.thinkhdi.com/files/pdfs/GettingStartedKCS.pdf

Cooper, V. A. (2007). Knowledge Transfer in Enterprise IT Support Provisions using Web-based Self-Service, unpublished PhD Thesis, Deakin University, Melbourne, Australia.

Cooper, V. A., Lichtenstein, S., & Smith, R. (2006a). Knowledge Transfer in Enterprise Information Technology Support using Web-based Self-service Systems. *International Journal of Technology Marketing, 1*(2), 145–170.

Cooper, V. A., Lichtenstein, S., & Smith, R. (2006b). Enabling the Transfer of Information Technology Support Knowledge to Enterprise Customers Using Web-based Self-service Systems: Critical Success Factors from the Support Organisations Perspective, in Proceedings of the Seventeenth Australasian Conference on Information Systems (ACIS 2006), Adelaide, Australia.

Cranefield, J., & Yoong, P. (2005). Organisational factors affecting inter-organisational knowledge transfer in the New Zealand state sector: A case study. *The Electronic Journal for Virtual Organisations and Networks, 7*, December.

Dahlberg, T., & Nyrhinen, M. (2006). A new instrument to measure the success of IT outsourcing. In *Proceedings of the 39th Hawaii International Conference on System Sciences.* Retrieved on November 30, 2010, from http://csdl2.computer.org/comp/proceedings/hicss/2006/2507/08/250780200a.pdf

Driedonks, C., Gregor, S., & Wassenaar, A. (2005). Economic and social analysis of the adoption of B2B electronic marketplaces: A case study in the Australian beef industry. *International Journal of Electronic Commerce, 9*(3), 49–72.

Ferdinand, J., & Simm, D. (2007). Re-theorizing external learning: Insights from economic and industrial espionage. *Management Learning*, *38*(3), 297–317. doi:10.1177/1350507607079030

Geib, M., Kolb, L., & Brenner, W. (2006). Collaborative customer management in financial services alliances. In Fjermestad, J., & Romano, N. C. (Eds.), *Electronic customer relationship management. New York, NY*. Armonk.

Gong, H., Tate, M., & Alborz, S. (2007). Managing the outsourcing marriage to achieve success. In *Proceedings of Hawaii International Conference on the System Sciences* (HICSS 2007. IEEE Society Press. Gutek, B., Bhappu, A. D., Liao-Troth, M. A., & Cherry, B. (1999). Distinguishing between service relationships and encounters. *The Journal of Applied Psychology*, *84*(2), 218–233.

Hall, M. (2008, February 14). Recession-proof MSP readies new service. *Computerworld*.

Holsapple, C. W., & Jin, H. (2007). Connecting some dots: E-commerce, supply chains and collaborative decision making. *Decision Line*, *38*(5), 14–21.

Kandampully, J. (2003). B2B relationships and networks in the Internet age. *Management Decision*, *41*(5), 443–451. doi:10.1108/00251740310479296

Khalifa, M., & Shen, N. (2005). Effects of electronic customer relationship management on customer satisfaction: A temporal model. In *Proceeding of the 38th Annual Hawaii International Conference on System Sciences* (HICSS'05). IEEE Society Press.

Knight, L., & Pye, A. (2007). In the search for network learning: Some practical and theoretical challenges in the process research. In Gossling, T., Oerlemans, L. A. G., & Jansen, R. (Eds.), *Inside networks: A process view on multi-organisational partnerships*. UK: Edward Elgar Publishing Limited.

Kumar, K., & van Dissel, H. G. (1996). Sustainable collaboration: Managing conflict and cooperation in interorganisational systems. *Management Information Systems Quarterly*, *20*(3), 279–300. doi:10.2307/249657

Kurnia, S., & Johnston, R. (2002). A review of approaches to EC-enabled IOS adoption studies. In *Proceedings of the 35th Annual Hawaii Conference on System Sciences* (HICSS'02), IEEE Society Press.

Lei, D. T. (1997). Competence-building, technology fusion and competitive advantage: The key roles of organizational learning and strategic alliances. *International Journal of Technology Management, 14*(2/3/4), 208-37.

Markus, M. L., & Christiaanse, E. (2003). Adoption and impact of electronic marketplaces. *Information Systems and E-business Management*, *1*(2), 139–155. doi:10.1007/s10257-003-0001-7

Mayring, P. (2000). Qualitative content analysis. *Forum Quality Social Research, 1*(2). Retrieved on November 30, 2010, from http://www.qualitative-research.net/fqs-texte/2-00/2-00mayring-e.htm

McCarthy, J., & Matzke, P. (2010). *Coming upheaval in tech services*. Forrester. Retrieved on November 27, 2010, from http://www.forrester.com/rb/Research/coming_upheaval_in_tech_services/q/id/56279/t/2

Nambisan, S., & Baron, R. A. (2010). Different roles, different strokes: Organizing virtual customer environments to promote two types of customer contributions. *Organization Science*, *21*(2), 554–572. doi:10.1287/orsc.1090.0460

Negash, S., Ryan, T., & Igbaria, M. (2003). Quality and effectiveness in Web-based customer Support systems. *Information & Management*, *40*(8), 757–768. doi:10.1016/S0378-7206(02)00101-5

Neuman, W. L. (2006). *Social research methods: Qualitative and quantitative approaches* (6th ed.). Boston, MA: Allwyn & Bacon.

Nieminen, H. (2005). Organizational receptivity – Understanding the inter-organizational learning ability. *Electronic Journal of Knowledge Management, 3*(2), 107–118.

Pan, G. S. C. (2005). Information Systems project abandonment: A stakeholder analysis. *International Journal of Information Management, 25*(2), 173–184. doi:10.1016/j.ijinfomgt.2004.12.003

Peppard, J. (2003). Managing IT as a portfolio of services. *European Management Journal, 21*(4), 467–483. doi:10.1016/S0263-2373(03)00074-4

Peppers, D., & Rogers, M. (2001). *One to one B2B: Customer development strategies for the business-to-business world.* New York, NY: Currency Doubleday.

Pujari, D. (2004). Self-service with a smile? Self-service technology (SST) encounters among Canadian business-to-business. *International Journal of Service Industry Management, 15*(2), 200–219. doi:10.1108/09564230410532510

Ragsdale, J. (2010). Service and support technology adoption and spending: The 2010-11 outlook. Retrieved on November 27, 2010, from http://searchcrm.techtarget.com/news/2240024630/Service-and-support-technology-adoption-and-spending-The-2010-11-outlook

Rai, A., & Sambamurthy, V. (2006). Editorial notes – The growth of interest in services management: Opportunities for Information Systems scholars. *Information Systems Research, 17*(4), 327–331. doi:10.1287/isre.1060.0108

Reimann, M., Oliver, S., & Thomas, J. S. (2010). Toward an understanding of industry commoditization: Its nature and role in evolving market competition. *International Journal of Research in Marketing, 27*, 188–197. doi:10.1016/j.ijresmar.2009.10.001

Rockart, J. F. (1979). Chief executives define their own data needs. *Harvard Business Review, 57*(2), 81–93.

Roloff, J. (2008). A life cycle model of multi-stakeholder networks. *Business Ethics (Oxford, England), 17*(3), 311–325. doi:10.1111/j.1467-8608.2008.00537.x

Schultze, U., & Bhappu, A. D. P. (2005). Incorporating self-serve technology into co-production designs. *International Journal of e-Collaboration, 1*(4), 1–23. doi:10.4018/jec.2005100101

Shukla, A. (2010). IT managed services market in India to hit $US3.8 billion in 2013. Retrieved on November 27, 2010, from http://www.techworld.com.au/article/345074/it_managed_services_market_india_hit_us3_8_billion_2013/

Sigala, M. (2007). Investigating the Internet's impact on interfirm relations: Evidence from the business travel management distribution chain. *Journal of Enterprise Information Management, 20*(3), 335–355. doi:10.1108/17410390710740772

Simons, L. P. A., van Loon, J., de Koning, N. M., Kruse, J., & Bouwman, H. (2009). Increasing the loyalty effects of e-CRM across the service delivery cycle. In *Proceedings of Bled eConference – eEnablement: Facilitating an Open, Effective and Representative eSociety*, June 14 – 17, Bled, Slovenia.

Singh, M., & Byrne, J. (2005). Performance evaluation of e-business in Australia. *Electronic Journal of Information Systems Evaluation, 8*(2).

Skanning, C. (2005). The cost and benefits of customer self-service. *The Wise Marketer*. Retrieved on November 28, 2010, from http://www.thewisemarketer.com/features/read.asp?id=83

Solitander, M. (2006). Balancing the flows: Managing the intellectual capital flows in inter-organizational projects. *Electronic Journal of Knowledge Management, 4*(2), 197–206.

SSPA. (2004). *2005 support demand research series. Tech strategy partners and SSPA*. Retrieved on August 7, 2007, from http://www.thesspa.com/publications/pr/041012b.asp

Subramaniam, A. (2009). *Web self-service will make you great*. Retrieved on November 28, 2010, from http://www.destinationcrm.com/Articles/Web-Exclusives/Viewpoints/Web-Self-Service-Will-Make-You-Great-54931.aspx

Szulanski, G. (2000). The process of knowledge transfer: A diachronic analysis of stickiness. *Organizational Behavior and Human Decision Processes*, *82*(1), 9–27. doi:10.1006/obhd.2000.2884

Udo, G. J., Bagchi, K. K., & Kris, P. J. (2010). An assessment of customers' e-service quality perception, satisfaction and intention. *Information & Management*, *30*(6), 481–492. doi:10.1016/j.ijinfomgt.2010.03.005

Vlachopoulou, M., & Manthou, V. (2003). Partnership alliances in virtual markets. *International Journal of Physical Distribution & Logistics Management*, *33*(3), 254–267. doi:10.1108/09600030310471998

Walsham, G. (1995). The emergence of interpretivism in IS research. *Information Systems Research*, *6*(4), 376–394. doi:10.1287/isre.6.4.376

Wang, S., & Archer, N. (2007). Business-to-business collaboration through electronic marketplaces: An exploratory study. *Journal of Purchasing and Supply Chain Management*, *13*(2), 113–126. doi:10.1016/j.pursup.2007.05.004

Yin, R. K. (2003). *Case study research, design and methods* (4th ed.). Newbury Park, CA: Sage Publications.

APPENDIX

CSFs for Knowledge Transfer from Support Organization to Customer Organization using WSSs

Critical Success Factor	Description
CSF-1 Cost Effectiveness	The cost equation for providing/using Web-based self-service must be better, or at least not worse, than providing/using non-Web-based self-service.
CSF-2: Provision of Additional Services and Cross-Selling Opportunities	Current WSS transactions are used proactively as an opportunity to offer the customer organization additional advice and services
CSF-3: Critical Mass: Knowledge Content and Knowledge Contributors	A sufficient number of end-users must proactively contribute sufficient knowledge content to the WSS knowledge base, to encourage all parties to initially use, and to continue to use, the WSS as a means of resolving their support issues or information requirements
CSF-4: Usefulness: Provision of Knowledge Which Meets User Requirements	The WSS must provide the functionality and knowledge required to meet the objectives of all stakeholders. For example, for the end-user customer, it should resolve a specific technical or business problem, or provide other required knowledge resources.
CSF-5: Ability to Provide Efficiency	Use of the WSS to resolve a support issue or provide other knowledge resources must be perceived as efficient by all parties. This is inclusive of not only the performance of the WSS tool but the surrounding processes for using the WSS.
CSF-6: Access, Connectivity, Availability and Performance	The providing organization, relevant business partners and the customer organization must have sufficient technology infrastructure in place, to enable all parties to participate in Web-based self-service.
CSF-7: Effective Information Architecture and Search Engine	The WSS must have an effective Information Architecture and Search Engine such that the information system that organises and retrieves knowledge in the knowledge base is perceived as effective by end-users.
CSF-8: Security, Privacy and Assurance	All stakeholders using WSS must feel secure, private and confident in all aspects of WSS transactions including the stored data components of transactions. Issues surrounding information security and information privacy, and the need to keep confidential related company secrets (intellectual property) must be addressed.
CSF-9: Ease of Use/Usability	An end-user must perceive that use of the WSS does not demand excessive cognitive or ergonomic effort.
CSF-10: Early Positive Experience	The first few end-user experiences with the WSS must result in a positive outcome, where end-user needs are met and they feel valued, in order for the end-user to adopt WSS long term.
CSF-11: Positive Experience	Using the WSS on an ongoing basis must result in a positive outcome, where corporate customer needs and all types of end-users' needs are met and they also feel valued. A positive experience is closely related to customer organization/end-user satisfaction
CSF-12 Confidence in Solution	The customer organization/end-user must feel confident that the solution provided by the WSS will resolve their issue and will not result in further issues. They must also have self-confidence in their own ability to apply the offered solution.
CSF-13: Customer Focus: Understand Customer and their Requirements	The support organization (and relevant business partners) must understand the individual business and technical needs of individual customer organizations and their end-users. With this understanding, WSS must be tailored to meet those individual needs.
CSF-14: Positive Relationship	The relationship between the support organization, business partners and the customer organization must be one which supports open communication and trust. This positive relationship should exist at both the corporate and end-user levels.

Critical Success Factors for Knowledge Transfer from an IT Support Organization to a Customer Organization using Web-based Self-service Systems (Part A).

Critical Success Factor	Description
CSF-15: Provision of Additional Support: Education & Training	Additional assistance, or education and training in respect to how to use the WSS must be provided by the support organization as requested by end-users.
CSF-16: Employee Focus	Management within the support organization, business partner and customer organizations must have an understanding of the work processes and conditions which will affect the ability and willingness of employees to adopt the WSS and associated strategies. With this understanding, management must focus on meeting the needs of their employees where possible, in order to maximise employee productivity and the benefits received from the WSS strategy.
CSF-17: Culture	The support organization should foster an environment that recognises that WSS is part of the way it wants to conduct business. In addition, an open, sharing culture is needed. The culture should extend to customer organizations and business partners.
CSF-18: Marketing and Awareness of Web-based Service	Marketing programs which raise awareness of and support for, the adoption of WSS, must be in place.
CSF-19: Knowledge Creation, Capture and Re-Use	Knowledge capture processes to ensure that valuable knowledge is created and captured into the WSS knowledge base by end-users must be in place. Knowledge reuse processes to ensure that this knowledge is subsequently accessed and re-used by end-users, must also be in place.
CSF-20: Knowledge Validation	Processes must be in place to ensure the accuracy of the knowledge which is captured into the WSS knowledge base and to ensure that once it is captured, it is frequently reviewed and updated for currency.
CSF-21: Knowledge Storage/Retrieval	Processes must be in place to ensure that the structure and format of captured knowledge facilitate findability.
CSF-22: Presentation of Knowledge	The knowledge must be presented in a form which maximizes the understanding acquired by end-users.
CSF-23: Measurement & Feedback of WSS	Sufficient measurement and feedback methods for assessing the effectiveness of the WSS strategy must be in place.
CSF-24: Alignment and Integration	There must be alignment and integration between the WSS and other channels' support processes, as well as with related business processes, in the context of the business/industry environment.
CSF-25: WSS Override and Recovery	The capability for the end-user or WSS to over-ride transactions initially made via the WSS, must be in place, whereby if an end-user is not finding a satisfactory resolution via the WSS, the transaction is directed to an alternative mode of service delivery (e.g. a chat session or telephone call).
CSF-26: Ease of Re-initiation	A process must be in place whereby an end-user can easily re-initiate a support transaction to re-locate a previously retrieved resolution or other knowledge resource.
CSF-27 Top Management Support	Top management must provide ongoing support and commitment to the WSS and associated strategies.

Critical Success Factors for Knowledge Transfer from an IT Support Organization to a Customer Organization using Web-based Self-service Systems (Part B).

Chapter 4
E-Services and Australian Accounting Practices

John Breen
Victoria University, Australia

Stephen Burgess
Victoria University, Australia

ABSTRACT

This research investigated the website use of Australian accounting practices in metropolitan Melbourne, Australia in the period 2006-2010. This occurred through an investigation of the websites of 100 practices and interviews involving selected practices at the start of the period and a further investigation of the websites at the end of the period. The analysis revealed a heavy emphasis towards website features that identify the businesses and the services that they offer, which is consistent with other website studies involving professional services. More than other businesses, accounting practices rely on keeping their clients up to date and thus the provision of information such as taxation news updates, changes to superannuation and regular newsletters to clients through their websites was identified as being an important tool for many practices. None of the accounting practices offered online order or payment features over the period. Again this is consistent with other studies of similar professional services businesses. Although some of the practices had log-in features in 2006, only two contained features that could perhaps be classed as 'e-services'. However, the 2010 review found an increase in the number of accounting practices with a log-in feature on their website. This suggests that the issues of trust in the technology are being addressed and that they could be moving towards a situation where e-services may become more commonplace in the future.

INTRODUCTION

Advances in information technology (IT) and the use of the Internet have brought forth new business models for professional services, thus enabling firms to operate more competitively on a wider and more efficient scale (Kotler et al. 2002). Firm-client relationships supported by the Internet will, by many predictions, define the successful accounting firm of the future (Bhansali

DOI: 10.4018/978-1-60960-607-7.ch004

2005; Schlageter 2005). Adopters of these new accounting practice models report significant benefits as they move into more productive and satisfying relationships with their clients. These benefits include a reduction in paperwork, the delivery of more responsive services, reduction in errors, and the incorporation of additional services (McCausland 2004). Firms have achieved these benefits by installing online accounting systems through which services are delivered using private and secure Web sites accessible to their clients.

This research investigates the use of the Internet among accounting practices located in metropolitan Melbourne, Australia (and its surrounding suburbs) by comparing and contrasting the changes in website content of 100 Australian accounting practices in the period 2007-2010 and, secondly, reporting on interviews with partners in 20 accounting practices conducted in 2007 to identify the level of e-service delivery in Australian accounting practices.

DELIVERING SERVICES ON THE INTERNET

The power of information and communication technology in improving business performance is underpinned by the versatility of the Internet as an information medium. Through their websites, firms have effectively reached out to their customers in a variety of ways, achieving greater efficiencies and in some cases offering added value. In the late 1990's, Angehrn (1997) identified four Internet 'dimensions' that a business could employ when developing a strategy to engage customers via the Internet. These dimensions are information (dissemination), communication (with customers), distribution (of digital goods or services) and (online) transactions (ICDT). More recently, Jelassi and Enders (2005) have suggested that these dimensions could be used by a business to help them select which Internet features to employ when engaging with customers. They describe Angehrn's dimensions as:

- **Virtual information space:** this includes online advertising and posting business information.
- **Virtual communication space:** is where the business engages in a two-way online communication with its customers, typically through email, chat facilities or bulletin boards.
- **Virtual distribution space:** allows for the delivery of digital goods (such as online books, software) and services (such as financial advice). This is an area where innovation in services delivery could provide benefits both to businesses that offer professional services and their clients.
- **Virtual transaction space:** allows for the acceptance of online orders and payments.

The four dimensions provide a useful framework from which to map out existing and potential value-adding activities and are particularly relevant to the subject of this study, accounting practices. In particular, the four dimensions will be used to classify the different e-service strategies of Australian accounting practices. Before this can be carried out, it is necessary to spend some time discussing e-services.

Stafford (2003) suggests that experts have not yet come upon a clear definition of what e-services actually are. One reason for this, as Stafford argues, could be attributed to the range of specialities involved in the delivery of e-services. From a marketing viewpoint, Bolton (2003) suggests that e-services are those services that can take place without buyer involvement. For instance, automatic collection of road tolls as a car passes through a tolling booth is an example of this. E-services can also be related to particular industries. For instance, Laffie (2005) discusses the US Internal Revenue Service (IRS) and describes their e-services as being a suite of online products that allow tax professionals and payers to conduct business electronically with the IRS. Another broad definition of e-services is offered by Hofacker, Goldsmith, Bridges and Swilley (2007:17), who define e-services as:

an act or performance that creates value and provides benefits for customers through a process that is stored as an algorithm and typically implemented by networked software

The authors go on to describe three different types of e-services: complements to existing offline services and goods; substitutes for existing offline services, and uniquely new core services. In addition, the authors also suggest four 'properties' that can help to distinguish them from traditional services. These are:

1. **Intangibility**. This can occur when an e-service replaces a previously tangible item, such as a theatre ticket or even a delivery of a package.
2. **Heterogeneity**. As many e-services are delivered in an automated fashion, there is less likelihood that the quality of service will vary.
3. **Inseparability of Production and Consumption**. The example provided by the authors is related to music, which can be performed online (production of service) and then downloaded and listened to (consumed) later via an mp3 player.
4. **Perishability**. As some online services (such as music) can be stored online and copied for repeat consumption, they are not, as such, perishable.

However, other authors suggest a more narrow definition of e-services. For instance, Hsu and Chiu (2004) describe e-services as "highly specialized electronic services" (p.359) for users and suggest that they include "support services such as consulting, outsourcing, website design, electronic data interchange, payment transfer and data storage backups" (p.359). Such complex e-services are sometimes known as web services.

E-services can occur between government, business and consumer and, as Chui et al (2005) suggest, from business to business. Despite the potential for complexity in e-service developments, Chui et al claim that the majority of existing e-services are available through interactive web pages in HTML (the well-known Internet formatting language). It has reached the stage where the *quality* of e-service delivery in some areas is becoming an issue when buyers are considering using the services of a business.

"As with other services, e-service customers need to be able to assess the quality of e-services in order to make informed buying decisions".

Field, Heim and Sinha (2004, p.291)

For the purposes of this chapter we will only be examining the relationship between business and consumer (now known as B2C). In relation to professional services businesses, such as accounting practices, we will take a broad definition of e-services. We define e-services as those professional services, traditionally performed face-to-face, which are now predominantly delivered online. These might include those services where at least one of the parties (business or consumer) is not required to actively participate for the service to be delivered or, at the less-sophisticated end, where the technology facilitates the delivery of the service.

ACCOUNTING FIRMS AND THE INTERNET

Information technology has played a critical role in the services provided by the accounting industry (Banker, Chang and Kao, 2002). The transition to e-commerce provides new opportunities to interconnect various organisational units as well as across different agencies (Gebauer and Shaw, 2002). At the same time, it demands from organisations and individuals new sets of behaviours. Chaston and Mangles (2001) used data from 174 small UK accounting firms to determine that entrepreneurial accounting practices believe that e-commerce offers major new trading opportunities, can deliver competitive advantage and improves internal efficiencies. They concluded that entrepreneurial accounting firms could be expected to be more involved in the adoption of new electronic communications technology before their non-entrepreneurial counterparts.

Further to this, Chen, Tseng and Chang (2005) collected data from 595 accounting firms in Taiwan and found that despite the fact that the impact of the Internet on accounting firms differed in degrees, they were all positive. Douglas (2004, p. 25) also remarks that "an accounting firm with no Web site or a lousy one is at a distinct disadvantage". However Fink (2007) suggests that trust of clients is important to the successful deployment of professional e-services.

The need for access to different levels of accounting information is likely to vary markedly across different individuals and groups. For instance, the financial reporting needs of many are currently met by reports in 'Adobe Acrobat' form, but these present limitations relating to file size and download time (Lymer and Debreceny 2003). The power of the Internet is multi-fold and its capacity to integrate with other information and communication technologies, such as databases and with the support of hypermedia presentations, enhances its capacity for financial reporting (Lymer et al 1999).

For most firms, early Internet reporting merely duplicated traditional 'hard copy' reporting (Bury 1999) and was treated as an additional way to disseminate the same information. There was limited evidence to suggest that the information available through the Internet was, in fact, useful for the users (Taylor 1998). A number of studies have reported problems relating to Internet reporting, such as difficulties associated with locating specific website and information items (Debreceny and Gray 2001); information omissions (Ettredge, Richardson and Scolz 2002) and the difficulty of distinguishing between audited and non-audited information (Bury 1999). In fact, recent studies in the area have tended to concentrate upon the financial reporting aspects of the Internet. For instance, Mohamed et al (2009) examined Internet reporting by 142 Omani businesses, Gutierrez-Nieto et al (2008) examined Internet reporting in businesses that provide microfinance and Yu and Rijia (2009) compared Internet financial reporting in China and the US.

Despite these problems, the Internet does have a number of important advantages over the traditional media used for financial reporting. It has the capacity to provide a whole range of additional, non-financial information in annual reports (Green and Spaul 1997). Lymer et al (1997), for instance, identified the advantages of the Internet for financial reporting as being cheap, quick, dynamic and flexible. Xiao, Jones and Lymer (2005) identify the advantages as including the use of digital technology, enhancements through technology integration, the capacity for real time communication, hypermedia presentations and limitless boundaries.

Lymer and Debreceny (2003) argue that the rate of development in technology is such that simply replicating paper reports into Acrobat format is nothing more than a simplistic response to information technology pressure for greater and wider dissemination of financial information. Therefore, they maintain that the professional responses to Internet financial reporting are grossly inadequate in relation to advances in technology developments and that accounting bodies have "a public responsibility to promulgate the most effective and efficient distribution of accounting data via the web" (p 118). Xiao, Jones and Lymer (2005), recognise that when it comes to financial reporting the Internet could, on the one hand, enable

Table 1. Examples of services and potential for e-services classified by virtual space

Virtual Space	Example	Potential for e-services
Information	Online advertising and posting business information	There is little potential for delivering *personalised* services as information is basically delivered one-way (business to consumer). The provision of generic advice could, however, be offered
Communication	Email, chat facilities or bulletin boards	Here there is potential for the delivery of enhanced personalised services as individual contact can be made with customers.
Distribution	Delivery of digital goods (such as online books, software) and services (such as financial advice)	There is some potential here, but it is difficult to determine whether or not it is possible to offer a highly personalised service, such as financial advice, in an automated fashion.
Transactions	Acceptance of online orders and payments	If customers are willing and able to pay for business services online and/or set up appointments on an automated basis, then there is potential here.

or facilitate new and innovative ways of reporting financial information, and, on the other hand, has the potential to unlock databases and provide real time reporting. This suggests that there is a potential for a whole range of services to be delivered by accounting practices over the Internet.

The next section introduces a study of Australian accounting practices to examine specifically the level of services that they are currently providing.

THE STUDY

In 2006, the Institute of Chartered Accountants in Australia (ICAA) funded a study that examined the website practices of Australian accounting practices in the following manner:

1. An analysis of the content of the websites of Australian accounting practices to describe the state of play regarding the use of websites among local accounting firms.
2. In-depth interviews with the relevant accounting practice staff to uncover drivers and inhibitors for the adoption of Internet technology in accounting firms as well as an investigation of techniques employed to determine the success of existing online initiatives.

This chapter reports on the results of both phases of the study, the analysis of website content of Australian accounting practices and subsequent interviews with 20 Australian accounting practices. In addition, a further website analysis was carried out in 2010 and this is reported to show how website features evolved over the (almost) four years between website investigations. The chapter focuses on the level of services delivered online in those practices and specifically how many of those services could be classified as *e-services* according to our definition.

The analysis of websites was initially carried out in November-December 2006, with the subsequent analysis occurring in September 2010. For this chapter, the website content was analysed according to Angehrn's (1997) four dimensions of space that businesses could employ to develop a strategy for dealing with customers on the Internet. Table 1 lists each of the virtual spaces and includes some examples provided earlier, along with some of their applications and where the authors see the potential for e-services.

The analysis technique used to identify website features was developed by Burgess (2002) and has been subsequently used in studies by Burgess and Zoppos (2002), Shackleton et al (2003) and, more recently, in Burgess, Sellitto and Wenn (2005) and Burgess, Bingley and Sellitto (2005). For this type of content analysis, Henderson and

Table 2. Main demographics of interview practices

Business type (number of employees)	Number of Interviews	Ave years of Operation	Typical no of clients		% with IT staff
			Individual	Business	
Micro (1-10)	6	19	100	100	17
Small (11-20)	5	40	500-1000	500-1000	20
Medium (21-50)	7	27	1000	500-1000	43
Large (51+)	2	26	1000+	1000+	100

Cowart (2002) quote McMillan (2000) in suggesting specific steps that should be followed in selecting a sample number of websites. In our case, the selection process involved the random selection of 100 firms listed on ICAA's website that were either located in metropolitan Melbourne or its surrounding suburbs. Of these, 85 websites were investigated in 2006 (the other 15 not being found in searches in standard search engines). Of these 85 websites, 79 were still in existence for the 2010 study. It is assumed that with the six missing websites the businesses closed, merged or were taken over. In order to determine some of the strategies used by accounting practices when developing their websites, 20 practices were selected for in-depth interviews in 2007 in order that their use of the Internet could be explored further. The stratified sampling technique employed in this study was able to generate a useful spread of accounting practices. The technique identified *strata* for practices within different categories in relation to the number of features on their websites. In other words, the sample included three types of businesses, those with: few features on their websites; an average number of features; and many features. In each case, the interviews were conducted at the premises of the practices with a partner of the practice; although, in a few instances, dedicated information technology employees were interviewed. Table 2 shows the main demographics of the interview sample.

Most of the practices interviewed were micro, small or medium sized practices. They had all been in operation for a number of years, ranging from 10 years (a medium sized practice) to 100 years (a small practice). Virtually all of the practices had a combination of individual and business clients. Not unexpectedly, the smaller practices were less likely to employ a dedicated information technology employee. Where dedicated IT employees were not to be found, the role was often carried out by a partner in addition to their assigned accounting duties or was shared between partners. In many of these instances, support also came from external consultants. There were also instances where external support was used when the business had dedicated IT employees. This was especially the case in relation to the design and hosting of websites.

RESULTS

Virtual Information Space

Features related to the virtual information space provided the first dimension from which to assess firms' website strategy. These features generally provide the Web surfer with information about how to contact or find the organisations, some basic information about the organisations, information about their products and services and news about the organisation or the industry. Alonso Mendo and Fitzgerald (2005) further classify the information provided on websites into features related to *identification and image* and those related to *promotion and contract*. We know from previous studies of small business websites (Burgess,

Bingley and Sellitto 2005; Burgess, Sellitto and Wenn 2005) that most small businesses include a number of contact features. In this part of the study, we examined websites to see if they had the following features:

- Identification and Image
  - Physical address; Email address; Fax number; Telephone number; Location map; Information about the firm; Information about the ICAA; ICAA logo*; Policies and/or Mission statement; Information about affiliates*. (*These features were added in during the study when it was noticed that they were prominent).
- Promotion and Contract
  - Information about services offered; Information about how to engage the services of the business; News/ press releases; Customer or member reviews.

The percentage of practices that had identification and image features on their websites across the two time periods is shown in Figure 1. In 2006, all accounting practice websites had telephone number, email address, and fax number. All but one had their physical address listed on their website. These results differed from other studies, including the analysis of business services websites (Burgess, Bingley and Sellitto 2005), where most - but not all - small businesses had these features. The 2010 results show a slight drop in the ratio of businesses that had these features, but not a substantial drop. In 2006 one in four accounting businesses had a location map on their website, whereas none of the business services practices in the earlier Burgess, Bingley and Sellitto (2005) study had a location map. The 2010 analysis showed that the location map was the one specific identification and image feature that was adopted by more accounting practices, with over 40% of the businesses having that feature. This could be a reflection of the ease in which location maps can be placed on websites and also the 'copycat' effect, where businesses note a feature

Figure 1. Identification and image features

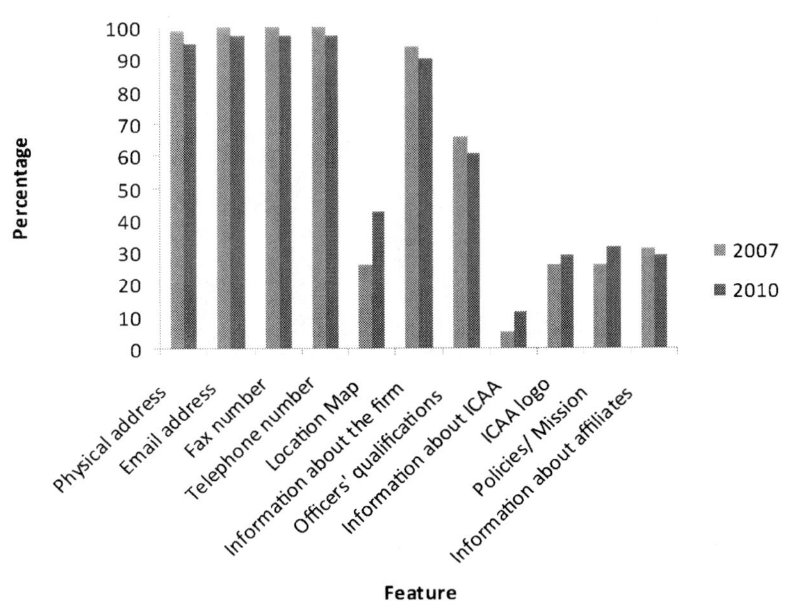

on a competitor website and replicate it on their own. When the physical address is listed on the website it can be used to locate the business, but also for other purposes (such as to send mail). A location map is primarily put on the website to help somebody *geographically or visually* locate the business. It might be typically employed in a situation where businesses may feel their customers need some extra assistance to find the business. In 2006 only 5% of the practices had information about their professional association and one in four practices actually had their association logo on their website. At the time the authors expected that a higher proportion of websites would have one or both of these features. In 2010 a slightly higher proportion of businesses had these features. There was a similar trend in the proportion of accounting practices that had their mission or policy statements on their website, with this proportion showing a slight increase in 2010.

Figure 2 shows the percentage of accounting practices that had promotion and contract features.

Some 95% of the accounting practice websites examined in 2006 had information about the services offered by the practices. There was a slight drop in 2010, but the overall result was also consistent with the earlier business services study (Burgess, Bingley and Sellitto 2005), where 93% of these businesses had information about the services offered.

An important feature of accounting practice websites was the inclusion of the officers' qualifications in two-thirds of the websites in 2006 (refer Figure 1). Again, there was a slight decrease in 2010. Also, in 2006 some 46% of the accounting practice websites listed the clients that were catered to (refer Figure 2). In 2010, 8% fewer businesses provided this list. Whilst these decreases are not substantial, it would have been expected that there would be an increase in these features over time as practices attempted to attract more clientele.

In the previous business services study only a small number (around 10%) had news or press releases about the business. In comparison, the accounting practice websites were more likely to have some type of news or newsletter available on their websites, with similar levels recorded in 2060 and 2010. Testimonials from clients were not a prominent feature of accounting practice websites, but the proportion of websites that listed

Figure 2. Promotion and contract features

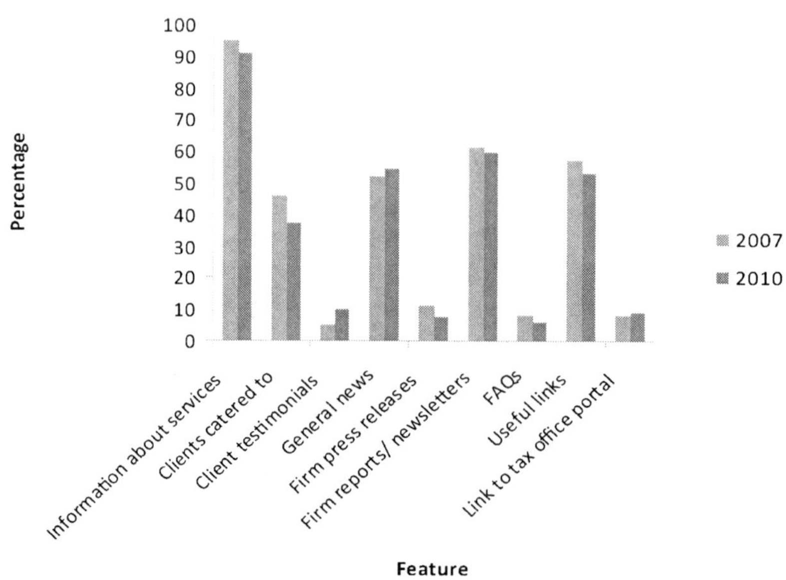

them did double from 2006 to 2010 (albeit from a low base).

In relation to the data collected during the interviews, many of the practices interviewed were of the opinion that the website has two main objectives: to convey information about the business to the general public and to deliver information from the business to clients (which is consistent with the website analysis).

In this regard, especially for those firms that cited the website as primarily a client resource, the website is used to distribute information such as taxation news updates, changes to superannuation and regular newsletters to clients. For one firm that was interviewed, the most important feature of the website is considered to be "content – nothing else" whilst for another firm, importance is placed on the website's capacity "to convey information about the staff, their qualifications and specialisation".

Overall, the interviews confirmed that the websites are primarily used for the dissemination of information. For example, in one firm that aimed to be technologically efficient, it was pointed out that all documents are either scanned or converted into 'acrobat' format and uploaded onto their website. It is important to note that, within the context of eservices delivery, email (refer next section) is sometimes used for the 'one-way' delivery of information, such as news bulletins.

Another important source of information for many clients is via the secure log-in feature, which one in four of the websites we analysed provided. Further information about the types of information accessed via this feature is in the 'Distribution Space' section.

Virtual Communication Space

This space covers the use of Internet technologies to provide or support communication between a business and its partners. From the website analysis, the only form of two way communication between business and client in 2006 and 2010 was via an email address, which was evident on virtually all of the websites analysed. The interviews confirmed that by far the most dominant form of Internet-based communication was via email, between employees themselves and between employees and clients. Almost all the businesses interviewed mentioned the importance of email, especially when dealing with clients that were some distance away. Many described it as 'vital' or 'critical'; with others suggesting that it is a useful supplement to their normal operations. The difference between face-to-face and email communications was an interesting discussion point that arose on a number of occasions. In this regard, a few practices conveyed their concerns about the increased use of email insofar as that there is less opportunity for face-to-face dealings and the chance to build relationships with clients.

The convenience and immediacy of reply is considered to be particularly beneficial in the management of interstate and overseas clients in different time zones. Likewise, the immediacy of reply is an important efficiency gain for practices.

Other beneficial reasons cited for use of email included:

- The capacity to send out group emails and information, such as newsletters, to specific (targeted) groups,
- Use as a quality assurance mechanism as there is documentary evidence of the communication,
- There is minimal work interruption to answer simple questions,
- Staff can respond quickly even if they are not in the office (as many firms offer flexible working conditions for their employees, clients are encouraged to email employees rather than call for a faster response).

It is also important to note that, although the firms interviewed regard email as an indispensable business tool, there are different opinions underpinning the firm's use of email communication. In one large firm, for example, "email is regarded as a facilitator only - not a relationship management tool. Major issues are conducted face-to-face". Another firm interviewed also described email as "a supplement, rather than a replacement, to the ways we deliver our services". For example, clients have the option of emailing their data files before booking a face-to-face appointment.

The interviews also revealed concerns associated with the growing use of email. Concerns included:

- The business etiquette or appropriateness pertaining to the (non) use of email, especially with larger or more important clients
- Unsolicited and junk email (SPAM) is becoming an issue
- The difficulty of maintaining relationships which are conducted virtually
- The problems associated with the construction of email correspondence insofar as less formal and more ad hoc decisions are likely to be made compared to the care usually taken with traditional letters.

As will be discussed in the next section, many businesses use email to 'swap' data files as attachments.

Virtual Distribution Space

In Table 1, we indicated that the virtual distribution space might show the most promise for the delivery of e-services (as we have defined it) by accounting practices. The virtual distribution space allows for the delivery of digital goods and services using Internet technologies. Alonso Mendo and Fitzgerald (2005) extend this by introducing the notion of website features that can *enhance the relationship between the business and its clients* (refer Figure 3). In this analysis, website features can provide simple services (such as tax, investment or other financial calculators), forms that the business needs to complete as part of its compliance or restricted areas on the website that can provide access to personalised areas (such as to investment portfolios). The website analysis provided little evidence in 2006 as to any automated e-service delivery occurring via the websites, unless it was occurring via the secure log-in areas (which could

Figure 3. Relationship Enhancement features

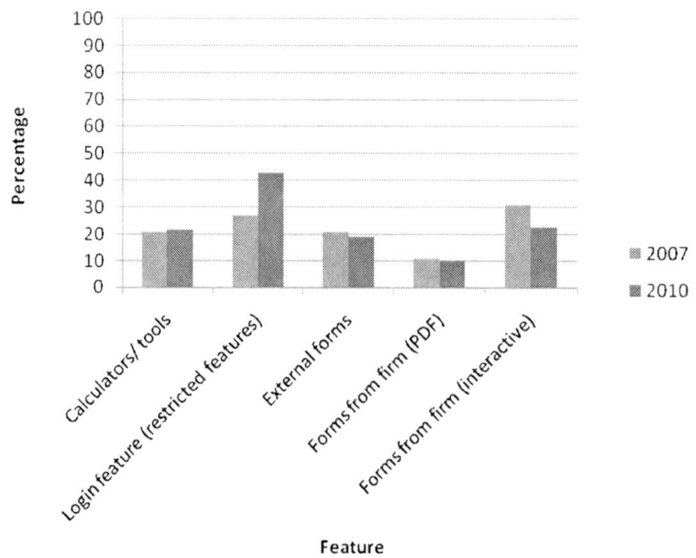

not be accessed). However, the 2010 analysis show a marked increase in the proportion of practices that provided log-in access to restricted areas. This suggests that there is a growing realisation of the importance and potential of e-services.

In the previous section we indicated that many of our interviewees transferred files over the Internet using email attachments. About half of the firms interviewed at the time carried out this practice, but none of these were large businesses. This suggested that, at least for larger firms, the emailing of attachments was not considered to be a suitable work practice. Three other businesses employed the use of file transfer protocol (FTP) to switch files between themselves and clients. Two practices used both FTP and email attachments to transfer files. Seven of the practices that were interviewed had a log-in feature. These provided access to special (but not personally tailored) accounting or financial information (3 firms), regular newsletter or reports (2 firms) and access to portfolio accounts (2 firms). Further studies will need to be carried out to determine if the increased proportion of websites with a log-in feature in 2010 were offering these, or even newer, services.

Could any of these be considered to be e-services? Do we have any instances where traditional face-to-face professional services are being delivered online? In relation to the transfer of data files and reports, in 2006 the websites were primarily being used as conduits to *transfer* the results of the professional services (already carried out by a partner or other employee) from business to clients. Certainly, email and FTP features facilitated this process. In relation to the log-in service, most were used for information provision and were therefore not necessarily tailored to the client. These could not be classed as e-services. Perhaps the one feature that could be classed as e-services, at least partially, is where practices allow clients to log-in and access their investment portfolios. In cases where clients are able to manipulate funds it could be argued that this is a professional service offered online. It is interesting to note that a number of practices that did not have this feature indicated that they might look to adopt it in the future as a way of enhancing the delivery of their professional services.

Virtual Transaction Space

The virtual transactions space allows for the online acceptance or orders and/or payments. From previous studies it is known that certain types of businesses are more likely to encourage ordering/booking (and sometimes payments) online (for example: Burgess, Bingley and Sellitto 2005; Burgess, Sellitto and Wenn 2005). For instance, art galleries are not likely to have any of these features. Bed and breakfast establishments will typically have a facility for customers to book via a web form that sends an email to the business or have the option to print out a booking form that can then be faxed to the business. Small manufacturing businesses are more likely to have interactive order/payment facilities on their website.

The Burgess, Bingley and Sellitto (2005) study of business services websites, however, showed no transactional interactivity features on the website. This study of accounting practice websites is consistent with this finding – with none of the websites having any of these features in 2006 or 2010. Once again, this reflects the nature of the business and the type of service that is being offered. In other words, the business of accounting can be characterised by its reliance on the provision and delivery of accurate and timely information. The feedback from interviewees in 2007 was that there is not yet enough sophistication in the client market to drive the adoption of transactional facilities. More specifically, the concerns in relation to security and low client demand for such services take precedence over their implementation.

DISCUSSION

Chaston and Mangles (2001) postulated that entrepreneurial accounting firms will be more likely to be involved in the adoption of new technologies. While there was no clear evidence of a measure of entrepreneurial attitude among the firms in this study, the more proactive practices that were interviewed in 2007 tended to be less conservative in their attitudes to technology adoption and more focused on building their practice.

All interviewees reported that they needed to be 'in the game' with respect to having a website. Such a finding is consistent with the views of Douglas (2004) who argued that accounting firms with no website were at a competitive disadvantage.

The finding that all accounting practices surveyed have several contact features among the information content on their website in both 2006 and 2010 is consistent with the business services sector studies conducted by Burgess, Bingley and Sellito (2005). However, there was a substantial increase in the proportion of websites that had a location map on their website in 2010 when compared with 2006.

The Burgess, Bingley and Sellito (2005) study found that none of the financial service businesses websites that were examined conducted transactions over the Internet and this finding was supported in this study of accounting practices, again in 2006 *and* 2010. To date, it is mainly information, communication and some relationship enhancement activity that is being offered over the Internet by accounting practices.

Chen, Tseng and Chang (2005) found that the impact of the Internet on accounting firms differed across firms but all impacts were positive. The findings of this study generally support those findings. There were, however, some negative aspects reported and these included setup and maintenance costs, time spent updating websites, dealing with SPAM and client concerns over security.

Xiao, Jones and Lymer (2005) argued that financial reporting using the Internet has the potential to unlock databases and provide real time reporting. While this study did not find substantial evidence to support their view, there were some examples of uses of the Internet that indicate that the practices are taking notice. These examples include clients' ability to access their investment portfolios in real time; and some practices having direct access to client details through the Australian Taxation Office portal. It is important to note that the 2010 review of websites showed a marked increase in the proportion of businesses that had a secure log-in feature, which provide the doorway for more businesses to potentially offer these features.

In relation to service delivery, Banker, Chang and Kao (2002) reported that IT plays a critical role in the services provided in the accounting industry and this study supports that argument. The findings from interviews indicated the use of computer accounting packages is almost mandatory among clients, IT supports an improved delivery of information, and email is now seen as enhancing the traditional face-to-face services provided. However there is little evidence of the use of websites for the delivery of automated e-services. At the moment, the Internet is primarily used to *support* the delivery of services rather than completely *automate* it. There was no evidence of activity on websites at the transaction end of the spectrum in 2006 or 2010.

The interviews conducted in this study provide some real-life context of the current attitudes and Internet practices of accounting practices in and around Metropolitan Melbourne. When combined with the analysis of accounting practice websites in 2006 and 2010, the study provides strong evidence that these businesses are very client-focussed and are willing to use the Internet and their websites to improve the delivery of services to their clients. However, at the moment only a few of these services could potentially be classified as *e-services*.

The authors encourage similar studies to be conducted in other parts of the world. Melbourne would be typical of other Australian capital cities in relation to its accounting practices activities and the level of Internet penetration. Also, Australian practices, whilst connected to the rest of the world electronically, are isolated from much of it in relation to geographic distance and time zone differences. It would be worthwhile to examine how accounting practices attitudes and use of the Internet in other countries would differ, especially where different government regulations, much less distance between neighbouring countries and different levels of Internet penetration were evident.

CONCLUSION

The analysis of Australian accounting practice websites reveals a heavy emphasis towards the identification of the business and its services in 2006 and 2010, which is consistent with other website studies that have been carried out. More than other businesses, accounting practices rely on the provision of various types of information to clients. The provision of news in various formats on their websites is also an important tool for many practices. One interesting finding was the increased number of practices in 2010 that included a location on the website. This could be a reflection of the ease in this feature can be placed on websites and/or where businesses note a feature on a competitor website and replicate it on their own. Interestingly, none of the accounting practices offered transactional interactivity features on their websites. The interviews revealed a number of strategies in relation to the use of IT and the Internet, ranging from conservative to entrepreneurial strategies. Following this, it was found that the websites were developed for a number of reasons. Although a number of the websites discussed during the interviews had log-in features, only two contained features that could perhaps be classed as 'e-services'. However, the increase in accounting practices with a log-in feature on their website in 2010 suggests that the issues of trust in the technology are being addressed and that they could be moving towards a situation where e-services may become more commonplace in the future.

NOTE

This research has been assisted by the Institute of Chartered Accountants in Australia through its Academic Research Grants Scheme.

The Institute of Chartered Accountants in Australia

REFERENCES

Alonso Mendo, F., & Fitzgerald, G. (2005). Theoretical approaches to study SMES e-business progression. *Journal of Computing and Information Technology*, *13*(2), 123–136. doi:10.2498/cit.2005.02.04

Angehrn, A. (1997). Designing mature internet business strategies: The ICDT model. *European Management Journal*, *15*(4), 361–369. doi:10.1016/S0263-2373(97)00016-9

Banker, R., Chang, H., & Kao, Y. (2002). Impact of Information Technology on public accounting firm productivity. *Journal of Information Systems*, *16*(2), 209–222. doi:10.2308/jis.2002.16.2.209

Bhansali, C. (2005, June). Successful firms will be online on time. *Accounting Technology*.

Bolton, R. N. (2003). Marketing challenges of e-services. *Communications of the ACM*, *46*(6), 43–44. doi:10.1145/777313.777337

Burgess, S. (2002). *Business-to-consumer interactions on the Internet: A model for small businesses*. PhD Thesis. Monash University, Melbourne Australia.

Burgess, S., Bingley, S., & Sellitto, C. (2005). A model for website development of micro tourism enterprises. *Illuminating Entrepreneurship: Proceedings of the Institute for Small Business and Entrepreneurship 28th National Conference*, Perlex Associates, UK.

Burgess, S., Sellitto, C., & Wenn, A. (2005). Maturity in the websites of Australian wineries: A study of varying website content. *International Journal of Electronic Business*, *3*(5), 473–490. doi:10.1504/IJEB.2005.008521

Burgess, S., & Zoppos, B. (2002). The features of selected Australian websites. *E-Conomy: From here to where? (CD Rom Proceedings)*, School of Management Information Systems, Edith Cowan University, Perth, December.

Bury, L. (1999, August 28-29). On line and on time. *Accountancy.*

Chaston, L., & Mangles, T. (2001). E-commerce and small UK accounting firms: Influence of marketing styles and orientation. *The Service Industries Journal*, *21*(4), 83–99. doi:10.1080/714005049

Chen, M., Tseng, C., & Chang, J. (2005). A survey investigation into the use of the Internet among accounting firms. *International Journal of Management*, *22*(4), 649–660.

Chiu, D., Kok, D., Lee, A., & Cheung, S. (2005). Integrating legacy sites into Web services with Webxcript. *International Journal of Cooperative Information Systems*, *14*(1), 25–44. doi:10.1142/S0218843005001006

Debreceny, R., & Gray, G. (2001). Financial reporting on the Internet and the Internet and the external audit. *European Accounting Review*, *8*(2), 335–350. doi:10.1080/096381899336078

Douglas, A. K. (2004). *Marketing your firm on the Internet. Tax Practice Management*. January-February.

Ettredge, M., Richardson, V., & Scolz, S. (2002). Dissemination of information for investors at corporate websites. *Journal of Accounting and Public Policy*, *21*, 357–369. doi:10.1016/S0278-4254(02)00066-2

Field, J., Heim, G., & Sinha, K. (2004). Managing quality in the e-service system: Development and application of a process model. *Production and Operations Management*, *13*(4), 291–306. doi:10.1111/j.1937-5956.2004.tb00219.x

Fink, D. (2007). Antecedents for building trust in professional e-services. *Integration and Innovation Orient to E-Society*, *2*, 41–49.

Gebauer, J., & Shaw, M. (2002). Introduction to the special section: Business-to-business electronic commerce. *International Journal of Electronic Commerce*, *6*(4), 7–17.

Gutierrez-Nieto, B., Fuertes-Callen, Y., & Serrano-Cinca, C. (2008). Internet reporting in microfinance institutions. *Online Information Review*, *32*(3), 415–436. doi:10.1108/14684520810889709

Henderson, K., & Cowart, L. (2002). Bucking e-commerce trends: A content analysis comparing commercial real estate brokerage and residential real estate brokerage websites. *Journal of Corporate Real Estate*, *4*(4), 375–385. doi:10.1108/14630010210811967

Hofacker, C. F., Goldsmith, R. E., Bridges, E., & Swilley, E. (2007). E-services: A synthesis and research agenda. *Journal of Value Chain Management*, *1*(1/2), 13–44.

Hsu, M., & Chiu, C. (2004). Predicting electronic service continuance with a decomposed theory of planned behavior. *Behaviour & Information Technology*, *23*(5), 359–373. doi:10.1080/01449290410001669969

Jelassi, T., & Enders, A. (2005). *Strategies for e-business: Creating value though electronic and mobile commerce*. UK: Pearson Education.

Kotler, P., Hayes, T., & Bloom, P. (2002). *Marketing professional services—Forward thinking strategies for boosting your business, your image and your profits* (2nd ed.). Paramius, NJ: Prentice Hall Press.

Laffie, L. (2005). From the tax adviser: Expanded e-services access. *Journal of Accountancy*, *199*(6), 98–98.

Lymer, A. (1997). The use of the Internet in company reporting: A survey and commentary on the use of the WWW in corporate reporting in the UK. *Journal of Financial Information Systems*.

Lymer, A., & Debreceny, R. (2003). The auditor and corporate reporting on the Internet: Challenges and institutional responses. *International Journal of Auditing*, *7*(2), 103–120. doi:10.1111/1099-1123.00063

Lymer, A., Debreceny, R., Gray, G., & Rahman, A. (1999). *Business reporting on the Internet*. London, UK: IASC.

McCausland, R. (2004, 7 September). Home, sweet home page. *Accounting Technology*.

McMillan, S. J. (2000). The microscope and the moving target: The challenge of applying content analysis to the World Wide Web. *Journalism & Mass Communication Quarterly*, *77*(1), 80–98.

Mohamed, E. K., Oyelere, P., & Al-Busaidi, M. (2009). A survey of Internet financial reporting in Oman. *International Journal of Emerging Markets*, *4*(1), 56–71. doi:10.1108/17468800910931670

Schlageter, M. (2005). Leveraging technology. *Journal of Accountancy*, *199*(6), 14.

Shackleton, P., Fisher, J., & Dawson, L. (2003). Town hall e-government: A study of local government electronic service delivery. *Proceedings of the 14th Australasian Conference on Information Systems*, Perth, Australia November.

Stafford, T. (2003). E-services. *Communications of the ACM*, *46*(6), 26–28. doi:10.1145/777313.777333

Taylor, S. (1998). *Web sites – A missed opportunity?* London: Business Briefing. Institute of Chartered Accountants in England and Wales.

Xiao, Z., Jones, M., & Lymer, A. (2002). Immediate trends in Internet reporting. *European Accounting Review*, *11*(2), 245–275. doi:10.1080/09638180020017087a

Yu, S., & Rijia, D. (2009). *Comparison of Internet financial reporting between China and United States*. 2009 International Conference on Information Management, Innovation Management and Industrial Engineering, (p. 211). Xian, China, IEEE.

Chapter 5
Online Self-Services:
Investigating the Stages of Customer-SST Systems Interaction

Călin Gurău
GSCM – Montpellier Business School, France

ABSTRACT

The introduction of self-service applications by online service providers provided an effective solution to the lack of direct personal contact over the Internet, and offered a quick way to expand business operations. However, the success of this customer empowerment method is influenced by the capacity of various clients to use and adopt self service technology. Despite the interest raised by this topic within academic and practitioners communities, the factors that determine the adoption of online self-service applications are still a matter of debate. Using a combination of qualitative and quantitative analysis, this study attempts to identify the main phases of customer-system interaction, as well as the quality dimensions perceived as important by SST users in each particular phase. The findings reinforce the interpretation of SST adoption as a gradual process of learning that presents different challenges for various types of customers depending on their level of online experience.

INTRODUCTION

In the last 20 years, self-service technology (SST) has become a familiar presence in the modern business environment. The introduction of effective self-service systems allow companies to modularize and automatize the repetitive elements of services, concentrating their resources and personnel on more personalised aspects of the company-customer relationship, and thus providing more added-value to their clients (Bobbitt & Dabholkar, 2001). On the other hand, SST systems give clients the feeling that they have a better control on the service choice and delivery. This is especially true for time consuming services that require a personalized research of information and the selection of the best alternative among several available offers.

The implementation of self-service technology (SST) has created new service channels and

DOI: 10.4018/978-1-60960-607-7.ch005

procedures. Nowadays, clients can conduct bank transactions through automated teller machines (ATM) or on the Internet (online banking), make reservations or purchase tickets through online kiosks, check-in automated hotels, or use self-scanning systems in retail stores. The integration of self-service technology with Internet applications has increased even more the convenience of information-rich services – the customers can now access online service from their homes or offices, 24 hours a day, without geographical limitations.

However, the implementation of this strategy requires more than the introduction of self-service applications. These self-service systems need to be tied-in with employee-related policies and procedures. These internal procedures must insure that the customer can rely on virtual assistance when using a self-service interface and feel comfortable in a range of transactions without human intervention (Kotler, Armstrong, Saunders, & Wong, 2002).

The introduction of online self-services has changed the way in which companies relate to their customers. This new technology eliminates firm's personnel from the service interface, replacing it with software applications that can be accessed through real-time Internet connections. On the other hand, the self-service system gives additional responsibilities to the customer, who will initiate, generate and consume the service interacting directly with software applications. However, this additional responsibility is not necessarily perceived as negative by the involved customers. In fact, existing studies show that customers, and especially online service users, enjoy having a greater degree of control over the service they access (Ding, Verma, & Iqbal, 2007). The self-service systems may also permit a better customisation of the online service, resulting in improved satisfaction for the user.

On the other hand, the lack of a direct relationship with an employee might represent a disadvantage for customers that prefer direct human interactions. The dialogue with an employee adds a personal quality to the service provided, and can provide quick and flexible solutions when things do not go as planned.

All these factors, together with the propensity of customers to adopt a new technology, can significantly influence the adoption and frequency of use of a specific online self-service.

The purpose of this exploratory study is to analytically define and describe the stages of customer–system interaction during the adoption of new self-service technology, and to identify the quality dimensions that influence customers' perceptions regarding online self-service systems in each of these interaction stages. On the other hand, considering that these interactions and perceptions can vary from one customer to another, the research attempts to compare the behaviour of high-frequency online SST users with that of occasional users.

After a comprehensive presentation of the main research streams that are relevant for the adoption of self-service technology, the chapter presents the research objectives of this research project, as well as the methodology applied to collect primary and secondary data. These results obtained from data analysis are then presented and discussed in direct relation with the defined research objectives. The chapter ends with a summary of the main findings and with propositions for future research.

THEORETICAL BACKGROUND

Customer empowerment through self-service technology is a new, but fertile field of studies. Previous research on this topic can be categorised into four main streams:

1. the innovation diffusion theory – although many self-services offered online were previously provided into the physical environment, the application of self-service technology can be considered as an innovation;

2. consumer readiness to adopt the online self-service technology;
3. the technology adoption model (TAM) and its development – the Unified Theory of Acceptance and Use of Technology (UTAUT); and
4. e-service quality and customer satisfaction - which studies the self-service dimensions that are considered essential for online service quality.

From an empirical point of view, these research streams complement well each other - the diffusion theory considers customers' perceptions about the characteristics of online self-services before and during the trial phase, the consumer readiness framework analyses the personal capacities of online users, the technology acceptance model and the unified theory of acceptance and use of technology attempt to identify and explain the elements influencing the psychological process of decision making and the subsequent behaviour of technology users, and, finally, the e-service quality-customer satisfaction theory analysis customers' perceptions during and after their interaction with self-service systems.

The Innovation Diffusion Theory

The innovation diffusion theory was formulated by Rogers (1962), being then applied in a large number of projects (Gerrard & Cunningham, 2003). Rogers (1995) defined five main characteristics of innovations that influence their adoption rate by the population:

1. Relative advantage – represents the additional benefit offered by the innovation in comparison with the existing market offers. The research conducted on the adoption of banking self-services identified as sources of relative advantage economic benefits (Black, Lockett, Wiklofer, & Ennew, 2001; Loudon & Delle Bitta, 1993; Polatoglu & Ekin, 2001), convenience (Black et al., 2001; Loudon & Delle Bitta, 1993; Polatoglu & Ekin, 2001), performance (Polatoglu & Ekin, 2001), and independence – eliminating the need to rely on others (Black et al., 2001).
2. Compatibility – is the consistency of the innovation with the existing values, past experiences and needs of the potential adopters. For example, familiarity with the Internet and online applications was identified as an essential factor for customers' willingness to adopt online self-service technologies (Black et al., 2001).
3. Complexity – the complexity of a new product/service will influence its speed of adoption. In the case of online self-service technology this dimension can be related either with the simplicity and/or friendliness of the web interface, or with the user's level of capability (Black et al., 2001; Hewer & Howcroft, 1999; Polatoglu & Ekin, 2001; Rotchanakitumnuai & Speece, 2003).
4. Trialability – represents the facility of interacting with the innovation. Trialability is essential to transmit first-hand information about self-service applications and to facilitate consumers' learning about the use of the new technology (Hewer & Howcroft, 1999).
5. Observability - is the degree of visibility within the society at large, or in a social/professional group. This dimension is not always applicable to characterize the adoption of online self-services, because in most cases the interaction between the customer and the online applications is not visible to other users (Black et al., 2001; Gerrard & Cunningham, 2003). However, the organisation that attempts to introduce a new self-service application on the Internet may increase its observability by actively promoting and demonstrating the use of the service.

Another determinant of customer's readiness to adopt innovations is the perceived level of risk (Holak, 1988; Labay & Kinnear, 1981; Lockett & Littler, 1997; Ostlund, 1974). Many studies have outlined that the online environment is considered more risky than the physical market (Pavlou, Liang, & Xue, 2007). This perception can represent an important deterrent for the adoption of self-services, especially when the online procedure includes payment or a transfer of confidential information.

Consumer Readiness

Consumer readiness to embrace the self-service technology is explained in the academic literature by three main variables:

1. Role clarity – since traditionally services were produced through the interaction between a customer and a specialised employee, the adoption of self-service technology requires customer empowerment and a modified consumer behaviour. Role ambiguity for staff or customers can create important challenges for the introduction of new services (Easingwood, 1986), and limits consumers' capacity to participate to the co-production of services (Larsson & Bowen, 1989). Meuter, Bitner, Ostrom and Brown, (2005) have also investigated the relationship between role clarity and the self-service system trial.
2. Motivation: Intrinsic and extrinsic motivation were identified as essential factors for the successful adoption of online self-service technology (Barczac, Ellen, & Pilling, 1997). Some customers welcome a higher degree of empowerment, because they find the participation to the co-production of services intrinsically attractive (Bateson, 1985; Dadholkar, 1996; Rogers, 1995; Schneider & Bowen, 1995). On the other hand, some customers might be motivated by specific extrinsic advantages, such as price discounts, time saving or convenience (Dadholkar, 1996; Schneider & Bowen, 1995).
3. Ability: The online environment in general and the self-service system in particular may require specific personal skills. A complex or unfriendly website interface can induce concern about customer's ability to interact with the system, and therefore prevent her/him to engage in a proactive behaviour (Jayanti, & Burns, 1998; Meuter et al., 2005).

Technology Acceptance Models

The technology acceptance model (TAM) represents an adaptation of the theory of reasoned action (Ajzen and Fishbein, 1975), which explains the development of users' attitudes and behavioural intentions by the perceived usefulness and the perceived ease of use of the new technology (Davis, Bagozzi, & Warshaw, 1989). Despite its predictive power, this model represents a specific research perspective which eliminates a series of other internal and external factors that influence the adoption of new technology (such as social pressure, previous user experience, or availability of technological support).

In an attempt to provide a general model of technology adoption, Venkatesh, Morris, Davis and Davis (2003) proposed the Unified Theory of Acceptance and Use of Technology (UTAUT), which explains users' propensity to adopt a technology by four categories of constructs: performance expectancy, effort expectancy, social influence, and facilitating conditions. On the other hand, a series of variables defining the demographic profile of users (gender, age, experience and voluntariness of use) were considered as mediators of this model. The main criticism of this theory was its complexity, Bagozzi (2007) arguing that a model that uses 41 independent variables to predict intentions and at least 8 independent variables for predicting behavior, leads the theory of technology adoption to a state of chaos.

E-Service Quality

The theories of e-service quality are based on the traditional SERVQUAL model (Parasuraman, Berry, & Zeithaml, 1988). This model is based on extensive research conducted by Parasuraman, Zeithaml and Berry (1985) who initially identified ten determinants of service quality: tangibles, reliability, responsiveness, competency, courtesy, communication, credibility, security, access, and understanding the customer. These dimensions were later reduced to five features, using factor analysis: tangibles, reliability, responsiveness, assurance and empathy.

The specific characteristics of the Internet have required a significant adaptation of these quality dimensions to online services. Zeithaml, Parasuraman and Malhotra (2001 and 2002) have identified a series of e-service quality dimensions: reliability, responsibility, access, flexibility, ease of navigation, efficiency, assurance/trust, security, price knowledge, site aesthetics and customisation/personalisation, developing the e-SERVQUAL model. Another study of Jun and Cai (2001), realized in the context of Internet banking services indicated six main dimensions of online systems quality: content, accuracy, ease of use, timeliness, aesthetics and security.

Curran and Meuter (2005) explained how customers decide whether to use or not a self-service technology. They identified four elements that can be used as predictors of attitudes toward self-service systems: ease of use, usefulness, need of interaction and risk. The study demonstrates that self-service systems should be useful and easy to handle, but also that these dimensions are not enough to ensure customer satisfaction. Analysing self-service systems such as ATMs and phone banking, the authors conclude that there is room for improvement in the design and functionality of self-service technology.

Bateson (1985) explored the factors that determine a customer to choose a do-it-yourself option (including self-service alternatives), or the traditional service delivery system. The findings indicated that time (consuming time), control (control over the received service), effort (the effort made to access the service) and dependence (depending or not on another person) are the most important elements that influence the decision to use a specific service system. The customers that preferred the self-service option emphasised that time and control represent essential factors for their choice. The importance of personal control was confirmed by Meuter, Ostrom, Roundtree and Bitner (2000) and Meuter, Ostrom, Bitner and Roundtree (2003) who demonstrated that many perceived benefits of the self-service technology can be linked to this quality dimension.

A study conducted by Lee and Allaway (2002) attempted to verify if the manipulation of personal control in relation to new self-service applications leads to predictable changes in the way consumers perceived the risk associated with this technology. An additional objective was to verify if the specific dimensions of personal control (predictability, controllability and outcome desirability) contribute equally to variations in perceived risk, perceived value and adoption intention. The results showed that an increased sense of control over the self-service technology reduces perceived risk, heightens perceived value and stimulate the intention to adopt and use self-service technology. This study also stated that marketers should evaluate six issues in relation to their self-service system: (a) whether the website is easy to understand for new customers, (b) whether potential adopters are able to predict the benefits before using the service, (c) whether customers can adapt the service to their personal preferences, (d) whether customers can adapt their usage level/involvement to the system, (e) how desirable are the benefits of using the self-service technology from the customers perspective and (f) if potential adopters have enough knowledge and skills to successfully use self-service systems.

Beatson, Coote and Rudd (2006) investigated the specific connection between personal service attributes, SST attributes, overall customer satisfaction, and customer commitment, attempting to understand the elements that determine not only SST use but also its adoption. As expected, both personal service and SST attributes contribute to overall satisfaction. However, the results identified a specific customer inertia which is introduced by the repeated use of SST applications. The customers seem to perceive the interaction with a specific SST application as a gradual learning process, which may prevent them to move to another SST system that offers similar services but with a different technology and design platform (Buell, Campbell, & Frei, 2010). On the other hand, a successful experience of personal service relationships does not have the same effect on customer commitment.

This study outlines an important characteristic of SST adoption that was not yet taken into account in the existing technology use/technology adoption studies – the interaction between customer and SST application is a gradual process of learning that comprises several stages. It is therefore necessary to make a direct link between the success of this learning process, customer satisfaction, and the subsequent use of the SST application. This link is even stronger on the Internet, since the user is alone in front of his/her computer, using the available online information to understand and became familiar with the effective functioning of the SST system. This premise explains the specific approach adopted in this study: first, the interaction between customer and online SST applications is considered as a multi-stage process; second, the factors influencing customers' perception regarding the online SST application are specific for each interaction stage; and third, the customers' level of experience is expected to influence the effectiveness of this interaction/learning process, as well as the specific quality dimensions perceived as important in every stage.

RESEARCH METHODOLOGY

The following research objectives have been defined for this study:

1. To identify the main of stages of interaction between the online customer and self-service systems.
2. To define the specific factors influencing the perception of online users regarding the quality of the self-service system in each interaction stage.
3. To investigate the way in which the level of experience of online users with self-service web applications influences their perception about the quality dimensions of online self-service systems.

The research was focused on only one type of self-service systems that involve online transactions and payments - hotel booking, which is perceived by respondents as having a low/medium level of risk.

In order to answer these research objectives, both secondary and primary data have been collected and analysed. In the first stage of the research process, a series of academic and professional articles on the topic of self-service systems, technology adoption and quality dimensions of online services have been accessed and consulted, in order to build a theoretical framework for research investigation and analysis.

On the basis of this secondary research (Bateson, 1985; Curran & Meuter, 2005, Davis et al., 1989; Lee & Allaway, 2002; Meuter et al., 2000 and 2003; Parasuraman et al., 1985 and 1988; Zeithaml et al., 2001 and 2002), the following dimensions have been identified:

- Perceived level of security: the level of risk perceived by a customer using an online self-service system.
- Perceived ease of use: the facility perceived by the online user to interact with

the self-service application and to successfully complete the transaction;

- Perceived usefulness: the perceived utility of the self-service application in terms of customer value, in comparison with other options of accessing the service;
- Flexibility: the possibility of the client to change options during his/her interaction with the self-service system without re-starting the whole process.
- Personalization: the possibility to adapt the self-service systems to personal aesthetic and procedural preferences;
- Information about pricing conditions and fees: the clarity and completeness of the information provided by the self-service website regarding the pricing conditions and fees of the service transaction.
- Tools for problem solving: the self-service website provides specific information/solutions to the users that encounter problems during the service transaction;
- Web page design: the attractiveness and logic of the self-service webpage interface;
- Personal contact: the possibility to contact an employee who can help in problem situations;
- Available demo before using the actual service: the self-service website provides a demonstration of how to interact with the self-service system, helping the user to understand SST functioning without concluding a real transaction.

The definitions of these ten quality dimensions have been refined during the interviews realized with SST users.

In the second stage of the research project 150 face-to-face interviews have been realized in the waiting area of Montpellier airport. The interviews lasted between 10 and 15 minutes, focusing on the following categories of information: the level of respondent's experience regarding the use of SST applications (high versus low frequency of use), the specific stages of customer-system interaction, and the specific quality dimensions considered as very important in each stage of interaction. The questions regarding the respondents' level of experience provided the opportunity to select an equal number of high-frequency and low-frequency online SST users (75 of each). A high-frequency use was defined as the access of at least one online SST application per week, while the low-frequency use definition was applied to all the other cases.

In what concerns the third research objective, the respondents were presented with the ten quality factors identified in the literature, and asked to evaluate their importance in every stage of their interaction with the self-service application, using a three items Likert scale (low, medium and high importance). In addition, two more open-ended questions were added, asking the respondents to indicate any other quality dimensions that were not included in the first list, as well as their respective level of importance for every interaction stage.

The recorded interviews were then transcribed and analyzed using discourse analysis techniques, which identified the specific patterns of answers. The results regarding the third research objective, that had a quantitative nature, were registered and analyzed using the SPSS software.

PRESENTATION AND ANALYSIS OF DATA

The Specific Stages of Customer - SST System Interaction

When accessing a new SST application, the respondents outlined the complexity of their interaction with the online system, indicating several stages of this learning/familiarisation process:

(1) **Initial contact**: describes the specific way in which respondents identify and access for the first time the online SSL application.

There are three main sources of information used by respondents to identify and reach the online self-service system: the recommendation/advice of trusted friends or family, individual search using web engines, and a referential obtained from another site or offline media channel (such as the weblinks displayed on associated websites, or the web address indicated in a TV, newspaper or poster advert).

The access to new SST applications as a result of other weblinks was indicated mainly by high-frequency users (81.3% of high-frequency users in comparison 56% of low-frequency users), while the two other sources of initial information are favoured mostly by the low-frequency users (77.3% for other media channels and 48% for the referral of friends/relatives, in comparison with 69.3% and, respectively, 34.7% for high-frequency users). These results possibly indicate the tendency of high-frequency SSL users to be more familiarized and influenced by the Internet that the low-frequency users. The following comment indicates well this trend: "I spend every day at least four-five hours on the Internet and I like to try different things. Many of the booking websites that I use now I discovered by reading recommendations posted on other websites" [V.M., 36 years, male].

Taking into account these preferences, it is possible to classify the respondents into three main categories: relationship-driven (relying on advice from friends or relatives), online savvy (identifying new sites during their daily web surfing), and prospectors/pragmatics (searching for SST applications that answer to a clear and momentary need).

(2) **Trial**: during the trial phase, the user navigates the SST website, attempting to identify the specific interaction procedure. Many low-frequency users are looking at this stage for a demo version of the SST application, while the high-frequency users rely mostly on their previous experience with similar SST systems. However, all respondents indicated the need for a minimum of information about the functioning of the self-service application, visible placed on the access webpage. Many of the relationship-driven people positively appreciate the possibility to contact by phone or by mail a personal advisor, who can provide guidance regarding their interaction with the SST application. However, this possibility became very rare, many personalized contacts being replaced with impersonal lists of answers to frequently-asked questions: "I do not understand why everybody is talking about customer service when, in reality, they only give you a stupid list of information, and they tell you to solve the problem yourself. Because of this I only use these online self-services when I am too tired or too busy to go to a normal tourist agency" [S.T., 43 years, female].

The success of customer's interaction with the SST application in the trial phase is paramount for the effective use and the subsequent adoption of this technology: "I am a busy person: if I do not understand how it functions in the first 10 minutes, or if I simply get lost on the website, it is not a good enough for me. Time is often more important than money" [F.G., 31 years, male].

(3) **Early use**: is defined as the first several operational interactions between the customer and the SST application, while the respondent is still learning the effective use of the technology. The number of these interactions varies largely depending on the previous user experience: usually, high-frequency SST users need only one interaction to understand the functional logic of the system, while some low-frequency users need up to 5-6 repetitions until they are completely familiarized with the self-service system: " I am a slow

learner, therefore I need to repeat every step several times in order to understand all the intricacies of the system,...and I am talking only about the basic usage. Sometimes a booking website is re-designed and becomes more complex – which confuses me a lot, and so I have to restart the learning process from zero. If the website has become too complicated – too many secondary functionalities that I will never use..., I simply search for a simpler booking site" [F.D., 51 years, female].

As the above comment indicates, there are several problems in this stage that can prevent the use and the subsequent adoption of the SST application: perceived complexity, lack of support, change of functional design, the addition of secondary functionalities perceived as useless or too complex. As in any learning process, the application must remain unchanged, at least during this stage, in order to reassure the user and reinforce his/her previous experience.

On the other hand, even if the users became familiarized with the SST system, a sudden change of design or functionality may determine an involution of the interaction process from the effective use stage to early use. In many cases this phenomenon can weaken or destroy the loyalty of the customer, unless the self-service application provides other clear differential advantages in comparison with other booking sites: "The site design changed twice in the last six months and I struggled to learn again the interactive procedures, because the prices are always correct and the comments very informative. I tried other sites but I was disappointed, so I stayed with this one" [M.S., 44 years, female].

(4) **Effective use**: this phase starts when the user is completely familiarized with the functioning of the website. The interaction become more efficient and many customers start to explore the additional functionalities offered by the self-service system. If they are very satisfied with the SST application and with the quality of service accessed though the website (hotels, restaurants, general tourist experience, etc.), many respondents start to recommend the website to friends, relatives or colleagues.

The loyalty effect is reinforced every time the customer has a positive experience in relation to the website, and it takes several disappointing transactions to destroy this loyalty effect. From this perspective, the findings of this study reinforce the conclusions reached by Beatson et al. (2006) regarding the existence of a higher loyalty threshold for the customers reaching the effective use phase. In this stage, some respondents also appreciate the possibility to personalize the self-service application, defined as the possibility to customize the interaction in relation to their specific personal needs: "It is a nice touch to define your own interaction preferences, I feel this way that I am more in control of the whole transaction" [A.J., 33 years, male].

The Quality Dimensions Associated with Each Interaction Stage and the Effect of Customer's Level of Experience

Some of the quality dimensions associated with various interaction phases are strongly influenced by the level of consumer's experience. Tables 1 to 4 indicate the number and percentage of respondents from the two compared categories that indicated a specific quality dimension as highly important for different interaction stages. The Chi square indicates the statistical significance of the differences between the two categories of respondents.

The data displayed in Table 1 indicates that the usefulness of the web site represents a quality dimension considered as very important by a large percentage of both high-frequency and low-frequency users, followed by a quality dimension

Table 1. The differences in the perceived importance of quality dimensions of the SST system in the initial contact stage

Frequency of use / Quality dimensions	High-frequency users		Low-frequency users		Chi square	
	N	%	N	%		
Security	29	38.7	42	56	**4.520**	**p = 0.034**
Ease of use	32	42.7	47	62.7	**6.648**	**p = 0.036**
Usefulness	61	81.3	62	82.7	0.045	p = 0.832
Information about pricing and fees	42	56	51	68	**2.292**	**p = 0.130**
Design	35	46.7	41	54.7	0.960	p = 0.327
Accessibility	45	60	59	78.7	**6.145**	**p = 0.013**
Total	75	100	75	100		

Table 2. The differences in the perceived importance of quality dimensions of the SST system in the trial stage

Frequency of use / Quality dimensions	High-frequency users		Low-frequency users		Chi square	
	N	%	N	%		
Security	30	40	49	64	**8.654**	**p = 0.003**
Ease of use	39	52	61	81.3	**14.520**	**p < 0.0001**
Usefulness	57	76	74	98.7	**17.417**	**p < 0.0001**
Flexibility	23	30.7	47	62.7	**15.429**	**p < 0.0001**
Personalization	11	14.7	40	53.3	**24.985**	**p < 0.0001**
Information about pricing and fees	39	52	51	68	**4.000**	**p = 0.046**
Tools for problem solving	27	36	58	77.3	**26.090**	**p < 0.0001**
Design	54	72	56	74.7	0.136	p = 0.712
Personal contact	45	60	59	78.7	**6.145**	**p = 0.013**
Demonstration	38	50.7	51	68	**4.669**	**p = 0.031**
Total	75	100	75	100		

that is not considered in the existing literature – accessibility. Accessibility was defined by respondents as the facility to identify and access the self-service system in the Internet environment, as well as the degree of compatibility of the SST application with the operating system of their personal computer.

There are statistically significant differences between the answers provided by high- and low-frequency users regarding the level of perceived security, the ease of use, the information provided about pricing and fees, and the level of accessibility. These factors are evaluated by analyzing the information displayed on the homepage of the website, and have a strong impact on the subsequent decision to use or not the self-service application. On the other hand, there are several quality dimensions that were considered as non-relevant for this interaction stage: flexibility, personalization, tools for problem solving, personal contact and demonstration.

In the trial phase, all quality dimensions defined in the literature become relevant, while the added dimension of accessibility disappears (see

Table 3. The differences in the perceived importance of quality dimensions of the SST system in the early use stage

Frequency of use / Quality dimensions	High-frequency users		Low-frequency users		Chi square	
	N	%	N	%		
Security	32	42.7	54	72	**13.190**	**p < 0.0001**
Ease of use	33	44	64	85.3	**28.039**	**p < 0.0001**
Usefulness	59	78.7	74	98.7	**14.927**	**p < 0.0001**
Flexibility	29	38.7	47	62.7	**8.642**	**p = 0.003**
Personalization	23	30.7	41	54.7	**8.830**	**p = 0.003**
Information about pricing and fees	42	56	51	68	2.292	p = 0.130
Tools for problem solving	21	28	69	92	**64.000**	**p < 0.0001**
Design	55	73.3	59	78.7	0.585	p = 0.444
Personal contact	26	34.7	68	90.7	**50.266**	**p < 0.0001**
Demonstration	13	17.3	35	46.7	**14.828**	**p < 0.0001**
Total	75	100	75	100		

Table 4. The differences in the perceived importance of quality dimensions of the SST system in the effective use stage

Frequency of use / Quality dimensions	High-frequency users		Low-frequency users		Chi square	
	N	%	N	%		
Security	32	42.7	56	74.7	**15.836**	**p < 0.0001**
Ease of use	26	34.7	62	82.7	**35.630**	**p < 0.0001**
Usefulness	62	82.7	75	100	**14.234**	**p < 0.0001**
Flexibility	17	22.7	44	58.7	**20.142**	**p < 0.0001**
Personalization	31	41.3	37	49.3	0.968	p = 0.325
Information about pricing and fees	51	68	52	69.3	0.031	p = 0.860
Tools for problem solving	12	16	65	86.7	**74.960**	**p < 0.0001**
Design	54	72	59	74.7	0.897	p = 0.344
Personal contact	16	21.3	66	88	**67.253**	**p < 0.0001**
Demonstration	5	6.7	20	26.7	**10.800**	**p = 0.001**
Total	75	100	75	100		

Table 2). The comments provided by respondents explain this phenomenon: "Of course, accessibility is important for the first contact, but then, if I like the website I am saving it as a favourite which greatly simplifies accessibility" [D.J., 47 years, male]. "Accessibility is for me like a sort of filter – if I have problems accessing the website in the first instance I look for other applications, since I don't have time to waste" [M.R., 23 years, male].

With the exception of design, which represents a quality dimension considered important by both high- and low-frequency SST users, for all the other quality dimensions, the low-frequency users are significantly more concerned, requiring higher levels of security, ease of use, usefulness, flexibility, personalization, commercial information, tools for problems solving, personal contact and online demonstration. This effect is probable induced by the lower level of experience of these respondents, which determine a greater need for support and reassurance.

In the third interaction phase – early use – the hierarchy of needs is again modified (see Table 3). The information about pricing and service fees becomes important for a larger number of high-frequency users – concerned by the commercial advantages offered by the self-service application. Another reason for this trend is the comparison made by several customers between similar SST systems (in this case hotel booking sites): "It is important for me to feel that I have the best value for money, or time, if you want. As soon as I understand the functioning of the website, I try to compare the commercial conditions offered by several service providers, and then I choose the most convenient offer".

It is interesting to note that in this phase, the number of experienced respondents slightly decreases in comparison with the previous phase, for a number of quality dimensions, while the number of less-experienced users increases. We can identify this phenomenon for ease of use, tools for problem solving, design, personal contact and demonstrations. The explanation provided by respondents outlines that experienced users become more independent and enter more quickly than the less-experienced customers into a routine phase, mainly because their previous experience helps them to learn quicker the functioning of the self-service application. Since many low-frequency SST users are still in a process of gradual learning, their need for support increases during this stage.

When customers enter in the effective use phase, the need for support significantly decreases (see Table 4). However, this need decreases quicker for the experienced users than for the less experienced ones (e.g. in what concerns ease of use, flexibility, tools for problem solving, personal contact and demonstration) – which validates the behavioural differences between these two categories of SST users. On the other hand, a larger number of high-frequency customers become interested in this phase in personalization and commercial information, attempting to increase the efficiency of their interaction with the SST application.

The needs of low-frequency users for personalized support, tools for problem solving, ease of use and security remain high even in this stage, which can be explained by their low familiarity with SST applications in general. It might also be a problem of personal attitude and confidence, since at this stage the preliminary learning process should be already finished for both categories of investigated users.

The results presented and commented above reinforce the interpretation of self-service technology use and adoption as a gradual learning process that raises specific challenges for various categories of online users, in each interaction stage. This statement represents an important change of perspective, since most of the theoretical models presented in the existing literature do not take into account the complexity and the temporal dimensions of this learning process. It also indicates possible areas of future research and cross-fertilization between the technology adoption/innovation diffusion theories, and practical models regarding the learning process and the types of learners (e.g. Kolb models).

CONCLUDING REMARKS

This chapter enriches the theoretical literature on the use and adoption of SST applications by identifying the specific phases of customer-system interaction, as well as the main quality dimensions used for evaluating the effectiveness of the system in every stage. These dimensions vary in relation to the specific profile of SST users. Although this study has evidenced only the effect of user's experience on the perceived quality dimensions, further research is necessary to identify and analyze the influence of other socio-demographic variables.

Interpreting this research project from the perspective of a value co-creation process (Prahalad & Ramaswamy, 2004), it is possible to build a meta model, which states that not only the participation to the co-creation of services can represent a source of satisfaction and commitment for customers, but also the way in which this participation is organized (by the company) and experienced (by the customer).

The findings of this study have both an academic and practical significance. At academic level, the results reinforce the idea, already expressed by other researchers (Beatson et al., 2006), that the adoption of SST can be equated to a gradual learning process, which has specific effects on customer satisfaction and loyalty. At practical level, this conclusion should be translated in a practical approach to enhance the learning process for various categories of SST users, by supporting and guiding them during the various phases of the interaction process. At this level, however, there is still much to be done. As some respondents have emphasized in their comments, the use and adoption of a specific SST application is not determined only by the isolated evaluation of that particular website, but rather through a process of dynamic assessment and comparison among several competing self-service systems. It can be expected that very few customers develop an exclusive commitment towards one SST application, the rule being rather a shared loyalty between several similar websites. All these research propositions need to be properly formulated and validated by future studies.

REFERENCES

Ajzen, I., & Fishbein, M. (1975). *Understanding attitudes and predicting social behaviour*. New Jersey: Prentice-Hall.

Bagozzi, R. P. (2007). The legacy of the technology acceptance model and a proposal for a paradigm shift. *Journal of the Association for Information Systems*, *8*(4), 244–254.

Barczac, G., Ellen, P. S., & Pilling, B. K. (1997). Developing typologies of consumer motives for use of technologically based banking services. *Journal of Business Research*, *38*(2), 131–139. doi:10.1016/S0148-2963(96)00032-X

Bateson, J. (1985). Self-service consumer: An exploratory study. *Journal of Retailing*, *61*(3), 49–76.

Beatson, A., Coote, L. V., & Rudd, J. M. (2006). Determining consumer satisfaction and commitment through self-service technology and personal service usage. *Journal of Marketing Management*, *22*(7-8), 853–882. doi:10.1362/026725706778612121

Black, N. J., Lockett, A., Wiklofer, H., & Ennew, C. (2001). The adoption of Internet financial services: A qualitative study. *International Journal of Retail & Distribution Management*, *29*(8), 390–398. doi:10.1108/09590550110397033

Bobbitt, L. M., & Dadholkar, P. A. (2001). Integrating attitudinal theories to understand and predict use of technology-based self-service. *International Journal of Service Industry Management*, *12*(5), 423–450. doi:10.1108/EUM0000000006092

Buell, R. W., Campbell, D., & Frei, F. X. (2010). Are self-service customers satisfied or stuck? *Production and Operations Management*, *19*(6), 679–697. doi:10.1111/j.1937-5956.2010.01151.x

Curran, J., & Meuter, M. (2005). Self-service technology adoption: Comparing three technologies. *Journal of Services Marketing*, *19*(2), 103–113. doi:10.1108/08876040510591411

Dabholkar, P. A. (1996). Consumer evaluations of new technology-based self-service options: An investigation of alternative models of service quality. *International Journal of Research in Marketing*, *13*(1), 29–51. doi:10.1016/0167-8116(95)00027-5

Davis, F. D., Bagozzi, R. P., & Warshaw, P. R. (1989). User acceptance of computer technology: A comparison of two theoretical models. *Management Science*, *35*, 982–1003. doi:10.1287/mnsc.35.8.982

Ding, X., Verma, R., & Iqbal, Z. (2007). Self-service technology and online financial service choice. *International Journal of Service Industry Management*, *18*(3), 246–268. doi:10.1108/09564230710751479

Easingwood, C. J. (1986). New product development for service companies. *Journal of Product Innovation Management*, *3*(4), 264–275. doi:10.1016/0737-6782(86)90005-6

Gerrard, P., & Cunningham, J. B. (2003). The diffusion of Internet banking among Singapore consumers. *International Journal of Bank Marketing*, *21*(1), 16–28. doi:10.1108/02652320310457776

Hewer, P., & Howcroft, B. (1999). Consumers distribution channel adoption and usage in the financial services industry: A review of existing approaches. *Journal of Financial Services Marketing*, *3*(4), 344–358.

Holak, S. L. (1988). Determinants of innovative durable adoption: An empirical study with implications for early product screening. *Journal of Product Innovation Management*, *5*(1), 50–69. doi:10.1016/0737-6782(88)90032-X

Jayanti, R. K., & Burns, A. C. (1998). The antecedents of preventative healthcare behavior: An empirical study. *Journal of the Academy of Marketing Science*, *26*(1), 6–15. doi:10.1177/0092070398261002

Jun, M., & Cai, S. (2001). The key determinants of Internet banking service quality: A content analysis. *International Journal of Bank Marketing*, *19*(7), 276–291. doi:10.1108/02652320110409825

Kotler, P., Armstrong, G., Saunders, J., & Wong, V. (2002). *Introduction to marketing* (2nd ed.). London, UK: Pearsons Education.

Labay, D. G., & Kinnear, T. C. (1981). Exploring the consumer decision process in the adoption of solar energy systems. *The Journal of Consumer Research*, *8*(3), 271–278. doi:10.1086/208865

Larsson, R., & Bowen, D. E. (1989). Organization and customer: Managing design and coordination of services. *Academy of Management Review*, *14*(2), 213–233.

Lee, J., & Allaway, A. (2002). Effects of personal control on adoption of self-service technology innovations. *Journal of Services Marketing*, *16*(6), 553–572. doi:10.1108/08876040210443418

Lockett, A., & Littler, D. (1997). The adoption of direct banking services. *Journal of Marketing Management*, *13*, 791–811. doi:10.1080/0267257X.1997.9964512

Loudon, D. L., & Della Bitta, A. J. (1993). *Consumer behaviour: Concepts and applications* (4th ed.). New York, NY: McGraw-Hill.

Meuter, M., Ostrom, A., Roundtree, R., & Bitner, M. (2000). Self-service technologies: Understanding customer satisfaction with technology-based service encounters. *Journal of Marketing*, *64*(3), 50–64. doi:10.1509/jmkg.64.3.50.18024

Meuter, M. L., Bitner, M. J., Ostrom, A. L., & Brown, S. W. (2005). Choosing among alternative service delivery modes: An investigation of customer trial of self-service technologies. *Journal of Marketing*, *69*(2), 61–83. doi:10.1509/jmkg.69.2.61.60759

Meuter, M. L., Ostrom, A. L., Bitner, M. J., & Roundtree, R. I. (2003). The influence of technology anxiety on consumer use and experience with self-service technologies. *Journal of Business Research*, *56*(11), 899–906. doi:10.1016/S0148-2963(01)00276-4

Ostlund, L. E. (1974). Perceived innovation attributes as predictors of innovativeness. *The Journal of Consumer Research*, *1*(2), 23–29. doi:10.1086/208587

Parasuraman, A., Berry, L. L., & Zeithaml, V. A. (1988). Servqual: A multiple-item scales for measuring consumer perceptions of service quality. *Journal of Retailing*, *64*(1), 12–40.

Parasuraman, A., Zeithaml, V. A., & Berry, L. L. (1985). A conceptual model of service quality and its implications for future research. *Journal of Marketing*, *49*(4), 41–50. doi:10.2307/1251430

Pavlou, P. A., Liang, H., & Xue, Y. (2007). Understanding and mitigating uncertainty in online exchange relationships: A principal-agent perspective. *Management Information Systems Quarterly*, *31*(1), 105–136.

Polatoglu, V. N., & Ekin, S. (2001). An empirical investigation of the Turkish consumers' acceptance of Internet banking services. *International Journal of Bank Marketing*, *19*(4), 156–165. doi:10.1108/02652320110392527

Prahalad, C. K., & Ramaswamy, V. (2004). *The future of competition: Co-creating unique value with customers*. Boston, MA: Harvard Business School Press.

Rogers, E. M. (1962). *The diffusion of innovations* (1st ed.). New York, NY: The Free Press.

Rogers, E. M. (1995). *The diffusion of innovations* (4th ed.). New York, NY: The Free Press.

Rotchanakitumnuai, S., & Speece, M. (2003). Barriers to Internet banking adoption: A qualitative stud among corporate customers in Thailand. *International Journal of Bank Marketing*, *21*(6/7), 312–323. doi:10.1108/02652320310498465

Schneider, B., & Bowen, D. E. (1995). *Winning the service game*. Boston, MA: Harvard Business School Press.

Venkatesh, V., Morris, M. G., Davis, G. B., & Davis, F. D. (2003). User acceptance of Information Technology: Toward a unified view. *Management Information Systems Quarterly*, *27*, 425–478.

Zeithaml, V. A., Parasuraman, A., & Malhotra, A. (2001). *A conceptual framework for understanding e-service quality: Implications for future research and managerial practice*, (pp. 1-49). (MSI Working Paper Series, No 00-115), Cambridge, MA.

Zeithaml, V. A., Parasuraman, A., & Malhotra, A. (2002). Service quality delivery through websites: A critical review of extant knowledge. *Journal of the Academy of Marketing Science*, *30*(4), 362–375. doi:10.1177/009207002236911

Chapter 6
Sponsored Search as a Strategic E-Service

Roumen Vragov
Baruch College, CUNY, USA

ABSTRACT

In addition to helping consumers navigate the web search engines offer a new kind of e-service known as sponsored search. A search engine provides an auction interface to advertisers that allows them to be included in a list of sponsored links. Consumers then search the list to find the best deal. This paper uses a version of traditional consumer price search models from economic theory to evaluate the extent to which consumers and advertisers can rely on sponsored search as an effective strategic infomediary. The article also describes steps that the manager of the search engine can take to improve the sponsored search service. The findings are relevant to web advertising as well as to mobile advertising.

INTRODUCTION

With about 213 million queries per day web search engines are the most often visited locations in cyberspace (Sullivan, 2006). They are the places where most Internet customers start searching for information available on the web. Many customer searches conclude in a purchase, so in effect, search engines provide an important global e-service by facilitating transmission of commercial information from producers to consumers. While the search engines' service was relatively unbiased but not of high quality at the beginning of the Internet age, its nature has gradually changed over the last decade. Search engines have improved the relevance of their search results and, in addition to their regular search service, they have started offering sponsored search services by allowing advertisers to directly interface with the search engines' own systems. Google, Yahoo and MSN let advertisers bid in on-line auctions for spots on the search engines' list of sponsored links (Cu-

DOI: 10.4018/978-1-60960-607-7.ch006

sumano, 2005) and then collect revenue for each click. The list of sponsored links is shown at the top of the regular search results and/or in a column to the right of the regular search results. In spite of reports (see Jansen & Resnik, 2005) showing that users have strong biases against sponsored links, search engine advertising revenues have been increasing dramatically over the last few years. Google has indicated $10 billion revenue from advertising for the 2006 fiscal year. Moreover, Google has recently started another similar program, AdSense, which also uses auctions to show business ads to customers who browse anywhere on the web.

Sponsored search is becoming increasingly important in two other recent areas of interest: mobile advertising and location-based services. Once search engines are readily accessible through mobile devices, they will be able to deliver ad messages based on location as well as search terms and other customer preferences. According to recent reports (Kharif, 2008) 58 million US wireless users viewed an ad on their device in February 2008, however, linking the ad to a specific location and persuading the users to act on the ad is still a challenge. Because of its increasing popularity, online and more recently mobile advertising through auctions will have a lasting impact on E-commerce and M-commerce.

Advertising is the main source of revenue for the big search engines. There are many factors that contribute to the success in collecting advertising revenue. One of them is the number of users who click on the sponsored links. In the long run, this number depends on the relevance of the regular search results shown and the relevance of the list of sponsored links. More relevant regular search results will attract more visitors to the search engine's web site and more relevant lists of sponsored links will encourage a bigger proportion of users to click on them. Less clear is the importance of the auction mechanism used to allocate the spots in the list of sponsored links even though analytical attempts to compare some currently used auction rules and strategic bidder behavior are reported in Edelman and Ostrovsky (2007), and Varian (2006). Since the start of the sponsored search service, search engines have occupied the delicate strategic position of a direct infomediary between advertisers and consumers. This article investigates the link between the search engines' revenue generating process and the usefulness of sponsored search to advertisers and consumers. The paper offers a theoretical model of the search engine's revenue-generating process based on the concept of a pure strategy Nash equilibrium by taking into account the strategic interactions between the main three interested parties: the search engine, the consumers, and the advertising firms. A specific auction algorithm rather than the more abstract oligopoly competition solution concept is used to represent the relationship between the three parties. In the context of the model we investigate several more specific questions and claims that have arisen recently in the scientific and popular press. First, the article estimates the collusive pressure that advertisers face and its relationship to the number of advertisers present in the advertising auction; Second, it investigates the differences between requiring advertisers to pay their bid versus requiring them to pay the bid of the advertiser below them on the list of sponsored links. Third, it examines how participants' profits change when the search engine considers the click-through rate (the number of times that a link has been clicked in the past) when it ranks the advertisers' bids. Using the model the article evaluates the extent to which sponsored search is useful to consumers and advertisers and describes steps that the manager of the search engine can take to improve the sponsored search service.

RELATED LITERATURE

Search engines have already received some attention in the E-services literature because, from a consumer's point of view, web search engines possess the first inherent characteristic of e-Services: information service (Rust & Lemon, 2001). From an advertiser's point of view, search engines are collaborative value-chain partners that the advertiser uses to deliver one's value proposition to customers. That is why, within the common framework of web services described in Singh et. al. (2003), search engines can be classified as infomediaries that serve the e-business need for exchange of knowledge and information with customers. There is already one study by Kumar et. al. (2004/2005) which established that search engine capability and user experience are two important factors that determine overall search cost and performance. This paper adds to this line of research by investigating how the strategic interactions between the search engine, advertisers, and consumers influence the quality of price information provided by the search service. Thus, the study is also part of the latest e-Services research framework described in Rowley (2006), in which information quality and customer perception of it are some of the central problems.

A good starting point for our investigation is a long tradition of previous literature in economics dealing with optimal consumer price search. The first theoretical observations are attributed to Bertrand (1883). He observed that if consumer price search is costless, producer prices converge to the perfectly competitive price. The area of consumer price search flourished in the 1970's and 1980's starting with Diamond (1971) who stated that if search costs are positive, producer prices will converge to the monopoly price. To show that, Diamond used a simple model based on a pure strategy Nash equilibrium. The theoretical focus later shifted to finding intermediate outcomes (price dispersion) which are supportable as mixed strategy equilibria. Consumer search models started dividing consumers into two types: those that search with zero cost, and others who have positive search costs (see Braverman, 1980; Varian, 1980 and Stiglitz, 1979). The idea of sequential search with or without recall and the usage of dynamic programming were added as other ways to support price dispersion (see Burdett and Judd, 1983; Carlson and McAfee 1983; Stiglitz, 1987, and Stahl, 1989). Even though most of these models are mathematically complex they are still based on the unrealistic assumption that all producers have the same costs. The first models that considered producers with different product costs are described in Reinganum (1979) and Rob (1985) and they also support price dispersion using mixed strategy equilibria. Most recently theory has shown that intuitive sub-game perfect equilibria exist also in the case of ordered search when firms have constant marginal cost normalized to 0. In these equilibria sellers' prices and profits decline in the order of search, and consumers with lower search costs search longer and obtain better deals (Arbatskaya, 2006). The link between consumer price search and producer price advertising is analyzed extensively in Robert and Stahl (1993). The authors explore the effects of an advertiser's entry and the effects of changes in the advertising costs and search costs on equilibrium prices and quantities.

The model presented here borrows some of the basic assumptions about consumer types and optimal consumer search behavior from the aforementioned authors to create a modified version of sequential search with perfect recall. The analysis becomes more specific when a search engine is introduced to act as an intermediary between consumers and advertisers. Consumers are still homogeneous in terms of their value for the product sold, however advertisers are heterogeneous in terms of production costs. We also introduce a specific auction mechanism through which the theoretical equilibrium is implemented.

Because of its success Google's advertising auction mechanism has already received some attention in scientific literature. Most notably Edelman et. al. (2005), Varian (2006) and Liu et. al. (2005) provide an interesting game theoretic analysis of the interaction between advertisers and the rules of the advertising auction. They show that the auction mechanism used by Google does not have an equilibrium in dominant strategies and is not strategically equivalent to the value-revealing generalized multi-unit version of a Vickrey auction as claimed in the popular press (Coy, 2006). These studies do not consider the indirect interaction between consumers and advertisers through the ultimate buying decision. Feng et. al. (2003) use simulations to compare Google's auction mechanism to that of Yahoo. (See http://www.google.com/ads/ for a description of Google's auctions, http://smallbusiness.yahoo.com/marketing/sponsoredsearch.php for a description of Yahoo's auctions, and http://advertising.microsoft.com/advertise/search/introducing-adcenter for a description of MSN.com's AdCenter). They report that ranking paid placement links by the product of the advertiser's bid and past click-through rates is better, in most cases, than ranking by advertiser's bid alone, which performs best only when the correlation between the advertiser's past click-through rates and bid is large. However, both of these studies fail to fully consider the impact of advertisers' prices on consumers' search and buy behavior. The model presented here demonstrates that under certain conditions the advertising auctions have a pure strategy Nash equilibrium in weakly dominant strategies. We also demonstrate that Feng et. al.'s results do not hold if product price competition between advertisers is considered.

Worth noting is another study by Animesh et. al. (2006) that describes a way to use sponsored search auctions as a segmentation tool. There are some other articles that demonstrate fruitful practical approaches to dealing with search engine advertising auctions. For example, Mehta et. al. (2005) use an algorithmic approach to optimizing the revenue from AdWord auctions. Kitts and LeBlanc (2004) build a simulated trading agent platform and report results showing the success of different bidding strategies, and Sarukkai (2005) provides an accounting framework that can be used by advertisers to determine the worth of a keyword.

THEORETICAL MODEL

This model represents the strategic interactions between consumers, a search engine, and the advertising company in a market for one product of interest (Product X). We assume that search results are 100% relevant so when a consumer uses a description of the product as a search term, the search engine displays all firms participating in the advertising auction that offer the product of interest in the list of sponsored links. To maintain logical continuity with previous search models we will assume initially that advertisers are charged by the search engine based on consumer purchases. Tracing customer purchases can be done with the help of cookies, however, this is still not a completely reliable method. However, to keep in tune with reality we will relax that assumption later in the model, so that we can see the implications of charging advertisers per consumer click instead of per customer purchase.

Consumer Characteristics

We assume that there are two groups of consumers that visit the web site of a search engine. The first group is made up of consumers who come to retrieve information. These consumers do not have any purchase intentions and therefore have no commercial significance. The second group is made up of consumers who are visiting the search engine to look for and purchase a product of interest. Let us mark by N the number of consumers in the second group and suppose that all of them are interested in purchasing product X. All consumers value product X at v. Suppose further that there are many types of consumers. Type 1 consumers prefer not to search all the sponsored links shown. They click on the first link in the list of sponsored links and buy product X from the corresponding web site. Generally type i consumers click on the first i links in the list of sponsored links, and choose to buy product X from the advertiser who offers X at the lowest price (A graph of consumer click patterns on MSN.com is available in Debmczynski et. al, 2008). Let us mark by a_i ($0 \leq a_i \leq 1$) the proportion of consumers of type i. It is easy to imagine that a consumer's decision to search or not to search is motivated by opportunity cost, search cost, or beliefs about the way the search engine's list of sponsored links operates. Generally, we assume that the proportions a_i are a result of utility maximizing behavior on the part of the consumers given the information and beliefs that they have. This is justified because it is the result of thorough theoretical analysis presented in many of the previous search models mentioned in the section above. Without loss of generality we also assume that if product prices are the same, consumers purchase product X from the top advertiser in the list of sponsored links.

Basic Auction Procedures

All companies, which want to advertise their product with the help of a search engine, have to submit a list of key words that closely match the product that they are selling and a short ad describing the product or service. These ads are displayed to every user who searches for the keywords chosen by the company. In our model, every advertiser who makes product X participates in the advertising auction and submits the same product description as keywords. The advertisers also submit a bid which expresses how much they are willing to pay to the search engine for every user who purchases product X from the company's web site (note that later in the analysis we look at pay-per-click mode versus pay-per-purchase mode). Companies are listed in the list of sponsored links according to their bids. The advertiser who has the highest bid is shown at the top of the list of sponsored links and is charged his/her bid for every consumer who buys the product from that advertiser. Advertisers are given some time to submit their bids, which are private. At the end of this period, advertisers are ranked from top to bottom as described and this list is shown to all consumers who start shopping for product X at the same time. We assume that ties between advertisers' bids are broken by the time of arrival of the bids into the advertising auction before the list of advertisers is shown simultaneously to all consumers.

Advertising Companies' Characteristics and Behavior

Suppose that there are $M > 1$ risk-neutral advertisers that can produce product X and offer it for purchase over the Internet. Every advertiser i is characterized by a cost of c_i per unit of X. All costs are greater than 0 and less than v. They are public information and for simplicity, we will label the lowest cost c_1, the second lowest cost c_2, etc. ($c_1 < c_2 < c_3 ... < c_M$) We assume that every advertiser has enough capacity to satisfy completely all existing demand for product X and knows the consumers' maximum willingness to pay for product X. Since consumer values are equal to v, the cost requirement guarantees that there is a possibility for non-negative gains to trade. All M advertisers decide to use a search engine to advertise their product through sponsored links. We are modeling a decision-making process, in which all advertisers submit the same key word(s). As mentioned earlier the advertiser with the highest bid per click gets the first slot in the list of sponsored links; the advertiser with the second highest bid gets the second slot and so on. Each advertiser knows how consumers behave and can influence one's profit in two ways. The advertiser can manipulate the price that he/she charges for product X and the bid that he/she submits in the advertising auction.

Observation 1: *If* $a_M = 1$ *(all consumers search the entire list of sponsored links), the advertising* *auction game has a Nash equilibrium in pure strategies in which all advertisers bid 0 and charge their costs while the advertiser with the lowest cost charges* $p_1 = c_2 + \varepsilon$ *(where* ε *is a very small number close to 0).*

Proof: *Given the strategy profile described above we can show that every advertiser's profit is greater than or equal to zero and no one has an incentive to deviate. Under this strategy profile the advertiser with the lowest cost receives positive profits and all other advertisers gain 0. Also note that changing the bid does not affect anyone of the advertisers' profits because consumers search the whole list of sponsored links and choose the advertiser with the lowest price. If the winning advertiser charges a higher price than* $c_2 + \varepsilon$*, s/he is replaced by the advertiser with the second highest cost and thus will loose the auction. If the winning advertiser charges a lower price, s/he will still win the auction but will receive a lower profit. All remaining advertisers will earn negative profits if they charge less than the winning advertiser.* ■

Since consumers search all sponsored links, the first spot on the list has no value. As long as the advertiser charges the lowest price for product *X*, he/she will be able to capture all consumers. This result corresponds to that of Bertrand (1883) who noted that, when consumers are fully informed about the prices that stores are charging, the resulting equilibrium price is the Walrasian price. Note that under these equilibrium conditions, the advertiser's profit is $N(c_1 - c_1 + \varepsilon)$, the total consumer surplus is $N(v - c_2 - \varepsilon)$, and the search engine's revenue is *0*.

Observation 2: *If* $a_1 = 1$ *(all consumers decide not to search for the advertiser with the lowest price and buy product X from the first link in the list of sponsored links) the advertising auction game has a Nash equilibrium in pure strategies in which every advertiser* $j > 1$ *charges a price of* $p_j = v$ *and bids* $b_j = v - c_j$ *while the advertiser with the lowest cost charges a price of v and bids* $b_1 = v - c_2 + \varepsilon$*. (j is used as the cost rank index)*

Proof: *Given the strategy profile described above we can show that everyone's profit is greater than or equal to zero and no one has an incentive to deviate. Under this strategy profile the winner of the top spot in the list of sponsored links receives positive profits and all other advertisers gain 0. Increasing the price reduces everyone's earnings because no purchases will be made. Decreasing the price can only lead to less earnings. For the winning advertiser bidding lower than* $v - c_2 + \varepsilon$ *will result in 0 profits because the bid will leave room for the advertiser with the second lowest cost to win the first spot in the list of sponsored links and therefore all consumers. Bidding higher than* $v - c_2 + \varepsilon$ *will result in lower profits for the winning advertiser.* ■

Since consumers do not search the list of sponsored links for the advertiser with the lowest price advertisers cannot compete in terms of price. All value is concentrated at the first spot of the list of sponsored links. The purpose of all advertisers will be to win this spot. This result corresponds to that of Diamond (1971) who noted that when search is so costly that consumers cannot search for the lowest price, the resulting equilibrium price is the monopoly price. Note that under these equilibrium conditions, the winning advertiser's profit is $N(c_1 - c_1 + \varepsilon)$, the total consumer surplus is *0*, and the search engine revenue is $N(v - c_2 - \varepsilon)$.

The analysis so far suggests that there is an inherent "conflict of interest" played out between the searching consumer and the search engine. The more consumers search for the lowest price, the less valuable is the first spot in the list of sponsored links. The advertisers have less of an incentive to compete for the first spot on the list of sponsored links, and search engine advertising revenues drop. If consumers do not search for the lowest price and click only on the top link, then advertisers will not compete in terms of the price of product *X*; they will compete for the top spots in the list of sponsored links. This will drive the revenue of the advertising engine up and consumer surplus down.

Although useful as an illustration the extreme cases presented in Observations 1 and 2 are hard to imagine under naturally occurring circumstances. It is more realistic to assume that sponsored links are extremely rarely clicked if they are beyond the first page of the search results. Let us look at the case in which the first *k* links are clicked a non-zero number of times and

$$a_1 > a_2 > a_3 .... > a_k. \quad (1)$$

This is a standard result from many of the models of consumer search behavior mentioned in the previous section and confirmed by observation.

Proposition 1: *If*

$$c_2 > \max\left\{\frac{v+c_1}{2}, \frac{v+(1-a_1)c_1}{2-a_1}\right\}, \quad (2)$$

the advertising auction game has a Nash equilibrium in pure strategies in which $b_1 = v - c_2 + \varepsilon$, $p_1 = c_2 - \varepsilon$, *and for all* $j > 1$ $p_j = c_j$ *and* $b_j = v - c_j$.

Proof: *Let us mark by* $\pi_j(p_j, b_j)$ *the profit of advertiser j if s/he submits a bid of* b_j *and charges a price of* p_j. *Let* p_j* *and* b_j* *be advertiser j's equilibrium strategies. Nash equilibrium in weakly dominant strategies requires that, given the strategies of all advertisers,*

$$\pi_j(p_j^*, b_j^*) \geq \pi_j(p_j, b_j) \text{ for all j, } p_j \text{ and } b_j, \text{ and} \quad (3)$$

$$\pi_j(p_j^*, b_j^*) \geq 0 \text{ for all j, } p_j^*, b_j^* \quad (4)$$

It is obvious that, given the strategies of all other players, for every $j > 1$, *any* p_j *and any* b_j, $\pi_j(p_j, b_j) \leq 0$. *This is because no one can profitably outbid or out-price the advertiser with the lowest cost. Given the strategy profile* $\pi_j(p_j^*, b_j^*) = 0$ *for all* $j > 1$. *This means that criteria (3) and (4) are satisfied for every* $j > 1$.

We know that in the equilibrium above the profit of the advertiser with the lowest cost is $\pi_1(p_1^*, b_1^*) = N(2c_2 - v - 2\varepsilon - c_1)$. *Let us see under what conditions inequality (3) is satisfied for this advertiser. If the advertiser lowers the bid, s/he can loose the first spot in the list of sponsored links to the advertiser with the second highest cost, and if s/he increases the bid, his/her payoff decreases. Therefore changing the bid can never improve the profit. Decreasing the price also results in lower payoff, however, the advertiser can raise the price to the monopoly level for the* a_1N *consumers who only search the first spot and ignore the rest of the consumers. To guarantee that this is not a more profitable strategy we need*

$$c_2 > \frac{v+(1-a_1)c_1}{(2-a_1)},$$

which constitutes the first part in inequality (2).

To complete the proof we only have to discover the conditions under which (4) is satisfied for the advertiser with the lowest cost. It is easy to see that $\pi_j(p_j^*, b_j^*) \geq 0$ *if* $2c_2 - v - c_1 - 2\varepsilon \geq 0$, *which is equivalent to*

$$c_2 > \frac{v+c_1}{2}$$

because ε *is a small number greater than 0. This constitutes the second part in inequality (2). We also have to check that the price charged is less than or equal to the consumers' value v. This translates into* $p_1 = c_2 - \varepsilon \leq v$ *or* $c_2 < v$, *which is true by assumption.* ■

The Nash equilibrium described in Proposition 1 is not unique. In fact, there are many bids and prices submitted by anyone of the *j > 2* advertisers that will constitute a Nash equilibrium. This is because these advertisers are never visited by consumers and their strategies do not interfere with the equilibrium strategies of the lowest cost and the second lowest cost advertisers. Search engine revenues, advertiser profits, and consumer surplus are the same for all these equilibria. Under the assumptions of our model we can show that the pure strategy Nash equilibrium in proposition 1 is also unique in terms of the strategies of the two advertisers with the lowest costs. (see Appendix)

Under the equilibrium conditions described above there is only one advertiser who has the highest bid and the lowest price at the same time. The winning advertiser's profit is $N(2c_2 - v - c_1 - 2\varepsilon)$, total consumer surplus is $N(v - c_2 + 2\varepsilon)$, and the search engine revenue is $N(v - c_2 + 2\varepsilon)$. In the following sections we look at some alternative model specifications and their strategic implications.

Cooperative Gaming

When conditions for Nash equilibrium in pure strategies are not satisfied, one has to resign to either using other more sophisticated equilibrium concepts[2] or to finding cooperative equilibria (see next section for a brief review of relevant experimental litarature). Because of the structure of the advertising auction, it might become apparent to all players that, if the top $z \leq k$ advertisers collude, they can increase their combined profits and help attain a stable point for the auction mechanism. The advantage of this approach is that the total payoff to successful cooperation increases with the number of bidders who decide to join the winning coalition and the conditions for cooperation are less restrictive than (2). If advertisers decide to cooperate, they can set $p_1 = c_{z+1} - \varepsilon$ and $b_1 = v - c_{z+1} + \varepsilon$ for the coalition member with the lowest cost. The winner of the top spot in the list of sponsored links can pay some amount of money to the remaining *z-1* bidders to not participate in the auction. The total earnings for the winning coalition will be $N(2c_{z+1} - v - 2\varepsilon - c_1)$ which can be split in some way among the members of the coalition. The only requirements here are that the revenue has to be positive and the price charged $p_1 = c_{z+1} - \varepsilon$ should be less than v. One cannot fail to notice that the above conditions are much less strict than inequality (2). The earnings for the search engine in this case are $N(v - c_{z+1} - \varepsilon)$ which is less than when advertisers do not cooperate. Clearly, consumer surplus is also less.

Usefulness of the Pure-Strategy Nash Equilibrium Concept

Experiments with human subjects have shown that the concept of a pure strategy Nash equilibrium is powerful and intuitive (see Gunthorsdottir et. al, 2007). Many other solution concepts used in previous models of consumer search behavior or consumer auction behavior prove to be either too complex or too confusing for humans to understand and use. For example Muller et. al. (2002) and Cason et. al (2003) show that consumers and producers under monopoly and oligopoly competition behave very differently than prescribed by theory. Many studies have confirmed that humans have trouble mixing strategies (e.g. Walker & Wooders , 2001; Brown & Rosenthal, 1990; Erev & Roth, 1998) or applying backward induction to find sub-game perfect equilibria (Rapaport et. al, 2003). As the previous and following sections demonstrate, the concept of a pure strategy Nash equilibrium can give us a basic idea about the strategic interaction between three categories of players in the sponsored search auction game and help us derive some useful basic results on the influence of the auction mechanism on profit and surplus distribution.

Using Consumer Click Histories

Suppose that the search engine uses the past click through rate of a sponsored link to determine its position on the list of sponsored links. There are many different ways to do this. To simplify the analysis we will assume that the search engine displays in the top spot in the list of sponsored links the company which had the highest click through rate in the previous period no matter what its bid is in the current period as long as the bid is strictly positive. This rule will affect the strategic power of new entrants in an auction for a spot on the list of sponsored links. Let us assume that company W is the only company displayed as a sponsored link in period 0 for a certain keyword. In period one, another company E with lower cost is considering to participate in the auction for the same keyword together with company W. To guarantee the existence of a pure strategy Nash equilibrium we still hold that inequality (2) is satisfied.

If the search engine does not use the click history to determine the winner of the top spot, the new entrant sets $b_E = v - c_W + \varepsilon$ and $p_W = c_W - \varepsilon$, and the incumbent sets $b_W = v - c_W$ and $p_W = c_W$ in equilibrium. Therefore, the new entrant displaces the previous winner in round 1. The winning advertiser's profit is $N(2c_W - v - c_E - 2\varepsilon)$, total consumer surplus is $N(v - c_W + \varepsilon)$, and the search engine revenue is $N(v - c_W + \varepsilon)$.

If the search engine decides to use past click history as operationalized in the paragraph above, the new entrant sets $b_E = 0$ and $p_W = v - \varepsilon$ in equilibrium and the incumbent advertiser sets $b_W = \varepsilon$ and $p_W = v$ in equilibrium. The previous winner uses his/her strategic advantage to lure the consumers that are only looking at the top spot on the list of sponsored links and charges the highest possible price. The new entrant cannot compete successfully for the top spot in the list of sponsored links, however, s/he can lure the consumers by offering product X at a price slightly lower than the monopoly price. The resulting profit for the previous winner is $a_1N(v - c_W - \varepsilon)$, the resulting profit for the new entrant is $a_2N(v - c_E - \varepsilon)$. Total consumer surplus is $a_2N\varepsilon$, and the search engine revenue is $a_1N\varepsilon$. The search engine can use default initial click through rates that are different from 0 as discussed in Richardson, et. al. (2006). In that case the advantage of the incumbent company might be lowered or disappear completely. In the following section we investigate how the above results change if we relax some of the assumptions that we made in the beginning.

Paying Per Click Auction

So far we have seen how the search engine's advertising auction would work if advertisers are charged for every purchase originated through the search engine. In practice, it might be hard for search engines to charge advertisers based on the purchases. Currently both Google and Yahoo charge advertisers per consumer click, not per consumer purchase. Changing this assumption does not alter Observations 1 and 2. Under 1 all advertisers except the one with the lowest cost bid *0* and under 2 all consumers click only on the top link, so advertiser's equilibrium payments are not affected. The same does not hold true for Proposition 1. In fact, it is easy to show that if payments are based on consumer clicks, a Nash equilibrium in pure strategies does not exist.

Proposition 2: *If advertisers are required to pay their bids in the auction for every click on their links, the advertising auction has no Nash equilibrium in pure strategies.*

Proof: *Because of (1) we know that if an advertiser's link is on the first page of the search results, the link will be clicked a positive number of times. Because of (4) every advertiser shown on the first page has to bid 0 in equilibrium if the consumer clicks result in no purchases. Otherwise that advertiser will have a negative profit. Let us focus our attention on the second advertiser on the first page in the list of sponsored links. There are two possible cases to consider:*

- *Case 1:* His/her consumer clicks result in no purchases, so his/her bid in equilibrium is 0, his price is the highest, and the bids of all advertisers after him/her are 0. The bid of the advertiser in the first spot in this equilibrium should be ε, which is the minimum necessary to keep the first spot. However, once this is in place, the advertiser in the second spot can always bid 2ε, get the first spot, charge v, and make a positive profit. Therefore the case discussed is unstable and a Nash equilibrium in pure strategies does not exist.

- *Case 2*: His/her consumer clicks result in some purchases, so his/her price p_2 and bid b_2 are lower than these of the advertiser in the first spot, and his/her current profit is non-negative. The optimal strategy of the advertiser in the first spot will be to charge v, and bid $b_2 + \varepsilon$, and for the advertiser in the second spot to bid b_2 and charge $v - \varepsilon$. However, once these bids and prices are in place, the advertiser in the first spot of the list can unilaterally charge $v - 2\varepsilon$ and increase his/her profit. Therefore the case discussed is also unstable and a Nash equilibrium in pure strategies does not exist. ■

This result is similar to some of the examples mentioned in Edelman et. al (2005) although the model discussed there does not include price competition between the advertisers. The authors of that article also provide a chart of actual bidder behavior at Overture (one of the first companies to start auctioning sponsored links) that directly supports Proposition 2.

Pay-the-Next-Lowest-Bid Auction

In this section we examine the effect of requiring advertisers to pay the next lowest bid in the auction instead of their own bid. The idea stems from Vickrey's seminal article on auctions published in 1961 (see Vickrey, 1961). It is easy to see that this new payment rule only slightly affects the advertisers' Nash equilibrium strategies described in observations 1 and 2, and proposition 1 above when payment is based on purchase because, eventually, only the first spot in the list of sponsored links is important. Proposition 1 in the model presented here guarantees that the theorem of revenue equivalence (see Myerson, 1981 and Riley & Samuelson, 1981) between the pay-your-bid and the pay-the-next-lowest bid auction holds. Changing the pricing rule in this way does not change the search engine revenue, the advertisers' profits, and consumer surplus. Under both pricing rules it is a weakly dominant strategy for all advertisers but the one with the lowest cost to reveal their costs in the prices that they charge. Their bids are also functions of their true costs in equilibrium.

It is interesting to note that if a pay-the-next lowest bid auction is used, the game has a Nash equilibrium in pure strategies under some conditions even when advertisers are required to pay *per click* instead of per purchase.

Proposition 3: *When advertisers pay per click, if*

$$c2 > c1 + a1(v - c1), \qquad (5)$$

there is a Nash equilibrium in pure strategies in which $b_1 = v - c_2 + \varepsilon$, $p_1 = c_2 - \varepsilon$, *and for all* $j > 1$ $p_j = c_j$ *and* $p_j = 0$.

Proof: *Given the strategy profile above, no one can profitably outbid or out-price the advertiser with the lowest cost, so (3) is satisfied for j > 1. Also only the advertiser with the lowest cost makes positive profit and everyone else receives 0, so (4) is satisfied for j > 1. The profit of the lowest-cost advertiser is* $\pi_1(p_1{}^*, b_1{}^*) = N(c_2 - \varepsilon - c_1)$ *which is positive by assumption, so (4) is satisfied for j = 1 too. We only have to show that the strategy profile also satisfies (3) for j = 1. Clearly just changing the bid does not affect the profit of the winning advertiser. However the winning advertiser can choose to raise his price to the monopoly level and give up his price leadership. The profit in that* *case will be* $\pi_1(v, b_1{}^*) = a_1N(v - c_1)$. *(3) is satisfied for j = 1 if his/her profit under proposition 3 is greater than or equal to his/her monopoly profit. This requirement results in* $c_2 > c_1 + a_1(v - c_1)$. ■

Under the equilibrium conditions described above there is only one advertiser who has the highest bid and the lowest price at the same time. The winning advertiser's profit is $N(c_2 - c_1 - \varepsilon)$, total consumer surplus is $N(v - c_2 + \varepsilon)$, and the search engine revenue is *0*.

Google's Auction Rule

Google uses an alternative rule to determine the advertisers' payments. According to the advertiser instructions posted on Google's web site, an advertiser has to pay the bid of the advertiser below him/her in the list of sponsored links plus \$0.01. This means that all advertisers on the first page of the search results have to pay a minimum of \$0.01 per click. As mentioned earlier equilibrium conditions require that all advertisers on the first page have positive or *0* profits, so purchases will have to occur from all advertisers because the minimum that any advertiser will need to pay is \$0.01 per click. As discussed in the previous paragraph, this would not be possible under the current assumptions of the model. Alternatively, we can assume that all advertisers will get some benefit in the future from the consumers who are currently visiting their web sites without making a purchase. Suppose that there is a number *d*, which represents the *future* benefit per click of any advertiser who wins a spot in the list of sponsored links but does not sell any items in the current period. This happens because some consumers might be interested in buying the same or similar product from that advertiser in the future. Also, in many cases advertisers consider their presence in the list of sponsored links as a branding or marketing instrument. With this new addition, it becomes easy to see that an equilibrium similar to the one described in proposition 3 is possible if $d > 0.01$ for all advertisers on the first page.

According to the advertising auction instructions posted on Google's web site, Google uses consumer click histories to rank the advertisers' bids similar to section 3.5 in this article. Google has not revealed the exact formula used in the ranking of the bids, however, Feng et. al. (2003) hint that Google might be multiplying the advertiser's bid by the past click-through rate. In that case the strategic advantage of the incumbent advertiser still exists although it is less extreme than described in section 3.6. When advertisers are charged per click, the rule has one obvious benefit. It discourages competing advertisers from clicking on each other's links in order to increase charges without raising revenues. The rule does not seem to benefit consumers at least according to this model. However, it is possible to analyze an alternative formulation of the model where the relevance of search results has a direct impact on consumer utility. In that case the rule might have a different effect (see Vragov, 2006).

IMPLICATIONS

Before discussing the implications of the theoretical model described above we report the results of a small experiment that was conducted as a class homework assignment in November 2005. During the experiment undergraduate students performed 175 searches on Google and Yahoo to compare the prices of various products offered by web sites in the list of sponsored links. We found that most of the time we could purchase the product of interest from at least one of the web sites shown in the list of sponsored links. However, more than 35% of the sponsored links shown were irrelevant. This means that the sponsored search service was still not used that well by advertisers and had not reached the state in which all sponsored links shown are relevant. In addition, in 72 cases we found that the product's total price (including shipping and handling) listed on the web-site ranked on top of the list was lower than the price of the web site in the second spot. In 41 cases the prices were approximately equal, and in 40 cases the price was higher. In the remaining 22 cases the product of interest could be purchased from only one of the sponsored links. We can say that only 40 out of 175 search results contradict our model even though the relevance of sponsored links is much below 100%.

The model has many important implications for practice. The model demonstrates that there are two major conflicts of interest when the search engine serves as a price infomediary to consumers through sponsored search. The first one arises because search engine profits increase when customers search less, so it is in the interest of the search engine to discourage customers from visiting many of the sites listed in the list of sponsored links. The second one arises because search engines use past click history to rank the advertisers in the list of sponsored links. This policy gives unfair advantage to well-established dominant advertisers in a particular industry and makes it harder for newly emerging small companies to advertise their products through sponsored search.

The model suggests that if one is a manager of a search engine, one should concentrate efforts on improving the sponsored search service rather than maximizing sponsored search revenue. This is not just the only option when new competing search engines emerge but this is also the only way to remove the two conflicts of interest mentioned above and make both consumers and advertisers more satisfied with the sponsored search service. There are specific steps that the manager can take to improve the service. One of them is to find new ways in which payments from advertisers can depend on actual purchases rather than just on clicks (perhaps by using cookies). This will simplify the advertisers' task and will increase the search engine revenue without hurting consumers. Advertiser collusion hurts both search engine revenues and consumers, so some techniques can be developed to monitor advertisers' bids, especially when there is only a couple of advertisers bidding on one keyword.

The auction mechanism can also be adjusted to improve the service. Instead of pay-your-bid auction or pay-the-next-highest-bid auction, the search engine can use a demand revealing mechanism that is much simpler to implement, would not negatively impact consumers, and would be easier for advertisers. The search engine can let advertisers bid for *only one* of the slots in the list of sponsored links. If there are ten spots available on the first page of the results then the advertiser's with the 10 highest bids get the slots on the first page and are displayed to the consumer. The search engine then charges all advertisers on the first page the same price per click which is equal to the value of the 11th highest bid every time when a consumer clicks on the link of the respective advertiser. Since the amount of the bid that every advertiser submits only decides whether the advertiser wins a slot or not, but does not determine the amount that the advertiser pays, it is a weakly dominant strategy for every advertiser to bid their true benefit from a consumer's click (Vickrey, 1961). If there are K less than 10 advertisers bidding for a given keyword, then the search engine can list the top K-1 advertisers and charge them the price of the last K^{th} bid.

The sponsored search e-service has an interesting property with respect to consumers. The usefulness of the service to consumers depends on their behavior. If consumers search through more of the sponsored links, they make the market for spots on the list less monopolistic, and consumer satisfaction with the sponsored search results increases.

LIMITATIONS

The model presented above relies on several assumptions that are common in theoretical modeling but not completely valid in practice. Full information about product costs is rarely available at the beginning of the auction process. It is easy to imagine, however, that with the flow of time advertisers true costs will become known at least within some range and at least within the respective industry. At the same time it is easy to show that a version of Proposition 1 will apply also in cases when the advertisers' costs are unknown but their distribution is known. We only need to replace c_1 and c_2 with their expected values. The conditions of Proposition 1 are admittedly too strict to be satisfied in a large number of cases, however, establishing an equilibrium prediction for the auction result allows us to answer the questions that we posed. It is also natural to expect that product search relevance is much below 100% shortly after the auction starts. It will take some time before companies find out for which key words they need to compete. Another problem is the mere calculation of relevant costs and prices. For example, as mentioned in Hansell (2005), Google's largest advertising auction consumer in the beginning was eBay which offers a variety of products at a variety of prices that change in real time. Finally, in this model, the use of click-through rates in bid rankings is treated in an extreme fashion to illustrate better the strategic advantage that this rule creates for incumbent advertisers.

CONCLUSION

This paper offers a theoretical model of the on-line sponsored search e-service based on the concept of pure strategy Nash equilirium. The model extends previous consumer search models to include explicitly an on-line search engine as an infomediary between consumers and advertisers. The search engine uses an auction to create a list of sponsored links that is displayed next to the usual search results. Consumers search the list to find the best deal. The paper shows how the classic Bertrand's and Diamond's results extend to our new model. The paper also demonstrates that many useful conclusions about possible search engine strategies to improve service can be derived without the complexities inherent in calculating mixed-strategy or other more sophisticated equilibria which are hard to implement in practice. For example, the model can inform us about the effect of advertiser collusion and the effect of some modifications in the auction rules on advertisers' revenues, search engine revenue and consumer surplus. The model describes the limits of the currently available sponsored search service and offers actionable advice to managers of search engines on how sponsored search can be improved.

FUTURE RESEARCH

The model presented here can be extended to incorporate more features of sponsored search auctions. Future models can include the possibility of interaction effects between different positions in the list of sponsored links, the interplay between product price and quality, the effect of existing advertiser reputation on consumer search patterns, the effect of current consumer visits on future consumer visits, and dynamic delays in discovering optimal strategies. When mobile advertising and location-based services are considered an additional important variable to be considered is the physical distance to a location of a business being advertised. Efficiency would require that the distance should be incorporated as a parameter in the auction mechanism.

Another promising way to continue exploring the topic is to conduct experiments with economically-motivated human subjects that probe the effect of different auction rules on profits and surplus under more complex parameter structures. I am currently considering experimental designs in which customers have combinatorial preferences for products and variable search costs while advertisers have combinatorial preferences for slots and are allowed to submit combinatorial bids.

REFERENCES

Animesh, A. , Ramachandran, V., Viswanathan, S. "Quality Uncertainty and the Performance of online Search Markets: An Empirical Investigation", working paper, University of Maryland, College Park, 2006

Arbatskaya, M. 2005 "Ordered Search" working paper, Emory University, February, 2005

Bertrand, J. 1883 "Theorie Mathematique de la Richesse Sociale" *Journal De Savants*, 1883, 499-508

Braverman, A. 1980 "Consumer Search and Alternative Market Equilibria" *Review of Economic Studies*, April, 47, 487-502

Brown, J. and Rosenthal, R. 1990 Testing The Minimax Hypothesis: A Re-Examination Of O'neill's Game Experiment. *Econometrical, 58*, (5) 1065-1081.

Burdett, K. and Judd, K. 1983 "Equilibrium Price Dispersion" *Econometrica*, July, 51, 955-70

Carlson, J. and McAfee, P 1983 "Discrete Equilibrium Price Dispersion" *Journal of Political Economy*, June, 91, 480-93

Cason, T., Gangadharan, L., and Duke, C. "Market Power in Tradable Emission Markets: A Laboratory Testbed for Emission Trading in Port Phillip Bay, Victoria," *Ecological Economics* (46:3) 2003, pp 469-491.

Coy, P. "The Secret to Google's Success: It Innovative Auction System Has Ad Revenues Soaring" *Business Week*, March 6, i3974, 2006, pg42

Cusumano, M. "Google: What It is and What it is not" *Communications of the ACM*, v48 n2, February 2005, pg. 15-17

Debmczynski, D., Kotlowski, W. & Weiss, D. "Predicting Ads Click-through Rate with Decision Rules" paper presented at 17th WWW2008 Conference in Beijing, China, April 21-25, 2008

Diamond, P. 1971 "A Model of Price Adjustment" *Journal of Economic Theory*, June, 3, 156-68

Edelman, B, Ostrovsky, M. & Schwartz, M. 2005 "Internet Advertising and the Generalized Second Price Auction: Selling Billons of Dollars Worth of Keywords" *Research Paper Series*, RP 1917, Graduate school of Business, Stanford University, October

Edelman, B & Ostrovsky, M. 2007 "Strategic Bidder Behavior in Sponsored Search Auctions" *Decision Support Systems*, v43 n1, February 2007, 192-198

Erev, I. , & Alvin Roth. 1998. "Predicting how people play games: Reinforcement learning and experimental games with unique, mixed strategy equilibria." *American Economic Review,* 88: 848-881.

Estes, William "Probability Learning." In A.W. Melton (ed.), *Categories of Human Learning*. New York: Academic Press. 1964

Feng, J., Bhargava, H. & Pennock, D. 2003 "Comparison of Allocation Rules for Paid Placement Advertising in Search Engines" *Proceedings of ICEC*, Pittsburg, PA,

Hansell, S. 2005 "Your Ads Here (All of Them): Google Wants to Dominate Madison Avenue, Too" *New York Times*, October, sec 2, pg1

Jansen, B. & Resnik, M. 2005 "Examining Searcher Perceptions of and Interactions with Sponsored Results" *Proceedings of the ACM Conference on Electronic Commerce*, Vancouver, Canada, June 5-8,

Kharif, O. Mobile Ads: Slow to Take Off, *Business Week Online*, 00077135, 3/11/2008

Kitts, B. and LeBlanc, B. 2004 "A Trading Agent Simulator for Keyword Auctions" *Proceedings of AAMAS*, July 19-23, New York, NY

Kumar, N., Lang, K. & Peng, Q. 2004/2005 "Consumer Search Behavior in Online Shopping Environments" *E-Service Journal*, v3 n3, Summer 2004/2005, pg. 87

Liu, D, Chen, J. & Whinston, A. "Ex-Ante Information and Design of Keyword Auctions", working paper, University of Kentucky, Lexington

Mehta, A., Saberi, A., Vazirani, U. & Vazirani, V. 2004 "AdWords and Generalized On-line Matching"

Muller, R., Mestelman, S., Spraggon, J., and Godby, R. "Can Double Auctions Control Monopoly and Monopsony Power in Emissions Trading Markets," *Journal of Environmental Economics and Management* (44:1) 2002, pp 70-92.

Myerson, R. 1981 "Optimal Auction Design" *Mathematics of Operations Research*, 6, 58-73

Rapoport, A., Stein, W.E., Parco, J.E. and Nicholas, T.E. (2003). Equilibrium play and adaptive learning in a three-person centipede game. Games and Economic Behavior. 43, 239-265.

Reinganum, J. 1979 "A Simple Model of Equilibrium Price Dispersion" *Journal of Political Economy*, August, 87, 851-58

Richardson,M., Prakash, A. & Brill, E "Beyond PageRank: Machine Learning for Static Ranking" Proceedings of 15th WWW2006 Conference, Edinburgh, Scotland, May 23-26, 2006

Riley, J & Samuelson, W. 1981 "Optimal Auctions" *American Economic Review*, 71, 381-02

Rob, R. 1985 "Equilibrium Price Distributions" *Review of Economic Studies*, July, 52, 457-504

Robert, J. and Stahl, D. 1993 "Informative Price Advertising in Sequential Search Model" *Econometrica*, v61 n3, May, 657-86

Rowley, J. 2006 "An Analysis of the E-Service Literature: Towards a Research Agenda" *Internet Research*, v16 n3, 339 - 359

Rust, R. & Lemon, K. 2001 "E-service and the Consumer" *International Journal of Electronic Commerce*, v5 n3, 83-99

Sarrukai, R. 2005 "How much is a Keyword Worth?" *Proceedings of WWW 2005*, May 10-14,Chiba, Japan

Singh, R., Ier, L, and Salam, A. 2003 "Web Service for Knowledge Management in E-marketplaces" *E-Service Journal*, v3 n1, Fall 2003, pg. 32

Stahl, D. 1989 "Oligopolistic Pricing with Sequential Consumer Search" *American Economic Review*, Sep, v79 n4, 700-

Stiglitz, J. 1979 "Equilibrium in Product Markets with Imperfect Information" *American Economic Review*, May 69, 339-45

Stiglitz, J. 1987 "Competition and the Number of Firms in a Market: Are Duopolies more Competitive than Atomistic Markets?" *Journal of Political Economy*, 95, October, 1041-61

Sullivan, D. 2006 "comScore Media Metrix Search Engine Ratings" *The ClickZ Network*, August 21, 2006 last accessed on August 03, 2007 at http://searchenginewatch.com/showPage.html?page=2156431

Varian, H. 1980 "A Model of Sales" *American Economic Review*, September, 70, 651-59

Varian, H. 2006 "Position Auctions", working paper, California Institute of Technology, November 2006

Vickrey, W., 1961, Counterspeculation, Auctions, and Competitive Sealed Tenders," *Journal of Finance*, 16, 8–27

Vragov, R. 2006 "Flexible Advertising on the Web" working paper, Baruch College, CUNY, April 30, 2006

Walker, M. & Wooders, J. "Minimax Play at Wimbledon." *American Economic Review*, 91(5): 1521-1538, 2001

ENDNOTES

1 A preliminary version of this paper was presented at the workshop on Targeting and Ranking for Online Advertising, which was held as part of the 17th International World Wide Web Conference in Beijing, China on April 21-25, 2008

2 The auction game described here always has at least one mixed-strategy Nash equilibrium, however, mixed strategy equilibria have been shown to have little practical significance (see next section) and are therefore omitted from this discussion. Edelmann et. al (2005) on the other hand use the concept of locally envy-free equilibria in their model of adword auctions. These are equilibria in which an advertiser cannot improve his/her payoff by exchanging bids with the advertiser ranked one positions above him/her. Their model, however, does not include the stage in which advertisers make pricing decisions.

APPENDIX

Proposition A1: *The equilibrium described in Proposition 1 is a unique pure strategy Nash equilibrium in terms of the strategies of the two advertisers with the two lowest costs.*

Proof:

a. Suppose that there is a Nash equilibrium in pure strategies in which the advertiser in spot $z > 1$ has positive profit. Let us find out the characteristic features of this equilibrium. Because of (1) we know that, for every $j < z$, $p_j > p_z$ because otherwise no one will purchase product X from the advertiser in the z spot. Inequality (3) guarantees that increasing the price p_z with ε(a very small number close to zero) and keeping everything else the same should result in a new profit which is less than or equal to the current profit. This happens only if there is a p_i $(i < z)$ such that $p_j = p_i - \varepsilon$, which means also that all advertisers in the spots between i and z have no purchases and therefore zero profits. We have to further consider two possible cases:

   Case 1: Advertiser i has a positive profit. In this case there is a contradiction in our setup because, keeping everything else the same, the advertiser in the i spot can decrease his/her price with ε, capture all consumers of the advertiser in the z spot, and increase his profit.

   Case 2: Advertiser i has zero profits but positive purchases. So $p_i - b_i - c_i = 0$. Since i is before z, $b_z < b_i$ and $p_i - b_z - c_i > 0$. In this case there is also a contradiction in our setup because advertiser i can increase his profit by submitting a lower bid.

   These two cases show that a Nash equilibrium in pure strategies does not exist when anyone except the advertiser owning the first spot has positive profits.

b. Suppose that there is a Nash equilibrium in pure strategies in which the advertiser with lowest cost does not own the first spot. In that case the advertiser with the lowest cost must have zero profits, otherwise, as shown in a) above, an equilibrium does not exist. We will now show that the lowest cost advertiser can always have positive profits and thus this Nash equilibrium does not exist. Let us mark by p_w, c_w, and b_w the price, cost, and bid of the winner of the first spot. Since the number of purchases is always positive in the first spot, we have $p_w - b_w - c_w \geq 0$ or $p_w - c_w \geq b_w \Rightarrow v - c_w \geq b_w \Rightarrow v - c_2 \geq b_w \Rightarrow v - c_2 + \varepsilon > b_w$. This means that anyone who bids $v - c_2 + \varepsilon$ can win the top spot in the list of sponsored links. The advertiser with the lowest cost can always bid that amount, charge v, and get positive profit equal to $a_1 N(c_2 - c_1 - \varepsilon)$. Therefore Nash equilibrium in pure strategies is not possible if the advertiser with the lowest cost does not own the first spot.

c) Suppose that there is a pure strategy Nash equilibrium in which the lowest cost advertiser owns the first spot but does not charge the lowest price. This means that there is another advertiser i who has the lowest price in the first page of the search results and the highest spot compared to all other advertisers that charge the same price. Because of a) that advertiser should have zero profits so $p_i - b_i - c_i = 0$. Inequality (3) requires that some other advertiser z before i has a price of $p_z = p_i + \varepsilon$ otherwise i can profitably deviate from his/her equilibrium strategy. We have to consider further two possible cases:

   Case 1: $p_z = p_1$. The lowest cost advertiser can always capture i's consumers by charging $p_1 - 2\varepsilon$, keep everything else the same, and thus increase his/her profits.

   Case 2: $p_z < p_1$. Because of a) advertiser z must have 0 profits, so $p_z - b_z - c_z = 0$. Because of the zero profit requirements for z and i, it follows that $b_z < b_i$ which means that in order for an equilibrium to exist $b_{z+1} = b_z - \varepsilon$ and $p_{z+1} = p_z - \varepsilon = p_i$, which contradicts our assumption that i

has the lowest price in the first page and owns the highest spot compared to all other advertisers that charge the same price.

Sections a), b), and c) together show that a Nash equilibrium in pure strategies exists only if the lowest cost advertiser owns the first spot in the list of sponsored links and charges the lowest price while all other advertisers earn zero profits. To guarantee that the first spot is incontestable we need $b_1 = v - c_2 + \varepsilon$, and to guarantee that the price leadership is incontestable, we need $p_1 = c_2 - \varepsilon$. Lastly, to satisfy (3) and (4) for the top advertiser we need the conditions shown in proposition 3. ■

This work was previously published in International Journal of E-Services and Mobile Applications, Volume 1, Issue 1, edited by Ada Scupola, pp. 21-37, copyright 2009 by IGI Publishing (an imprint of IGI Global).

Section 2
E-Services:
Adoption, Perceptions and User Values

Chapter 7
The Adoption of "E-Banking" by Lebanese Banks:
Success and Critical Factors[1]

Elissar Toufaily
Université du Québec à Montréal (UQÀM), Canada

Naoufel Daghfous
Université du Québec à Montréal (UQÀM), Canada

Roy Toffoli
Université du Québec à Montréal (UQÀM), Canada

ABSTRACT

This article examines the organizational, structural, and strategic factors that can speed up or slow down the adoption of E-banking innovations by financial institutions in the Lebanese market. A conceptual model is developed based on a review of the major innovation adoption theories and other research findings, and tested with a survey administered through a census of the population of Lebanese banks. Findings revealed that the organizational variables of bank size, presence of functional divisions, technical staff and infrastructure, as well as the degree of international experience and tolerance of risk of the decision makers exert a significant impact on the adoption of E-banking. Two structural characteristics, the internal technological environment, and relative advantage of the innovations were also found to impact adoption. The two strategic factors related to the banks' degree of international operations were also found to positively influence the adoption of E-banking. Recommendations were provided on ways of accelerating the rate of adoption of E-banking in an important developing Middle Eastern country.

DOI: 10.4018/978-1-60960-607-7.ch007

INTRODUCTION

A decade after the introduction of the first commercial Internet sites, e-commerce is still captivating both university researchers, as well as business communities (Bergeron, 2001; Ngai & Walt, 2002; Cyr, Bonanni, Bowes & Ilsever, 2005). A number of surveys show that approximately 97% of consumers who make on-line transactions will continue adopting this type of behaviour, with a majority indicating that they intend to significantly increase their number of transactions over the next 12 months (Ernst & Young, 2001). In general, consumers appear to be satisfied with e-commerce as an efficient, fast, and cheaper way of conducting business, and their online purchases are growing in a diverse range of products and services.

Among the industries influenced by e-commerce, the financial institutions are the most affected by the technological revolution. The financial industry is based on the sharing of information, which itself relies heavily on information and communication technology (ICT) in order to acquire, analyse and deliver data to all relevant users. Financial institutions innovate continuously and update their marketing strategies in order to closely meet the desires and demands of their customers in a safe and confidential environment. These institutions must, on the one hand, equip themselves with cutting edge technologies in order to adapt, and, on the other hand, bring forward plans for a relationship-oriented marketing strategy. The end result will be the establishment of a culture which prioritizes the customer, generates a climate of trust, and promotes increased knowledge and customer loyalty (Perrien and al., 1993; Ricard and Perrien, 1999). ICT helps the financial organization to improve the efficiency and efficacy of their business processes and to improve their competitive positions (Rust & Kannan, 2002).

Therefore, in an era where different media (e.g. computers, Mobile, Internet, TV, cable etc.) complement and overlap one another, the physical location of the bank is no longer considered to be critical for the customer. The customer wants more flexibility without paying more, and his demands are clear: Make transactions wherever, whenever, and whichever way he wants. Hence, the development of the concept of electronic financial services, commonly known as "E-banking". This umbrella term embodies a variety of different services, including the common automatic teller machine (ATM) services, telephone banking, direct deposit, automatic bill payment (ABP), the use of debit cards, Internet banking and computer banking (PC banking), etc. (Kolodinsky, Hogarth & Hilgert, 2004; Hernandez &Mazzon, 2006). The use of E-banking technologies has grown rapidly in industrial markets like the USA (ABA Banking Journal, 2001), but it is still in its infancy in emerging markets like Lebanon - the object of the present study.

The organisational consequences of the introduction of ICT, specifically E-banking, can initially take the form of strategic changes to management processes, the reorganization of the management network and even the reformulation of the company's purposes and business portfolio (Venkatraman, 1995). According to Badoc, Lavayssière and Copin (1994), banks must include E- banking into their processes to develop new activities and to approach new markets, more specifically, those of virtual financial services. E-banking seems a disruptive technology that doesn't support current bank-based servuction practices (Walsh et al, 2002). E-banking can also be considered as a disruptive innovation since it generates service and/or physical products with different performance attributes that may not be valued by existing customers (Ibid, 2002). This is in agreement with Christensen & Raynor (2003) who claim that disruptive technologies do not necessarily need to be radically new from a technological point of view, but that it could contain superior "performance trajectories" along critical dimensions that customers value. However, the adoption of an innovation within an organization in general or, more specifically, that of "E-banking" within financial institutions, is not always an easy task. Although some factors will speed it up, many others can slow it down.

The purpose of this article is to analyze the organizational, structural and strategic factors that can speed up or, conversely, slow down the Implementation of E-banking by financial institutions. By E-banking, we refer to the use of various electronic media as channels for conducting financial transactions. The empirical context of this study will be the Lebanese banking sector, an emerging market where e-banking is only in its initial stages. This article will also analyze the importance and the potential that each form of electronic channel represents for Lebanese banks.

The article proceeds as follows. The first section will review existing theories with respect to the adoption of E-banking. The second section proposes a theoretical model of the factors which influence the adoption of E-banking. Section 3 describes the methodology used for testing the above model. Section 4 presents the results of the statistical data analysis, while the last section discusses the study's implications for research and practice and presents the study's limitations and some directions for future research.

LITERATURE REVIEW

E-Banking: Definition and Advantages

Literature on banking technology has developed in a major way only as of the late 1980s and the early 1990s, with the emergence of new technologies that simplified remote access to banks (Hernandez & Mazzon, 2007). According to Insley, Al-Abed and Fleming (2002), "Electronic banking is an umbrella term for the process by which a customer may perform banking transactions electronically without visiting a brick-and-mortar institution". The following terms all refer to one form or another of electronic banking: personal computer (PC) banking, Internet banking, Mobile banking, TV banking and phone banking. PC banking and Internet or online banking are the most frequently used designations. It should be noted that electronic banking is a larger concept than banking via the Internet (Karjaluoto et al. 2002). The Internet is one of the principal delivery channels for electronic banking (Karjaluoto et al., 2002). The movement towards E-banking is a natural process for credit institutions in the electronic age.

E-banking is attractive for customers as it makes it possible to conduct banking transactions anytime and anywhere, faster and with lower fees compared to using traditional bank branches. From the bank's perspective, the adoption of electronic businesses offers major opportunities in terms of competitive advantage: it allows the bank, on the one hand, to improve efficiency and operational effectiveness within the company; while on the other, it enables it to develop a stronger and more durable business relationship with its customers. The main benefits of electronic banking are cost savings, reaching new segments of the population, efficiency, cross selling, third-party integration, and customer satisfaction (Hiltunen et al., 2004). Wah (1999) remarks that the success of banks operating via the Internet depends on their ability to attract and keep customers. Despite all the apparent advantages for banks, adoption rates of this innovation vary across countries. In general, constraints imposed by regulations and technology (for instance the requirements for signatures for mortgages) as well as the personal perception of customers about electronic banking and transactions can slow the adoption of e-banking by banks.

Adoption of New Technologies

Several competing theoretical approaches have been used to investigate the determinants of individual acceptance and use of information technology (Venkatesh et al., 2003). One line of research focuses on the factors that predict the behavioural intention of the consumer (intention to adopt a new technology) or on the behaviour itself (actual adoption of new technology). The most important of these studies are centered on the Theory of Reasoned Action (TRA) (Fishbein and Ajzen, 1975), on the Theory of Planned Behaviour (TPB) (Ajzen, 1985), and on the Technology Acceptance Model (TAM) (Davis, 1989). Among the adoption determinants upon which these models are based can be found: beliefs, subjective norms, attitudes, perceived usefulness, perceived ease of use and perception of behavioural control.

A second line of research has considered adoption of new technologies from the perspective of the Innovation Diffusion Theory (Rogers, 1983). Rogers (1983) proposed a model of the adoption of innovations that included five product or service characteristics postulated to influence consumer acceptance of an innovation. These are defined as follows:

- **Relative advantage:** This refers to the degree to which consumers perceive a new product or service as different from and better than its substitutes.
- **Trialability:** Refers to the ability of consumers to experiment with a new innovation and evaluate its benefits. It is the extent to which consumers can try out the new technologies and products before purchasing (Kolodinsky, Hogarth &Hilgert, 2004).
- **Compatibility:** This is the extent to which the innovation is perceived as being in line with the values, needs, beliefs and experiences of prospective adopters. In the case of E-banking, it means the degree to which a given technology fits in with the banking behaviour of the institutions and their customers; specifically, the way in which they have historically managed their finances.
- **Complexity:** Refers to the extent to which the innovation is perceived as difficult to understand and use. For consumers without previous computer experience, and for whom that e-banking is difficult to use, adoption of the innovation may be frustrated (Kolodinsky, Hogarth &Hilgert, 2004).
- **Observability:** It is the extent to which an innovation is visible and communicable to consumers. In the case of E-banking, some channels are more observable than others.

These are collectively important factors for customers in their decisions to adopt E-banking (Sayar &Wolfe, 2007). Several researchers have incorporated some of the dimensions of Rogers' model in empirical work that examined technological innovations (Raju, 1980; Dabholkar, 1996; Daniel, 1999; Tan & Teo, 2000Lee et al., 2003).

It should be pointed out that Rogers' (1983) theory was conceived for use by a decision making unit, a unit which could just as well be an individual consumer or an organization.

From the perspective of the financial organizations, the results of the study on the drivers and inhibitors of the adoption of E-banking show that the decision by banks to adopt E-banking is driven mostly by "external factors", such as the number of other banks adopting, competitive forces, consumer demand and technological availability (Bradley and Stewart, 2002). According to Sayar and Wolf (2007), the decision to adopt is mostly effected by new revenue potential, cost reduction, security and access through other distribution channels. The main inhibitors of E-banking were identified as PC ownership, authentification issues on the Internet and resistance to change. Mols (2001) conducted a survey among the managers of banks in Denmark, which shows that management support and future orientation are the most important factors driving the introduction of a new electronic distribution channel. For bank managers the efficiency and enhancement of customer services, through faster, easier and more reliable service and improvement of the competitive position (Aladwani, 2001) are considered as the main advantages of offering E-banking services. Size and cost-effectiveness (Bughin, 2003) are also two critical factors for a bank's adoption decision.

Other dimensions that can affect the adoption of E-banking will be described in the conceptual framework section.

Profile of the Lebanese Banking Sector

Compared to its geographic and economic dimensions, Lebanon's banking sector plays a disproportionate role in its economy. The Lebanese banking sector is well-integrated into the domestic economy, as it contributes to around 4.8% of the GDP, whereas it only employs 1.2% of the total Lebanese labor force, which confirms the high productivity of this sector (ABL, 2004). It is, in fact, the major financial intermediary and provider of funds to the domestic economy, in both its private and public sectors. The share of banks' assets invested in the Lebanese economy reached 80.4% of total assets at the end of September 2005 (i.e. 83,138 billion L.L.), against 19.6% for investments outside Lebanon. The total assets of banks represented 316% of GDP, in 2004. This ratio is a record high in the region and is among the highest ratios in the world (BankData, 2004). In fact, the banking sector has helped Lebanon to rebuild itself after its devastating civil war, and it also helps to shield its economy from the political turmoil that it continues to face.

It is worth highlighting the rapid rise lately in the exports of services on the part of the Lebanese banking sector, after its modernization and diversification, to currently include almost all the services provided by large international banks. This has been supported by the modernization and restructuring of the banking institutions, be it with regards to working methods and procedures, human resources, capital base or adherence to international banking industry norms. After the expansion of the network of branches on the domestic level and the promotion of activities in the European countries and other economies where their presence goes far back, Lebanese banks have benefited recently from the opening of some banking markets in the neighbouring Arab countries. The Lebanese banking sector expanded during the last two years to include Syria, Jordan, Sudan, Algeria, Egypt, and the Emirates. This was also done in the hope of reaching other markets, like Iraq, Libya and maybe Saudi Arabia, in preparation for the time when restrictions would be lifted and conditions would become favourable. Many banks are also aiming at reaching markets where an important Lebanese Diaspora exists, as in some African countries, Australia and Latin America.

The penetration of computers and Internet access in Lebanon is small relative to that of East Asia and the Organisation for the Economic Cooperation and Development (OECD) countries mainly because of high access costs in the Arab World. Another reason for the low penetration is that people in the Arab region are not well trained to use these new technologies. In spite of that, the Lebanese banking sector has a strong reputation and is a leader in the region and has outperformed other sectors in the economy in the 1990s (Infropo, 2003). Electronic Delivery channels have been expanding as banks try to provide customers with new and innovative access methods to their accounts. In fact, ATMs, phone banking, wireless banking and even TV banking are enabling customers to have a quick and permanent access to their funds. Electronic banking is being offered in all it flavours on the Lebanese market (ABL, 2003). But not all banks are engaging in this new type of banking as many factors can affect the decisions to offer these services. Some of them are macro dimensions, related to the political, cultural and judicial environments of the bank; while others are related to the organisational culture, structure, and capabilities of the bank itself. Some barriers to E-banking adoption by the Lebanese customers are lack of awareness and trust, the absence of rules and regulations, the complexity of such services, the fear arising from security issues, and the costs of PCs and Internet connections. Also, adoption barriers can be related to the customers' mentality as they prefer the personal contact with

Figure 1. The conceptual model of the adoption of e-banking

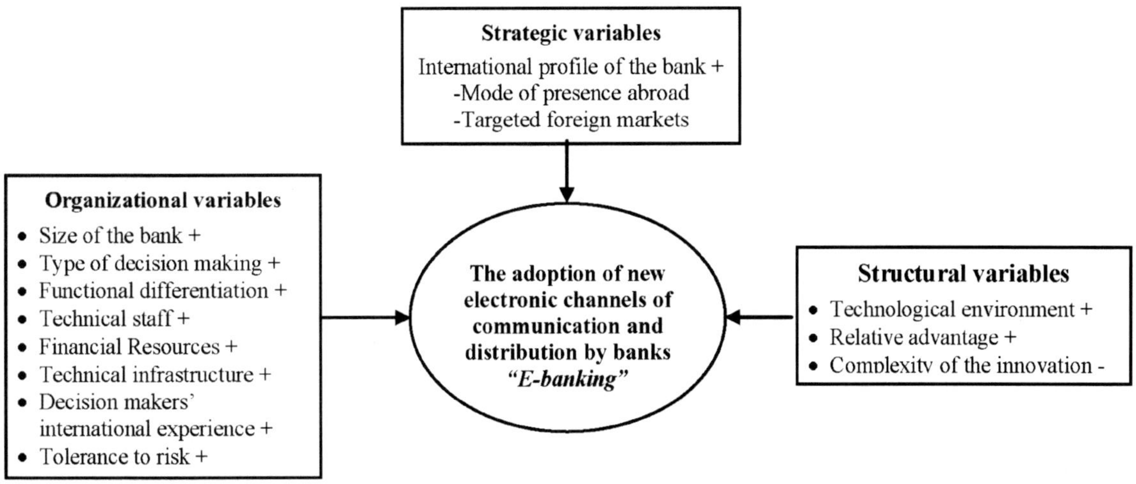

branch managers, especially for customers aged above 50 and who are not familiar with the use of these technologies.

In this context, and because of the above factors and quasi total lack of studies on the factors affecting E-banking adoption in the Middle East in general, and Lebanon in particular, we decided to undertake this study, which we feel should make an important contribution to better understanding the factors influencing adoption of new technologies in the region.

CONCEPTUAL FRAMEWORK

According to Rogers (1983), "An innovation is an idea, practice or object that is perceived as new by an individual or other unit of adoption". The adoption of E-banking by a financial institution, like any other innovation, isn't an easy process. Rather, it depends on many factors, some of them are organizational; others are related to structural factors of the innovation, and others arise from the strategy of the unit that adopts it.

The conceptual framework presented in Figure 1 summarizes the variables that are retained and that will be used as the fundamental basis of our empirical study.

Organizational Variables

The analysis of the relationship between the organizational dimensions and the adoption of E-banking is an important component of our study. Marketing authors have shown that the effect of organizational characteristics is a critical factor in the reaction to an innovation (Skhiri, 2000). As such, bank size, type of decision-making, functional differentiation, the presence of technical staff, financial resources, technical infrastructure, decision makers' international experience and tolerance to risk are all factors that can speed up or slow down a bank's adoption of E-banking.

Despite conflicting and divergent opinions about the relationship that exists between company size in general and the adoption of innovations, authors such as Kimberley and Evanisko (1981), Coren and Klepper (1996) and Bughin (2003) have shown that large businesses are most likely in the best position to benefit from the adoption of new technologies. Consequently, we think that large institutions are more likely to introduce E-banking because they have a greater likelihood of being exposed to innovations, and to possess the necessary resources to acquire and implement them (financial funds, social capital, processes, expertise, distinctive capabilities, etc). This leads us to our first hypothesis

H1.a: *E-banking is more likely to be adopted by large institutions.*

Decision-making methods refer to the way decisions are made inside the organization, whether centralized or decentralized. Authors such as Rogers and Shoemaker (1971) note that the relationship between centralization and the adoption of innovations is negative. Other researchers (e,g, Zmud, 1982) concluded that centralization had a variable effect on adoption. Zmud (1982) for instance, found that the impact of this variable differed according to the phase of development of the innovation and to the degree of compatibility of the innovation with the interests of the decision makers within the organization. Given that one of the main functions of banks is the transfer and sharing of information, any technology which facilitates these processes would be congruent with the mission of the bank. Consequently we feel that centralized banks would have a greater interest in adopting new technologies. In this sense we are taking the approach of Zmud (1982) and his collaborators. This is recast as the following hypothesis:

H1.b: *In the institution where decision-making is centralized, the probability of adopting E-banking is higher.*

Functional differentiation refers to the degree in which a bank is divided into subunits (i.e. the organization of a company in several distinct departments according to different functions of management like finance, marketing, human resources, production, etc). Balbridge & Brunham (1975) state that the administrative complexity of the organization increases the probability of adoption of the innovation. This means that the more an organization is subdivided into distinct units or departments which are able to evaluate technological innovations, the higher the likelihood of the adoption of these innovations (Gatignon & Robertson, 1989). This leads us to the following hypothesis:

H1.c: *The adoption of E-banking is more prevalent in institutions which have multiple functional units.*

Langley and Truax (1994) state that the presence of qualified staff members in the company having appropriate expertise in new information and communication technologies facilitates for banks the choice of these technologies and speeds up their adoption. Dixon and Nixon (2000) point out that financial institutions must have the internal resources, technical qualifications, and control of the technology in order to be able to adopt the Internet as a new distribution channel of products and services. Therefore, a bank with a high number of technically and scientifically skilled personnel will be more able to deal with risk and uncertainty brought on by the adoption of E-banking, and more likely to ultimately adopt the new technology. This is recast as the following hypothesis:

H1.d: *The adoption of E-banking is more prevalent in banks with more highly qualified technical staff members.*

Langley and Truax (1994) also emphasize that the possession of funds is a key factor when companies introduce technology. Our research demonstrates that the adoption of E-banking technology, with its high costs, requires serious financial justification, not only to cover the costs of introducing the new technology, but also for learning about the new management model, technology support, application of new procedures, and the training of existing staff members. Thus, organizations which have greater financial resources and/or which have easier access to capital are more likely to introduce new electronic distribution and communication channels. Our hypothesis therefore becomes:

H1.e: *The degree of adoption of E-banking is positively correlated with the institution's financial resources.*

The integration of new electronic channels within the bank requires the presence of telecommunication, multimedia and network infrastructures, electronic platforms, advanced programming languages (e.g. IP, HTML, Java, XML) and technical architecture with secure data sharing (e.g. The Public Key Infrastructure which enables the bank to communicate with its customers and partners in a secure and confidential way, while also protecting data). This corresponds to the technical infrastructure of the bank. Thus, we think that the more the technical infrastructure of the bank is developed, the more the probability of adoption of e-banking is high. This leads us to the following hypothesis:

H1.f: *The adoption of E-banking is more prevalent in institutions with more advanced technical infrastructures.*

According to Skhiri (2000), the degree of international experience of the decision-makers has considerable impact on the adoption of innovations. In fact, it is expected that a manager who has international experience within an institution, from training or professional activities, will have a more positive attitude regarding the adoption of innovations, and would be more likely to introduce and integrate them into the services of the institution. This is especially true when the decision maker operates in a country where the level of technology is less advanced. In such a case, experience abroad will most likely expose the manager to new technologies and make him aware of their positive impacts within the institution, even before introducing them in the country where the bank operates. This leads to the following hypothesis:

H1.g: *E-banking is more likely to be adopted when the decision maker has experience abroad.*

Risk tolerance refers to the level of uncertainty that decision makers can tolerate when introducing organizational change like the adoption of new electronic distribution and communication channels. It's a multidimensional construct (Bauer, 1960) composed of five distinct dimensions: tolerance to financial risk, performance risk, psychological risk, physical risk and social risk (Cox, 1967; Jacoby & Kaplan, 1972). For many specialists, perceived risk is a key factor determining the behaviour vis-à-vis the adoption of innovations (Arndt, 1967; Ostlund, 1969; Shandh, 1968). Thus, the rapid adoption of an innovation would be attributable to, among other things, low perceived risk or high tolerance to risk (Gatignon and Robertson, 1985; Webster, 1969). More specifically, some authors point out that security and confidentiality risks on the Internet are serious barriers to the adoption of e-commerce (Bhimani, 1996; Cockburn and Wilson, 1996; Quelch and Klein, 1996). In our research, we think that the more bank managers tolerate risk from innovations, the more open they will be towards change in general and, specifically, towards the introduction of new electronic distribution and communication channels. Based on this, we present the following hypothesis:

H1.h: *The degree of adoption of E-banking is positively correlated with the degree of risk tolerance of decision-makers.*

Structural Variables

In this category of variables are included both the product and service characteristics of the innovation, (Rogers, 1962; 1983; Rogers E.M. and Shoemaker, 1971; Robertson, 1971), as well as the firm's internal technological environment. Most writers in marketing are keenly interested in the relationship between structural factors and the adoption of innovations and state that these factors generally put pressure on the company at every level. This pressure becomes particularly great when changes are introduced within the organization. In the present study we analyse the impact of the internal technological environment of the bank and two features of the innovation itself, relative advantage and complexity, from the point of view of the managers.

Miller (1985) analyzed the impact of the technological environment on the adoption behaviour of technological innovations by organizations. He states that the more the firm has innovative manufacturing technologies, the higher is the probability of adopting other innovations. In essence, this means that the more the firm is used to adopting technologies, the greater its propensity to adopt more. This is a form of organizational innovativeness (Hirschman, 1980). With respect to the adoption of E-banking, we believe that the propensity of adopting new distribution channels would therefore be greater if the bank is used to introducing new information and communication technology as part of its business. This leads to the following hypothesis:

H2.a: *The level of adoption of E-banking in a bank is higher if the bank continuously adopts different technological innovations.*

Rogers (1983) posited that the perceived relative advantage of an innovation is positively related to its level of adoption. In studying the factors influencing the adoption of Internet banking, Tan and Tao (2000) have shown that the perceived relative advantage of E-banking is positively related to its level of adoption by consumers. Webster (1969) mentions that the firms that are likely to adopt the innovations more quickly than the others are those who perceive the greatest relative advantage from the innovation. Based on these assertions, we think that the degree of adoption of E-banking is higher if the bank believes that this innovation might enhance its image, improve its efficiency and operational effectiveness, and increase its performance, in terms of market share, customer satisfaction, and costs. Based on these arguments, we present the following hypothesis:

H2.b: *The more a bank perceives that an innovation provides a higher relative advantage, the greater its propensity to adopt it.*

Previous studies have shown that substantially complex innovations require greater technical and operational abilities and effort in order to increase their chance of being adopted (Cooper and Zmud, 1990 ; Dickerson and Gentry, 1983). On the organizational level, Gauvin & Sinha (1991) indicate that the complexity of an innovation is associated with the concept of Inertia. In fact, the more the innovation is complex, the more the decision about its introduction within the organization is important. As E-banking increases in complexity, its adoption requires major changes within the bank, in terms of management and operations processes. This leads to the following hypothesis:

H2.c: *The degree of adoption of E-banking correlates negatively with the perceived complexity of the innovation.*

Strategic Variables

The strategic factors retained in this study are related to two variables that determine the bank's international profile: The type of presence in foreign markets and the number of markets targeted.

One of the critical decisions that a firm must make as it internationalizes its operations is the type of presence that it will have on the foreign markets (see, for e.g., Calof & Beamish, 1995). The different modes include licensing or franchise agreements, indirect or direct exporting, sales branches, joint ventures or wholly owned subsidiaries (Calof &Beamish, 1995; Cho & Padmanabhan, 1995). In this sense, and in the context of this study, two strategic modes of presence abroad are retained, namely direct foreign investment and strategic alliance or joint venture. Each strategy is related to commitment, risk and potential benefit by the firm (Kotler, 1991).

In fact, the international environment in which the bank finds itself offers a wide opportunity to learn and to evaluate the importance of technological innovations and stimulates the adoption of E-banking. The international presence allows the company to follow the tendencies of its competitive environment, to better serve customers in its global markets and to be informed in a continuous way. Possessing the right information at the right time constitutes an undeniable key of success in an international market. As well, the more the bank progresses in its internationalization process, the more it faces a large number of competitors from different origins. When it targets many markets at the same time, the more the adoption of electronic channels of E-banking becomes a necessity. Based on these arguments, we present the following hypothesis,

H3.a: *The degree of adoption of E-banking is greater as the institution expands internationally.*

The number of foreign markets targeted is defined by the number of foreign markets the bank wants to conquer internationally. The higher the number, the more likely is the possibility of introducing new electronic channels of distribution and communication. In these conditions the need for banks to serve dispersed customers throughout a number of countries implies the necessity of cheaper and more efficient channels. This is even more so as the competitive environment intensifies. This enables us to propose the following hypothesis:

H3.b: *The degree of adoption of E-banking by a bank correlates positively with the number of markets targeted internationally.*

METHODOLOGY

Mohr (1982) suggests that two fundamental types of theoretical approaches can be used to investigate organizational phenomena: "process" and "variance" models. While process models aim to understand the sequence of events leading to some result over time, variance models focus on correlations between groups of variables and a specific outcome. In this study, we will be following the latter approach.

Sampling and Data Collection

The validation of the research model is based on a quantitative method of data collection, specifically a survey administered to Lebanese banks. First, a pre-test was administered successfully in seven banks. This led either to the elimination of some ambiguous questions, the improvement of others, or the addition of new items. Following this, the main survey was administered. It consisted of a fifty-minute personal interview questionnaire either with E-banking managers or information technology managers of banks on the official list of institutions operating in the Lebanese market. The questionnaire consisted of both close-ended and open-ended questions, with three major sections.

The first section assessed the bank's attitude towards electronic channels of distribution. The objective was to measure the degree of knowledge, utilization and adoption of E-banking innovations by banks. We examined every type of electronic distribution and communication media, namely: Phone banking, Mobile banking, ATM, TV banking, PC banking, Extranet and Internet banking.

The second section tested the impact of organizational, structural and strategic variables on the reaction of the banks towards the adoption of E-banking.

The third section captured the profile of the banks and the respondents. Our purpose here is to test the impact of the "international profile" variable on the adoption of E-banking innovations.

The list of respondents was obtained from the Central Bank of Lebanon (CBL), and represents the total number of banks that operate in this country. Among the 57 banks chosen, 31 of them operate internationally and 26 are strictly local. In total, 53 banks accepted to participate in our study, which represents 93% of our population. Because of this high rate, we estimated that there was no representativeness problem, since this was essentially a census of the population.

Description of the Sample

The Banks' profiles are measured by their mission, nationality, their date of creation, and their size in terms of their total assets. The respondents' profile was indicated by job position and by foreign experience. The profile of the banks is shown in detail in appendix A.

Among the 53 banks that participated in our study, ¼ are foreign banks operating in the Lebanese market and ¾ are of Lebanese origin. 40% of the banks have international activities and 60% operate locally. Among these foreign banks, more than half have their head office in Asia (53.8%), followed by Europe (23.1%), Africa (15.4%) and America (7.7%). The sample varies according to the date of creation. The first banks to operate on the Lebanese market were the foreign banks. Local banks are relatively younger. Most banks from the sample are commercial or retail banks (73.6%). This is also true for both foreign (53.8%) and Lebanese banks (80%). 58.5% of the banks have one to ten agencies in Lebanon, and this is true for international and local banks as well. Banks employ an average of 278 individuals, between managers and support employees, with the mean being higher among the international banks (an average of 387 individuals) compared to 146 for local banks. The distribution of bank branches among the Lebanese regions, which also corresponds to the geographic distribution of the economic activities, is as follows: Beirut and its suburbs (455 branches), Mount – Lebanon (145 branches), North Lebanon (80 branches), South Lebanon (86 branches) and the Bekaa (59 branches).

Among the banks with international operations, 89.7% have overseas subsidiaries, 34.5% have strategic alliances, and some of the banks operate both ways. 57.5% of international banks in our sub-sample target between one to five foreign markets, while 24.1% target between six to ten. The first continents targeted by banks with international activities are Asia and Europe (72.4% each), followed by America and Africa (44.8% each). However, Australia is only targeted by a minority of the banks mentioned in our sample.

With respect to the respondents' personal characteristics, it should be noted that the majority of respondents occupy information technology management positions (67.9%) whereas 20.8% are E-banking managers. In fact, with respect to the international experience of managers, 44.8% of respondents had extensive foreign experience, compared to only 8.4% for local banks.

MEASURES

E-Banking Adoption Rate

The degree of adoption of E-banking by each bank is measured by the rate of use of seven types of electronic distribution and communication channels of financial services. These channels are "PC banking" "Phone banking", "Mobile banking", "Internet banking", "TV banking", "Automated teller machine" and "Extranet". These seven were chosen to represent different types of e-banking technologies at different stages in their development and that might attract different types of

Table 1. Degree of adoption of new electronic channels "e-banking"

E-banking Adoption rate	Local banks N= 40		Foreign banks N= 13 (%)	Total sample N= 53 (%)
	International activity (%)	Completely local activities (%)		
Less than 33%	43.8	95.8	61.5	71.7
33 to 66%	37.5	4.2	30.8	20.8
More than 66%	18.8	0	7.7	7.5
Total	16	24	13	53

users. Appendix B gives the definition of each type of E-banking system retained in this study. The analysis of the reliability of the measure of E-banking adoption displays a Cronbach alpha equal to 0.7035, which indicates that this variable has an adequate internal consistency for this type of research (Nunnally 1978, Churchill 1979). In this study adoption was measured on a two point nominal scale. Managers were asked if each of these seven technologies was adopted or not by their bank. With these choices of answers, a score of 0 was attributed to the bank that does not use a given electronic channel, whereas a score of 1 for the one that uses it.

Following the recoding of answers, the rate of adoption was calculated by making the summation of the scores obtained for each bank according to the 7 selected channels, divided by 7. This was than converted to a percentage, which represented the adoption rate. This percentage varies between 0% (none of the technologies were adopted) to 100% (all the technologies were adopted). This measurement of the adoption rate was inspired by Damanpour (1987).

Independent Variables

The measures of the independent variables are given in appendix C. These are largely single item measures with scales ranging from ratio to nominal. For the variables of relative advantage and complexity, 5 point –Likert interval type scales were used. For the type of presence abroad, 2 binary items were used for the measure. All the items were developed for this study according to the definitions given in the literature (e.g. Rogers, 1983; Hernandez and Mazzon, 2007; Bradley and Stewart, 2002).

RESULTS

Descriptive Statistics

Table 1 illustrates the degree of adoption of E-banking by the banks used in our sample, depending on whether they are foreign or local, and whether they operate internationally or locally.

These results show that our total sample is formed primarily of banks which adopt E-banking at a lower level (71.7%), and secondly, of banks which adopt E-banking at a moderate level (20.8%). A bivariate cross tabulation was carried out between the type of bank and the degree of adoption to test for association. The Chi square test was found to be significant (p =.006) showing that for the whole population, there is a significant association between the adoption of E-banking and the different types of banks. The analysis of the Cramer's V coefficient (0,368) shows that the strength of the interdependence between the two variables is of average intensity. This indicates that there is a relationship between the type of bank and the degree of adoption.

The banks which have high adoption rate represent only about 7.5% of our total sample. Foreign and national banks with international activities have a higher level of adoption, confirming that international bank are, on average, more willing to introduce new E-banking channels than banks with exclusively local activities. In fact, banks with international operations find themselves facing a great number of competitors operating in different socio-economic and cultural contexts. Also, banks must learn how to control new variables and adapt to them. Under these conditions, being informed, able to communicate, and able to satisfy the needs and the personal requirements of customers, twenty-four hours a day, from any location, has become a crucial requirement and an undeniable condition for performance.

The rate of adoption for each type of electronic channel is shown in appendix D. The results indicate that the most frequently used electronic channels are Automated Teller Machines (ATMs) (60.4%), Phone banking (30.2%) and Internet banking (26.4%). Nevertheless, the level of adoption is relatively higher among foreign and international banks for each type of channel studied. This affirms what is stated above: that international banks are generally more likely than local banks to adopt new E-banking channels.

Tests of Hypotheses

In order to test the hypothesis about the relationship between the degree of adoption of E-banking channels and the organizational, structural and strategic variables, bivariate associative tests are used; that is, Pearson correlation analysis when independent variables are of a continuous nature, analysis of variance (ANOVA Statistics) when independent variables are discreet or categorical, as well as Crosstabs analysis with Post Hoc analysis to examine all possible linear combinations between group means.

The Influence of Organizational Variables on the Adoption of E-Banking

The relationships between the eight organizational variables and the degree of adoption of E-banking were analyzed using Analysis of Variance. Their results are shown in Table 2.

As can be observed in the above table, six organizational variables have a significant effect on the degree of adoption of E-banking by Lebanese banks.

Table 2 shows that the largest banks (i.e. those with $10 billion or more in assets) have the highest mean level of adoption, followed by the midsize banks (i.e., those with between $1 billion and $ 10 billion in assets). The small banks (i.e. those with under $100 million in assets) have the lowest rate of adoption of E-banking. So, our hypothesis based on the affirmation of Kimberly and Evanisko (1981) is supported: company size is one of the most important indicators of innovation, and bank size correlates positively to the degree of adoption of E-banking (Diniz E., 1998). Consequently, the largest banks are more likely to adopt E-banking innovations because they have more opportunities to become exposed to the innovations and have the necessary resources to acquire and implement them. *Hypothesis H1.a is thus confirmed.*

The relationship between the type of decision and the adoption of E-banking innovation is not significant and thus *hypothesis H1.b is not confirmed.* This result is in line with the assertions of Kimberly and Evanisko (1981) and Zmud (1982) that stipulate that the nature of the relationship between the type of decision making and the adoption of IT innovation varies depending on the context in consideration, where other variables are involved and can moderate the relationship between the two variables.

Hypothesis H1.c, which stipulates that the adoption of E-banking innovations is greater in banks divided into functional units, is *confirmed.* This indicates the importance of the functional

Table 2. ANOVA results for the relationship between organizational variables and the degree of adoption of "e-banking"

Variable	Mean of adoption	Fisher analysis (sig)	Intensity of the relation W^2
Size[2] (H1.a)			
< $ 100 million (n=7)	.0414		
$100 million - $ 1 billion (n=26)	.1784	**8.657 (.000)****	0.23
$1-10billion (n=13)	.4300		
> $10 billion (n=3)	.4569		
Kind of decision-making (H1.b)			
Centralized (n=47)	.2128	2.112 (.152)	0.02
Decentralized (n=6)	.3571		
Functional division (H1.c)			
Little (4 to 9) (n=18)	.1429		
Average (10 to 15) (n=29)	.2463	**3.327 (.044)***	0.08
High (16 to 21) (n=6)	.4048		
Level of expertise of technical staff (H1.d)			
Low (n=12)	.0238	**66.322 (.000)****	0.71
Average (n=25)	.1429		
High (n=16)	.5179		
Financial resources (H1.e)			
0-30% (n=20)	.2571		
31-60% (n=5)	.1429	1.767 (.196)	0.06
61-100% (n=7)	.5714		
Technical infrastructure (H1.f)			
Under developed (n=14)	.0204	**77.696 (.000)****	0.74
Moderately developed (n=24)	.1607		
Highly developed (n=15)	.5333		
Decision makers' international experience (H1.g)			
No experience (n=5)	.1143		
Little experience (n=14)	.0816	**17.310 (.000)****	0.48
Average experience (n=19)	.1654		
Strong experience (n=15)	.4857		
Tolerance of risk (H1.h)			
Never (n=24)	.0536		
Sometimes (n=18)	.2381	**-100.342 (.000)****	0.79
Often (n=11)	.5974		

* Significant with a probability of error ≤0.05

** Significant with a probability of error ≤ 0.01

division of the company into specialized sub-units capable of evaluating the benefits of incorporating technological innovations in order to simplify tasks and communication between the bank and its internal and external partners.

The results of the ANOVA in table two demonstrate a significant mean difference in adoption of E-banking due to the different levels of expertise of the technical staff at the bank, as well as with the degree of international experience of decision makers. In fact, the table shows that the average rate of adoption increases as the level of expertise of the bank staff increases. Furthermore, the degree of adoption rises significantly with the level of international experience of the decision makers. Such experience enables the exposure of decision makers to different types of innovations in information and communication technology, thus enabling them to better estimate their importance. This may contribute to the creation of a more favorable attitude towards innovations in electronic distribution and communication channels, thereby enhancing the chance of integrating these technologies into the bank. *Therefore, hypotheses H1.d and H1.g are confirmed.*

Results do not confirm H1.e. No significant relationship between financial resources and the level of adoption of E-banking innovations was found. It appears that the innovations of E-banking are considered by the majority of banks as a necessary tool of marketing and promotion, required to attract customers and create a good image about the technological development of the bank. Thus, the bank finds that even if the adoption requires large additional financial resources, they still adopt these innovations so as to satisfy their customers, provide an added value on the quality of their services and products and to enhance their reputation.

The results of Table 2 show that a positive and significant relationship exists between the technical infrastructure of the bank and the degree of adoption of E-banking. This supports the argument that a bank with a more developed technical infrastructure, enabling higher level of secure and private communication with clients will be more likely to introduce e-banking. The ANOVA shows that the mean levels of the adoption are different, according to the level of development of the technical infrastructure in the bank: low, average or high. *Hypothesis H1.f is thus confirmed.*

Hypothesis H1.h, emphasizing that the level of adoption of E-banking innovations correlates positively with the level of tolerance of risk by decision makers was confirmed. This result is in line with the opinion of several authors like Webster (1969), Arndt (1967), Ostlund (1969), Sheth (1968) Gatignon and Robertson (1985), who believe that the tolerance of risk correlates positively with the propensity to adopt innovations. In fact, the more decision makers are ready to take risks, the less reticent they are in facing change like the adoption of E-banking innovations (Bhimani, 1996; Cockburn et Wilson, 1996). Results from the analysis of variance show significant differences in the means of the degree of adoption, *confirming hypothesis H1.h.*

The Influence of Structural Variables on the Adoption of E-Banking

The results of the relationship between the degree of adoption of E-banking innovations and the structural variables (i.e. the technological environment of the bank, relative advantages and the complexity of the innovation), are presented in Tables 3.a and 3.b.

Fisher's statistical analysis demonstrates that there is a highly positive significant relationship between a bank's internal technological status and the rate of adoption of E-banking channels, *confirming Hypothesis H2.a.* Thus, the more the bank is used to regularly introduce new technologies into its organizational structure, the greater its propensity to adopt E-Banking innovations; hence confirming Miller's assertion (1985).

Table 3.b shows that the correlation between perceived relative advantage from E-banking and

Table 3a. ANOVA results of the relationship between structural variables and the rate of adoption of "e-banking"

Variable	Mean of adoption	Fisher analysis (sig)	Intensity of the relation W^2
Regular adoption of new technologies (H2.a)		**16.952 (.000)****	0.23
Yes (n=18)	.3889		
No (n=35)	.1469		

Table 3b. Correlations between structural variables and rate of adoption of "e-banking"

Variable	Coefficient de Pearson (sig)	Intensity of the correlation
Perceived relative advantage from the innovation (H2.b)	**.384(.004) ****	r=.384
Technological complexity of the innovation (H2.c)	-.121 (.389)	r=-.121

** Significant at $\alpha \leq 0.01$

the level of adoption is positive and significant (r=.384, p=.004) supporting the affirmations of Rogers (1983) and Tan and Teo (2000) that the perceived relative advantage of an innovation is related positively to the adoption of the innovation. Therefore, the bank adopts the electronic channels of E-banking if it perceives that these innovations could increase its performance level with respect to increased market shares, cost reduction, customer satisfaction, etc. *Hypothesis H2.b is therefore confirmed.*

Hypothesis H2.c, which postulates that the degree of adoption of E-banking is negatively related with the complexity of adoption of the innovation, *was disproved.* Based on the results of the Pearson correlation shown in Table 3.b, the relationship between the complexity of the innovation and the level of its adoption is in fact negative, but not significant. There is, however, at least directional support for the hypothesis. In fact, we believe that the most likely reason for this result is that 67.9% of the bank managers with whom we dealt with are ICT directors and 20.8% are E-banking managers. Consequently, they are experts in technology and electronic channels and do not see the adoption of these channels as complex. The level of adoption of innovations in the bank is influenced by other factors, not by the complexity of the innovation. In this way, the decision to adopt E-banking is a strategic one based on bank resources, market demands, and the existence of a specialized "back office"—not on the technological complexity of the channels.

The Influence of Strategic Variables on the Adoption of E-Banking

With regard to the relationship between the strategic variables and the degree of adoption of E-banking innovations, our results supported our assumptions that the international profile of a bank has a significant impact on the degree of adoption of E-banking. These results are given in Table 4.

In fact, the two strategies that display a positive and significant relationship with the organizational adoption of electronic channels are the strategic alliance to one or more foreign partners and foreign direct investment in the form of subsidiaries. These results were predictable: In the case of a strategic alliance, the bank is in a direct relationship with its correspondents and foreign partners and, consequently, has more opportunities to be exposed to E-banking innovations. The bank is directly influenced by the internal tech-

Table 4. ANOVA results of the relationship between strategic variables and the degree of adoption of "e-banking"

Variable	Mean of adoption	Fisher statistical analysis (sig)	Intensity of the relation W2
Mode of presence abroad (H3.a)			
• Strategic alliance			
Yes (n=10)	.4000	**7.556 (.008)***	0.11
No (n=43)	.1894		
• Foreign Direct Investment			
Yes (n=26)	.3297	**11.579 (.001)****	0.17
No (n=27)	.1323		
Targeted Foreign markets (H3.b)			
None (n=24)	.1131	**5.985 (.005)***	0.16
"1 to 5" (n=15)	.2857		
"6 and up" (n=11)	.3506		

* Significant at $\alpha \leq 0.05$, ** Significant at $\alpha \leq 0.01$

nological status of its partners and must align itself with them. Also, as the bank sets up more foreign subsidiaries, it will be faced with a larger number of competitors of different origins, especially when targeting several markets at once. And it is also exposed to an increasing number of different innovations. The adoption of E-banking becomes necessary in order to maintain a competitive advantage and to continue to innovate. This *confirms hypothesis H3.a.*

Fisher's analysis, presented in Table 4, shows that the relationship between the numbers of foreign markets targeted and the degree of adoption of E-banking is positive and significant. In fact, as banks target more foreign markets, the importance of satisfying customers in these markets increases. This results in a higher probability that the new E-banking channels will be adopted, *confirming Hypothesis H3.b.*

DISCUSSION

This study examined the adoption of seven different E-banking channels by Lebanese banks and the factors that affect the bank's adoption of these innovations. The results of the analyses show that the six organizational variables: bank size, functional divisions, technical staff, technical infrastructure, tolerance of risk and decision makers' international experience exert a significant impact on the adoption of E-banking. Also, among the structural characteristics, the results reveal that the internal technological environment of the bank is a very important factor in determining the adoption of E-banking. In fact, a bank that is continuously using technologies as part of its operations, will have the experience, the resources and the management to introduce E-banking channels. As well, the relative advantage seen in adopting such technologies, in terms of the bank's performance, is a deciding factor in whether to adopt or not these innovations. As for the bank's international profile, the level of adoption of electronic E-banking channels is higher with international banks than with exclusively local banks.

Some of the hypotheses were not confirmed. For example, the relationship between the type of decision making, and the adoption of E-banking. The nature of this relationship can vary depending on the type of innovation considered and its compatibility with the interests of the organization's members (Evanisko, 1981; Zmud, 1982). Future research should consider the effects of moderator variables on the relationship between centralization/decentralization and the adoption of E-banking innovations. Another example is the absence of a significant relationship between the complexity of E-banking and the degree of adoption. Although not significant, the direction of the relationship between the complexity and the adoption was as hypothesized, which is in line with the literature (Cooper & Zmud, 1990; Dickerson & Gentry, 1983). The lack of significance is most likely due to the fact that the respondents were managers of ICT services. These tend to be experts in E-banking channels and would not perceive these technologies as challenging.

Based on the above results, the profile of a bank which adopts E-banking innovations and operates on the Lebanese market can be deduced. As such, it is an international bank operating in several markets at once, by implementing subsidiaries or through strategic alliances with one or more foreign partners. It is a medium or large bank (with at least $ 1 billion in total assets), divided into different functional units, with a highly competent electronic and technical staff and a highly developed technical infrastructure. Such a bank also regularly introduces different types of technological innovations, and considers that the adoption of these innovations provides it with a relative advantage. An innovative E-bank has decision makers who can tolerate risks related to the introduction of these media within their organization, and who have international experience (through professional activities or education). These characteristics must be taken into account by new banks which want to invest in ICT in an emerging market.

The results also indicate that the most frequently used electronic channels are Automated Teller Machines (ATMs) (60.4%), Phone banking (30.2%) and Internet banking (26.4%). These statistics can help bank managers develop the market for their better known and used electronic channels, and to reconsider their marketing strategies for those in which the degree of adoption remains weak. Of these delivery channels, Internet banking deserves special attention as having the potential to offer many more services than any other channel. Banks need to visibly demonstrate concern for security, privacy, reliability and liability with concrete solutions to reduce or eliminate costs to customers in case transactions fail or are processed inaccurately.

IMPLICATIONS, LIMITATIONS AND FUTURE RESEARCH

These results have important implications for policy makers, managers of financial institutions, suppliers of E-banking systems, and academic researchers in the field. This is especially important in the Lebanese context since we have identified a relatively low level of adoption of E-banking channels (71.7% of banks adopt E-banking at a lower level). This shows that the acceptance of electronic financial transactions is very limited. However, there are many ways to encourage E-banking.

Recommendations for Policy Makers

The major challenge to electronic banking in Lebanon is the lack of an adequate legal framework. In fact, Lebanon is behind in legal issues as there are no laws against hacking into other people's systems, the ATM print-out is not considered legal, and there is no electronic signature law. Thus, banks cannot build a case against any fraud and any action taken through electronic banking. Lebanon is currently in the early stages of developing e-commerce law. E-banking has to be legalized and a digital signature law has to be passed, and this is the role of policy makers and the central bank. A legal support for privacy protection is also an issue that policy makers must consider. Also, banks do not have to wait for the Lebanese government to catch up here; they can take the lead in devising policies, and enforceable electronic banking contracts, which customers think are fairer than much of current practice. Policy makers and banks should, therefore, give more attention to the laws and regulations concerning E-banking.

There is also a need for cheaper communication and Internet access, what seems another macro factor that affects the adoption of these technologies.

For Bank Managers

A bank which wants to innovate in E-banking should employ decision makers who believe that the adoption of E-banking represents less risk and more benefits, thereby increasing the rate of adoption. It should employ managers who have international experience from training or professional activities, and it should develop training programs with international scope to bring the expertise in ICT from developed countries into the bank. Furthermore, there is a need to spread awareness among people so that they can change their culture and mentality. Awareness can be achieved by encouraging corporations to trade in a B2B environment, by spreading e-education through advertising and seminars, by providing free access to kiosks for communities not having access to the Internet, by effectively dealing with security and privacy issues for the customers, and finally by providing value-added services at the corporate and retail levels.

Banks already offering electronic banking need to get their customers to increase their use of these channels by increasing customers' consciousness through marketing campaigns, increasing the rate of usage, making demonstrations on these services and by providing training to corporate users. The need for staff training in the bank is therefore essential. There must be someone in each branch trained to handle common electronic problems, precisely Internet problems, and who know where to go within the bank for more complex problems. Banks should also target the right people to use the E-banking facilities: for instance, individuals with Internet and PC access and expatriates. They should increase customers' trust by clearly explaining security-related issues. Often, security and privacy concerns are not purely technical issues, but rather are sometimes related to process design, or even to customer psychology and beliefs, which may or may not be consistent with the actual technology and system. To encourage the usage of E-banking channels, banks should provide incentives such as better rates and lower transaction costs.

All banks need to have the basic electronic banking services: ATMs, Internet cards and an informational web site. Banks not engaging in these channels should include them in their IT strategic plans.

Academic Contribution

This is the first study, to the best of our knowledge, that examined the adoption process and Rogers' concepts for an E-banking innovation in the context of the Middle East. It has confirmed the validity of the North American adoption model in the Lebanese market and, thus, has helped to extend the boundaries of this model. It has also given researchers interested in the concept of the adoption of innovations a new framework with which to compare this phenomenon in developed and developing countries.

Suppliers of E-Banking Systems

The results also have implications for suppliers of E-banking systems. It supports previous findings about the importance of developing systems which possess significantly better advantages than what existed previously (Webster, 1969; Tan & Tao, 2000; Gourville, 2003). Moreover, they have to communicate and highlight these advantages to the banks. They also have to provide adequate technical training to compensate for poor technical knowledge on the part of bank staff.

Despite the encouraging results, the study has some limitations that should be addressed in the future. The first limitation relates to the validity of our variables. Because many of the variables were measured by single items, their reliabilities and validity could not be assessed. Future research should include multi-items scales to permit validation. The second limitation relates to the analysis of the data. In fact, in order to reduce the complexity of the interpretations, we did not examine the potential interactions between the variables of the model.

The study was conducted in the micro-context of a decisional unit: a bank. Pertinent macro variables such as the socio-economic and cultural dimensions were left aside. Future studies incorporating the macro perspective, like the competition, market costs of telecommunication, legislation and regulations and their influence on the adoption of E-banking, should be carried out.

The study can be repeated not only from the perspective of the bank, but also from its customers' point of view, with the aim of better understanding their perceptions, interests and opinions regarding electronic distribution and communication channels.

Lastly, more comparative studies should be carried out between different countries or cultures with the aim of discovering the organizational, structural, strategic, and cultural factors which are at the root of the similarities and differences in the adoption of E-banking from one country to another.

REFERENCES

ABA Banking Journal (2001). E-banking 2001: where are we headed? 93(1)- 52-53.

ABL Association des Banques au Liban. (2004), Activity and Performance of the Lebanese banking sector. Available online at :http://www.abl.org.lb/ABL/Pictures/Part%20IV-2004-2005%20eng.pdf

Ajzen, I. (1985). From intentions to actions: a theory of planned behaviour, in Kuhl, J. and Beckmann, J. (Eds), *Action control: From cognition to Behaviour,* Springer, Heilderberg. 11-39.

Aladwani, A. M. (2001). Online banking: A field study of drivers, development challenges and expectations. *International Journal of Information Management, 21*, 213–225. doi:10.1016/S0268-4012(01)00011-1

Arndt, J. (1967). Role of product- related conservations in the diffusion of a new product. *JMR, Journal of Marketing Research, 4*, 291–295. doi:10.2307/3149462

Badoc, M., Lavayssiere, B., & Copin, E. (1994). *E-marketing de la banque et de l'assurance*. Edition Organisation.

(2004). *BankData*. Beirut, Lebanon: Bilan-banques.

Bauer, R. A. (1960). Consumer Behaviour as Risk Taking. In Hancock, R. S. (Ed.), *Dynamic Marketing for a Changing World.* Chicago: American Marketing Association.

Bergeron, J. (2001). Les facteurs qui influencent la fidélité des clients qui achètent sur Internet [The factors influencing the loyalty of online purchasers]. *Recherche et Applications en Marketing, 16*(3), 39–53.

Bhimani, A. (1996). Securing the commercial Internet. *Communications of the ACM, 39*(6), 29–35. doi:10.1145/228503.228509

Bradley, L., & Stewart, K. (2002). A Delphi study of drivers and inhibitors of internet banking. *International Journal of Bank Marketing, 20*(6), 250–260. doi:10.1108/02652320210446715

Bughin, J. (2003). The diffusion of Internet banking in western Europe. *Electronic Markets*, *13*(3), 251–258. doi:10.1080/1019678032000108329

Calof, J. L., & Beamish, P. W. (1995). Adapting to foreign markets: Explaining internationalization. *International Business Review*, *4*(2), 15–31. doi:10.1016/0969-5931(95)00001-G

Cho, K.R. & Padmanabhan, P. (1995). Acquisition versus new venture: The choice of foreign establishment mode by Japanese firms. *Journal of International Management*, Fall, l(3), 255-85.

Christensen, C. M., & Raynor, M. E. (2003). *Innovator's Solution: creating and sustaining successful growth* (p. 303). Harvard Business School Press.

Churchill, G. A. (1979). A paradigm for developing better measures of marketing constructs. *JMR, Journal of Marketing Research*, *14*, 64–73. doi:10.2307/3150876

Cockburn, C., & Wilson, T. D. (1996). Business Use of the World-Wide Web. *International Journal of Information Management*, *16*(2), 83–102. doi:10.1016/0268-4012(95)00071-2

Cooper R. B. & Zmud, R. W. (1990), Information technology implementation research: a technological diffusion approach, *Management Science*, February 36, 2, 123-139.

Cox, D. F. (1967). *Risk taking and information handling in consumer behaviour*. Harvard University press.

Cyr, D., Bonnanni, C., Bowes, J., & Ilsever, J. (2005). Beyond Trust: Web site design preferences across cultures. *Journal of Global Information Management*, *13*(4), 24–52. doi:10.4018/jgim.2005100102

Dabholkar, P. A. (1996). Consumer evaluations of new technology- based self service options: an investigation of alternative models of service quality. *International Journal of Research in Marketing*, *13*, 29–51. doi:10.1016/0167-8116(95)00027-5

Damanpour, F. (1987). The adoption of technological, administrative, and ancillary innovations: Impact of organizational factors. *Journal of Management*, *13*, 675–688. doi:10.1177/014920638701300408

Daniel, E. (1999). Provision of electronic banking in the UK and the Republic of Ireland. *International Journal of Bank Marketing*, *17*(2), 72–82. doi:10.1108/02652329910258934

Davis, F. D. (1989). Perceived usefulness, perceived ease of use, and user acceptance of computer technology. *Management Information Systems Quarterly*, *13*(3), 319–340. doi:10.2307/249008

Deise, M. V., Nowikow, C., King, P., & Wright, A. (2000). *Exexutive's Guide to E-business: From Tactics to Strategy* (p. 272). Pricewaterhousecoopers L.L.P.

Dicherson, M. D., & Gentry, J. W. (1983). Characteristics of adopters and non-adopters of home computers. *The Journal of Consumer Research*, *10*(2), 225–235. doi:10.1086/208961

Diniz, E. (1998). Web banking in USA, *Journal of Internet Banking and E-commerce*, 3(2), online at: http://www.arraydev.com/commerce/JIBC/9806-06.htm

Dixon, M., & Nixon, B. (2000). *E-banking: Managing your money and transactions online*. SAMS publishing, pp.244.

Ernst and Young. (2001), Global online retailing, *An Ernst and Young Special Report*, pp.142.

Fishbein, M., & Ajzen, I. (1975). *Belief, Attitude, intention, and Behaviour: An introduction to theory and research*. Reading, MA: Addison-Wesley.

Furst, K., Lang, W.W. & Nolle, D.E. (2002). Internet Banking, *Journal of Financial services Research,* 22(1,2), 95-117.

Gatignon, H., & Robertson, T. S. (1985). A proposal inventory for new diffusion research. *The Journal of Consumer Research, 1*(March), 249–267.

Gauvin, S., & Sinha, R. (1991). Innovativeness In industrial Organisations: A two-Stage Model of Adoption». Working paper, University of Laval. Canada.

Gourville, J. T. (2003). *Why consumers don't buy: the psychology of new product adoption.* Harvard Business School Publishing.

Hernandez, J. M. C., & Mazzon, J. A. (2007). Adoption of Internet banking: proposition and implementation of an integrated methodology approach. *International Journal of Bank Marketing, 25*(2), 72–88. doi:10.1108/02652320710728410

Hiltunen, M., Heng, L., & Helgesen, L. (2004). Personalized Electronic Banking Services. In C.-M. Karat, J. Blom & J. Karat (Eds.), *Designing Personalized User Experiences in eCommerce* (5), 119-140. Netherlands: Kluwer Academic Publishers.

Hirchman, E. C. (1980). Innovativeness, Novelty Seeking And consumer Creativity. *The Journal of Consumer Research, 7,* 283–295. doi:10.1086/208816

Infopro (2003). *The Business Handbook.* Beirut. Lebanon.

Insley, R., Al-Abed, H., & Fleming, T. (2002), What's the definition of E-banking, Journal of Online Banking, available at: http://www.bankersonline.com/technology/gurustech081803d.html. Consulted in March 2006.

Jacoby, J., & Kaplan, L. (1972), The components of perceived risk. *Advances in Consumer Research,* ed. Venkatesan M., Association for Consumer research, 287-291.

Karjaluoto, H., Mallita, M., & Pento, T. (2002). Factors underlying attitude formation towards online banking in Finland. *International Journal of Bank Marketing, 20*(6), 261–272. doi:10.1108/02652320210446724

Kimberly, J. R., & Evanisko, M. J. (1981). Organization innovation: the influence of individual, organizational, and contextual factors on hospital adoption of technological and administrative innovations. *Academy of Management Journal, 24*(4), 689–713. doi:10.2307/256170

Kirsner, D., & Balbi, D. (1997). Les besoins des banques. *Banque, 586,* 38–39.

Kolodinsky, J. M., Hogarth, J. M., & Hilgert, M. A. (2004). The adoption of electronic banking technologies by US consumers. *International Journal of Bank Marketing, 22*(4), 238–259. doi:10.1108/02652320410542536

Kotler, P. Dimaulo, McDougall G.G. & Armstrong G. (1991), *Le Marketing de la théorie à la pratique* [Marketing : From theory to Practice]. Boucherville, Québec: G. Morin, 1991.

Kotler, P. (1991). *Marketing Management: Analysis, Planning, Implementation and Control* (7th ed.). Englewood Cliffs, New Jersey: Prentice-Hall International Inc.

Langley, A., & Truax, J. (1994). A Process Study of new Technology Adoption. In *Smaller Manufacturing firms.* Press of University of Quebec in Montreal.

Lee, E., Lee, J., & Eastwood, D. (2003). A two-step estimation of consumer adoption of technology-based service innovations. *The Journal of Consumer Affairs, 37*(2), 256–282. doi:10.1111/j.1745-6606.2003.tb00453.x

Lejeune, A., Préfontaine, L., & Ricard, L. (2001). Les chemins vers la performance : L'approche relationnelle et la transformation des entreprises [The path of performance : Relational approach and organisational transformation]. *Gestion*, *26*(3), 45–51. doi:10.3917/riges.263.0045

Lemaître, P. (1997). Les enjeux de la banque à distance [The issues of home banking]. *Banque*, *587*, 63–65.

Miller, A. (1985). Technological thinking: Its impact on environmental management. *Environmental Management*, *9*, 179–190. doi:10.1007/BF01867074

Mohr, L. (1982). *Explaining Organizational Behaviour*. San Francisco: Jossey-Bass.

Mols, N. P. (1998). The behavioural consequences of PC banking. *International Journal of Bank Marketing*, *16*(5), 195–201. doi:10.1108/02652329810228190

Mols, N. P. (2001). Organising for the effective introduction of new distribution channels in retail banking. *European Journal of Marketing*, *35*(5/6), 661–686. doi:10.1108/03090560110388150

Munos, A. (1999). Technologies & Métier de service [Technologies and service professions]. *Décisions Marketing*, *17*(August), 55–65.

Nunnally, J. C. (1978). *Psychometric Theory*. New York: McGraw-Hill.

Ostlund, L. E. (1969). The role of Product Perceptions in innovative behaviour, in ed. P.R. Mac Donald, *Marketing Involvement in Society and the Economy*, Chicago: American Marketing Association, 259-266.

Perrien, J., Filiatrault, P., & Ricard, L. (1993). The implementation of relationship marketing in commercial banking. *Industrial Marketing Management*, *22*, 141–148. doi:10.1016/0019-8501(93)90040-E

Quelch, J. A., & Klein, L. R. (1996). The Internet and International Marketing. *Sloan Management Review*, (Spring): 60–75.

Raju, P. S. (1980). Optimum stimulus level: its relationship to personality, demographics, and exploratory behaviour. *The Journal of Consumer Research*, *7*, 272–282. doi:10.1086/208815

Ricard, L., & Perrien, J. (1999). Explaining and evaluating the implementation of organizational relationship marketing in the banking industry: Client's perceptions. *Journal of Business Research*, *45*, 199–209. doi:10.1016/S0148-2963(97)00226-9

Riddle, D. I. (2001). Cadre d'analyse de rentabilisation pour l'adoption des affaires électroniques dans les petites entreprises, available at : www.strategis.ic.gc.ca/epic/site/sbrp-rppe.nsf/vwapj/framework_f.pdf/$FILE/framework_f.pdf - Consulted in octobre 2003.

Riggins, F. J., & Rhee, H.-S. S. (1998). Toward a Unified View of Electronic Commerce. *Communications of the ACM*, *41*(10), 88–95. doi:10.1145/286238.286252

Robertson, T. S. (1971). *Innovative behaviour and communication*. New York: Holt, Rinehart and Winston.

Rogers, E. M. (1983). *Diffusion of innovations*. New York: The free press.

Rogers, E. M., & Shoemaker, F. F. (1971). *Communications of Innovations*. New York: Free Press.

Rust, R. T., & Kannan, P. K. (2002). *E-Service: New Directions in Theory and Practice*. Armonk, New York, NY: ME Sharpe.

Sathye, M. (1999). Adoption of Internet banking by Australian consumers: an empirical investigation. *International Journal of Bank Marketing*, *17*(7), 324–334. doi:10.1108/02652329910305689

Sayar, C., & Wolfe, S. (2007). Internet banking market performance: Turkey versus the UK. *International Journal of Bank Marketing, 25*(3), 122–141. doi:10.1108/02652320710739841

Shandh, J. N. (1968). A factor analytic model of brand loyalty. *JMR, Journal of Marketing Research, 5*, 395–404. doi:10.2307/3150264

Skhiri, S. (2000). Adoption des innovations en technologie de l'information par les entreprises internationales [The adoption of IT innovations by international companies]. Unpublished Master's thesis in Business Administration, University of Quebec in Montreal, Canada.

Tan, M., & Teo, T. S. H. (2000). Factors Influencing the Adoption of Internet Banking. *Journal of the Association for Information Systems, 1*(5), 1–44.

Venkatesh, V., Morris, M. G., Davis, G. B., & Davis, F. D. (2003). User acceptance of information technology: Toward a unified view. *Management Information Systems Quarterly, 27*(2), 425–478.

Venkatraman, N. (1995). Reconfigurations d'entreprises provoquées par les technologies de l'information. In *L'entreprise compétitive au futur* (pp. 151–195). Les Éditions d'Organisation.

Villates, D. (1997). Demain, la banque à distance [The Distant bank of tomorrow]. *Banque, 585*, 68–70.

Wah, L. (1999). Banking on the Internet. *American Management Association, 88*(11), 44–48.

Walsh, S. T., Kirchhoff, B. A., & Newbert, S. (2002). Differentiating Market Strategies for Disruptive Technologies. *IEEE Transactions on Engineering Management, 49*, 341–351. doi:10.1109/TEM.2002.806718

Webster, F. E. J. (1969). New Product Adoption in Industrial Markets: A Framework for Analysis. *JMR, Journal of Marketing Research, 33*(3), 35–39.

ENDNOTES

1 The authors acknowledge the contribution of the two anonymous reviewers that has significantly improved earlier versions of the article.

2 The size of the banks is stratified by asset size, based on the ranking by total assets of the Bilanbanques (2004). The categorization of small, medium and large banks is adapted from Furst, Lang & Nolle (2002).

3 An intranet is a private network that is contained within an enterprise

APPENDIX A

Table A1.

Profile of all the banks			
Variable		**Frequency**	**%**
Nationality	Lebanese	40	**75.5**
	Foreign	13	24.5
Head office of the foreign bank	Asia	7	**53.8**
	Europe	3	23.1
	Africa	2	15.4
	America	1	7.7
	Australia	0	.0
Date of creation	1950 and earlier	11	25.0
	1951 to 1970	12	27.3
	1971 to 1900	13	**29.5**
	1991 and later	8	18.2
Function	Retail	39	**73.6**
	Corporation	3	5.7
	Private	3	5.7
	Merchant	8	15.0
Size	Small < $ 100 million	7	6,1
	Medium $ 100million-$1billion	26	26,5
	$ 1- $ 10 billion	13	**53,1**
	Large > $ 10 billion	3	14,3
Profile of the subgroup of international banks N = 29			
Variable		**Frequency**	**%**
Nationality	Lebanese	16	**55.2**
	Foreign	13	44.8
Kind of presence abroad	Strategic alliance	10	34.5
	Foreign direct investment	36	**89.7**
Targeted foreign markets	1 to 5	15	51.7
	6 and greater	11	38.0
Targeted Continents	America	13	44.8
	Europe	21	**72.4**
	Asia	21	**72.4**
	Africa	13	44.8
	Australia	6	20.7

APPENDIX B

Table A2. Definition of various types of e-banking

E-banking channel	Definition
PC banking	Is the fastest growing E-banking technology, calls for perhaps the most consumer involvement, as it requires the consumer to maintain and regularly interact with additional technology (a computer and an Internet connection) (Kolodinsky, Hogarth & Hilgert, 2004). PC banking is a form of E-banking that enables customers to execute bank transactions from a PC via a modem. In most PC banking ventures, the bank offers the customer a proprietary financial software program that allows the customer to perform financial transactions from his or her home computer. The customer then dials into the bank with his or her modem, downloads data, and runs the programs that are resident on the customer's computer. Currently, many banks offer PC banking systems that allow customers to obtain account balances and credit card statements, pay bills, and transfer funds between accounts.
Phone banking	Represents an active technology that has undergone an evolution, from calling the bank and talking in person to a customer service representative to calling an automated account management system. Unlike PC banking, however, phone banking requires no equipment or services other than a touch-tone phone line. It allows clients to manage most of their banking needs using a touchtone phone.
Mobile banking	Allows mobile phone users to perform secure e-commerce transactions using the next generation of WAP (Wireless Application Protocol) phones that link mobile devices with the Internet. Bank customers can check their account balances, pay bills and transfer funds while traveling in a taxi or relaxing in the park simply by following message prompts on the handset screen rather than cumbersome and more expensive mediums such as voice-prompted telephony or the Internet (Anderson, Jones, Butler, 1999). With the Mobile banking, customers can access their bank and account via SMS (Daniel, 1999).
Internet banking	Also known as online banking or web banking is an outgrowth of PC banking. Internet banking uses the Internet as the delivery channel by which to conduct banking activity, for example, transferring funds, paying bills, viewing checking and savings account balances, paying mortgages, and purchasing financial instruments and certificates of deposit. An Internet banking customer accesses his or her accounts from a browser— software that runs Internet banking programs resident on the bank's World Wide Web server, not on the user's PC. Internet banking is a more flexible solution than PC Banking. It doesn't require a software installation for the customer who uses his Web navigator to connect himself to the server the bank.
TV banking	Is a digital, interactive TV that allows customer to shop, bank, email or even surf the net all from his TV. TV banking is the use of satellite or cable to deliver account information to the TV screens of customers (Daniel, 1999).
Automated Teller Machine (ATM)	Probably the most widely used means of electronic funds transfer. ATM is a machine that permits customers to gain access to their accounts through the use of a magnetically encoded plastic card and by pushing appropriate buttons on a computer terminal. ATMs dispense cash, transfer funds from one account to another and perform other functions and are available 24 hours a day.
Extranet	A private network that uses Internet technology and the public telecommunication system to securely share part of a business's information or operations with suppliers, vendors, partners, customers, or other businesses. An extranet can be viewed as part of a company's Intranet[3] that is extended to users outside the company. You can access an extranet only if you have a valid username and password, and your identity determines which parts of the extranet you can view.

APPENDIX C

Table A3. Measurement scales of the independent variables

Variable	Measurement scale
Financial resources	Scale variable that varies from 0 to 100% evaluating the percentage of the annual budget devoted to the introduction and the improvement of E-banking services and channels
Size of the bank	Ordinal variable of 4 categories, stratified by asset size, according to the international standards of the size of the bank: Small < $ 100 million Medium $ 100million-$1billion $ 1- $ 10 billion Large > $ 10 billion
Functional division	Scale variable of n answers, transformed into a discrete variable of three categories evaluating the number of departments within the organization (four to nine departments/Ten to fifteen departments/sixteen departments and more)
Kind of decision- making	Discrete variable with 2 categories (often centralized/ Decentralized decision-making)
Level of expertise of technical staff	Discrete variable with three categories, evaluating the level of expertise of technical staff in general (low expertise/ average expertise / high expertise)
Technical infrastructure	Discrete variable with three categories, evaluating the degree of development of the technical infrastructure of the bank (under developed infrastructure/ Moderately developed / Highly developed)
Tolerance of risk	Discrete variable with three categories, evaluating the degree of tolerance of risk by the IT managers when introducing the E-banking innovations concerning the adoption of E-banking (never /sometimes /often)
Decision makers' international experience	Discrete variable with four categories, evaluating the international experience of the decision-maker (No experience abroad/a little experience of at least six months, an average experience from 1 to 3 years, strong experience of more than 3 years)
Regular adoption of new technology	Binary variable, evaluating if the bank regularly adopts the recently developed innovations (yes/no)
Perceived relative advantage from the innovation	Scale variable measured on five points Likert scale, evaluating the perception of the relative advantages related to the adoption of E-banking (not at all important/ little important/ fairly important /very important/extremely important) Explain why:
Complexity of the innovation	Scale variable measured on five points Likert scale, evaluating the extent to which the adoption of E-banking is perceived difficult to introduce within the bank (not at all complex/ Little complex/moderately complex/very complex/ extremely complex). Explain what are the reasons of this complexity:
Mode of presence abroad	Binary variable with two items measured with two categories each, which evaluates the international presence of the bank by two modes: a strategic alliance (yes/no) or/and the implementation of subsidiaries (yes/no).
Number of foreign markets targeted	Scale Variable with n answers, transformed to discrete variable with 3 categories, evaluating the number of market that the bank is targeting abroad (none /1 market to 5/ 6 and more)

APPENDIX D

Table A4. Bank categorization according to the type of channel adoption

	Local banks N = 40				Foreign banks N =13		Total sample N = 53	
	International activity N= 16		Local activity N= 24					
	freq	%	freq	%	freq	%	freq	%
Phone Banking	9	56.3	2	8.3	5	38.5	16	30.2
Mobile Banking	3	18.8	1	4.2	3	23.1	7	13.2
ATM	14	87.5	12	50.0	6	46.2	32	60.4
TV Banking	1	6.3	1	4.2	0	.0	2	3.8
PC Banking	4	25.0	1	4.2	3	23.1	8	15.1
Extranet	3	18.8	0	.0	3	23.1	6	11.3
Internet Banking	9	56.3	2	8.3	3	23.1	14	26.4

This work was previously published in International Journal of E-Services and Mobile Applications, Volume 1, Issue 1, edited by Ada Scupola, pp. 67-93, copyright 2009 by IGI Publishing (an imprint of IGI Global).

Chapter 8
The Impact of Online Product Presentation on Consumers' Perceptions:
An Experimental Analysis

Carlos Flavián
University of Zaragoza, Spain

Raquel Gurrea
University of Zaragoza, Spain

Carlos Orús
University of Zaragoza, Spain

ABSTRACT

The website design and specifically the visual aspects have been revealed as key factors for achieving a successfully e-service website. This research examines the effects of the visual product presentation on online users' perceptions. We carried out an experiment to analyze the impact of the size, the quality and the movement of the product picture on the perceived usability, satisfaction and purchase intention of users. We also analyzed the moderating role of user experience and product familiarity. The results suggest the importance of visual aspects, emphasizing the quality of the picture as the most salient factor. This study also indicates that users' satisfaction is influenced to a great extent by these attributes.

INTRODUCTION

In the last years, the diffusion of the Internet as a new retail and distribution channel has undergone a great growth. The possibilities of exchange are almost infinite if we take into account that there are 1,400 million people (Internet World Stats, 2008) and more than 70,000 million of *dotcom* (Domain Tools, 2008). Sales through the web are up to 65 billon US$ in United States in the first half of 2007, which means an increase of 20% (U. S. Census Bureau, 2008). It is thought that in Europe sales outcomes will reach 197 billion

DOI: 10.4018/978-1-60960-607-7.ch008

US$ by 2008, which supposes an increase of 37% related to 2006 (Emarketer, 2008).

In this context, a special interest has arisen focused on the analysis of the factors that affect the success of e-commerce and e-services websites. In this line, the specialized literature has identified the website design as one of the main factors for the development of a good interface for satisfying the consumer needs. A good design is relevant for companies to survive in the extremely competitive World Wide Web (Ranganathan & Ganapathy, 2002; Tan & Wei, 2006). Thus, the web design is important for getting higher levels of satisfaction with the website (Kim & Eom, 2002) or for increasing the online purchase intention of the consumer.

More specifically, visual aspects, such as images, graphics or icons, are being considered as one of the main components of the website design (Kim & Stoel, 2004; Park, Lennon & Stoel, 2005; Ranganathan & Ganapathy, 2002). In this line, the appearance of a store influence the consumer's perceptions and behaviours, which has been demonstrated in the traditional context (Donovan, Rossiter, Marcoolyn & Nesdale, 1994), as well as in the online context (Eroglu, Machleit & Davis, 2001; Kim & Eom, 2002; Vrechopoulos & Siomkos, 2002). However, very little research has been focused on studying how the specific features of visual elements could influence online consumers.

The aim of this study is therefore, to examine the effect of the online product presentation on user's perceptions and behaviours. In particular, this research analyzes the effects of three of the features most exploited by e-firms: the size, the quality and the movement of the product picture. We analyze three dependent variables related to consumer behaviour, basing on the model of Consumer Response System proposed by Holbrook and Hirschman (1982): one cognitive variable –perceived usability–, one affective variable –satisfaction with the website– and one behavioural variable –online purchase intention–. In this way, this research offers an integrated global overview about the users' perceptions in order to better examine their inner states. This approach integrates the information processing perspective and the experiential view, which is very useful for our study to acquire a more complete knowledge of the real online behaviour under this display of visual aspects of the product presentation.

The article is structured as follows. In the next section, we review the literature regarding the relevance of website design for the success of the electronic commerce and e-services. After that, we formulate hypotheses about the effects of different visual product presentations on online consumers' inner states, and the role played by users' experience with the Internet and familiarity with the type of product. We then discuss the methodology and present our findings. Lastly, we introduce the main conclusions and managerial implications, as well as the research limitations and future research lines.

THE IMPORTANCE OF WEBSITE DESIGN

With the development of the Internet, electronic commerce and e-services, most of the research carried out in the field of web design has been focused on identifying the key factors that could have an influence on website degree of acceptation and success (e,g. Kim & Stoel, 2004). On one hand, the usability defines what sort of features should the website have in order to make it easier for users to navigate. In this sense, Nielsen (1994) defines the usability of a website as the ease with which the user can learn to manage the system and memorise the basic functions, the efficiency of design of the site, the degree of error avoidance and the general satisfaction of the user. Usability can also be understood as a tool for measuring the quality of a website (Ranganathan & Ganapathy, 2002). Thus, the ease of use of a system could improve a more complex learning and a higher ability to anticipate how that system will perform. In fact, adequate usability levels improve the best understanding of the contents and tasks that the consumer has to know for the achievement of a goal (e.g. to place an order), which in turn reduces the probability of error and improves the level of trust. As a consequence, the best usability, in the sense of a comfortable atmosphere, can create a positive bias in the consumer. These aspects could lead to a higher level of satisfaction, trust and loyalty towards a specific website (Flavián, Guinalíu & Gurrea, 2006; Kim & Eom, 2002).

In a similar way, the marketing literature has studied how the factors that define the store environment influence the consumer's mood and purchase behaviour. Design factors related to visual cues (e.g., layout, colour) are important for getting a positive response of the consumer and to facilitate consumer goals (Donovan, Rossiter, Marcoolyn & Nesdale, 1994). This issue has been adapted to the web environment. Dailey (2004) defines web atmospherics as the intended design of web environments to generate positive effects (cognitive and affective) on the consumer to increase positively the consumer responses. Among these responses we can find a major number of visits or a major time browsing in the website, in order to turn the user of a website into a client. In this sense, Van der Heijden and Verhagen (2003) examined the image elements in a virtual store that could improve the purchase intention in the user, identifying the most important factors of design. In addition, Kim and Stoel (2004) demonstrated that the factors of web appearance can lead to higher consumer satisfaction. The development of a good website design turns into a key question, since it will determine the entry of the consumer to the store.

We could also mention Dailey (2004) and Eroglu, Malcheit and Davis (2001) research, that analyzed the influence of web atmosphere on consumers. These authors demonstrated that the insights of this atmosphere affect the cognitive and affective states of the consumer, and as a consequence, it affects the purchase behaviour towards the website too. Tan and Wei (2006) proposed that the appearance of a website is an instrumental factor that improves the perception of information in order for subjects to perform better cognitive mapping and assessment of decisions for execution. They also established that graphical representation, such as icons, colours, images and animations, is used to launch the website with more attractiveness, which increases the user satisfaction and improves the navigation. Consequently, the specialized literature has revealed that the visual aspects play an important role for the websites' success.

THE ROLE PLAYED BY THE VISUAL PRODUCT PRESENTATION HYPOTHESES

Traditionally, the marketing literature has studied the influence of consumers' perceptions on their purchasing attitudes and behaviours about the commercial establishments. Aspects such as the layout of the store, the setting of the establishment, the placing of the products, the colours or the background music are characteristics that influence on how consumers perceive the establishment and on their purchasing behaviour.

In the online context, there is not possibility either to "touch" the product or to examine it physically before the purchase. Thus, the importance of a good product image is critical for the consumer, because it let users identify and know the product. In this sense, Vrechopoulos and Siomkos (2002) emphasized that image is one of the main dimensions of the web atmosphere. They found aspects as the colour, the brightness, the size and the form of the pictures. In a similar way, Kim and Eom (2002) identified a proper size of the image and a good quality of the picture as critical factors to achieve an effective design of a website.

Visual representations show the information briefly, making the decision easier. Besides, the interactive visualization tools (e.g. motion and animations) could create a better match between the task and the environmental decision, which improves the quality of the information and reduces the effort required. In the same line, the use of pictures and the movement-dynamism of products could help to better know the product and therefore to consider its purchase (Lee & Benbasat, 2003).

Therefore, the designers have to take into account the importance of the image of the product when developing the website, since the product presentation could affect the user's perceptions and behaviours. The size and the movement of the product picture are resources very often manipulated by designers in their websites, showing the products in different modes (e.g. big/small or static/dynamic images). However, there is a lack of studies focused on analysing the effects of the product visual presentation on the consumer behaviour. The management of these features has to be considered carefully, since the speed of navigation could be diminished if there is an overload in the use of the images with a large size, too high quality and/or very sophisticated animations.

Influence of the Visual Presentation of the Product on Perceived Usability

In a website, usability reflects the perceived ease of navigating the site or making purchases through the Internet (Flavián, Guinalíu & Gurrea, 2006). In this way, the design is the main tool for managing the website and making it easy to use. A good design in terms of a good appearance should improve the ease of use and navigation through the website, make it more enjoyable and pleasant, and simplify the decision making (Tan & Wei, 2006).

In this sense, fast presentations, organized screens and easy search paths make the navigation more efficient, supporting a pleasurable experience and diminishing the cognitive effort (Park, Lennon & Stoel, 2005). Thus, the use of graphics, images, colours and animations are resources very often used in order to give a more efficient navigation through the website.

In fact, the image of a product can affect the degree of perceived usability, since images are demonstrated to be remembered more easily than text information (Hong, Thong & Tam, 2004). This could facilitate consumer's decision, improving the perceived usability of the website.

So, we propose that an adequate size and a high quality of the product picture will lead to a higher level of perceived usability. Furthermore, the availability of movement or animation of the image of the product may offer a higher interactivity between the consumer and the website, and also the possibility to see the product from different views, so it favours a better knowledge of the product:

H1a: *The size of the visual presentation of the product will have a positive and significant effect on the degree of perceived usability for the Internet user.*

H1b: *The quality of the visual presentation of the product will have a positive and significant effect on the degree of perceived usability for the Internet user.*

H1c: *The movement of the visual presentation of the product will have a positive and significant effect on the degree of perceived usability for the Internet user.*

Influence of the Visual Presentation of the Product on User Satisfaction with the Website

In general terms, satisfaction is defined as an affective consumer condition that results from a global evaluation of all the aspects that make up the relationship between the firm and the client (Anderson & Sullivan, 1993). Satisfaction is a sign of the favourable attitude of the person. In a virtual context, satisfied users may stay longer time in a website, may revisit the site, and may recommend it to others.

Marketing research has dealt widely with the establishment of routines to guide the web design to achieve the user satisfaction. Though a good website design may not guarantee consumer satisfaction (there are other factors) it has a direct impact (Flavián, Guinalíu & Gurrea, 2006; Muylle, Moenaert & Despontin, 2004). In a more specific way, it has been proved that graphical representation (e.g. icons, colours, images and animations) is used to launch the website with more vividness and so to increase the satisfaction. In the same line, Kim and Stoel (2004) identified the web appearance as a main dimension of the website quality, wherein the size and fidelity's image of a product played an important role. In addition, the study carried out by Muylle, Moenaert and Despontin (2004) developed a construct about the Internet user satisfaction. The design of the website was identified as one of its main dimensions. Besides, it was possible to demonstrate that several factors, such as colours, textures, graphic elements, and the new manners of animation, led to a higher satisfaction in goal-oriented consumers.

Therefore, a large size and a good quality of the image of the product could evoke more positive feelings when navigating through the website, because the product is perceived more clearly and it helps users to know the product better and recall it easier (Lee & Benbasat, 2003; Park, Lennon & Stoel, 2005). Likewise, the movement of the images could provide an entertaining experience. If we also consider the more information and interactivity that the motion offers in the presentation of the product, we expect that the addition of movement could imply higher level of consumer satisfaction with the website:

H2a: *The size of the visual presentation of the product will have a positive and significant effect on the degree of satisfaction with the website of the Internet user.*

H2b: *The quality of the visual presentation of the product will have a positive and significant effect on the degree of satisfaction with the website of the Internet user.*

H2c: *The movement of the visual presentation of the product will have a positive and significant effect on the degree of satisfaction with the website of the Internet user.*

Influence of the Visual Presentation of the Product on Online Purchase Intention

In a traditional purchase context, it is possible to find a great amount of studies that highlight the importance of visual aspects to provoke purchase responses of the consumer (Swinyard, 1993). When assessing multiple choices, consumer often needs to consider a lot of attributes related to the specification of the product, such as price, the quality of the materials and properties, the colour or the size. Thus, the way in which a product is displayed in a store may increase the purchase intention (Swinyard, 1993).

In the virtual environment, the elements related to visual appearance of a website could have an important impact on the consumer purchase intention. It seems reasonable that most of the online consumers will visit and make purchases through the Internet more probably in well-designed websites, which provides the necessary tools for facilitating the decision (Kim & Eom, 2002). In this way, Then and DeLong (1999) demonstrated that a good presentation of images can generate higher consumer purchase intention and, finally increase the sales of the firm.

Therefore, it seems logical to propose that if the visual presentations of the product displayed with a large size and a high quality in order to evoke positive feelings, the consumer's purchase intention will increase. Images with an adequate size and high quality could also offer a better knowledge of the product, which could lead to consider its purchase. In addition, a major knowledge of the product could augment the purchase intention. Furthermore, Ariely (2000) showed that the quality of the decision and attention of the users get better with high levels of interactivity. As we have pointed out, the movement of the product allows the consumer to examine it from multiple perspectives and interact with it, which also favours the user's purchase intention:

***H3a:** The size of the visual presentation of the product will have a positive and significant effect on the Internet consumer purchase intention.*

***H3b:** The quality of the visual presentation of the product will have a positive and significant effect on the Internet consumer purchase intention.*

***H3c:** The movement of the visual presentation of the product will have a positive and significant effect on the Internet consumer purchase intention.*

Moderator Effects

In the specialized literature, some authors have argued that the consumer characteristics are important for determining the acceptance of online shopping (e.g. Lian & Lin, 2008). In this line, several studies have analyzed the major role played by a consumer's experience as a variable moderating certain relationships or behaviours (e.g. Forbes & Rothschild, 2000). In addition, some researchers have suggested that the users' perceptions about the website qualities may differ depending on the type of product (Kim & Stoel, 2004; Zhou & Nakamoto, 2007), since they may have different degrees of knowledge about it. Thus, it seems interesting to analyse the moderator effect that the degree of internet experience and product knowledge could have in the relationships proposed.

Regarding the first aspect, consumers could make different judges depending on the levels of expertise with the use of a website. In this way, dynamic or visual properties may influence on people's judgements depending on the degree of experience. The perceptions of people who navigate a website for the first time could differ from the perceptions of experienced users. Thus, we could note a strong learning effect when the users revisit an e-store, since the knowledge of the layout of the website could enhance their perceived control in the navigation. This enhanced control could allow users to pay more attention to other features such as the visual aspects of the website, leading them to different perceptions with respect to more novice users.

In addition, product familiarity could affect the way in which users collect and process the information, affecting product learning and evaluation. In this way, consumers who have higher knowledge of a product could have a set of expectations about it, such as the characteristics more considered and its importance, or the relationships between the attributes. Nevertheless, non-familiarized consumers with little prior product experience could entail a more difficult and complex learning, because of their lack of knowledge structures (Zhou and Nakamoto, 2007). Thus, different levels of product knowledge could generate different outcomes in the inner states of the consumers. In this way, visual aspects associated to the product in the interface could imply different responses on users depending on their degree of familiarity with it.

Therefore, we could expect that Internet user experience and familiarity with the type of product play a moderating role in the previous proposed hypotheses.

***H4a:** The degree of Internet experience will moderate the influence of visual product presentation on user's perceived usability, satisfaction with the website and online purchase intention.*

Table 1. Representative nature of the sample

	eScope (2007)	Eurostat (2008)	AECEM (2008)	AIMC (2008)	Our Research
Sex (Male)	50%	58%	51.5%	57%	52.4%
Age (23-45 years)	69%	64%	68.1%	50.3%	60.2%
Online shopping experience	N.A.	N.A.	55.6%	80%	62.3%

***H4b:** The degree of product familiarity will moderate the influence of visual product presentation on user's perceived usability, satisfaction with the website and online purchase intention.*

METHODOLOGY

An experiment was carried out in order to test the hypotheses proposed. This method has been widely used in marketing and web design literature (Park, Lennon & Stoel, 2005; Van der Heijden & Verhagen, 2003). Subjects were randomly assigned to a between-subjects factorial design with 2 (image size: large versus small) x 2 (image quality: high versus low) x 2 (image movement: static versus dynamic) conditions. A total of 176 graduate and undergraduate students participated in the experiment in a session on February 2008. The specialized literature has frequently employed students in this type of studies (e.g. Hong, Thong & Tam, 2004), since they are more likely than other consumers to be potential Internet shoppers. Thus, we were satisfied with the sample profile achieved because it showed similarities to recent samples in some European studies (Asociación Española de Comercio Electrónico y Marketing Relacional –AECEM-, 2008; Asociación para la Investigación en Medios de Comunicación –AIMC- 2008; eScope, 2007; Eurostat, 2008) of Internet users (see Table 1). The majority of the respondents were male and young (between 18 and 34) and, on average, they had some online shopping experience (62.3% had purchased at least one product or service, and 30.9% had purchased more than 4 products or services through the web).

To carry out the experiment, 8 websites were created, which simulated an electronic products virtual store, named *Electroshop.com* (Figure 1), wherein the product visual presentation –digital video camera– was manipulated in the presentation to the consumer. We chose a digital video-camera mainly for two reasons: (1) it is an electronic product, and these products are especially attractive for the selected sample. A recent report shows that electronic goods are among the most purchased items online, as well as the most related information searched (EIAA, 2008); (2) digital video-camera is not a product so standardized as other products (e.g. mp4 player, digital camera), on which there is a generalized knowledge about its main characteristics; the characteristics of video cameras are not so common for the general public, and the consumer has to pay special attention to the information in order to know the product. In this way, the Internet could represent the most preferred medium for acquiring information and knowledge about this type of products (Burke, 2002), where the product specifications and detailed pictures are especially emphasized.

In this way, the size and the quality of the pictures were manipulated according to the idea of maximizing the difference (Lee & Benbasat, 2003) so that the participants could perceive such differences. Thus, large pictures were four times bigger (8,93cm x 7,61cm) than smaller ones (2,23cm x 1,90cm), and low quality pictures were at 25% of resolution with respect to high quality

Figure 1. Examples of experiment websites

(a) large size, high quality, static

(b) small size, low quality, static

(c) small size, low quality, dynamic

(d) large size, high quality, dynamic

images (28 pixels/cm). To provide with movement to the visual presentation, a flash animation was used to create a 360 degree view.

Before carrying out the experiment, a pilot test was conducted in a usability lab. A group of usability and marketing experts assessed the websites in order to detect possible errors in the design of the websites as well as in the questionnaire that participants had to fill in.

The subjects could interact with the product (enlarging the picture, seeing it from different views or accessing to more information about its characteristics) and navigated through the website for a few minutes. Once the time was over, the participants turned off the monitor and filled in a questionnaire.

In the final questionnaire, all the questions used a five-Point Likert scale. With reference to the measurement of the dependent variables, we used and adapted the scales widely validated by specialized literature for the perceived usability (USAB) (Nielsen, 1994; Roy, Dewit & Aubert, 2001), satisfaction with the website (SAT) (Flavián, Guinalíu & Gurrea, 2006, Janda, Trocchia & Gwinner, 2002; Smith & Barclay, 1997), online purchase intention (OPI) (Jarvenpaa, Tractinsky & Vitale, 2000; Van der Heijden & Verhagen, 2003) and familiarity with the type of product (FAM)

(Gefen, 2000; Gurrea, 2006) (see Appendix). Besides, we gathered subjects' demographics as well as user experience with Internet through one item which recorded three levels of experience (less than two years, from two to five years and more than five years) (Hong, Thong & Tam, 2004; Park, Lennon & Stoel, 2005).

RESULTS

We carried out an analysis of reliability and dimensionality to check the validity of the scales used in the questionnaire. Regarding the reliability of the scales, we based on Cronbach's alpha indicator and the item-total correlation between each element of the dependent variables. The results of the tests showed acceptable indices of internal consistency in the four scales considered, as we can see in Table 2. In the next step, we proceeded to evaluate the uni-dimensionality of the scales developed by carrying out a principal components analysis. The existence of uni-dimensionality is a very important issue in order to develop a quantitative analysis due to the fact that it allows calculating the average of the indicators that compose each construct. Consequently, it is possible to use a solely variable to represent each theoretic construct. Factorial analysis, with principal components and varimax rotation was carried out basing on the existence of eigen-values higher than 1. Moreover, it was required that factorial loadings were higher than 0.5 points and a significant total explained variance. Only one factor was extracted from each of the four scales: perceived usability, satisfaction with the website, online purchase intention and product familiarity.

Previous to the analyses for testing the hypotheses proposed, three variance analyses were developed in order to ensure that the manipulations in the visual presentations of the product were perceived significantly different. Thus, we con-

Table 2. Analysis of scale validity

		Internal reliability		Exploratory Factorial Analysis	
Construct	**Item**	**Cronbach α**	**Item-total Correlation**	**Factor Loadings**	**Variance Extracted**
Perceived Usability	USAB1	0.864	0.664	0.791	65.67%
	USAB2		0.738	0.846	
	USAB3		0.640	0.770	
	USAB4		0.675	0.796	
	USAB5		0.737	0.846	
Satisfaction with	SAT1	0.929	0.829	0.923	87.50%
the Website	SAT2		0.880	0.948	
	SAT3		0.852	0.935	
Online Purchase	OPI1	0.906	0.750	0.883	84.36%
Intention	OPI2		0.870	0.946	
	OPI3		0.823	0.925	
Product	FAM1	0.927	0.872	0.933	82.04%
Familiarity	FAM2		0.821	0.903	
	FAM3		0.877	0.935	
	FAM4		0.747	0.850	

Table 3. Analysis of variance for image manipulations[1]

Effect	Mean (stnd. deviation)		Mean diff.	*F*	sign.
	Large	Small			
Size	3.53 (1.06)	2.09 (0.90)	1.44	94.318	0.000***
	High	Low			
Quality	3.59 (1.01)	1.81 (0.89)	1.78	153.333	0.000***
	Dynamic	Static			
Movement	3.87 (1.94)	2.45 (1.17)	1.42	62.369	0.000***

firmed that participants significantly perceived the large image as larger than the small image (F = 15.693; p < 0.001), the high quality image as higher quality than the low quality image (F = 25.268; p < 0.001), and the dynamic presentation as more dynamic than the static presentation (F = 16.822; p < 0.001) (see Table 3).

To examine the hypotheses, a Multivariate Analysis of Variance (MANOVA) including all dependent and independent variables was developed. Firstly, we can state that the picture's size had a significant impact on the user's satisfaction with the website ($F = 3.516$; $p < 0.1$), supporting H2a. The user's satisfaction with the website increases with the employment of large product images. However, the other effects proposed for the image's size were not significant (see Table 4) so we must reject hypotheses H1a and H3a. In addition, we found the most significant effects for the quality of the product picture on the perceived usability ($F = 9.699$; $p < 0.01$), as well as the satisfaction with the website ($F = 34.671$; $p < 0.01$) and the online purchase intention ($F = 15.716$; $p < 0.01$). Thus, quality positively influences the users' perceptions (see Table 5), which supports hypotheses H1b, H2b and H3b.

Regarding the movement effect of the product visual presentation, significant differences were found for the user's perceived usability ($F = 2.863$; $p < 0.1$), satisfaction with the website ($F = 7.101$; $p < 0.01$) and online purchase intention ($F = 3.167$; $p < 0.1$). Participants showed a higher preference

Table 4. Hypotheses testing related to the image size[1]

Dependent variable	Mean (stnd. deviation)		Hypotheses testing		
	Large	Small	Mean diff.	*F*	sign.
Perceived usability	4.09 (0.70)	4.04 (0.61)	0.04	0.180	0.672
Satisfaction with the website	3.65 (0.79)	3.46 (0.86)	0.19	3.516	0.063*
Online purchase intention	2.87 (0.98)	2.70 (0.96)	0.17	1.603	0.207

Table 5. Hypotheses testing related to the image quality[1]

Dependent variable	Mean (stnd. deviation)		Hypotheses testing		
	High Quality	Low Quality	Mean diffs.	*F*	sign.
Perceived usability	4.24 (0.54)	3.93 (0.71)	0.31	9.969	0.002***
Satisfaction with the website	3.92 (0.67)	3.25 (0.83)	0.67	34.671	0.000***
Online purchase intention	3.09 (0.96)	2.52 (0.91)	0.57	15.716	0.000***

Table 6. Hypotheses testing related to the image movement[1]

Dependent variable	Mean (stnd. deviation)		Hypotheses testing		
	Dynamic	Static	Mean diff.	*F*	sign.
Perceived usability	4.15 (0.61)	3.98 (0.69)	0.17	2.863	0.092*
Satisfaction with the website	3.70 (0.74)	3.40 (0.89)	0.3	7.101	0.008***
Online purchase intention	2.90 (0.94)	2.65 (0.99)	0.25	3.167	0.077*

for the dynamic images than for the static ones (see Table 6). Therefore, these results give support to the hypotheses H1c, H2c and H3c. Nevertheless, we did not observe significant interaction effects of the independent variables.

Finally, we performed a MANCOVA test in order to detect possible moderator effects of user experience and product familiarity. We classified participants into three categories for each moderator variable: Internet user experience and product type familiarity. Thus, we used low experienced users, medium experienced users and high experienced users with the Internet for the user experience covariate. Besides, we categorized the product type familiarity covariate as low product familiarized, medium product familiarized and high product familiarized. The results of the test were evidence for significant moderator effects. Thus, Internet user experience had a moderator effect on perceived usability ($F = 2.837; p < 0.1$) as well as on satisfaction with the website ($F = 2.805; p < 0.1$). In addition, product type familiarity moderated the perceptions about usability ($F = 13.904; p < 0.01$) and satisfaction with the website ($F = 6.142; p < 0.05$). No moderator effects were found for the online purchase intention. Therefore, we can partially support the hypotheses H4a and H4b.

DISCUSSION

The results of the experiment allow us to state that the picture quality affects in a very significant way the three variables proposed: usability, satisfaction and purchase intention. Specifically, the quality with which the image is launched on the computer screen turns into a key feature to achieve a positive response of the online consumer. This result is concurrent with previous findings in the specialized literature (e.g. Lee & Benbasat, 2003; Zhang and Myers, 2005). In this way, we can affirm that users penalize the bad quality of the images, perceive a lower degree of usability in the website, a lower satisfaction with the website and diminish their online purchase intention.

With reference to the movement of the images, we found also positive and significant effects on the three dependent variables of our study. The possibility of product rotation, which provides different perspectives, implies users to perceive a higher interactivity with the site, which helps to increase the perceived usability of the website. Besides, this interactivity provokes positive feelings in the consumer (Jiang & Benbasat, 2007; Park, Lennon & Stoel, 2005), making the navigation more pleasant and therefore increasing the satisfaction with the website. Moreover, the availability of a 360 degrees view allows users to identify and know the product in a better way, which facilitates the decision making process and enhances the intention to buy the product.

Regarding the moderator effects, we found that Internet user experience moderated the effects of the relationships proposed. This finding supports previous studies which suggest that the degree of experience with the Internet influences consumers' behaviour in the online environment (e.g. Frambach, Roest and Krishnan, 2007) In this way, the more experienced the users were the

Figure 2. Users' perceptions depending on Internet experience

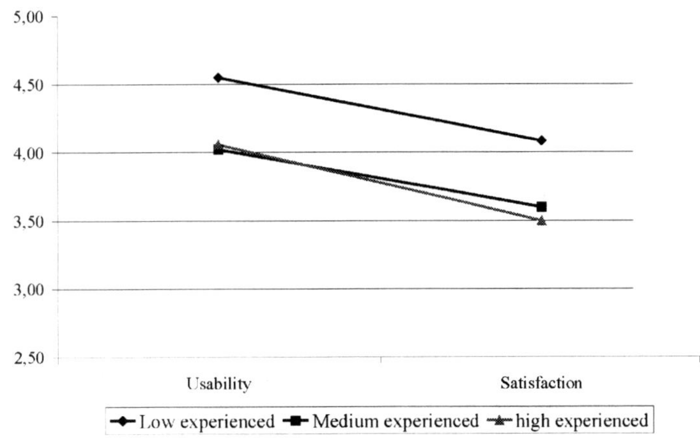

lower scores they rated for the perceived usability and satisfaction with the website (see Figure 2). Thus, as users become more experienced with the Internet, they become more demanding, since they could have a major knowledge about what they can find on the Internet and they could expect more and better features in a given website. In addition, the product familiarity also affected the intensity of the relationships analyzed. When participants were more familiarized with the type of product, they perceived higher levels of usability and satisfaction with the website. This result could be explained because of the information related to the product characteristics we provided in the websites. If the user was familiarized with this type of product, he noted that all information he needed was displayed on the screen, and therefore his perceptions positively increased (see Figure 3).

On the contrary, the effects of the picture size were only significant on the degree of user's satisfaction with the website. In this sense, large images help to enhance positive experiences browsing the website, evoking positive feeling during the navigation. If the user sees a product image with an adequate size in the same page where the product is presented, he does not have

Figure 3. Users' perceptions depending on product familiarity

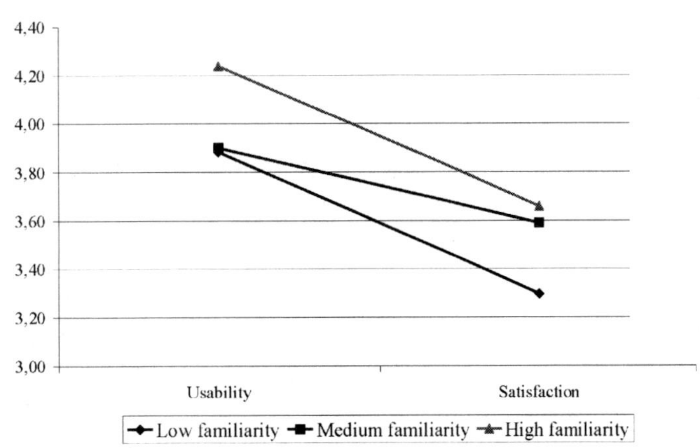

to do more clicks than necessaries to enlarge it and know the product better. This reduction of the required effort could lead to a higher level of satisfaction with the website.

However, launching big-size images did not affect the degree with which users perceived the website was ease to understand and manage, that is, perceived usability. This finding could be explained by the fact that users could consider other aspects more important, such as the location of the picture (Chau, Au & Tam, 2000; Hong, Thong & Tam, 2004). Besides, in the real world most of the websites offer the possibility of seeing the product pictures in different sizes, displaying a button link to enlarge o reduce that images. Because of that, the size of the initial picture that encloses the product is not especially valued. Besides, and in accordance with previous results (e.g., Park, Lennon & Stoel, 2005), online purchase intention was not affected by the size of the picture. This finding could indicate that, from a certain limit, the picture of the product offers the same information to the users, irrespective of its size.

Finally, the online purchase intention was not moderated by either the Internet user experience or product type familiarity. Though these variables affected significantly the users' perceptions, the purchase intention could be influenced by other variables or circumstances, such as the real necessity of buying that product or if it is appealing for the user.

CONCLUSION AND MANAGERIAL IMPLICATIONS

Visual aspects have been revealed as a key factor for the achievement of successfully interfaces in e-services. Because of that, this study has focused on analyzing the effects of the size, the quality and the movement of the visual presentation of the product on users' perceptions and behaviours in the website.

Thus, the literature review and the empirical tests allow us to state the importance of web design in order to get positive outcomes in the e-service context. Specifically, the findings of the experiment stress the importance that visual aspects have on online consumers, influencing their perceptions depending on the way the product is launched in the interface. In this way, this research demonstrates that the size, the quality and the movement of the product picture have an impact on the degree of perceived usability by the user, his satisfaction with the website and his online purchase intention. Moreover, the consumer's characteristics, such as the experience with the Internet or e-service and the product familiarity, moderate these relationships.

The results of our study have also important implications for people in charge of the design management of the websites and/or e-services. Thus, visual aspects should be considered when a website is being developed (Cao & Zhang, 2005; Dailey, 2004; Zhang and Myers, 2005). Designers not only have to take into account the costs associated to the implementation and maintenance of these visual elements, but also those derived from possible losses in the download and/or navigation speed because of the use of design tools. In this sense, many e-services websites tend to spend a great amount of resources in developing sophisticated modes of presenting their products irrespective of its utility or functionality for users (Helander and Khalid, 2000).

Anyhow, it seems logical to propose that, when products are presented in the interface, designers should consider the maximization of the quality of the images with respect to other features, since this attribute influences significantly the users' perceptions and behaviours towards a website. Given that online consumers only can examine the product in a virtual way, the fidelity with which they see the product has to match the reality as much as possible. In the same line, the possibility of movement affected users' perceptions in a great extent, since it allows users to interact with the product and know it better. Therefore, these two features have to be salient for bringing consumers near the real physical product.

Bearing in mind all above, it is possible to conclude that, when designers pursue to enhance the users' cognitive states in their websites, they must provide a comfortable atmosphere through the use of high quality images and the possibility of view the product from different views (Jiang & Benbasat, 2007; Park, Lennon & Stoel, 2005). This will provide users with high levels of interaction with the website and will increase their knowledge about the product, which will mean greater levels of perceived usability. Besides, to achieve great levels of users' satisfaction with the website, web designers must emphasize all the aspects related to the visual representation of the product, displaying pictures with an adequate size, excellent quality and with the possibility of interact with the product. Moreover, to encourage visitors to engage in an experience in the website with the final aim of turning them into clients, the quality and movement of the product images have to be considered, since the higher interactivity and the better knowledge will make the purchase decision easier.

Similarly, we did not observe any interaction effect among the variables manipulated in the experiment on the users' perceptions. Because of that, it is not possible to identify an optimal combination of the characteristics considered to achieve the best design. Thus, it is reasonable to state that there are different combinations which lead to an optimal visual design of the website, depending on the type of organization or the type of product offered by the firm.

In this way, we could consider the analysis of other variables related to the product presentation that could affect consumers, such as the way in which the information related to the product characteristics is presented on the computer screen. In the present study we have pointed out the importance of the visual aspects in the product presentation in the interface. However, the display of verbal information has to be considered in the same extent as visual information, since previous research has demonstrated that different modes of presenting text-based product information have different impacts on the consumers' perceptions (Flavián, Gurrea & Orús, 2008). Therefore, further research could examine the combined effects of visual and verbal information in order to find out an optimal website design.

Nevertheless, the results of this study should be taken with caution, since the sample profile limits the generalization of the findings. Although our sample characteristics are quite comparable to the general Internet users' profile, students represent a specific segment of online population and have some specific particularities that distinguish them from the rest of the potential base of online consumers. Besides, it should be noted that this study has examined only one type of product, a digital video camera. In this type of products, consumers may be guided more by utilitarian benefits, such as functionality and easy of use, than by other type of benefits related to hedonic cues. In this way, it would be interesting to analyse the effects of different product presentation modes on users' perceptions and behaviours with other type of products, where users could search for more hedonic benefits, such as pleasure, fun or playfulness. Furthermore, to overcome the limitations related to the sample profile and the product type, future research could be addressed to examine the online consumers' behaviours in a real shopping environment (Burke, 2002). Thus, an online survey could be convenient to analyse different scenarios with diverse types of consumers and multiple types of products.

Moreover, it could be interesting to consider other consumers' characteristics such as the degree of product involvement as a moderator factor in the relationships analyzed, since a product could be more appealing for some consumers than others, and therefore affecting their responses to visual stimulus.

REFERENCES

Anderson, E., & Sullivan, M. (1993). The antecedents and consequences of customer satisfactions for firms. *Management Science*, *12*(2), 125–143.

Ariely, D. (2000). Controlling information flow: on the importance of control in decision making and the value of information. *The Journal of Consumer Research*, *27*, 233–248. doi:10.1086/314322

Asociación Española de Comercio Electrónico y Marketing Relacional –AECEM-. (2008). Estudio sobre comercio electrónico b2c 2008. Retrieved November, 2008, from: http://www.red.es/media/2008-10/1224586013979.pdf

Asociación para la Investigación en Medios de Comunicación –AIMC-. (2008). Audiencia de Internet. abril-mayo 2008. Retrieved November, 2008, from: http://www.aimc.es/aimc.php

Burke, R. R. (2002). Technology and the customer interface: what consumers want in the physical and virtual store. *Journal of the Academy of Marketing Science*, *30*, 411–432. doi:10.1177/009207002236914

Cao, M., & Zhang, Q. (2005). Web site quality and usability in e-commerce. In Gao Y. (Eds) (2005): *Web System Design and Online Consumer Behavior* (pp. 107-124), Idea Group Publishing, Hershey.

Dailey, L. (2004). Navigational web atmospherics. Explaining the influence of restrictive navigation cues. *Journal of Business Research*, *57*(7), 1–9. doi:10.1016/S0148-2963(02)00364-8

Domain Tools. (2008). Domain counts & Internet statistics. Retrieved May, 2008, from: http://www.domaintools.com/internet-statistics/.

Donovan, R. J., Rossiter, J. R., Marcoolyn, G., & Nesdale, A. (1994). Store atmosphere and purchasing behaviour. *Journal of Retailing*, *70*(4), 283–294. doi:10.1016/0022-4359(94)90037-X

Emarketer (2008). European B2C e-commerce: spotlight on the UK, Germany and France. Retrieved March, 2008, from: http://www.emarketer.com/Report.aspx?code=emarketer_2000426.

Eroglu, S. A., Machleit, K. A., & Davis, L. M. (2001). Atmospheric Qualities of Online Retailing: a Conceptual Model and Implications. *Journal of Business Research. Special Issue on Retail Strategy and Consumer Decision Research*, *54*(2), 177–184.

eScope (2007). Overview of the European Internet landscape. Retrieved November, 2008, from: http://blog.boom.ro/documente/prezentari/EscapeReports.ppt#1

European Interactive Advertising Association. (2008). EIAA online shoppers 2008 – Report. Retrieved November, 2008, from http://www.eiaa.net/research/research-details.asp?SID=1&id=410&lang=6&origin=media-consumption.asp

Eurostat (2008). Internet as a data source. Retrieved November, 2008, from: http://epp.eurostat.ec.europa.eu/portal/page?_pageid=2973,64549069,2973_64554228&_dad=portal&_schema=PORTAL

Flavián, C., Guinalíu, M., & Gurrea, R. (2006). The Role Played by Perceived Usability, Satisfaction and Consumer Trust on Website Loyalty. *Information & Management*, *43*, 1–14. doi:10.1016/j.im.2005.01.002

Flavián, C., Gurrea, R., & Orús, C. (2008, July). *The relevance of visual aspects for the website store*. Paper presented at the 15th International Conference on Retailing and Consumer Services Science, Zagreb, Croatia.

Forbes, W., & Rothschild, M. (2000). *Toward an Understanding of Consumer Experience on the Internet: Implications for Website Design*. Paper presented at the 33rd Hawaii International Conference on Systems Sciences, Hawaii, Unitated States.

Frambach, R. T., Roest, H. C. A., & Krishnan, T. (2007). The impact of consumer internet experience on channel preference and usage intentions across the different stages of the buying process. *Journal of Interactive Marketing*, *21*(2), 26–41. doi:10.1002/dir.20079

Gefen, D. (2000). E-Commerce: The Role of Familiarity and Trust. *OMEGA: The International Journal of Management Science*, *28*, 725–737. doi:10.1016/S0305-0483(00)00021-9

Gurrea, R. (2006). *La Prensa Digital: Un Nuevo Escenario para el Análisis del Comportamiento del Lector de Periódicos*. Unpublished doctoral dissertation, University of Zaragoza, Spain.

Helander, M. G., & Khalid, H. M. (2000). Modelling the customer in electronic commerce. *Applied Ergonomics*, *31*, 609–619. doi:10.1016/S0003-6870(00)00035-1

Holbrook, M. B., & Hirschman, E. C. (1982). The experimental aspects of consumption: consumer fantasies, feelings, and fun. *The Journal of Consumer Research*, *9*, 325–345. doi:10.1086/208906

Hong, W., Thong, J. Y. L., & Tam, K. Y. (2004). Designing Product Listing Pages on E-Commerce Websites: An Examination of Presentation Mode and Information Format. *International Journal of Human-Computer Studies*, *61*, 481–503. doi:10.1016/j.ijhcs.2004.01.006

Internet World Stats. (2008). Home page. Retrieved May, 2008, from: www.internetworldstats.com.

Janda, S., Trocchia, P., & Gwinner, K. (2002). Consumer perceptions of Internet retail service quality. *International Journal of Service Industry Management*, *13*(5), 412–431. doi:10.1108/09564230210447913

Jarvenpaa, S. L., Tractinsky, N., & Vitale, M. (2000). Consumer trust in an Internet store. *Information Technology Management*, *1*(1), 45–71. doi:10.1023/A:1019104520776

Jiang, Z., & Benbasat, I. (2007). Investigating the influence of the functional mechanisms of online product presentations. *Information Systems Research*, *18*(4), 454–470. doi:10.1287/isre.1070.0124

Kim, E. B., & Eom, S. B. (2002). Designing Effective Cyber Store User Interface. *Industrial Management & Data Systems*, *102*(5), 241–251. doi:10.1108/02635570210428276

Kim, S., & Stoel, L. (2004). Apparel Retailers: Website Quality Dimensions and Satisfaction. *Journal of Retailing and Consumer Services*, *11*(2), 109–117. doi:10.1016/S0969-6989(03)00010-9

Lee, W., & Benbasat, I. (2003). Designing an Electronic Commerce Interface: Attention and Product Memory as Elicited by Web Design. *Electronic Commerce Research and Applications*, *2*, 240–253. doi:10.1016/S1567-4223(03)00026-7

Lian, J. W., & Lin, T. M. (2008). Effects of Consumer Characteristics on their Acceptance of Online Shopping: Comparisons among Different Product Types. *Computers in Human Behavior*, *24*, 48–65. doi:10.1016/j.chb.2007.01.002

Muylle, S., Moenaert, R., & Despotin, M. (2004). The Conceptualization and Empirical Validation of Web Site User Satisfaction. *Information & Management*, *41*, 543–560. doi:10.1016/S0378-7206(03)00089-2

Nielsen, J. (1994). *Usability Engineering*. San Francisco: Morgan Kaufmann.

Park, J., Lennon, S. J., & Stoel, L. (2005). On-Line Product Presentation: Effects on Mood, Perceived Risk, and Purchase Intention. *Psychology and Marketing*, *22*(9), 695–719. doi:10.1002/mar.20080

Ranganathan, C., & Ganapathy, S. (2002). Key Dimensions of B2C Web Sites. *Information & Management*, *39*, 457–465. doi:10.1016/S0378-7206(01)00112-4

Roy, M., Dewit, O., & Aubert, B. (2001). The impact of interface usability on trust in web retailers. *Internet Research: Electronic Networking Applications and Policy*, *11*(5), 388–398. doi:10.1108/10662240110410165

Smith, J., & Barclay, D. (1997). The effects of organizational differences and trust on the effectiveness of selling partner relationships. *Journal of Marketing*, *61*, 3–21. doi:10.2307/1252186

Swinyard, W. R. (1993). The effects of mood, involvement, and quality of store experience on shopping intentions. *The Journal of Consumer Research*, *20*, 271–280. doi:10.1086/209348

Tan, G. W., & Wei, K. K. (2006). An Empirical Study of Web Browsing Behaviour: Towards an Effective Website Design. *Electronic Commerce Research and Applications*, *5*, 261–271. doi:10.1016/j.elerap.2006.04.007

Then, N. K., & Delong, M. R. (1999). Apparel shopping on the web. *Journal of Family and Consumer Sciences*, *91*, 65–68.

U. S. Census Bureau. (2008). Quarterly retail e-commerce sales 4th quarter 2007. Retrieved February, 2008, from: http://www.census.gov/mrts/www/data/html/07Q4.html.

Van der Heijden, H., & Verhagen, T. (2003). Online Store Image: Conceptual Foundations and Empirical Measurement. *Information & Management*, *41*, 609–617. doi:10.1016/S0378-7206(03)00100-9

Vrechopoulos, A. P., & Siomkos, G. J. (2002). Store atmosphere in non-store retailing. *Journal of Internet Marketing, 3*(1), Retrieved May, 2008, from: http://www.arraydev.com/commerce/JIM/0207-03.htm.

Zhang, X., & Myers, M. (2005). Web design and e-commerce. In Gao, Y. (Ed.), *Web System Design and Online Consumer Behavior* (pp. 205–221). Hershey: Idea Group Publishing.

Zhou, K. Z., & Nakamoto, K. (2007). How do Enhanced and Unique Features Affect New Product Preference? The Moderating Role of Product Familiarity. *Journal of the Academy of Marketing Science*, *35*, 53–62. doi:10.1007/s11747-006-0011-3

ENDNOTE

1 * *p-value* < 0.1; ** *p-value* < 0.05; *** *p-value* < 0.01

APPENDIX

Table A1. Measurement Scales

Perceived Website Usability	
USAB1	In this website everything is easy to understand
USAB2	This website is simple to use
USAB3	It is easy to find the information I need from this website
USAB4	The structure and contents of this website are easy to understand (menus, display of information, etc.)
USAB5	It is easy to move within this website
Satisfaction With The Website	
SAT1	I am satisfied with the use of this website
SAT2	The experience that I have had with this website has been satisfactory
SAT3	In general terms, I am satisfied with the relationship with this website
Online Purchase Intention	
OPI1	If I had to buy a video-camera, I would probably use this website
OPI2	If I had to buy a video-camera, I would probably consider the purchase of a video-camera in Electroshop.com
OPI3	I will probably consider the purchase of a video-camera in Electroshop.com if I need this product
Product Familiarity	
FAM1	I am familiarized with this type of product
FAM2	I am familiarized with the characteristics of this type of product
FAM3	I am familiarized with the operation of this type of product
FAM4	I am regular user of this type of product

This work was previously published in International Journal of E-Services and Mobile Applications, Volume 1, Issue 3, edited by Ada Scupola, pp. 17-37, copyright 2009 by IGI Publishing (an imprint of IGI Global).

Chapter 9
A User Centered Innovation Approach Identifying Key User Values for the E-Newspaper

Carina Ihlström Eriksson
Halmstad University, Sweden

Jesper Svensson
Halmstad University, Sweden

ABSTRACT

We have studied the pre-adoption phase of the e-newspaper, i.e. a newspaper published with e-paper technology. The research question of this article is: In what way can a user centered innovation process contribute to identifying key values in mobile innovations? The aim of this article is threefold: firstly, to identify key values for the e-newspaper, secondly, to examine the intention to adopt a new possible innovation and thirdly, to explore user centered design processes ability to capture user values early in the innovation process. We have identified seven key user values: mobility, availability, personalization, usability, interactivity, readability and environment. Furthermore, we found that surprisingly many would consider exchanging their traditional newspaper with an e-newspaper in the future, indicating that the e-newspaper have a fair chance of becoming an important innovation. Finally, we propose a four phase model for identifying user values in a pre-adoption phase of an innovation.

INTRODUCTION

During the last decade we have experienced how the internet and mobile technologies have changed the way we live, work and communicate. This evolution has made it possible for us to have anytime, anywhere access to information and applications. There are three broad trends that are driving economy when it comes to mobile applications: changing customer priorities, new hardware/device innovation, and new infrastructure innovation (Kalakota & Robinson, 2001). The

DOI: 10.4018/978-1-60960-607-7.ch009

ongoing diffusion of personal computers, handheld devices, and mobile telephones (Lyytinen & Yoo, 2002) as well as the advent of new technologies such as the e-paper, enables new innovation possibilities in the mobile business (m-business) area.

However, despite all the innovation possibilities the adoption of mobile services/applications has not taken off as expected in Europe (Hammond, 2001; Carlsson *et al.*, 2006). A growing body of research into adoption of mobile services has sought to understand the reasons for this slow adoption rate (Sarker & Wells, 2003; Vrechopoulos *et al.*, 2003; Carlsson *et al.*, 2005; Mallat, 2006; Blechar *et al.*, 2006). Among the explanations as to why this adoption rate is low, usability factors, technological factors and business model factors are mentioned. As use patterns changes, new demands and expectations emerge, which lead to uncertainty about what people actually value and are willing to pay for (Tilson *et al.*, 2004)

To guarantee a successful m-business application the designers/managers need to have a complete understanding of the mobile business value including the experience of the user (Sims *et al.*, 2007). Kalakota and Robinson (2001) suggest several fundamental questions in elucidating customer experience, e.g. the relationship between customer experience and perceived value; what features is the customer looking for; and what is the customer willing to pay for the experience. The value-in use concept focuses on the experience perceived by a user interacting with products or services in use situations. According to this concept, the customers experience and perception are essential to be able to determinate user value (Kuusisto, 2008). Consumer value can also be defined in terms of the monetary sacrifice people are willing to make for a product (Boztepe, 2007). Indeed, identifying key user values early and emphasize these values in the innovation process could make the difference between a success and a failure.

Opening up innovation processes to involve users and other relevant stakeholders as innovators are recognized to be important to meet the user needs and leverage market value (von Hippel, 2005; Gassman, 2006; Lettl *et al.*, 2006). End users have specific domain knowledge that in many cases other stakeholders lack. Therefore, this group is in many cases an essential source for innovation. Involving users in the innovation process could increase the possibilities of identifying key values of products and services, which in turn enhance the odds to make it into the market and to be adopted by the users.

One example of an industry seeking new innovations is the newspaper industry. Lately, this industry has faced declining subscription and advertising revenues, forcing it to seek new alternative ways to profit and to attract new audiences (WAN, 2006; The Economist, 2006). The advent of e-paper technology has captured the interest of the industry (IFRA eNews initiative, 2006; The Swedish Newspaper Publishers' Association, 2007). The main advantage with e-paper technology is that it is reflecting, giving the same reader experience as paper and is thin, flexible and non-sensitive with WiFi connection, making it a device suitable for m-business applications. The potential replacement of the traditional newspaper with the e-newspaper, i.e. a newspaper published with e-paper technology, in the future would dramatically reduce production and distribution costs, making it an interesting prospect. Moreover, there is a belief within the industry that the e-newspaper could reach new audiences, e.g. young people.

The challenge for a successful launch lies within finding enough additional value for the users (readers) to choose an e-newspaper before the printed newspaper. Another challenge with the e-newspaper innovation is that a successful launch concerns both the adoption of new technology, i.e. the device, and the adoption of new services.

This article reports from two research projects (described in section 2) in which we have involved users in the process of designing and evaluating a future e-newspaper. The study focus on the pre-adoption phase, inspired by Karahanna *et al.* (1999), who argue the benefits of exploring this phase. The overall research question for this article is: *In what way can a user centered innovation process contribute to identifying key values in mobile innovations?*

The aim of this article is threefold: *firstly*, to identify key values for the e-newspaper, and *secondly*, to examine the intention to adopt a new possible innovation (i.e. the e-newspaper) and *thirdly*, to explore user centered design processes ability to capture user values early in the innovation process.

THE E-NEWSPAPER CONCEPT

One of the biggest advantages of using the e-paper technology is the readability of the display. Electronic paper (e-paper) is the common term for several different technologies that can be used to produce screens with a number of specific characteristics, of which many have properties that can be compared with print on paper. The technology works essentially in the same way, but instead of printing black dots on paper, the print is made electronically. The e-paper is reflecting, giving the same reader experience as paper (such as high contrast and the possibility to read in sunlight). Compared to transmissive displays such as LCD and TFT displays, e-paper has a reflective display technology, i.e. it has no back light. Transmissive displays that use backlight are in many cases almost unreadable in full sunlight. The e-paper is thin, flexible and non-sensitive. Many e-paper technologies have a high resolution that is comparable with the resolution used in traditional newspapers (170 dots per inch or more). This resolution is superior to the standard 72 dpi that is used on most of today's computer displays. The electronically made print also enables e-paper the same viewing angle as in any printed material. In addition, it does not require high battery performance – ultimately, the screen image is stable and fix even when there is no electrical voltage applied, power is only needed when the display is updated.

Today there exist a number of products using e-paper technology to support reading, so called eReaders. The Amazon Kindle [1] is Amazon's own product offering more than 225 000 books, top US and European newspapers and journals. It has a 6" diagonal E-Ink electronic paper display, 600 x 800 pixel resolution at 167 ppi with 4-level gray scale (Figure 1).

Other existing products with similar features are iRex iLiad [2], Sony Reader [3] and Bookeen Cybook [4]. Newspapers are also available in iRex iLiad, whereas the other products mostly focus on books.

The e-paper industry is expected to become big business, analysts from IDTechEX forecast for example that plastic electronics will be a $30 billion industry by 2015, and could reach as much as $250 billion by 2025 [5]. E-paper technology is also used in other products such as clocks and displays for mobile phones, Citizen, Seiko and Motorola have launched products based on e-paper displays.

An e-newspaper is a newspaper service published with e-paper technology (i.e. in an eReader), and is not to be mixed up with electronic versions of newspapers published online. The e-newspaper

Figure 1. Amazon Kindle[1]

is predicted to combine the readability and overview from the traditional newspaper with the possibilities of online media such as constant updates, interactivity and video (Ihlström *et al.*, 2004).

As online newspapers emerged in the mid-nineties many newspaper managers feared the possible cannibalization effect this could have on their main product, the traditional newspaper. But as time went by, the online newspaper genre has found its own content, form and functionality and became an important complement to the main product (Ihlström, 2004). The digitization of news media has led to the emergence of new opportunities but also to disruptive implications for traditional media (Christensen & Davis, 2006). Many newspaper organizations have met this changing media landscape by organizing themselves as media houses publishing in multiple channels such as print, online, PDAs, and mobile phones. The e-newspaper is believed to fit in with this mix of channels, bringing its own special features making it an complement at first, but is predicted to replace the printed edition in the long run due to the potential of the technology and the possibility of dramatically reducing printing and distribution costs (Ihlström 2005).

The Amazon Kindle 2.0 is about to be released. Amazons CEO Jeff Bezos have commented that because the Kindle allows you to buy texts wirelessly, it "*isn't a device, it's a service*". Editor Rosen at the journal New Atlantis meets this comment by writing:

The Kindle is a metaphor for the experience of reading in the twenty-first century. Like so many things we idolize today, it is extraordinarily convenient, technologically sophisticated, consumption-oriented, sterile, and distracting. The Kindle also encourages a kind of utopianism about instant gratification, and a confusion of needs and wants. Do we really need Dickens on demand? Part of the gratification for first readers of Dickens was rooted in the very anticipation they felt waiting for the next installment of his serialized novels ... They had time to connect to the story. (Shea, 2008).

We argue that an e-newspaper is a service delivered in a device, i.e. the eReader.

There is a need to differ between the device and the service provided when discussing the e-newspaper concept. There are several issues concerning the device, such as navigation and design of the hardware, whereas the service offered in the device has issues concerning navigation and design in the graphical user interface (GUI). There is also a conflict of interest between the producers of the eReaders and content providers (in this case the media houses), regarding e.g. business models and customer relations (Ihlström Eriksson *et al.*, 2008).

The empirical findings reported in this article are derived from two projects; DigiNews (ITEA 03015) and UbiMedia (Designing Ubiquitous Media Services through Action Research). The research started within the DigiNews project, which was a two year project including partners from Belgium, Spain, Netherlands, France and Sweden and consisted of several major technology firms, media houses and universities.

The DigiNews project aimed at defining, architecting and demonstrating a solution of a digital newspaper (i.e. the e-newspaper). The solution included all parts needed to produce, distribute and consume digital news, i.e. all steps from publisher to reader. The project also explored ways to maximize the chances of a successful introduction to the market; this included the creation of different business plans and strategies. To reach the project goal and to ensure acceptance of the proposed digital newspaper system by the various stakeholders, three points-of-views were taken: user, business, and technology. Our part of the project mainly concerned the user and business views, and this article is concentrated around the user view.

The DigiNews project had the ambition to create an e-newspaper acceptable to a wide audience, by exploiting the potential of this digital media. Since the e-paper terminal used in the project had a much smaller display than the traditional printed newspaper, one challenge was to make this smaller format acceptable to the reader and to provide a user experience that was similar to reading a printed newspaper. Another challenge was to create a user interface that was easy to use, not only by computer-experienced people, but for everybody. Our part in this project was to work with design and business related aspects. We mainly worked with seven Swedish newspaper partners, as well as with users/readers and advertisers. All our results derived from this project have been reported to the other project partners. The Swedish Newspaper Publishers' Association, who was also a project partner, was active in negotiating our findings to iRex during their development phase.

After the DigiNews project ended in mid-year 2006, the research continued with the real life tests using iRex iLiads (described in the research design section) within the UbiMedia project, which was a Swedish project with partners from 9 Swedish newspapers, the Swedish Newspaper Publishers′ Association and Stampen (a parent company for several newspapers, printing houses and distribution companies). This two-year project, which ended in mid-year 2008, targeted the challenge of designing ubiquitous media services for a multitude of devices and contexts to be consumed anytime and anywhere.

There are no exact figures available about the market of e-newspapers today. Le Monde, one of the newspaper partners in the DigiNews project published their newspaper in the Amazon Kindle initially but is no longer available in that device. None of the DigiNews partners choose to publish on the iRex iLiad, and the Swedish newspaper partners are still waiting for a better eReader with color capabilities to enter the market. However, there are many things happening in this industry today. For example, the newcomer Plastic Logic has announced the first content partners for their Plastic Logic Reader of which you can find the Financial Times and USA Today (Davis, 2009). The Plastic Logic Reader has a large e-paper display and unconfirmed wireless connectivity, in a device measuring around 8.5 x 11 inches. It's expected to go into trials and previews during the second half of 2009, with a full commercial launch following in 2010. The ideal format, a flexible display that could be rolled or folded like a newspaper, is still years off, according to E Ink (Taub, 2009), but they foresee color displays with moving images and interactive clickable advertising coming in only a few more years. Fujitsu has announced a full color e-ink display that is three times bigger than the Kindle (Eaton, 2009). The size, foldability and color possibilities may indeed influence the e-newspaper take off.

RELATED RESEARCH

In recent years the trends of user centered (von Hippel, 2005) and co-creative (Sawhney *et al.*, 2005) innovation processes has emerged. The concept of user centered innovation ranges from innovations created by the user to involvement of users in the innovation processes (von Hippel, 2005). Even if the customer always are important in a business setting, these new trends focus on a systematic approach in identifying customer needs where skills from social sciences are applied in the mapping of customer experiences and customer needs (Rosted, 2005).

Technology-driven innovation and user centered innovation often complements each other. However, the competences needed differ as well as the resources. User centered innovation requires resources for analyzing and assessing customer needs, while technology-driven innovation will require resources for new technology development (Rosted, 2005).

User centered innovation implies that companies are striving to deliver innovations that generate user value or user experience unmatched by the competitors. This requires a profound understanding of user needs and user values as well as the ability to translate these into unique products and experiences (von Hippel, 2005).

There is a growing recognition that providing superior value for users is a key aspect for business success (Boztepe, 2007). One way to mitigate competition and open up entirely new markets is by focusing on creating advances in customer value (Kim & Mauborgne 2005). One key attribute that distinguishes breakthrough products from their closest followers, is according to Cagan and Vogel (2002), the significant value they provide for users.

According to Kuusisto (2008) the concept of value adding services or products imply that value is contained in the product or the service. The value is created and offered by the producer. Another perspective is the value-in use concept that focuses on the experience perceived by a user interacting with products or services in use situations. This concept implies that the customer is always a co-creator of the value. According to this concept, the customers experience and perception are essential to be able to determinate user value (Kuusisto, 2008).

Consumer value can also be defined in terms of the monetary sacrifice people are willing to make for a product. The primary focus here is on the point of exchange where money is seen as an index of value. According to this perspective, the assumption is that at the moment of purchase, the consumer makes a calculation and evaluation of what is given (value) in respect to what is taken in terms of money (Boztepe, 2007). Consumer value and consumer needs are also important aspects of adoption and diffusion theory.

Technology diffusion and adoption has been a key area of research in the IS discipline since the influential work of Tornatzky and Klein (1982) and Davis (1989), and research has increased massively ever since and is today a major topic in m-business research.

Technology diffusion can be categorized based on supply-push and demand-pull theories which display the interaction between technology and the adopting unit (Damsgaard & Lyytinen, 1996). Supply-push theories assume that the diffusion of a technology is based on its functionality. The adoption decision is then a rational problem choosing between an existing and a new technology. The supplier of the technology tries to make the adoption choice easier for the adopting unit by providing a technological problem-solver (Tscherning & Damsgaard, 2008).

Demand-pull theories on the other hand are based on the users' rational choice (Rogers, 1995). Based on this, companies should work on topics which are perceived as problems on the demand side. This would increase the probability of a technology being adopted and diffused by forming it against the needs of the adopting unit (Tscherning & Damsgaard, 2008).

Some of the most well-cited theories and models for adoption are the theory of Diffusion of Innovations (Rogers, 1995), the Technology Acceptance Model (Davis, 1989) and the Unified Theory of Acceptance and Use of Technology (Venkatesh *et al.*, 2003). There are several studies that have applied these frameworks on how mobile services are adopted. Among others, studies on drivers of mobile service adoption (Anckar and DÍncau, 2002), factors influencing adoption (Khalifa and Cheng, 2002; Mallenius *et al.*, 2007; Damsgaard & Kelleher, 2007), patterns of adoption (Carlsson *et al.*, 2005; Malhotra and Segars, 2005), and attitudes towards using mobile services (Knutsen, 2005).

Karahanna *et al.* (1999) makes a distinction between the pre-adoption and post-adoption phase of an innovation and found differences in the attitudes towards the innovation in the two stages, e.g. they found that the pre-adopters had a much wider array of perceived characteristics influencing their intention to adopt the technology. They argue for the importance of further research concerning the distinction between these two phases. Davis and Venkatesh (2004) have also investigated user acceptance early in the development process by using TAM on product concepts with the help of mock-ups, suggesting that mock-ups can successfully be used to assess the usefulness of an indented system in a pre-prototype phase.

RESEARCH DESIGN

This research is an example of a multi-method approach (Mingers & Gill, 1997; Mingers, 2001). The key advantages of the multi-method approach are:

Firstly, "the use of multiple research methods is especially important in exploratory and explanatory studies. When knowledge is crude and limited [...], multiple methods permit a wider and more complete understanding of the phenomenon studied" (Pinsonneault & Kraemer, 1993, p. 88). Exploratory studies are appropriate "in new fields of study where little work has been done, [...] and little is known about the nature of the phenomenon" (Patton, 2002).

Secondly, "different research methods focus on different aspects of reality and therefore a richer understanding of a research topic will be gained by combining several methods together in a single piece of research" (Mingers, 2001, p. 241).

Other arguments in favor of a multi-method approach are e.g. "a research study is not usually a single, discrete event but a process that typically proceeds through a number of phases" (Mingers, 2001, p. 243).

The study presented in this article concerns an exploratory approach to user centered innovation processes with the aim to identify key user values relevant for the innovation adoption, and reports from four phases of the innovation process, during 2005 and 2006. The four phases in our research design includes multiple data collection activities which also are in line with the recommendations for m-commerce research by Lehner and Watson (2001). The design phase resulted in three e-newspaper prototypes that were tested and used for input for the discussions of the e-newspaper concept in the second phase. The results from the second phase formed the base for the validation phase and also informed the real life test. The validation phase was deepened in the real life test to get a rich picture of key user values.

Data Collection Activities

The data collection activities were conducted according to the four phases described above, i.e. the design phase, the prototype test, the validating and the real life test (Figure 2).

In *the design phase* we started with 8 initial three hour workshops, four with readers (23 persons) and four with newspaper designers (21 persons). The readers were reached by an advertisement in four online newspapers; Sydsvenskan, Göteborgs-Posten, Hallandsposten and Sundsvalls Tidning. The aim of the selection was to get a representative group of readers as well as a dynamic group for the workshop. It was an even distribution concerning gender and the distribution of age was in the interval between 20 to 70 years. These workshops were carried out as future workshops (Jungk & Müllert, 1996; Kensing & Halskov Madsen, 1991). We modified the original future workshop technique into a shorter more

Figure 2. Data collection activities

Phase	Year	No. of respondents	Duration	Design			
Design	2004-2005	54	1 year	8 initial workshops followed by 10 focus groups meetings			
Prototype testing	2005	36	2 hours	test 75 minutes	structured interview 45 minutes	questionnaire	
Validating	2006	3757	5 days	online questionnaire with 127 questions in total			
Real life evaluation	2006	12	2 weeks	questionnaire	daily activities with online diary, specific questions and critical incidents	semi-structured interview 90 minutes	questionnaire

focused version (Ihlström *et al.*, 2005), with the aim to find opportunities and problems related to the e-newspaper. Based on the opportunities identified, the participants created scenarios and mock-ups to support both design and requirement specification. The workshops were all audio recorded and parts were also video recorded.

To continue the involvement from the users and to receive the publishers domain knowledge two different focus groups were initiated; one with users and one with newspaper designers. The *user focus group* consisted of two women and four men in the age of 29 to 82 years that had answered the call for interest in the local newspaper (Hallandsposten). The participants had different education and all of them were working except one, who was retired. Five three hour sessions with this user focus group were conducted during the first year of the project. The *design focus group* consisted of four newspaper designers from Swedish newspapers (Aftonbladet. Göteborgs Posten, Norrköpings Tidningar and Sundsvalls Tidning). Five sessions, ranging from 3-8 hours, with this design focus group were held to discuss and create design ideas. In these sessions the prototypes were developed iteratively and input from the user focus group was discussed. The three prototypes illustrated e-newspapers of Aftonbladet, a national evening press (Figure 3), Göteborgs-Posten, a major regional newspaper (Figure 4), and Sundsvalls Tidning, a local morning paper (Figure 5). All focus group meetings were audio recorded.

The *prototype testing phase* was conducted with 36 users in the autumn of 2005. Each respondent tested two different e-newspaper prototypes. The selection of respondents was made to secure a diversity of gender and age. There was an even gender selection, six respondents were in the ages between 15-24, 12 between 25-44, 12 between 45-64 and six respondents were over 65 years.

The prototype test was designed to test different design solutions as well as to investigate user attitudes towards the e-newspaper concept and consisted of a 75 minutes test, a 45 minutes

Figure 3. The Aftonbladet prototype in A4 format

KÖP TV-TIDNINGEN I DAG

AFTONBLADET

TORSDAG 16 JUNI 2005

AFTONBLADET AVSLÖJAR

JÄTTE-BLUFF MED KÖTT

Så luras du köpa utländskt

LIZA MARKLUND: LEDARNA FÖR ROKS har varit galna i tio år

FREDRIK VIRTANEN: SOSSAR TAR DÖD på radions klassiker

KARIN AHLBORG: DUBBELT SÅ DYRT att köpa svensk mat

Kylie: "Jag har bröst-cancer"

Möt dom i film och bildspel

Zlatan: "Jag kan mycket bättre"

← Meny Innehåll Nyheter →

Figure 4. The Göteborgs-Posten prototype in A5 format

Göteborgs-Posten

Fredrik tror på jobb i Sverige

Bevakad dyrgrip på museet

Kristine, årets värvning

TORSDAG 14 APRIL 2005

Charterplan i flygdrama

Stigsson anklagas för fler brott

Vaccin mot ärftlig bröstcancer testas

Ny metod kan användas inom fem år

Sävehof klart för SM-final

Föräldrar lär sig sätta gränser

i stället

INDEX

Menu

Index

Hide

CLAS OHLSON

ALLTID AKTUELLA ERBJUDANDEN - KLICKA HÄR

Figure 5. The Sundsvalls Tidning prototype in A5 format

structured interview, and was ended with a questionnaire. As the e-paper technology was not available at this time in the project, we chose to test the prototypes on tablet PCs, which enabled the use of color and multimedia not possible in the real life test where actual e-paper devices were used.

The respondents were asked to think aloud (Dumas, 2003) during the *tests*, which were both audio and video recorded. At each test at least two members of the research team were present, of which one functioned as test leader while test protocols were filled out by another member of the research team. The test itself mainly focused on navigational and design related issues which have been reported in Svensson and Hakeröd (2006) and these results are not included in this article. However, the test itself made the respondents familiar with the prototypes and the e-newspaper concept, which was essential for the following interview and the questionnaire.

A structured *interview* approach (Patton, 2002) with 47 questions divided into five themes was used. The structured approach was chosen to ensure answers to all questions by all test subjects, although giving them the opportunity of elaborating on specific questions. The themes were: a) user experience, b) e-newspaper versus other media, c) preferred e-newspaper services, d) willingness to pay/ business models, and e) advertisement preferences. At each interview two researchers were present, one leading the interview and one taking notes. The interviews were all audio recorded. The *questionnaire* after the interview consisted of 19 questions. These questions mostly concerned adoption, preferences and media habits.

In the *validation phase* we chose to use online questionnaires because that allowed us to provide information for the respondents to obtain an understanding of the e-newspaper concept. Since we assumed that the e-newspaper concept was unknown to most of the potential respondents we provided them with the possibilities to:

- read more about e-paper technology,
- watch three concept videos, and
- explore the corresponding e-newspaper prototypes.

The concept videos available in conjunction to the questionnaires illustrated the potential of the e-newspaper by focusing on the benefits for three different personas: the business women, the student and the senior citizen (identified as potential target groups in the DigiNews project).

Another reason for choosing online questionnaires was that web samples can be as representative, or even more representative than traditionally collected samples because of the heterogeneity of the online population (Buchanan and Smith, 1999). Admittedly, there are inherent problems in controlling who responds to online questionnaires. Control for cases with multiple submis-

Figure 6. Newspapers hosts for questionnaires and number of respondents

Newspaper	URL	Unique visitors per day	No. of respondents
Aftonbladet	www.aftonbladet.se	1.200.000	3757
Göteborgs-Posten	www.gp.se	41.500	135
Sundsvalls Tidning	www.st.nu	14.500	447

sions from the same IP number was excluded in the data collection.

We presented the questionnaire at the news sites of the three Swedish newspapers that we had collaborated with in developing the e-newspaper prototypes (Figure 6), i.e. Aftonbladet, Göteborgs-Posten and Sundsvalls Tidning. Aftonbladet is a tabloid with the most visited news site in Sweden, Göteborgs-Posten is a regional morning paper covering Göteborg (the second largest city in Sweden) and its surroundings, and Sundsvalls Tidning is a local morning paper in the north of Sweden. The questionnaire was available at each site for about five days in April 2006.

In total, 4339 respondents answered the questionnaire. The respondents that had given an age under 15, those who did not complete or answered the questions included in this study contradictorily were excluded from the data set, resulting in a dataset containing 3626 respondents (Figure 7).

The questionnaire consisted of 127 questions in total, divided in 4 different parts, i.e. background questions, business models, e-newspaper preferences and preferences of mobile services. Many of the questions where of multiple choice nature, however, several open questions were included as well.

In *the real life evaluation phase*, the evaluation of the e-newspaper concept was conducted with 10 families who tested an early version of an e-newspaper published on an actual e-paper device (in this case the iRex iLiad) in real-life settings over a two week period in the autumn of 2006. The iRex iLiad has an 8.1-inch display with a resolution of 768x1024 pixels (160 dpi) with 16 levels of grayscale and supports WLAN and LAN connections.

The e-newspaper of Sundsvalls Tidning (Figure 8) was published twice daily, at 6 pm and 1 am, and was downloadable via Internet. The respondents were foremost selected to represent

Figure 7. Demographic and background data of the questionnaire respondents

<table>
<tr><th colspan="6">Demographic data</th></tr>
<tr><td colspan="3"></td><td>All</td><td>Men</td><td>Women</td></tr>
<tr><td colspan="3">No of</td><td>3626</td><td>2216</td><td>1410</td></tr>
<tr><td colspan="3">%</td><td>100</td><td>61,1</td><td>38,9</td></tr>
<tr><td colspan="3">Mid age</td><td>37,1</td><td>37,9</td><td>35,7</td></tr>
<tr><th colspan="6">Background data in percentage of total data set</th></tr>
<tr><td rowspan="7">Occupation</td><td>Full time</td><td>55</td><td rowspan="3">Income</td><td>Low</td><td>30</td></tr>
<tr><td>Part time</td><td>7,7</td><td>Medium</td><td>41,9</td></tr>
<tr><td>Unemployed</td><td>5,8</td><td>High</td><td>28,1</td></tr>
<tr><td>Senior citizen</td><td>6,3</td><td rowspan="4">Education</td><td>Elementary</td><td>9,1</td></tr>
<tr><td>Student</td><td>19</td><td>Grammar</td><td>45,5</td></tr>
<tr><td>Sick leave</td><td>3,6</td><td>University</td><td>44</td></tr>
<tr><td>Other</td><td>2,6</td><td>Other</td><td>1,5</td></tr>
<tr><td></td><td>Newspaper subscriber</td><td colspan="2">Read online news</td><td>Possession of mobile phone</td><td>Use mobile services</td></tr>
<tr><td>Yes</td><td>48,8</td><td colspan="2">99,4</td><td>97,5</td><td>53,8</td></tr>
<tr><td>No</td><td>51,2</td><td colspan="2">0,6</td><td>2,5</td><td>46,2</td></tr>
</table>

Figure 8. E-newspaper prototype in iRex iLiad

different types of households such as singles, couples, families with children, and senior citizens, to secure different use patterns, but we also tried to get differences in gender, ages, occupation and education. In two of the families both adults participated in the evaluation, resulting in a total of 12 respondents, consisting of three females and nine males in the age between 21 and 69.

The real life test was designed to more specifically explore pros and cons with an early version of an e-newspaper, as well as the users' preferences and intentions to adopt. This real life test consisted of a two week evaluation, two questionnaires (one before and one after the test), daily online activities (diary, questions and critical incidents), and a 90 minutes semi-structured interview.

The two-week evaluation started with a meeting in Sundsvall, where the respondents were introduced to the device, and got a *questionnaire* about their media and reading habits. The test design was explained and they signed an agreement, assuring their active participation through *daily online activities*, e.g. writing a diary, answering three specific questions and reporting critical incidents daily on the test web site. The answers to the questionnaire and the daily activities were all stored in a database.

After two weeks of e-newspaper use, the respondents were visited in their homes (Figure 9) for a 90 minute interview about their experiences and preferences of the e-newspaper. A semi-structured *interview* approach (Patton, 2002), with an interview guide consisting of 5 themes were used. The themes corresponded to the themes in the first evaluation. At this point we chose to use a semi-structured interview approach due to the nature of the test. The respondents had evaluated the e-newspaper concept for two weeks and were likely to have more opinions on the subject. We used an interview guide to facilitate the consistency of data collected between sites and interviewees. While allowing individual perspectives and experiences to emerge, the interview guide provided a systematic way of delimiting issues discussed in the interview (Patton, 2002). Our set-up for the interview, i.e. one researcher asking the questions and the others listening and taking notes, allowed for a detailed discussion within the research team after each interview. The interviews were all audio recorded.

Finally, the respondents received a questionnaire consisting of 14 questions which matched the questionnaire of the prototype test. This data was handled in the same way as in the test.

Data Analysis

The data from the first part of the *design phase* was used as input to the second part. For example, the produced mock-ups were studied and analyzed twice. The first analysis was done together with the design focus group where the mock-ups were studied together with the video presentations that explained the mock-ups. The primary goal of this analysis was to identify new innovative ideas.

Figure 9. Home setting

In the second analysis of the mock-ups, done by master students and researchers, the different design elements and navigational support where documented and a general list of different design elements where produced. This data functioned as a base for the design of the prototypes. We chose to make the three prototypes different in terms of navigational support and design elements in order to find out which worked the best.

The data from the *prototype testing phase* was analyzed as follows. The protocols and recordings from the tests have not been used in this study as described earlier. The structured interview recordings were stored in a database, question by question, and were thereafter transcribed by members of the research team. A coding technique (Miles & Huberman, 1994) was used when analyzing the answers of each question separately, where similar answers were clustered into categories. The answers from the questionnaire were also added to the database described above. Only mean scores were derived from this material to be used for comparison to the results of the other phases.

The *validation phase* results were analyzed using SPSS v16.0. Some of the questions were statements with a 7-grade Lickert scale and others were multiple choice questions. The analysis focused on calculation of mean scores and standard deviations for each statement and on frequencies and percentages for multiple choice questions. Comparisons of the results by gender, age and education were also done.

The transcriptions (~310 pages) from the interviews in the *real life evaluation phase* were transcribed by a professional audio typist and were later checked and corrected by the authors. In order to categorize the data collected, patterns were identified in the transcribed material (Miles & Huberman, 1994). Excerpts in the transcribed material were marked with assigned colors, facilitating data categorization according to corresponding themes. Within these themes, we scanned the data material for similarities and differences, providing input for re-examining the initial themes. Key quotes from different empirical activities are used in the article to illustrate our reflections of significant aspects of the themes.

The results from the questionnaire were used for comparisons with results from the other phases using mean scores.

EMPIRICAL FINDINGS

In this section we present the result corresponding to the three aims of the study. First we present findings concerning key user values, second we present findings regarding the respondents' intention to adopt and third we present our reflections on the user centered innovation process.

Key User Values

During the analysis of the empirical data seven key user values emerged; mobility, availability, personalization, usability, interactivity, readability and environment. We have chosen to structure this sub-section according to these key values, and present data from all relevant phases in conjunction to each key value. The key value can be seen as an umbrella term representing several values. In Figure 10, we present these values according to when they first appeared in the empirical data.

Mobility

We have identified "localization" and "portability" that together form the key user value "Mobility".

"Localization": In the design phase the possibility of geo-localized services was one of the topics discussed in the workshops. Ads and offers based on where you are positioned was one way of leveraging the right ads at the right time and situation according to many workshop participants. One common scenario generated by the participants was the case when a traveler or an individual living abroad during the winter month receives his/her local news via the e-newspaper.

In the prototype testing phase the value regarding localization was mentioned by several respondents who talked about being able to get news and content anywhere you like, e.g. possibilities of receiving local news on vacation, or relevant ads from the location you find yourself in. This is illustrated by a quotation from one of the respondents:

I see it as an advantage to bring it with me when I am travelling to the US or somewhere else, and have the possibility of receiving Göteborgs-Posten when it is fresh.

Figure 10. Key user values

Key user values	Design phase	Prototype testing phase	Real life evaluation phase
Mobility	Localization	Portability	
Availability	Accesibility Updates		
Personalization	Layout Content		
Usability	Navigation		Ease of use Efficiency
Interactivity	Functionality	Communication	
Readability	Format		Comfortability
Environment	Less trash	Less energy	

In the real life evaluation phase localization services where not incorporated in the e-newspaper. In one way the news was based on localization just as a local newspaper is, but geo-localized services were not incorporated and therefore the localization aspects where not dynamic. A few respondents stated that they would appreciate hyper local news about their own neighborhood, other than that, localization where not a frequent subject in the interviews.

There was only one respondent that traveled abroad and brought the e-newspaper with him. This respondent had hoped to be able to read his local news during his visit in Belgium:

*I brought it with me when I went away to Brussels and thought it would be neat to sit at the hotel in the morning and read my local newspaper, but then... then there was trouble logging on to the internet... so in the end I had to read old newspaper at my hotel *laughter*.*

"Portability": In the prototype testing phase many of the respondents mentioned aspects of portability, e.g. the fact that you can bring the e-newspaper with you and still get updated news when you are on the move. The possibility to bring the e-newspaper with you while commuting or during travels or visit abroad was often mentioned as an important aspect. This also concerns that the e-reading device is mobile and has a smaller format compared to laptops and tabloid newspaper. As one of the respondents said:

I think that when I travel by bus or train to work and I have my bag and a thing like this in a suitable size, it would be an advantage. And I avoid the printing ink.

In the real life evaluation phase the most common mentioned problem that the e-newspaper solves regarding the consumption of news is that you can bring the newspaper with you when you travel and still get local and updated news every day. Overall the respondents thought the size of the eReader was acceptable and some of the respondents read the news while they commuted on the bus. Some thought it was too small, while two respondents thought the eReader could be even smaller:

it doesn't fit your pocket, I mean, a newspaper you whirl and put in your pocket... you need a more flexible screen, that would have been excellent....

However, more than half of the respondents felt the size was quite perfect and that this smaller size enabled portability, as illustrated by one of the respondents:

I liked this format, even if it initially seemed a little small, it was quite good... I had it with me on the airplane and on the bus, and then this size was perfect.

In the validating phase values connecting to mobility was rated as rather important with means ranging from 4 to almost 6 (Figure 11). There were only minor differences between gender, age and education. For example, younger people thought it was more important with access to news while travelling than older people.

Availability

We have identified "accessibility" and "updates" that together form the key user value "Availability".

"Accessibility": In the design phase there were several discussions relating to accessibility of news. When a newspaper reader does not receive his/her newspaper in the morning or when it is delayed, it seems to generate a lot of negative emotions. Therefore one of the most important values with the e-newspaper were the possibility to always have access to the newspaper, even though some identified problems with the e-newspaper

Figure 11. Mobility issues

Mobility		Total	Gender		Age				Education			
			Male	*Female*	*15-24*	*25-44*	*45-64*	*65-*	*Elementary*	*Grammar*	*University*	*Other*
I want to have access to news while traveling or commuting												
7-grade Likert scale where 1 represented not interesting and 7 very interesting	Mean	4,81	4,87	4,56	5,15	4,85	4,32	3,98	4,55	4,68	4,89	4,11
N=3560	Std. Deviation	2,035	2,067	2,109	1,896	2,048	2,213	2,163	2,198	2,107	2,038	2,107
Mobile news offer the most value while I travel												
7-grade Likert scale where 1 represented do not agree and 7 agree	Mean	4,89	4,97	4,71	4,75	4,94	4,9	5,04	4,62	4,88	4,96	4,53
N=1356	Std. Deviation	1,866	1,811	1,981	1,866	1,859	1,875	1,8	2,105	1,931	1,758	2,348
I find it important to have news adapted to where I am												
7-grade Likert scale where 1 represented do not agree and 7 agree	Mean	4,14	4,09	4,27	4,23	4,16	4,03	4,15	4,36	4,16	4,09	4
N=1352	Std. Deviation	1,764	1,744	1,806	1,703	1,762	1,819	1,848	1,882	1,777	1,731	1,751
I find it valuable when content is adapted to the place where I am												
7-grade Likert scale where 1 represented do not agree and 7 agree	Mean	4,07	4,04	4,13	4,12	4,09	3,97	4,04	3,82	4,13	4,04	4,29
N=1342	Std. Deviation	1,695	1,7	1,683	1,68	1,686	1,747	1,509	1,708	1,731	1,654	1,724
To be able to recieve the newspaper anywhere												
7-grade Likert scale where 1 represented not important and 7 very important	Mean	5,81	5,85	5,74	5,72	5,94	5,66	5,54	5,75	5,76	5,88	5,29
N=2947	Std. Deviation	1,75	1,734	1,774	1,771	1,649	1,874	1,924	1,906	1,776	1,685	1,901

as well, such as power shortage or problems with the digital distribution.

In the prototype testing phase getting the news and content delivered to your eReader were also considered as important by many respondents, e.g. not missing out on your morning paper due to weather conditions etc. Other mentioned the importance of getting your news fast. One respondent illustrated this by:

The accessibility is important, I believe that more persons would read the news if they could have them delivered directly to their living rooms.

In the real life evaluation phase many respondents highlighted that they felt really annoyed when the traditional newspaper get delayed or do not arrive at all in the morning due to for example distribution or printing problems. The accessibility of the e-newspaper felt better, even though problems with the digital distribution (e.g. internet connection problems) are of course possible. Several respondents had minor problems updating their newspaper at home and the importance of accessibility seems evident in quotes such as:

Due to all the hassle to start and update the e-newspaper you needed to plan for your newspaper reading and start everything up a couple of minutes before you wanted to read... you don't want to do other stuff and go around waiting for the newspaper to be ready for you to read... it is like you need to put on the television 10 minutes before you start watching.

Another value with the e-newspaper relating to accessibility that was commonly mentioned is connected to not having to go out in the morning and fetch the newspaper. One user stated for example:

*... the absolutely best thing with this is not the reading itself, but the fact that I don't need to go down to my mailbox 100 meters from here *laughter*... this is the one main benefit for me at last, that you don't need to go out in the morning when it is cold and the weather is miserable and so on.*

"Updates": In the design phase one of the opportunities with the e-newspaper that the workshops participants recognized was the possibility to receive updated content similar to the online newspaper. There was no consensus if the updated content should be pushed out continuously or in editions. Some participants wanted to always have updated news, others wanted to be able to finish their newspaper just like a traditional newspaper and instead get updated editions during the day.

In the prototype testing phase many respondents commented on the value of updates. But also in this phase the opinions differed regarding how this should be done. Several respondents stated that they would like to be able to finish the newspaper before it changed with updates. Some respondents suggested a special section for updated news. But many respondents found the possibility of getting updates during the day as essential.

In the real life evaluation phase many respondents really appreciated the updated editions of the newspaper, a couple of them read the morning news late in the evening when the late edition where distributed to the e-newspaper. However, the preference about how the newspaper should be updated differs between different respondents. Some of the respondents felt stressed by constantly updated news and really appreciated that you where able to "finish" the newspaper without something changing during your time reading it. One user stated for example that:

*... if there would not be editions it would be more like... like a online newspaper because it would change constantly... then it feels like you need to read it all the time, you get a beep when something new arrives and then you need... now I get it in a package and need to chill down *laughter*, it won't change until next time that I want it to.*

However, the same respondent says later on in the interview that during certain circumstances you really want to have constant updates:

normally I would update it once a day and that would be in the morning... But suppose something happens, like 9/11 or a storm or something, then you need to....

Some of the respondents would like to have constantly updated news and did comparisons to the web. Finally, most of the respondents would like a special section that where updated so you could easily check what news that where new since you last read your edition of the e-newspaper.

In the validating phase questions about availability received a relative high mean score (Figure 12) which indicates the importance of the key value "availability". There were no obvious differences between gender, age and education.

Personalization

We have identified "layout" and "content" that together form the key user value "Personalization".

"Layout": In the design phase there were some discussions about personalization of the layout of the newspaper. Some individuals sought after customization options regarding the layout and appreciated their personal layout of the e-newspaper. However, a majority of the participants did not see this as an important aspect of the e-newspaper

Figure 12. Availability issues

Availability		Total	Gender		Age				Education			
			Male	*Female*	*15-24*	*25-44*	*45-64*	*65-*	*Elementary*	*Grammar*	*University*	*Other*
I want to have access to news the whole day												
7-grade Likert scale where 1 represented not interesting and 7 very interesting	Mean	5,77	5,75	5,63	5,75	5,77	5,56	5,6	5,55	5,7	5,75	5,61
N=3508	Std. Deviation	1,576	1,67	1,717	1,65	1,627	1,827	1,718	1,829	1,683	1,668	1,626
I want to have access to multiple newssources on the same time												
7-grade Likert scale where 1 represented not interesting and 7 very interesting	Mean	4,65	4,58	4,61	4,44	4,67	4,55	4,62	4,78	4,55	4,59	4,69
N=3561	Std. Deviation	1,887	1,975	1,891	1,855	1,917	2,034	2,1	1,969	1,886	1,996	1,861
I need to have access to news in real time												
7-grade Likert scale where 1 represented not interesting and 7 very interesting	Mean	4,77	4,67	4,77	4,8	4,76	4,54	4,65	4,57	4,74	4,69	4,92
N=3506	Std. Deviation	1,799	1,886	1,824	1,784	1,875	1,91	1,755	1,975	1,843	1,871	1,492
It is absolutely central that I have access to news at all times												
7-grade Likert scale where 1 represented not interesting and 7 very interesting	Mean	4,21	4,22	4,07	4,17	4,12	4,17	4,63	4,11	4,1	4,23	4,24
N=3481	Std. Deviation	1,918	1,928	2,011	1,909	1,955	2,035	1,785	2,028	1,963	1,951	1,863
My news source needs to update the news several times a day												
7-grade Likert scale where 1 represented not interesting and 7 very interesting	Mean	5,45	5,47	5,24	5,3	5,48	5,28	5,25	5,13	5,38	5,44	5,21
N=3499	Std. Deviation	1,701	1,721	1,891	1,801	1,751	1,851	1,832	1,919	1,785	1,77	1,833
I find it important to have services available at the right time												
7-grade Likert scale where 1 represented do not agree and 7 agree	Mean	5,8	5,84	5,71	5,65	5,86	5,79	5,92	5,78	5,7	5,9	5,81
N=1340	Std. Deviation	1,358	1,296	1,489	1,417	1,314	1,419	1,129	1,246	1,411	1,322	1,167
It is important that content is adapted to time of day												
7-grade Likert scale where 1 represented do not agree and 7 agree	Mean	3,92	3,96	3,82	4,09	3,88	3,87	3,85	3,81	4,06	3,79	4,12
N=1331	Std. Deviation	1,83	1,794	1,911	1,877	1,791	1,867	2,014	1,968	1,83	1,804	1,728
Constantly updated news												
7-grade Likert scale where 1 represented not important and 7 very important	Mean	6,18	6,11	6,31	6,2	6,2	6,17	5,96	6,19	6,27	6,09	6,29
N=2719	Std. Deviation	1,249	1,295	1,158	1,212	1,215	1,304	1,466	1,345	1,158	1,309	1,178

and instead talked about the importance of layout connected to newspaper feeling.

In the prototype testing phase some of the respondents mentioned the possibility of making adjustments to make a more personalized layout of your news, e.g. to decide where to put the updates and not to include any ads at all.

In the real life evaluation phase relatively few mentioned possibilities to personalize the layout of the e-newspaper with some exceptions. Many discussions were about the existing layout, but then the respondents had comments aimed to the newspaper handling the layout. These respondents did not mention that they wanted to customize the layout based on their own preferences. However some respondents mention that they lacked personalization options:

I don't mean that my way is the way people should read newspapers, but I miss this with the possibility to configure it so it fits me.

"Content": In the design phase personalization of content was important for a majority of the participants. The possibility to put together their own newspapers with sections from different newspaper brands was seen as an interesting possibility with the e-newspaper. Also personalized ads were in general something the participants were positive to, however, quite often the discussion ended up in the information value of both editorial and advertisement content.

Some participants argued that you want to be surprised and informed when you read a newspaper, therefore you should not exclude sections in a newspaper. The common knowledge that the newspaper delivers generates a base for discussion at work. If everyone read their personal newspaper they won't have a common ground for discussion. Several participants also said that they liked to read advertisement and classed this as valuable information. Therefore they did not want to have personalized ads that also removed the surprise factor discussed about editorial content.

In the prototype testing phase the possibility of getting personalized content was regarded as positive by many of the respondents. Being able to buy different parts from different newspapers was one option that was mentioned. But also here the opinions differed:

Yes, I would like to have personalization options, but I would not like to totally miss out on "show me something I didn't know-effect" – it is a hard balancing act.

Also in the real life evaluation phase the majority of the respondents wanted to be able to personalize the content of the e-newspaper in one or another way. Mainly the personalization was about possibilities to buy different sections from other newspapers to complement their local newspaper. Most of the respondents also wanted to personalize the advertisement to better fit their profiles with some exclusion such as:

I don't really like this. I find it... I could imagine that I tell them what ads that I want, but I don't want to connect that to my subscription, or to my name. I find it... I don't like it when they monitor me and try to find out who I am....

To exclude editorial content based on a profile to personalize the news were not that interesting with the exception of a few respondents. Most of the respondents either stated that they get all editorial content today from the newspaper and should also receive this in the e-newspaper. Another argument was that you never really know what you will find interesting. One user stated for example while discussing this:

It doesn't bother me that much (to exclude material), but I personally would like to have the full newspaper, that is what I pay for....

The same user later stated while discussing personalization of content that:

Figure 13. Personalization issues

Personalization		Total	Gender		Age				Education			
			Male	*Female*	*15-24*	*25-44*	*45-64*	*65-*	*Elementary*	*Grammar*	*University*	*Other*
Personalized newspaper, i.e. be able to choose the content you want to payfor												
7-grade Likert scale where 1 represented not important and 7 very important	Mean	5,3	5,24	5,39	5,37	5,43	5,07	4,69	4,96	5,37	5,3	4,95
N=2925	Std. Deviation	1,935	1,996	1,827	1,885	1,871	2,021	2,176	2,097	1,901	1,924	2,155
Personal information (e.g. phonebook, calendar, economy)												
7-grade Likert scale where 1 represented not important and 7 very important	Mean	3,52	3,76	3,12	3,66	3,56	3,36	3,38	3,52	3,59	3,46	3,24
N=2916	Std. Deviation	2,202	2,217	2,118	2,217	2,212	2,159	2,253	2,233	2,195	2,205	2,1
It is important that services are adaptable to my own interests												
7-grade Likert scale where 1 represented do not agree and 7 agree	Mean	5,35	5,31	5,44	5,24	5,42	5,26	5,5	5,43	5,38	5,32	5,06
N=1351	Std. Deviation	1,553	1,553	1,552	1,514	1,487	1,739	1,503	1,734	1,541	1,544	1,144

I don't really know what I want to read, you don't have a clue about what will be interesting in the newspaper the next day.

In the validating phase personalization regarding content and services aimed at your own interests was considered relative important, whereas the ability for the device to host personal information was not considered as important (Figure 13). The mean scores are evenly spread between gender, age and education.

Usability

We have identified "navigation", "ease of use" and "efficiency" that together form the key user value "Usability".

"Navigation": In the design phase it became quite evident that it was important to quickly be able to get an overview over the newspaper content. The participants designed their own navigational solutions and the preferences differed quite a lot. Three basic conceptual designs where generated based on the navigation in a traditional newspaper, the online newspaper and hybrid versions of these two.

In the prototype testing phase many of the respondents commented on the importance of navigation support and overview of the content. One of the respondents compared the e-newspaper with the traditional newspaper and stated:

In an ordinary newspaper you feel its weight and how much information it contains.

But several of the respondents felt that they got a better overview of the content due to indexes and navigation support.

In the real life evaluation most of the respondents felt that the device controlled their reading and that they had to adjust their reading to the navigation system and to the device. It was generally problems with the navigation that negatively affected the feeling of control, as illustrated with the following quotation:

... but this is the part that is stressful. I did not feel one hundred percent in control. It was not intuitive, it do not do what I feel is right... when you read you do not want to feel like the distribution of presents on Christmas when you never really know what you will get.

For a few respondents these issues where not perceived that problematic and some felt that the only problem was the trouble to move between the different levels of information (i.e. the first page, the sections and the articles), as one respondent pointed out:

Well I could not move back to the first page so in that sense the device controlled my reading. It was not as efficient that you would like.

"Ease of use": In the real life evaluation phase more than half of the respondents thought it were hard to learn how to find information in the newspaper. Some of the respondents stated that they gained control and understood the device after a couple of days. However, even after the test was concluded and we conducted our interviews, some perceived that they still could not achieve control over the e-newspaper. One of the respondents said:

it is a little bit like a psychology test, you do not know what will happen, this eventually leads to that you do not really dare to press on something cause you do not now what will happen.

The most commonly mentioned problem was the fact that you could not move between the different levels of information, there were no shortcuts to the sections or the first page. Some also mentioned the lack of a working pagination as a reason.

However, a minority of the respondents thought that it was quite easy to find what they sought after and stated that it took them three to four days to learn how they should act to find what they wanted.

"Efficiency": In the real life evaluation phase approximately half of the respondents mentioned negative aspects of the e-newspaper relating to efficiency compared to the traditional newspaper. The most commonly mentioned problems were that the e-newspaper was slower and harder to get an overview compared to the paper edition. The most commonly negative aspect mentioned compared to the online version was that the eReader had a slow refresh rate, as one of the respondents remarked:

The visual part is just fine, the rest is like being in a hurry and get behind a tractor on the highway.

In the validating phase the importance of the key value "usability" is evident based on the high mean scores (Figure 14). The value "ease of use" is considered as most important. It is interesting to notice that younger people rate the format and layout more important than older people, whereas older people prefer easy technology compared to much functionality as more important than younger people.

Interactivity

We have identified "functionality" and "communication" that together form the key user value "Interactivity".

"Functionality": In the design phase there were functions that came up as extra important for the e-newspaper. These also related to added value primarily compared to the traditional newspaper but also in some cases to the online newspaper. The functions that where extracted as important where the possibility to save and print content, text size settings, archive function, personal settings and to be able to read e-mail.

Figure 14. Usability issues

Usability		Total	Gender		Age				Education			
			Male	*Female*	*15-24*	*25-44*	*45-64*	*65-*	*Elementary*	*Grammar*	*University*	*Other*
That it is easy to use and easy to learn												
7-grade Likert scale where 1 represented not important and 7 very important	Mean	5,94	5,71	6,32	5,77	5,88	6,15	5,99	6,08	6,06	5,78	6,17
N=2722	Std. Deviation	1,518	1,629	1,224	1,593	1,532	1,407	1,578	1,478	1,403	1,623	1,224
That I can depend the technology always working												
7-grade Likert scale where 1 represented not important and 7 very important	Mean	6,35	6,29	6,45	6,25	6,37	6,39	6,26	6,33	6,41	6,29	6,54
N=2722	Std. Deviation	1,125	1,177	1,026	1,198	1,067	1,159	1,259	1,192	1,081	1,161	0,741
Easy to find what you look for or want to read												
7-grade Likert scale where 1 represented not important and 7 very important	Mean	6,27	6,18	6,43	6,14	6,34	6,26	6,16	6,21	6,34	6,22	6,17
N=2716	Std. Deviation	1,224	1,261	1,144	1,321	1,124	1,309	1,337	1,359	1,141	1,259	1,671
I prefer easy technology compared to much functionality												
7-grade Likert scale where 1 represented do not agree and 7 agree	Mean	4,38	4,25	4,6	3,68	4,22	5,01	5,54	4,56	4,29	4,42	4,67
N=2720	Std. Deviation	1,93	1,932	1,907	1,857	1,879	1,864	1,659	1,974	1,932	1,919	1,867
It is important to quickly get an overview of the content												
7-grade Likert scale where 1 represented do not agree and 7 agree	Mean	5,97	5,89	6,14	5,72	6,04	6,01	6	5,74	5,88	6,09	6
N=1360	Std. Deviation	1,299	1,295	1,295	1,44	1,201	1,376	1,296	1,507	1,395	1,147	1,275
It is important that it is easy to find what I am looking for												
7-grade Likert scale where 1 represented do not agree and 7 agree	Mean	6,24	6,16	6,41	6,08	6,27	6,28	6,44	6,25	6,24	6,23	6,47
N=1342	Std. Deviation	1,126	1,146	1,061	1,236	1,07	1,166	0,87	1,275	1,162	1,073	0,717

In the prototype testing phase many of the respondents also mentioned added functionality they would like to have, for example possibility of enlarging text, sharing content, printing, saving and searching. Other preferred functionality mentioned was multimedia functionality, calendar and payment transactions. Some respondents mentioned that it needed to be "breakfast proof", i.e. that it would not be ruined if someone spilled coffee on it.

In the real life evaluation phase a majority of the respondents had ideas regarding adding functionality to the e-newspaper, specifically a save and a print function where you could save and print articles and for example obituaries. The same functions also came up when problems with the e-newspaper was discussed, the most frequently mentioned problem was the lack of a save and print function. Another functionality that leveraged value in the e-reading device was the possibility for a dedicated digital reading device. Several respondents commented on the fact that they could handle all their reading with this device including different brands of newspapers and books and magazines. One user discussed for example this value in the student life context: *...it would be much better to not have to carry several kilos of books in your backpack" and another user said: "this would be the dream where newspapers were updated on the fly and I would be able to carry pdf files and other documents with me, or maybe even comic strips or whatever... this would be a reading device that I could carry with me and fill with content. Then it would be just the same as the portable video and music players.*

"Communication": In the prototype testing phase several respondents mentioned communication possibilities as added value. Communication with for example journalists and advertisers but also with other readers was something that was mentioned by some. Examples mentioned were the possibility of commenting on articles, respond directly on an ad and chat possibilities with other readers.

In the validating phase different values in the key value "interactivity" were considered more or less important (Figure 15). From Figure 15 we can discern that "functionality" issues are considered more important than "communication" issues. Older people seemed more in favor of feedback

Figure 15. Interactivity issues

Interactivity		Total	Gender		Age				Education			
			Male	*Female*	*15-24*	*25-44*	*45-64*	*65-*	*Elementary*	*Grammar*	*University*	*Other*
I want to be able to interact with editors and journalists												
7-grade Likert scale where 1 represented not interesting and 7 very interesting	Mean	3,72	3,75	3,56	3,9	3,68	3,5	3,5	3,68	3,67	3,69	3,43
N=3561	Std. Deviation	1,905	1,962	1,891	1,903	1,936	1,938	2,011	1,949	1,936	1,937	1,909
I want to be able to communicate with advertisers												
7-grade Likert scale where 1 represented not interesting and 7 very interesting	Mean	3,15	3,13	3,09	3,09	3,1	3,14	3,35	3,28	3,17	3	3,8
N=3502	Std. Deviation	1,856	1,906	1,831	1,767	1,868	1,966	2,004	2,005	1,883	1,883	1,946
Transactions (e.g. e-business, ordering, booking)												
7-grade Likert scale where 1 represented not important and 7 very important	Mean	4,24	4,32	4,11	3,97	4,36	4,18	4,48	4,1	4,33	4,18	3,89
N=2900	Std. Deviation	2,118	2,131	2,091	2,095	2,083	2,167	2,23	2,198	2,092	2,127	2,083
That I am offered more functionality such as chat, multimedia, interactivity												
7-grade Likert scale where 1 represented not important and 7 very important	Mean	3,97	4,05	3,85	4,24	3,96	3,86	3,69	4,14	4,2	3,74	3,43
N=2725	Std. Deviation	2,074	2,072	2,072	2,073	2,047	2,119	2,005	2,217	2,083	2,017	1,914
I would like to communicate more with other readers												
7-grade Likert scale where 1 represented do not agree and 7 agree	Mean	2,71	2,84	2,49	2,97	2,65	2,6	3,02	2,82	2,76	2,65	2,11
N=2715	Std. Deviation	1,842	1,866	1,781	1,984	1,812	1,769	1,911	1,988	1,871	1,795	1,388
I would like to access information from other readers												
7-grade Likert scale where 1 represented do not agree and 7 agree	Mean	3,34	3,41	3,18	3,5	3,35	3,19	3,44	3,62	3,45	3,2	3,06
N=1365	Std. Deviation	1,747	1,719	1,803	1,822	1,672	1,828	2,043	1,876	1,748	1,717	1,611
The opportunity to give feedback concerning services is important												
7-grade Likert scale where 1 represented do not agree and 7 agree	Mean	4,63	4,63	4,61	4,74	4,6	4,53	5,12	4,6	4,72	4,53	4,88
N=1342	Std. Deviation	1,871	1,843	1,94	1,907	1,812	1,979	1,883	2,069	1,851	1,862	1,616
It is important with multimedia in mobile services												
7-grade Likert scale where 1 represented do not agree and 7 agree	Mean	4,21	4,17	4,3	4,73	4,16	3,87	4,08	4,25	4,41	4,03	3,71
N=1359	Std. Deviation	2,002	2,009	1,985	1,93	1,976	2,059	1,853	2,135	2,032	1,938	1,829
I appreciate services that puts me in contact with others												
7-grade Likert scale where 1 represented do not agree and 7 agree	Mean	3,18	3,21	3,11	3,42	3,1	3,13	3,65	3,45	3,34	2,97	3,29
N=1331	Std. Deviation	1,756	1,739	1,794	1,817	1,71	1,765	2,058	1,893	1,766	1,693	2,054

possibilities than the other age groups, whereas the younger people were more in favor of chat and multimedia functionality than the other age groups.

Readability

We have identified "format" and "comfortability" that together form the key user value "Readability".

"Format": In the design phase aspects relating to format and specifically the size of the envisioned eReader was one of the most commonly discussed subjects. A very strong majority of the participants in the workshops argued that an e-newspaper could not be smaller than an A4 format. Mainly the arguments for the A4 format related to retaining readability and the newspaper feeling. Only a few individuals reflected on that this bigger format affected the portability negatively, however, the mockups generated in the workshops where all based on ideas of flexible e-paper.

In the prototype testing phase the format was discussed by almost all respondents, where most stated that an A4 size was the minimum size acceptable. About one fourth of the respondents were in favor of a smaller size – about an A5. One respondent said:

It is important that it is portable and has a good overview.

In the real life evaluation phase many respondents discussed the existing layout and pointed out some parts that made them lack a newspaper feeling. The way indexes were built was one example of layout related aspects that felt more like a book than a newspaper. Several respondents related newspaper feeling to typefaces and said that the e-newspaper felt more like an traditional newspaper due to the similarities in the typefaces and that the letters looked and "felt" just like in the normal newspaper:

I think it is important that it looks good to make the content feel more serious, I mean like a consequence of typefaces and, eh, packaging… I mean the format affects the feeling of the newspaper.

"Comfortability": In the real life evaluation phase the all respondents said that the e-paper provided a really good and comfortable reading experience, especially compared to a computer screen. However, many respondents also complained that the e-paper had the same problem as print on normal paper; you needed illumination to be able to read:

It (the e-paper display) is very comfortable for the eyes but on the same time it is a little bit annoying that you need light to be able to read, maybe this is annoying because I am used to read on a computer screen, you don't think about this when it is a traditional newspaper….

A majority of the user also stated that the e-newspaper provided a much more laid-back reading experience compared to the online newspaper. Many respondents said that the reading experience where more relaxed and similar to reading a traditional newspaper. The e-newspaper where often read in the sofa or in the bed which felt relaxed and laid back. The answers also includes that the interaction felt more relaxed compared to reading an online newspaper. One user said:

… I think this is better with this (the eReader) because you're more flexible. You can sit as you want like this or like this or more laid back but you can't do that with a computer or with a screen.

Another user stated:

…then I can lie down in the bed and read some in this. A regular newspaper is a bit inconvenient then and troublesome to read, but this is… practical.

Figure 16. Readability issues

Readability		Total	Gender		Age				Education			
			Male	Female	15-24	25-44	45-64	65-	Elementary	Grammar	University	Other
The format and layout of the e-paper and the content												
7-grade Likert scale where 1 represented not important and 7 very important	Mean	5,2	5,2	5,2	5,51	5,33	4,85	4,22	4,78	5,25	5,23	4,69
N=2727	Std. Deviation	1,886	1,852	1,943	1,661	1,787	2,072	2,206	2,205	1,836	1,857	2,069
The format of a service is more important than the performance												
7-grade Likert scale where 1 represented do not agree and 7 agree	Mean	3,41	3,41	3,41	3,46	3,33	3,56	3,68	3,65	3,5	3,27	4,24
N= 1340	Std. Deviation	1,672	1,67	1,681	1,59	1,657	1,771	1,749	1,769	1,716	1,587	2,047
The format of a service is more important than the content												
7-grade Likert scale where 1 represented do not agree and 7 agree	Mean	2,95	3	2,82	3,23	2,9	2,82	2,92	3,51	3,03	2,78	2,88
N=1338	Std. Deviation	1,525	1,523	1,523	1,587	1,487	1,527	1,605	1,722	1,556	1,433	1,525

In the validating phase the format and layout of the e-newspaper were rated as quite important, however, it was not more important in relation to the content and the performance of the service according to the mean scores (Figure 16). The mean scores were relatively evenly spread between gender, age and education.

Environment

We have identified "less trash" and "less energy" that together form the key user value "Environment".

"Less Trash": In the design phase during discussions about the problems with traditional newspapers one aspect with the e-newspaper leveraging user value compared to the traditional newspaper became evident. Most of the participants thought it was tedious to handle all the paper trash generated by the traditional newspaper and paper advertisement. Some participants also talked about the environmental problems a traditional newspaper generated both concerning the production and the distribution of the newspaper.

In the prototype testing phase this value also came up. Several respondents commented on the trouble to handle and throw away paper which will be avoided by replacing the printed edition by the e-newspaper. This is illustrated by the following quotation:

The biggest problem is to go and through away all the newspapers that fills half the house.

On the negative side there were comments about the sensitive technology and the risk to misplace or break the device, which in turn will lead to more electronic waste.

Also in the real life evaluation phase most respondents mentioned the benefits with the e-newspaper concerning less trash. These respondents stated the importance of the fact that they didn't need to handle and throw away paper. Most of the respondents reflected upon the inconvenience relating to carrying newspapers and paper advertising to the paper recycling stations. Some respondents identified this value as their one primary argument for adopting an e-newspaper:

I would like to have an e-newspaper because... last week I carried out an IKEA bag with paper, I think it was paper from two weeks back and this is just very disturbing and annoying.

"Less energy": In the prototype testing phase many respondents mentioned positive environmental benefits with the e-newspaper compared to environmental problems related to the traditional newspaper, i.e. the production and distribution of the paper. One respondent said:

It has to be basic respect for the environment, one avoids the printing on paper and the distribution of paper waste.

In the real life evaluation phase a majority of the respondents mentioned environmental benefits with the e-newspaper compared to the traditional newspaper. The aspects concerning the fact that less energy was needed related to both the production and the distribution of the traditional newspaper. Some respondents also commented that they needed to use the car to drive to the paper recycle stations which also negatively affect the environment:

I think this is more durable and that you consume less energy regarding the environmental aspect, most people comment on the negative aspects of the newspaper itself, but I mean today we went some miles (a mile equals 10 km in Sweden) just to be able to leave our newspapers… there is also someone who sits in a car and distribute it every night and that put restrains on the environment….

The environmental issues were more important for some respondents compared to others, which is shown in this quotation:

I would much rather have all advertisement sent to me in the e-newspaper so I can keep the ads that I want. From my point of view I think you should legislate against sending out paper advertisement because I think it's just such a waste of resources….

In the validating phase the environmental issues were considered as relatively important by the respondents (Figure 17). Female and younger respondents found these issues somewhat more important than the other respondents.

Intention to Adopt

In this section we present data from the questionnaires used in the prototype testing phase, the validation phase and the real life evaluation phase. We have analyzed this data according to gender, age and education to further illuminate similarities or differences.

The first set of questions regarded if the respondents ever would consider exchanging the traditional newspaper with an e-newspaper, and in which timeframe. In Figure 18, we give an overview of the answer of the total population, and in Figure 19-21 we present the figures according to gender, age and education.

Figure 17. Environmental issues

Environmental		Total	Gender		Age				Education			
			Male	*Female*	*15-24*	*25-44*	*45-64*	*65-*	*Elementary*	*Grammar*	*University*	*Other*
Environmental reasons, improving the environment in the future *7-grade Likert scale where 1 represented not important and 7 very important*	Mean	4,26	4,07	4,57	4,48	4,36	3,93	4,02	4,24	4,26	4,25	4,27
N=2953	Std. Deviation	2,243	2,258	2,185	2,244	2,191	2,285	2,402	2,358	2,244	2,224	2,236
Positive effect on the environment (e.g. less cutting) *7-grade Likert scale where 1 represented not important and 7 very important*	Mean	4,83	4,55	5,3	5,09	4,89	4,57	4,67	4,87	4,97	4,7	4,74
N=2713	Std. Deviation	2,079	2,185	1,903	1,989	2,046	2,158	2,208	2,133	2,061	2,082	2,034

Figure 18. Intention and timeframe of the total population

Total *Percentage*	Prototype testing phase	Validation phase	Real life evaluation phase
Would you ever consider exchanging your traditional newspaper with an e-newspaper?			
	N=36	*N=2724*	*N=11*
Yes	83,3	75,3	100
No	16,7	24,7	0
Within what timeframe are you ready to read your newspaper on e-paper?			
	N=36	*N=2724*	*N=11*
Today	58,3	55,8	54,5
Within 2-5 years	27,8	25,1	45,5
Within 10 years	11,1	7,7	
Within 20 years	2,8	3,2	
Never	0	8,2	

Figure 19. Intention and timeframe divided by gender and phases

Percentage	Prototype testing phase		Validation phase		Real life evaluation phase	
	Male	Female	Male	Female	Male	Female
Would you ever consider exchanging your traditional newspaper with an e-newspaper?						
	N=36		*N=2724*		*N=11*	
Yes	87,5	80	78,4	70,1	100	100
No	12,5	20	21,6	29,9	0	0
Within what timeframe are you ready to read your newspaper on e-paper?						
	N=36		*N=2724*		*N=11*	
Today	68,8	50	61,3	46,7	62,5	33,3
Within 2-5 years	25	30	22,2	29,8	37,5	66,7
Within 10 years	0	20	6,3	9,9	0	0
Within 20 years	6,2	0	3,1	3,5	0	0
Never	0	0	7,1	10,1	0	0

Figure 20. Intention and timeframe divided by age and phases

Percentage	Prototype testing phase				Validation phase				Real life evaluation phase			
	15-24	25-44	45-64	65-	15-24	25-44	45-64	65-	15-24	25-44	45-64	65-
Would you ever consider exchanging your traditional newspaper with an e-newspaper?												
	N=36				*N=2724*				*N=11*			
Yes	100	75	86,7	75	73,5	79,9	68,5	71	100	100	100	100
No	0	25	13,3	25	26,5	20,1	31,5	29	0	0	0	0
Within what timeframe are you ready to read your newspaper on e-paper??												
	N=36				*N=2724*				*N=11*			
Today	80	58,3	53,3	50	50,5	60,3	53	44,1	0	57,1	50	100
Within 2-5 years	20	25	33,3	25	24,8	23,5	26,5	38,6	100	42,9	50	0
Within 10 years	0	16,7	13,4	0	10,6	6,8	7,5	5,4	0	0	0	0
Within 20 years	0	0	0	25	4,5	3	2,8	2,2	0	0	0	0
Never	0	0	0	0	9,6	6,4	10,2	9,7	0	0	0	0

Figure 21. Intention and timeframe divided by education and phases

Percentage	Prototype testing phase				Validation phase				Real life evaluation phase			
	Elementary	Grammar	University	Other	Elementary	Grammar	University	Other	Elementary	Grammar	University	Other
Would you ever consider exchanging your traditional newspaper with an e-newspaper?												
	N=36				*N=2724*				*N=11*			
Yes	50	100	82,6	100	77,6	74,8	75,8	57,1	100	100	100	0
No	50	0	17,4	0	22,4	25,2	24,2	42,9	0	0	0	0
Within what timeframe are you ready to read your newspaper on e-paper?												
	N=36				*N=2724*				*N=11*			
Today	0	57,1	65,3	100	55,9	52,9	59,2	36,1	0	20	100	0
Within 2-5 years	25	42,9	26,1	0	24,3	26	24,1	33,3	100	80	0	0
Within 10 years	75	0	4,3	0	7,9	9	6,2	11,1	0	0	0	0
Within 20 years	0	0	4,3	0	2,6	3,7	2,9	2,8	0	0	0	0
Never	0	0	0	0	9,3	8,4	7,6	16,7	0	0	0	0

In Figure 19, we can see that a very high percentage of the respondents stated that they would consider exchanging their traditional newspaper. It is interesting to notice that even with a lot of technological problem in the real life evaluation phase these respondents would all consider an exchange. The percentage of respondents that answered yes to this question was also higher in the prototype testing phase. There was a bit higher percentage of males than females that would consider the exchange.

In all phases the male respondents were more inclined to start reading their newspaper on e-paper already today than the females, although this option also had the highest percentages for females in the prototype testing phase and the validation phase. Only in the validation phase respondents stated that they would never read their newspaper on e-paper.

Looking at the same questions with an age perspective (Figure 20) reveals differences within the three phases. In the prototype testing phase the two younger age groups were most in favor of reading on e-paper today, this was also the case for the 25-44 age group in the validation phase. The age group 45-64 in the prototype testing phase and the 65+ group in the validation phase favored the 2-5 year perspective.

Taking an education perspective (Figure 21) on the same questions shows that the ones that have attended grammar school or university in the prototype testing phase were more positive to consider the exchange, whereas this were more evenly spread in the other phases. A higher percentage of university educated was ready to start reading on e-paper today than the other groups.

The next set of questions regarded the cost of the e-newspaper and how the eReader should be financed. In Figure 22, we give an overview of the answer of the total population, and in Figure 23-25 we present the figures according to gender, age and education.

Regarding the cost to exchange the newspaper (Figure 23), most respondents in the validation phase stated that it has to be cheaper than the traditional newspaper. Of these respondents more females preferred this option as well as in the real life evaluation phase whereas in the prototype testing phase there was a strong dominance of males. The females in the prototype testing phase favored the "same price" option, whereas in the other two phases this was more equalized. In all three phases there was a dominance of males choosing the option "may be more expensive if the e-newspaper delivers added value", although this was of lesser interest for the respondents in the validation and real life evaluation phases. There were only a few respondents in the validation phase that stated that cost was irrelevant.

The most common answer to this question regarding the payment of the eReader was that "the reader receives the eReader for free with the e-newspaper subscription". In both the prototype testing phase and the validation phase more females than males chose this option, whereas in the real life evaluation phase more males chose this option. Only males in the prototype testing phase and the real life evaluation phase wanted to pay a single fee for the eReader, also more males than females in the validation phase chose this option. In the real life evaluation phase some females wanted to purchase the eReader by installment payments, whereas there were more males than females that preferred this option in the other two phases.

Looking at the same questions from an age perspective (Figure 24) reveals mostly similarities

Figure 22. Cost and finance issues of the total population

Total / *Percentage*	Prototype testing phase	Validation phase	Real life evaluation phase
For you to exchange your newspaper, to what cost per newspaper edition?	*N=36*	*N=2942*	*N=11*
Cheaper than the traditional newspaper	27,8	72	45,4
Same price	52,8	12,6	36,4
May be more expensive if the e-newspaper delivers added value	19,4	9,6	18,2
Cost is not relevant	0	5,8	0
How should the e-reader be financed?	*N=36*	*N=2700*	*N=11*
The reader buy the e-reader by installment payments	11,1	10,4	9,1
The reader pays a single fee for the e-reader	11,1	26,3	9,1
The reader receives the e-reader for free together with the e-newspaper subscription	77,8	63,3	81,8

Figure 23. Cost and finance issues divided by gender and phases

Percentage	Prototype testing phase		Validation phase		Real life evaluation phase	
	Male	Female	Male	Female	Male	Female
For you to exchange your newspaper, to what cost per newspaper edition?						
	N=36		N=2942		N=11	
Cheaper than the traditional newspaper	50	10	68,6	77,6	37,5	66,7
Same price	25	75	13,8	10,6	37,5	33,3
May be more expensive if the e-newspaper delivers added value	25	15	11,3	6,7	25	0
Cost is not relevant	0	0	6,3	5,1	0	0
How should the e-reader be financed?						
	N=36		N=2700		N=11	
The reader buy the e-reader by installment payments	18,8	5	12,2	7,4	0	33,3
The reader pays a single fee for the e-reader	25	0	30,5	19,2	12,5	0
The reader receives the e-reader for free together with the e-newspaper subscription	56,2	95	57,3	73,4	87,5	66,7

Figure 24. Cost and finance issues divided by age and phases

Percentage	Prototype testing phase				Validation phase				Real life evaluation phase			
	15-24	25-44	45-64	65-	15-24	25-44	45-64	65-	15-24	25-44	45-64	65-
For you to exchange your newspaper, to what cost per newspaper edition?												
	N=36				N=2942				N=11			
Cheaper than the traditional newspaper	40	41,7	13,3	25	68,3	73,5	73,2	62,6	0	57,1	0	100
Same price	40	50	53,3	75	16,1	12,3	10,6	13,1	100	28,6	50	0
May be more expensive if the e-newspaper delivers added value	20	8,3	33,4	0	10,5	9,3	8,8	13,1	0	14,3	50	0
Cost is not relevant	0	0	0	0	5,1	4,9	7,4	11,2	0	0	0	0
How should the e-reader be financed?												
	N=36				N=2700				N=11			
The reader buy the e-reader by installment payments	20	16,7	0	25	10,3	10,9	9,6	9,7	0	0	0	100
The reader pays a single fee for the e-reader	0	16,7	13,3	0	30,2	26,8	22,6	28	0	14,3	0	0
The reader receives the e-reader for free together with the e-newspaper subscription	80	66,6	86,7	75	59,5	62,3	67,8	62,3	100	85,7	100	0

but also some differences. In the prototype testing phase the ages over 45 found it less important that it was cheaper than the traditional newspaper, and they also formed the largest percentage in that phase that stated that it may be more expensive. The age group of 65+ in the validation phase has the highest percentage of considering cost is not relevant. Since there were only 11 respondents in the real life evaluation phase of which one was in the age group of 15-24 and one in the 65+ group, the figures speak for themselves. No major differences regarding the finance of the eReader in the prototype testing phase and the validation phase were found.

The grammar school respondents in the prototype test (Figure 25) also favored the "same price" option, whereas all education categories in the validation phase favored the "cheaper" option.

The results regarding the financing of the eReader show that the elementary school respondents were least in favor of the installment payments option. Otherwise the figures were rather evenly spread among the groups and phases.

Figure 25. Cost and finance issues divided by education and phases

	Prototype testing phase				Validation phase				Real life evaluation phase			
Percentage	Elementary	Grammar	University	Other	Elementary	Grammar	University	Other	Elementary	Grammar	University	Other
For you to exchange your newspaper, to what cost per newspaper edition?												
	N=36				*N=2942*				*N=11*			
Cheaper than the traditional newspaper	50	0	34,8	0	75	73,3	70,2	68,3	0	80	20	0
Same price	50	85,7	39,1	100	10,5	12,3	13,5	7,3	0	20	60	0
May be more expensive if the e-newspaper delivers added value	0	14,3	26,1	0	5,6	9,2	10,5	14,6	100	0	20	0
Cost is not relevant	0	0	0	0	8,9	5,2	5,8	9,8	0	0	0	0
How should the e-reader be financed?												
	N=36				*N=2700*				*N=11*			
The reader buy the e-reader by installment payments	0	14,3	13	0	6,9	10,1	11,3	11,1	0	20	0	0
The reader pays a single fee for the e-reader	25	14,3	8,7	0	27,7	24,3	28,3	16,7	0	20	0	0
The reader receives the e-reader for free together with the e-newspaper subscription	75	71,4	78,3	0	65,4	65,6	60,4	72,2	100	60	100	0

User Centered Innovation Process

In the Diginews project we worked with user centered design activities iteratively during all of the stages in the innovation process. In the design phase we conducted workshops focusing on creating design suggestions and generate ideas of how the e-newspaper innovation should and could be formed including both design and functionality. These activities also served as requirement elicitation activities. However, these user centered activities also provided an opportunity to early identify user values by discussing opportunities as well as envisioned problems with the e-newspaper innovation.

The user created mock-ups and scenarios formed a base for the upcoming design activities together with the newspaper designers and this work was constantly discussed and validated with our user focus group which enabled more opportunities to explore what user values the e-newspaper innovation could and should bring to the market from a consumer perspective.

The designs solutions generated by these user centered innovation activities were incorporated in the prototypes that were evaluated and discussed with 36 respondents during the prototype testing phase. In this phase several new user values were identified in the interviews, and the values identified in the design phase was validated in this process.

The online questionnaire conducted in the project gave us an opportunity to validate the identified user values during the earlier stages. It also provided an opportunity to measure the importance of the key user values that was identified. The result of the validation phase indicate that by involving users early in the innovation process, important key user values can be identified and specified that can help guide the design of an innovation leveraging consumer value.

These results were further strengthen in the real life testing phase where 12 users tested the e-newspaper innovation during a longer time in a real life setting as opposed to the traditional tests that were conducted during the prototype phase. In this phase two new categories of user values were identified. These values related to the actual e-newspaper, i.e. the early version of the iRex iLiad but also to the e-newspaper service tested out by the content provider, in this case Sundsvalls Tidning. Many of the problems reported during the test did indeed correlate to the identified key user values in the earlier phases of the innovation process. However, due to technical and resource constraints, they could not be incorporated into the e-newspaper that was tested.

It became evident that several user values that we identified early in the process were not taken into consideration and incorporated into the e-newspaper innovation by the device producer, leading to problems for the content provider.

Figure 26. Descriptions of key user values of the e-newspaper

Key user values	Description	Device	Service
Mobility	*Localization* - content anywhere, geo-localized services	X	X
	Portability - convenience of handling the product anywhere	X	
Availability	*Accessibility* - content anytime	X	
	Updates - frequency of updated content		X
Personalization	*Layout* - design options and functionality	X	X
	Content - newspaper parts and ads of your liking		X
Usability	*Navigation* - overview and navigational support	X	X
	Ease of use - also includes learnability	X	X
	Efficiency - reading experience compared to other channles	X	X
Interactivity	*Functionality* - added functionality compared to other channels	X	
	Communication - discussion and feedback, sociability		X
Readability	*Format* - size, newspaper feeling		X
	Comfortability - laid back reading and display qualities	X	
Environment	*Less trash* - no need to handle paper trash	X	
	Less energy - environmental benefits compared to the production and distribution of traditional newspapers	X	

The iRex iLiad is according to us a device that focuses on delivering a functional e-paper solution not really aimed for newspaper reading. One example is the layout of the hardware (i.e. buttons, switches, etc.), that makes it complicated for the service providers to deliver a natural newspaper interface mapping to the hardware. The company seems to allocate their recourses on the technical problems and overlooking the usability aspects of the device.

DISCUSSION

In this paper we have identified seven key user values for the e-newspaper, i.e. mobility, availability, personalization, usability, interactivity, readability and environment. We have also examined the intention to adopt a new possible innovation in a pre-adoption phase. Furthermore, we have reflected on the user centered design process conducted within the project from a user value perspective.

From this study we can derive both practical and theoretical implications. The practical contribution regards implications for device producers and service providers based on the identified key user values and figures regarding the intention to adopt the e-newspaper innovation. The theoretical contribution is based on the empirical exploration of user centered design processes ability to capture user values in the pre-adoption phase of the innovation process.

The key user values for the e-newspaper in the pre-adoption phase is summarized and explained in Figure 26. As stated earlier it is essential to differ between what concerns the device (the eReader) and the newspaper service when referring to the e-newspaper. Taking this perspective on the values described above, Figure 26 also provides an overview of how the identified values map into the device or the service part of the e-newspaper.

The mobility value concerns both the device and the service. The localization aspect is connected to the service regarding for example to be able to read local news anywhere. This service is comparable to the online newspaper which also can be regarded as a service. However, to be able to publish geo-localized content, the device needs to send information about its position via for example GPS, positioning via WiFi or triangulation via the GSM network. The portability value mainly maps to the device, the eReader needs to be in a portable format to enable the mobility of the readers. One way to offer a bigger screen and still support the portability would be to use flexible displays.

The availability value also concerns both the device and the service. To be able to deliver content anytime (accessibility), the device needs to have access to a server with the content. The value concerning updates connects to the service regarding if the e-newspaper should be delivered in editions, constantly updated or a hybrid version with a special section of the edition that is constantly updated.

The personalization value primarily maps to the service part of the e-newspaper. However some users wanted to be able to personalize the device as well. This would for example mean that soft buttons could be used to change the functionality of the device. To be able to personalize the layout the content management system needs to support modules that can be selected by the reader. Specific functionality also put demands on the way the newspaper service is delivered and managed by the newspapers. The same requirements also apply to personalized content and ads. The content management system needs to deliver content based on reader profiles.

The usability value concerns both the device and the service. The device manufacturer needs to provide better hardware support for different navigation solutions. The service provider needs to provide a good overview of the content and deliver an efficient and functional navigation support for the reader. The ease of use value and the efficiency value connects both to the device and the service. The e-newspaper needs to be easy to use and easy to learn. Furthermore it needs to be as efficient as the traditional newspaper and the online newspaper concerning finding content that you are searching for.

The interactivity value is based on two values, functionality and communication. In the study, suggested added functionality for the device were for example to be able to save and print content, text size settings, archive function and personal settings, which would make the e-newspaper more comparable to existing newspaper channels. Communication values maps to the service side of the e-newspaper, e.g. to be able to communicate with editors, journalists and other readers. In this value the sociability aspect is also included.

The value of readability is also split between the device and the service. Even if the size is inherent in the device and affects the format, the value of format primarily affects the service side. In the empirical data, the value of format relates to retaining readability and the newspaper feeling. The value of comfortability relates to the device, this value primarily concerns the display qualities and the size and functionalities of the device which enables a laid back reading experience.

Finally, both the values connected to environment (less trash and less energy) are connected to the device. In the same way as an online newspaper is related to a computer and therefore put less constraint on the environment; this value of the e-newspaper is connected to the device. Less trash is seen as a value primarily for the reader's convenience of not having to handle as much paper trash. Less energy is seen in the light of more general environmental benefits such as not having to produce and distribute as many paper editions of the newspaper. On a side note, it could be argued that the waste of the electronic trash of computers and other devices put severe constraint on the environment as well, this was however not a common comment in our studies.

According to Kim and Mauborgne (2005) you need to focus on creating advances in customer value which is a way to mitigate competition and open up new markets. Based on this, the e-newspaper seems to be able to bridge several user values that today either is inherent in the traditional newspaper or in the online newspaper. It also has the possibility to bring new unique values which might be enough to make it competitive on the existing media market. As Cagan and Voguel (2002) argues, to be able to deliver breakthrough products that distinguishes themselves from the competitors, significant user value needs to be delivered, hence the importance of investigating user value.

In this study we tried to identify user value early and throughout the innovation process. As we have shown, many key user values were identified early and then validated and sometimes expanded during the later stages of the innovation process. One interesting remark is that the identified key values where found and validated independent of the actual usage of the e-newspaper. According to Kuusisto (2008) the concept of value can be seen from two perspectives, either as value contained in a product, or as value-in use. In our study all seven key values (9 of 16 values) where identified early in the innovation process, i.e. in the discussions in the workshops. In the prototype testing phase three new values where identified and in the real life evaluation phase, the last three values emerged. This indicates that the actual usage of prototypes and early products is of importance to be able to identify values that connects to the interaction with a product, i.e. the in-use value concept.

Interestingly, most of the respondents in all three phases would consider exchanging their traditional newspaper with an e-newspaper. The respondents in the real life evaluation phase were 100% positive. In the prototype testing phase 83,3% were positive compared to 75,3% in the validation phase. All respondents that answered the questionnaires had the opportunity to test either the prototypes or the e-newspaper in the final test. Noticeable is that these figures correspond to the amount of time the respondents spend with the services, even though there were more technical problems in the real life evaluation phase. As reported in the empirical findings section there were also some differences regarding gender, age and education. Furthermore, over 50% of the respondents in all three phases answered that they are ready to read their newspaper on e-paper already today. Moreover, in all three phases over 25% answered that they were ready within 2-5 years. These figures indicate that the e-newspaper have a fair chance of being adopted by a large population. However, this study was conducted in a pre-adoption phase and just as Karahanna *et al.* (1999) reports, there might be differences in attitudes towards the innovation in a pre-adoption phase compared to a post-adoption phase.

Nevertheless, taking these figures serious the media house management needs to start preparing for a forthcoming launch of their e-newspapers. Regarding the cost of the e-newspaper there were some differences between the respondents in the three phases, whereas there were a more even spread regarding the payment of the eReader. Many respondents in the validation phase (72%) and the real life evaluation phase (45,5%) answered that the e-newspaper should be cheaper than the traditional newspaper, while the respondents in the prototype testing phase (52,8%) answered that it could have the same price. Several of the respondents in the real life evaluation phase (36,4%) also found the same price acceptable. A majority of the respondents in all three phases favored the option where the reader receives the eReader for free together with the e-newspaper subscription.

From the above discussion, we summarize implications for further e-newspaper development for both device and service producers in Figure 27. From this reasoning we continue to formulate three major challenges for a successful e-newspaper launch.

Based on the detailed descriptions of the user values we have identified several practical implications for both the device producers as well as the service providers. It became evident in the real life evaluation phase that the e-newspaper in this early version did not succeed to deliver the

Figure 27. Practical implications

Key user values		Implications for device producers	Implications for service producers
Mobility	Localization	GPS or Wifi positioning needs to be incorporated	Content management systems that dynamicly adjusts content depending on device position and reader profile
	Portability	Flexible display allows bigger display, still being very portable	
Availability	Accessibility	Support more communication standards to enable "always connected"	If the e-newspaper is based on editions, these need to be delivered at the same time every day
	Updates		Special section, always updated The ability of updating just some pages, not the entire edition. Indication of news flashes when always connected
Personalization	Layout	Soft buttons needs to be incorporated to support personalization of functions	Module based content management system supporting personalization of the layout
	Content		Content management system and business models for handling personalization of the content
Usability	Navigation	Better hardware support for navigation solutions	Provide overview over the content. Need to incorporate an efficient and functional navigational support
	Ease of use	Better feedback to users, more intuitve user interface	Provide an service that is easy to use and easy to learn
	Efficiency	Faster refresh rate would reduce the "computer stress syndrome"	A service that is as efficient to use as existing newspaper channels
Interactivity	Functionality	Focus on added functionality that enables e.g. printing, saving, archiving etc	
	Communication		Possibility to make a contribution to the debate adds value Sociability possibilities with other readers adds value
Readability	Format		Providing a more traditional newspaper layout would increase "the newspaper feeling"
	Comfortability	The size of the reader affects the comfortability	
Environment	Less trash	This is an important marketing aspect that should be considered	
	Les energy	This is an important marketing aspect that should be considered	

full potential of the innovation. From a device producer viewpoint the most important implications concerns communication standards to support the accessibility value, this also relates to the mobility value concerning ways of implementing positioning technology such as for example GPS. What is also important to consider is the way the size of the eReader affects the values of mobility and readability. Furthermore, the usability issues that were identified in the real life evaluation phase needs to be addressed, e.g. include a better hardware support for different navigation solutions.

From a service provider viewpoint there are several implications that are important. To be able to fulfill the values concerning localization and personalization, content management systems needs to be developed that are flexible and dynamic. Furthermore the service providers need to be able to deliver the content in several different ways, including real time news feeds as well as content packaged in editions. The usability value also needs to be addressed by the service providers to be able to launch a successful e-newspaper innovation. Finally, there is a challenge to be able to keep the newspaper feeling in the e-newspaper. Even if the prototype test showed that it can be done, the real life evaluation did not succeed to do this with an automated publication process. Just as in any newspaper channel, the e-newspaper probably needs dedicated designers and editors that can customize the content for the e-newspaper.

The ongoing e-paper development with larger color displays with moving images and interactive clickable advertising coming in only a few more years will have impact on the device producers with increased competition. Since many of the newspapers are just waiting for this kind of e-paper to enter the market, the e-newspaper could have a promising future. However, for the media houses to succeed with their e-newspaper launch they have still to meet some challenges. Continuing the discussion from the practical implications we have formulated three main challenges for a successful e-newspaper launch.

- *Challenge 1: How to develop a layout that is acceptable to the audience?* By meeting this challenge the media houses need to consider usability, mobility and readability values. They will have to balance the readability issues like newspaper feeling with usability issues like navigation support. They will also have to balance update frequencies with layout issues.
- *Challenge 2: In what ways can content and additional services constitute added value?* This challenge mostly concerns the availability, interactivity and personalization values. Here the main challenge is to find ways to reduce manual work. Most of the issues raised within the corresponding values imply content management systems and system integration. This challenge is also about finding niches and reaching new audiences.
- *Challenge 3: How to develop business models that leverage user value?* This challenge includes the above challenges and values as well as the environment value as a marketing advantage. In this study the intention to adopt was high in the pre-adoption phase, but when it comes to actual decisions we do not know what happens. Most of the respondents preferred a payment solution for the eReader, in which it was incorporated in the e-newspaper subscription. However, the media houses have to consider all options in order to attract as many readers as possible. If the media houses manage to meet the two above challenges they may find a cost level that is suitable for the readers, given that production and distribution costs could be reduced.

We argue that the media houses could be guided in their choices by the results of this study, but we also believe that they need to continue involving users in their innovation processes for further development of their e-newspapers. We also argue that all key user values identified in this study are more or less applicable to other mobile applications as well. Given that some mobile phone producers, e.g. Motorola, already have started to use e-paper in the display of mobile phones, the readability value could be applicable here as well.

Moreover, we argue that involving users in many ways throughout the innovation process could increase the adoption rate of the product or service due to user values captured early in the process. To be able to deliver innovations that generate user value or user experience unmatched, von Hippel (2005) argues that there is a need for a profound understanding of these, as well as the ability to translate these into unique products and experiences. Rosted (2005) highlights that user centered innovation processes needs to allocate resources for analyzing and assessing customer needs which implies that these activities needs to be planned for and modeled into the innovation processes. There is a trend according to Rosted (2005) that focus on a more systematic approach in identifying customer needs where skills from social sciences are applied in the mapping of customer experiences and customer needs. We therefore suggest a model for identifying user values in a pre-adoption phase of an innovation (Figure 28), that we believe could work as guidance for other innovation processes as well.

To be able to identify key user values early in the innovation process users need to be involved

Figure 28. Model for identifying key user values

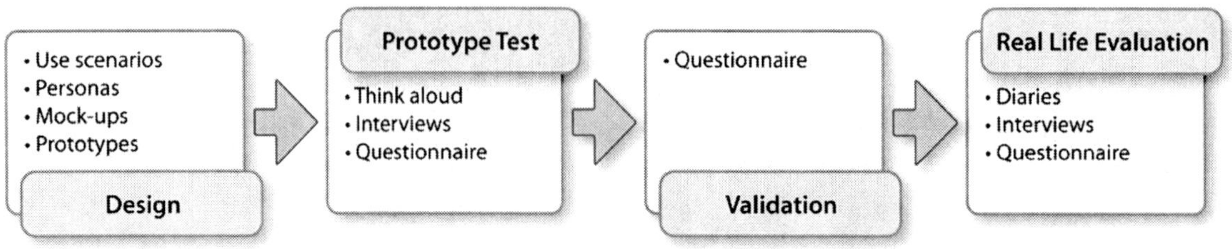

from the beginning. The values can be identified in discussions with focus groups, but we have found that by involving the users actively in the design work, more values were discussed and became tangible. Design methods and tools such as mock-ups, scenarios, personas and prototypes developed by the users themselves provides a rich picture regarding what user values that are important.

To be able to validate and identify values connected to in-use situations prototypes can be used early in the innovation process as a base for discussing user values. Instead of only focusing on traditional evaluations of prototypes, e.g. usability tests, we suggest that prototype evaluations is complemented with interviews aiming at investigating user values as well. Together with for example think aloud methods and questionnaires a good data set can be created concerning user values. In our study several new values emerged during the prototype test, but the test also served as a way to get a richer picture of the different key values of the innovation. This might prove a valuable source of information, helping developers to understand in which areas they should focus their resources.

In our proposed model we suggest that questionnaires could be used to validate identified key user values. A well designed questionnaire can be an efficient way of validating the findings before moving on and develop the final version of a product or service. But it is important for the respondents to have access to prototypes, sketches or other design solutions to be able to investigate their intention to adopt.

Traditionally many products and services are tested before they are shipped out to the market. However, many of these tests are done in labs during a quite short time span, focusing on usability issues and functionality. The data generated from these activities are often affected by the fact that the tests are done in an artificial environment where the users are observed and asked questions. In our study we have found that conducting evaluations in a real life setting during a longer time span provides a rich data set. This approach let the users have time to get over the first impressions. It was quite evident when studying the diaries and comparing them with the final interviews, that people might change their mind during a longer test period. In our study we found new values but several values also where complemented, again providing a richer picture of the key user values.

Based on our results, we argue that the proposed model could ensure that key user values are identified early in an innovation process, which in turn could influence the intention to adopt, helping developers to focus on the most important aspects when developing new innovations.

CONCLUSION

We argue that by taking a user centered innovation approach, key user values that could influence the intention to adopt, could be identified early in the process. Many of the values were identified in the first phase, and then complemented in the other phases, providing a rich picture of key user values. Furthermore, the use of product concepts, mock-ups and prototypes early in an innovation process seems to be powerful tools for identifying key user values.

To conclude, in this study we have contributed both with practical and theoretical insights relevant for further studies of pre-adoption, user centered design and innovation processes by:

- studying the pre-adoption of an innovation from both a technical as well as service-based perspective, revealing implications for further development for both device and service providers,
- empirically identifying and discussing user values influencing intention to adopt, that could also be of interest to other m-business applications, and
- proposing a four phase model for identifying user values in a pre-adoption phase of an innovation.

The limitations of the research lie within studying the pre-adoption phase of an innovation and not being able to confirm whether the user values identified also are valid in a post-adoption phase. Another limitation is that our proposed model is only tested within one specific case with its own contextual limitations.

Future research will include further studies on user and other stakeholder involvement in open innovation processes in a Living Lab setting. This will also enable us to test the model in other contexts and application areas.

REFERENCES

Anckar, B. & DÍncau, D. (2002). Value-creation in Mobile Commerce: Findings from a Consumer Survey. *Journal of Information Technology Theory and Application*, *4*(1), 43–64.

Blechar, J., Constantiou, I. D., & Damsgaard, J. (2006). Exploring the influence of reference situations and reference pricing on mobile service user behaviour. *European Journal of Information Systems*, *15*, 285–291. doi:10.1057/palgrave.ejis.3000618

Boztepe, S. (2007). User value: Competing theories and models. *International Journal of Design*, *1*(2), 55–63.

Buchanan, T., & Smith, J. L. (1999). Using the Internet for psychological research: Personality testing on the World-Wide Web. *The British Journal of Psychology*, *90*, 125–144. doi:10.1348/000712699161189

Cagan, J., & Vogel, C. M. (2002). *Creating breakthrough products: Innovation from product planning to program approval.* Upper Saddle River, NJ: Prentice Hall.

Carlsson, C., Carlsson, J., Hyvönen, K., Puhakainen, J., & Walden, P. (2006). Adoption of Mobile Devices/Services – Searching for Answers with the UTAUT. In *Proceedings of the 39th Hawaii International Conference on System Sciences, Hawaii.* (CD-ROM), Computer Society Press (10 pages).

Carlsson, C., Hyvönen, K., Repo, P., & Walden, P. (2005). Asynchronous Adoption Patterns of Mobile Services. In *Proceedings of the 38th Hawaii International Conference on System Sciences, Hawaii.* (CD-ROM), Computer Society Press (10 pages).

Christensen, C. M., & Davis, A. B. (2006). The Future of Newspapers. Retrieved September 29th, 2008. Available at:http://www.forbes.com/home/leadership/2006/10/10/leadership-newspapers-media-lead-innovation-cx_cc_1011clayton.html

Damsgaard, J., & Kelleher, C. (2007). *What drives the innovation, diffusion and adoption of mobile services? - An analysis of four alternative studies.* Los Angeles: Presented at LA Global Mobility Round Table.

Damsgaard, J., & Lyytinen, K. (1996). Government Strategies to Promote the Diffusion of Electronic Data Interchange (EDI): What we Know and What we don't Know. *Information Infrastructure and Policy*, *5*(3), 169–190.

Davis, C. (2009). Plastic Logic announce content partners: USA Today, Zinio and more. Retrieved February 15th, 2009. Available at: http://www.slashgear.com/plastic-logic-announce-content-partners-usa-today-zinio-and-more-0933577/

Davis, F. D. (1989). Perceived Usefulness, Perceived Ease of Use, and User Acceptance of Information Technology. *Management Information Systems Quarterly*, *13*(3), 319–340. doi:10.2307/249008

Davis, F. D., & Venkatesh, V. (2004). Toward Preprototype User Acceptance Testing of New Information Systems: Implications for Software Project Management. *IEEE Transactions on Engineering Management*, *51*(1), 31–46. doi:10.1109/TEM.2003.822468

Dumas, J. S. (2003). User-Based Evaluations. In Jacko, J. A., & Sears, A. (Eds.), *The Human-Computer Interaction Handbook* (pp. 1093–1117). Lawrence Erlbaum Associates.

Eaton, K. (2009). Forget Kindle 2: Fujitsu's E-Reader Screen is Bigger, and It's in Color. Retrieved February 15th, 2009. Available at: http://www.fastcompany.com/blog/kit-eaton/technomix/forget-kindle-2-fujitsus-e-reader-screen-bigger-color

Gassmann, O. (2006). Opening up the innovation process: towards an agenda. *R & D Management*, *36*(3), 223–228. doi:10.1111/j.1467-9310.2006.00437.x

Hammond, K. (2001). B2C e-Commerce 2000-2010: What Experts Predict. *Business Strategy Review*, *12*, 43–50. doi:10.1111/1467-8616.00165

IFRA eNews initiative (2006), retrieved August 14th, 2008. Available at:http://www.ifra.com/website/website.nsf/html/CONT_ENEWS?OpenDocument&ENW&

Ihlström, C. (2004). *The Evolution of a New(s) Genre*. Doctoral Dissertation, Göteborg University.

Ihlström, C. (2005). The e-newspaper innovation - converging print and online. In *Proceedings of the International Workshop on Innovation and Media: Managing Changes in Technology, Products and Processes, Stockholm*.

Ihlström, C., Åkesson, M., & Nordqvist, S. (2004). From Print to Web to e-paper - the challenge of designing the e-newspaper. In *Proceedings of ICCC 8th International Conference on Electronic Publishing, ELPUB 2004, Brasilia*, pp. 249-260.

Ihlström, C., Svensson, J., & Åkesson, M. (2005). Participatory Design of Future Every Day IT Artifacts - Engaging readers and publishers in designing the e-newspaper. In *Proceedings of the 28th Information Systems Research Seminar in Scandinavia, Norway*. CD-ROM.

Ihlström Eriksson, C., Åkesson, M., Bergqvist, M., & Ljungberg, J. (2009). Forming a value network - analyzing the negotiations between actors in the e-newspaper case. In *Proceedings of the 42nd Hawaii International Conference on System Sciences, Hawaii*. (CD-ROM), Computer Society Press (10 pages).

Jungk, R., & Müllert, N. (1996). *Future Workshops - How to Create Desirable Futures*. London: Institute for Social Inventions.

Kalakota, R., & Robinson, M. (2001). *M-Business: The Race to Mobility*. McGraw-Hill Companies.

Karahanna, E., Straub, W. D., & Norman, L. C. (1999). Information technology adoption across time: a cross-sectional comparison of pre-adoption and post-adoption beliefs. *Management Information Systems Quarterly*, *23*(2), 183–213. doi:10.2307/249751

Kensing, F., & Halskov Madsen, K. (1991). "enerating Visions: Future Workshops and Metaphorical Design, in J. Greenbaum and M. Kyng (eds.), *Design at Work*, 155-168. Hillsdale: Lawrence Erlbaum Associates, Inc.

Khalifa, M., & Cheng, S. K. N. (2002). Adoption of Mobile Commerce: Role of Exposure. In *Proceedings of the 35th Hawaii International Conference on System Sciences, Hawaii.* (CD-ROM), Computer Society Press (10 pages).

Kim, W. C., & Mauborgne, R. A. (2005). Blue ocean strategy: From theory to practice. *California Management Review*, *47*(3), 105–121.

Knutsen, A. (2005). M-Service Expectancies and Attitudes: Linkages and Effects of First Impressions. In *Proceedings of the 38th Hawaii International Conference on System Sciences, Hawaii.* (CD-ROM), Computer Society Press (10 pages).

Kuusisto, A. (2008). *Customer roles in business service production - implications for involving the customer in service innovation. Research report 195.* Lappeenranta: Lappeenranta University of Technology.

Lehner, F., & Watson, R. T. (2001). *From E-Commerce to M-Commerce: Research Directions.* Working paper. University of Regensburg: Chair of Business Informatics.

Lettl, C., Herstatt, C., & Gemuenden, H. G. (2006). Users' contributions to radical innovation: evidence from four cases in the field of medical equipment technology. *R & D Management*, *36*(3), 251–272. doi:10.1111/j.1467-9310.2006.00431.x

Lyytinen, K., & Yoo, Y. (2002). Research Commentary: The Next Wave of Nomadic Computing. *Information Systems Research*, *13*(4), 377–388. doi:10.1287/isre.13.4.377.75

Malhotra, A., & Segars, A. H. (2005). Investigating Wireless Web Adoption Patterns in the U.S. *Communications of the ACM*, *48*(10), 105–110. doi:10.1145/1089107.1089113

Mallat, N. (2006). *Exploring Consumer Adoption of Mobile Payments - A Qualitative Study.* Helsinki, Finland: Presented at Helsinki Global Mobility Roundtable.

Mallenius, S., Rossi, M. & Tuunainen, V. K. Factors affecting the adoption and use of mobile devices and services by elderly people – results from a pilot study. *Presented at LA Global Mobility Round Table, Los Angeles.*

Miles, M. B., & Huberman, A. M. (1994). *Qualitative Data Analysis.* California: Sage Publications, Inc.

Mingers, J. (2001). Combining IS Research Methods: Towards a Pluralist Methodology. *Information Systems Research*, *12*(3), 240–259. doi:10.1287/isre.12.3.240.9709

Mingers, J., & Gill, A. (1997). *Multimethodology: Theory and Practice of Combining Management Science Methodologies.* Chichester, U.K.: Wiley.

Patton, M. Q. (2002). *Qualitative Research & Evaluation Methods* (3 ed.). Sage Publications, Inc. California.

Pinsonneault, A., & Kraemer, K. L. (1993). Survey Research Methodology in Management Information Systems: An Assessment. *Journal of Management Information Systems*, *10*(2), 75–105.

Plastic Logic. retrieved February 8th, 2007. Available at:http://www.plasticlogic.com/news.php

Polymer Vision. retrieved August 14th, 2008. Available at: http://www.polymervision.com/

Rogers, E. M. (1995). *Diffusion of innovations.* New York: The Free Press.

Rosted, J. (2005). *User-driven innovation–Results and Recommendations.* FORA Report.

Sarker, S., & Wells, J. D. (2003). Understanding mobile handheld device use and adoption. *Communications of the ACM*, *46*(12), 35–40. doi:10.1145/953460.953484

Sawhney, M., Verona, G., & Prandelli, E. (2005). Collaborating to create: The Internet as a platform for customer engagement in product innovation. *Journal of Interactive Marketing*, *19*(4), 4–17. doi:10.1002/dir.20046

Shea, C. (2009). Kindling controversy. Retrieved February 15th, 2009. Available at: http://www.boston.com/bostonglobe/ideas/brainiac/2009/02/kindling_controversy.html

Sims, F., Williams, M.-A., & Elliot, S. (2007). Understanding the Mobile Experience Economy: A key to richer more effective M-Business Technologies, Models and Strategies. In *Proceedings of Sixth International Conference on the Management of Mobile Business* (ICMB 2007).

Svensson, J., & Hakeröd, J. (2006). Design of navigational support for task oriented reading in e-newspapers. In *Proceedings of the 29th Information Systems Research Seminar in Scandinavia, Denmark*.

Taub, E. A. (2008). New E-Newspaper Reader Echoes Look of the Paper. Retreived February 2nd, 2009. Available at:http://www.nytimes.com/2008/09/08/technology/08ink.html

The Economist. (2006). Who killed the newspaper? Retrieved September 29th, 2008. Available at: http://www.economist.com/opinion/displaystory.cfm?story_id=7830218.

The Swedish Newspaper Publishers' Association. (2008), retrieved August 14th, 2008. Available at: http://www.tu.se/index_target.asp?c1=1214

Tilson, D., Lyytinen, K., & Baxter, R. (2004). A Framework for selecting a Location Based Service (LBS) Strategy and Service Portfolio. In *Proceedings of the 37th Hawaii International Conference on System Sciences, Hawaii*. (CD-ROM), Computer Society Press (10 pages).

Tornatzky, L. G., & Klein, K. J. (1982). Innovation Characteristics and Adoption-Implementation. *IEEE Transactions on Engineering Management*, *29*(1), 28–45.

Tscherning, H., & Damsgaard, J. (2008). Understanding the Diffusion and Adoption of Telecommunication Innovations: What We Know and What We Don't Know. In IFIP, Volume 287, *Open IT-Based Innovation: Moving Towards Cooperative IT Transfer and Knowledge Diffusion*, eds. León, G., Bernardos, A., Casar, J., Kautz, K., and DeGross, J. (Boston: Springer), 41-62.

Venkatesh, V., Morris, M. G., Davis, G. B., & Davis, F. D. (2003). User Acceptance of Information Technology: Toward a Unified View. *Management Information Systems Quarterly*, *27*(3), 425–478.

von Hippel, E. (2005). Democratizing Innovation: The evolving phenomenon of user innovation. *Journal für Betriebswirtschaft*, *55*(1), 63–78. doi:10.1007/s11301-004-0002-8

Vrechopoulos, A., Constantiou, I., & Sideris, G. (2003). The critical role of consumer behaviour research in mobile commerce. *International Journal of Mobile Communications*, *1*(3), 329–340. doi:10.1504/IJMC.2003.003498

WAN. (2006). *World Association of Newspapers*. World Press Trends.

ENDNOTES

1. Amazon Kindle (2009), retrieved February 1st, 2009. Available at: http://www.amazon.com/Kindle-Amazons-Wireless-Reading-Device/dp/B000F82I73MA
2. iRex iLiad (2009), retrieved February 1st, 2009. Available at: http://www.irextechnologies.com/products/iliad

3 Sony Reader (2009), retrieved February 1st, 2009. Available at: http://www.sonystyle.com/webapp/wcs/stores/servlet/CategoryDisplay?catalogId=10551&storeId=10151&langId=-1&categoryId=8198552921644523779

4 Cybook (2009), retrieved February 1st, 2009. Available at: http://www.bookeen.com/ebook/ebook-reading-device.aspx

5 Oak Investment Partners invests in Plastic Logic, retrieved February 1st, 2009. Available at: http://www.idtechex.com/printelecreview/en/articles/00000399.asp

This work was previously published in International Journal of E-Services and Mobile Applications, Volume 1, Issue 3, edited by Ada Scupola, pp. 38-78, copyright 2009 by IGI Publishing (an imprint of IGI Global).

Section 3

Mobile Services:

Acceptance and Consumer Values

Chapter 10
The Influence of Content and Trust on Consumers' Intention to Accept Mobile Advertisements

Mika Westerlund
Helsinki School of Economics, Finland

Risto Rajala
Helsinki School of Economics, Finland

Tuure Tuunanen
The University of Auckland, New Zealand

Jari Salo
University of Oulu, Finland

ABSTRACT

A new model for investigating consumers' behavioral intention in mobile advertising is developed and related to mobile marketing theory and practice. Hypotheses are tested about the hierarchical structure and the effects of the factors that precede consumers' acceptance of mobile advertisement. The results suggest that attractive content and trust in advertisers are key predictors of mobile device users' behavioral intention towards mobile advertising. The findings lead to a modification and extension of Fishbein's behavioral intention model and its application in the research on advertising effectiveness.

INTRODUCTION

The growth of mobile advertising has opened a new area for research. Previously, advertising and marketing researchers paid considerable attention to consumers' attitude toward advertisements in the major media (Homer, 1990). The research has shown that consumers now ignore advertising more than they did in previous decades. At the same time, however, Internet advertising seems to generate positive consumer attitudes (Schlosser et al., 1999) and the rapid proliferation of mobile

DOI: 10.4018/978-1-60960-607-7.ch010

devices has created a new channel for marketing (Tsang et al., 2004). Hence, consumers' attitudes toward advertising in the mobile media have gained significant attention among both researchers and business practitioners (Bauer et al., 2005; Ferris, 2007; Karjaluoto and Alatalo, 2007; Lee et al., 2006; Okazaki et al. 2007).

Acceptance research has provided important insights in explaining the success of mobile advertisement (Bauer et al., 2005). This research stream has investigated a myriad of aspects that affect consumers' attitude toward mobile marketing. These attitudes reflect the degree to which consumers identify with the advertising (MacKenzie & Lutz, 1989). Schlosser et al. (1999) reveal that mobile device users' attitude towards mobile marketing is less resistant to change than are consumers' attitude towards advertising in general. For example, prior research indicates that context, credibility and subjective norm are positively related to consumers' intentions to participate in mobile marketing (Karjaluoto & Alatalo, 2007; Lee et al., 2006). However, little is known about the hierarchical structure and effects of the underlying factors that antecede consumers' behavioral intention to accept specific mobile advertising content.

The present study examines consumers' behavioral intentions in the context of mobile marketing. Extending the Theory of Reasoned Action (TRA) proposed by Fishbein and Ajzen (1975), the study responds to the recent call for research (e.g., Nantel & Sekhavat, 2008; Okazaki et al., 2007) on advertisement content and trust in the advertisers as the key predictors of consumers' willingness to participate in mobile advertising. In particular, by employing PLS structural equation modeling and the composition of higher-order constructs comprising their underlying variables, the study investigates (1) the hierarchical structure; (2) the relative effects of the factors that precede consumers' intention to accept mobile advertisement; and (3) provides evidence that validates the model and discusses whether such a conceptualization is generalizable across consumer groups.

The article is structured as follows. After this introductory section, we offer a literature review on the theoretical foundations of the study. Moreover, we establish a research model and formulate hypotheses based on previous literature. Thereafter, we present our research design, measures, data analysis and the results. Finally, we conclude the article by discussing the implications of the study.

THEORY AND HYPOTHESES

The literature suggests that intention towards action is the best predictor of individuals' behavior (e.g., Fishbein and Ajzen, 1975). Behavioral intention measures the strength of consumers' conscious plans to perform the target behavior, such as consumers' acceptance of mobile advertising. Nantel and Sekhavat (2008) argue that the content of marketing messages significantly influence the behavioral intentions evoked by mobile advertisement. They further include the concepts of credibility and trust in the advertiser as important dimensions of the source of these messages. Thus, content and trust need further investigation as the key determinants of consumers' perception on mobile advertising.

Advertisement Content

Mobile advertising depends upon consumers' acceptance. Rettie et al. (2005) show that although their initial attitudes toward mobile advertising are sometimes negative, many consumers are prepared to accept advertising if the content appeals to them. Similarly, Muk (2007) reports that the relevance of the advertising message content and the perceived value of the offering are of great importance in the success of advertising via mobile devices. According to Muk (*ibid.*), this is due to that these factors have the largest impact on attitude toward advertising by reducing the perception of intrusiveness and, thus, increasing the acceptance of mobile advertisements.

Bauer et al. (2005) distinguish two kinds of acceptance drivers: perceived utility and perceived risk. Consumers will only accept mobile marketing if they perceive a benefit in receiving advertising messages on their mobile phone and if they can avoid the potential negative consequences receiving those advertisements. Moreover, Carroll et al. (2005) show that the content and its relevance to the audience are critical to the acceptance of mobile marketing. These arguments are congruent with Kaas' (1990) claim that consumers' perception of the advertisement is subject to its marginal utility compared to the utility that results from an alternative activity. This implies that a consumer's attitude toward mobile marketing will be more positive if the content has a higher perceived utility. The content of mobile advertisements consists of several dimensions: the degree of personalization, advertisement context, informativeness of the message, and entertainment value (Barwise & Strong, 2002; Bauer et al. 2005; Heinonen & Strandvik, 2007; Lee et al., 2006; Tsang et al. 2004).

Personalization. Bauer et al. (2005) accentuate the role of personalization in mobile marketing, arguing that a mobile phone is rarely used by anyone but its owner. Moreover, Barwise and Strong (2002) claim that consumers expect highly personalized messages, because they receive the messages through their mobile phones which they consider an intimate device. Bauer et al. (2005) emphasize that most users maintain a very personal relationship with their mobile phone, and, that the Subscriber Identity Module (SIM) card allows for the exact identification of each mobile phone and its user. Using the mobile medium for communication therefore enables the advertiser to personalize the content. Leppäniemi and Karjaluoto (2005) posit that personalization is grounded on "building customer loyalty by building a meaningful one-to-one relationship." That is, they stress the importance of understanding the needs of each consumer, and, helping to satisfy a goal that efficiently and knowledgeably meets each consumer's need in a given context. Hence, they define personalization in terms of individual preferences, needs, mindsets and lifestyles, and both cultural as well as geographical differences. Jingjun Xu (2006-2007) shows that personalization can enable marketers to reach their potential customers in a very individual way while improving their relationships with the existing customers. Furthermore, Jingjun Xu (2006-2007) adds personalization as an essential means of influencing consumers' attitudes toward mobile advertising.

Context. Heinonen and Strandvik (2007) argue that consumers' willingness to receive and respond to marketing communication can be viewed as a function of the content and the context of the message. They contend that any channel can and should be evaluated according to consumer responsiveness in order to understand the effects and effectiveness of communication. In conformity with this view, Muk (2007) and Leppäniemi and Karjaluoto (2008) stress that the uniqueness of mobile advertising lies in its potential to target consumers in a specific context. In other words, mobile advertising should take into account the usefulness of information in regard to consumers' time, location, and other contextual attributes.

Informativeness. Consumers' propensity to search and use information is an important construct in the analysis and explanation of their behavior (Kroeber-Riel & Weinberg, 2003). Both Ducoffe (1996) and Barwise and Strong (2002) assert that consumers prefer informative messages and the informativeness of the content of advertisements is an important predictor of their value and one that is crucial to the effectiveness of advertising. Lee et al. (2006) have studied the informativeness of mobile advertisement content in terms of its helpfulness and factuality. To investigate informativeness, they use variables such as "mobile advertisements are interesting to me" and "mobile advertisements are a great source of timely information." Barwise and Strong (2002) suggest that highly relevant content can only be achieved from data obtained explicitly from the consumer at the time of obtaining permission, rather than by mining an existing customer database. They claim that if the messages are considered irrelevant by the target audience, there is a potential for negative reactions.

Entertainment. Tsang et al. (2004) and Lee et al. (2006) investigate the entertainment value of mobile advertisements through consumers' perceived enjoyment and entertainment of mobile advertisement. For this purpose, their variables include, "receiving mobile advertisements is enjoyable and entertaining" and "receiving mobile advertisements is pleasant." Ducoffe (1996) posits that the form, or entertainment, of the advertisement content is a vital determinant of their value and crucial to the effectiveness of advertising. Along with their entertainment value, Ducoffe (1996) states that irritation with advertisements also influences people's attitude. This finding is consistent with earlier research (Mitchell and Olson, 1981) that interesting and pleasing advertisements have a positive impact on consumers' attitudes. Barwise and Strong (2002) also maintain that consumers prefer short and entertaining messages.

Trust in the Advertiser

Consumers prefer to minimize risk rather than maximize utility. According to Bauer et al. (2005), a consumer's subjective risk perception can thus strongly determine his or her behavior. Bauer et al. (ibid.) suggest that in the context of mobile marketing, trust should be conceptualized as consisting of two constructs, brand trust and mobile advertising trust, both of which affect users' choice to push mobile advertising. They posit that consumers are likely to accept push advertisement only when they perceive both the medium and the content to be trustworthy and non-deceptive. Attributes of trust are identified in several ways. Traditionally, the relevance and credibility of advertisements have been considered important mediators of advertising effects (MacKenzie & Lutz, 1989). Trust becomes crucial when consumers make decisions or take action on the basis of unclear information. In particular, perceived trust can be seen as a reflection of an individual's determination of the veracity of the imparted information. However, Eastin (2001) states that it is difficult to distinguish more from less trustworthy information because much of the content of online information is not subject to governmental or ethical regulation. In the literature, the aspects of trust consist of attitude towards advertising, subjective norm, the perceived credibility of the advertiser, and consumer privacy issues (Bauer et al., 2005; Nantel & Sekhavat, 2008; Okazaki et al., 2007; Tsang et al., 2004).

Attitude. Bauer et al. (2005) posit that the more positive one's attitude toward advertising in general the more positive his or her attitude toward mobile marketing. Zanot (1984) reveals that attitudes toward advertising in the major media became increasingly negative after the 1970s. Internet advertising, however, seems to generate positive consumer attitudes (Schlosser et al., 1999). Karjaluoto and Alatalo (2007) show that the more favorable the consumers' attitude towards mobile marketing, the higher their intention to participate in mobile marketing. Jun and Lee (2007) add that advertising attitudes are positively related with consumers' past behavioral experience. This study considers the experience-based attitude as an important antecedent of trust in the advertisers.

Subjective norm. The theory of reasoned action suggests that subjective norm, i.e. social influence, affects a person's intention to perform a behavior. It is defined as the person's perception that people think that they should or should not perform the behavior in question (Ajzen, 1991; Davis et al., 1989; Fishbein & Azjen, 1975). According to Ajzen and Fishbein (1980), subjective norm is a function of the product of one's normative belief and motivation. It is the person's belief that the salient referent thinks he/she should (or should not) perform the behavior and his/her motivation to comply with that referent. In the information systems research, the effect of subjective norm is shown to be more salient for consumers with little or no experience (Venkatesh et al., 2003). Thus, subjective norm may be considered influential in the emergence of trust.

Credibility. The term "credibility" is used to refer to the traits of the communicator (whether an individual or a company). It includes expertise, trustworthiness, attractiveness, and power (Lee et al., 2006). Mackenzie and Lutz (1989) and Drossos et al. (2007) show that credibility strongly influences a consumer's attitude toward the advertiser, which in turn is an important predictor of the consumer's behavioral intention toward the advertisement. Sternthal et al. (1978) noted that a source with high credibility has a greater power of persuasion than a source with low credibility. Furthermore, Choi and Rifon (2002) define the credibility of the advertiser as "the extent to which consumers believe that a firm can design and deliver products and services that satisfy customer needs and wants." Moreover, Nantel and Sekhavat (2008) point out that credibility of a source is connected with its power to inspire trust in the advertiser.

Privacy. The perceived risk of losing one's privacy is also related to a consumer's trust in the advertiser. The risk associated with mobile marketing is mainly perceived as one of data security. The conception of privacy is vital in mobile advertising; Bauer et al. (2005) emphasize that new media service users tend to have concerns about data manipulation, unauthorized data access, and unwanted tracking of usage patterns. According to Bauer et al. (2005), another security issue is the protection of consumers' privacy. By using the mobile medium it is possible for marketers to reach consumers at any time and in any place. This characteristic provides the basis for high-potential, personalized mobile marketing on one hand, but also accounts for consumers' fear of privacy violations on the other. In addition, Rappaport (2007) cites the urgency of treating consumers' personal information with integrity. According to him, this concern is framed in terms like the "privacy issue" as an essential foundation for brand-consumer relationships and brand demand. It pertains to consumers' belief that advertisers are reputable and responsible, and that they will treat consumers' personal information with care.

The Theory of Reasoned Action in Mobile Advertising

The purpose of the Theory of Reasoned Action (TRA) is to predict and understand the factors influencing an individual's behavior in a specific context. Originating in social psychology, it was developed by Fishbein and Azjen (1975). The theory and its subsequent variation, the Theory of Planned Behavior (TPB), have been applied to research in a variety of fields. TRA provides a theoretical foundation for the linkage among four constructs: behavior, intention, attitude, and belief. Through exposure to an object, people link the object with its attributes in varying strengths. Its premise is that the totality of a person's belief is the informational base that ultimately determines his or her attitudes, intentions, and behaviors (Fishbein & Ajzen, 1975).

Bauer et al. (2005) show that the TRA is a means of measuring consumers' acceptance of mobile advertisements. According to their study, acceptance should be forecasted by measuring an individual's intention to accept. As such, it is of major relevance to both theory and practice to develop a model for testing mobile marketing acceptance. The basic assumption of Ajzen and Fishbein's theory is that individuals consciously decide to perform or not to perform a specific behavior. In this, they consider and evaluate various criteria pertaining to the behavior before actually performing it. Therefore, for a reliable forecast of the acceptance of mobile marketing, it is necessary to examine the combination of acceptance and use decisions.

The fundamental proposition of the TRA is that behavior (B) is determined by behavioral intention (BI). The behavioral intention is in turn postulated to be a function of the individual's attitude toward the act (A_{act}) and toward social norms (SN). Whether the attitude toward the act or the social norms exerts the greater influence on the behavioral intention depends on the individual and the decision object (Ajzen and Fishbein, 1980; Fishbein and Ajzen, 1975). This relationship is originally written as:

$$B \sim BI = A_{act}\, w_1 + SN\, w_2 \qquad (1)$$

The parameters w_1 and w_2 are empirically determined standardized regression coefficients (Fishbein and Ajzen, 1975), each reflecting the strength of the relative impact of the attitude toward the act and the social norms on the behavioral intention. However, prior literature on the acceptance of mobile advertisement (e.g., Gopal & Tripathi, 2006) argues that the perceived content of the advertisement is important in an individual's decision to accept the advertisement (e.g., Ferris, 2007). Moreover, both the effect of A_{act} and SN on the behavioral intention (BI), and, ultimately, to the acceptance of mobile advertisement (B), are mediated by trust in the advertiser. Furthermore, trust is associated with other variables such as credibility (MacKenzie & Lutz, 1989; Okazaki et al., 2007) and privacy (Rappaport, 2007). Therefore, in the context of a consumer's acceptance of mobile advertisement, this relationship can be restated as:

$$B \sim BI_{Acc} = w_1 CTN + w_2 TRU, \qquad (2)$$

where behavior *(B)* is a reflection of a consumer's intention to accept advertisements *(BI_{Acc})*, which in turn, is subject to the consumer's perception of the content (*CTN*), and, beliefs about its consequences (C_i) and the evaluation of those consequences (E_i). That is:

$$\sum_{i=1}^{n} CTN = C_i \times E_i,$$
where i = number of consequences. (3)

Furthermore, the second level of explanation (specified above) includes the effect of trust (*TRU*), which is subject to the individual's perception of the consequences of trusting the advertiser. This can be written as:

$$\sum_{i=1}^{n} TRU = C_i \times E_i,$$
where i = number of consequences. (4)

Measuring the second level constructs, i.e., "perceived advertisement content" (*CTN)* and "perceived trust in advertisers" (*TRU)* will therefore be sufficient for this study. On the basis of these considerations the following hypotheses are formulated:

H_1: *Perceived advertisement content (CTN) has a direct positive effect on consumers' behavioral intention to accept (BI_{Acc}) mobile advertisements.*

H_2: *Perceived trust in advertisers (TRU) has a direct positive effect on consumers' behavioral intention to accept (BI_{Acc}) mobile advertisements.*

METHODOLOGY AND DATA

An online survey to test the hypotheses was conducted in 2007. The empirical inquiry was carried out among European mobile phone users hailing from Finland and the Netherlands. Using the competitiveness indicators, Finland and the Netherlands are the leading European information societies in the benchmark analysis of EU-15 countries (Ahokas & Kaivo-oja, 2003). Furthermore, according to the European Commission's Information Society and Media reports

(2008a; 2008b), in 2007 the mobile penetration rate in Finland exceeded 108% and in the Netherlands 111%. Thus, mobile market profiles in these two countries are highly similar, making them a feasible sampling frame. Moreover, these countries have high broadband penetration rates (over 34%), and thus, they attractive markets for mobile advertisers.

Respondents were recruited via an e-mail in which an invitation and a link to the survey were included in the message body. The total sample size consisted of 420 potential respondents and the questionnaire yielded 103 usable responses for analysis. Thus, the response rate equals 24.5% which is considered acceptable. Respondents' demographic information was collected for control purposes. The sample comprised 68% males and 32% females. The average age of the respondents was 32, with the youngest respondent being 21 and the oldest 63 years old. Overall, over 75% of the respondents were younger than 39 years old and more than half (51.5%) were younger than 30 years old. Most respondents had been exposed to mobile advertising; 79% of them had previously received mobile advertisements and 21% had not.

Multi-item scales were used to measure all constructs. The survey explored user experience, attributes of mobile advertising, and the consumer's intention to accept mobile advertising. All items were measured on a five-point Likert-type scale (1="strongly disagree" to 5="strongly agree"). The items were drawn from the literature, but the wording of the questionnaire was slightly modified in order to fit this context of mobile advertising (see Appendix 1 for the list of surveyed items).

Scale Validity and Reliability

To avoid the standard assumptions of multinormality and the necessity of a large sample size, Wold's (1982) method of partial least squares (PLS) was used for parameters estimation. To assess the reliability and validity of the constructs, composite reliability values (ρc) and average variance extracted values (ρv) were examined for each first-order latent variable. Construct reliability was assessed using composite reliability analysis as suggested by Fornell and Larcker (1981). Following Chin et al. (2003), it can be written using the calculation formula:

$$\rho_c = \frac{(\sum \lambda_i)^2 \operatorname{var} F}{(\sum \lambda_i)^2 \operatorname{var} F + \sum \Theta_{ii}}. \tag{5}$$

All composite reliability values were above the recommended level of.70 (Fornell & Larcker, 1981). A complementary measure to composite reliability is the average variance extracted, which is useful in examining convergent validity. Average variance extracted is the average variance shared between a construct and its measures (Hulland, 1999) and the equation is defined as:

$$\rho_v = \frac{(\sum \lambda_{yi})^2}{(\sum \lambda_{yi})^2 + \sum \operatorname{var}(\varepsilon_i)}. \tag{6}$$

It shows directly the amount of variance captured by the construct in relation to the variance due to measurement error. In our study, all constructs exceeded the recommended.50 benchmark (Diamantopoulos & Siguaw, 2000) except the second-order construct *TRU*, which, however, had an acceptable value of.46. Overall, the composite reliability values and average variance extracted values indicate that the scales perform adequately. In addition to these two measures, Table 1 shows the means, standard deviations, Cronbach's alphas for internal consistency, and correlations for the constructs.

Second-Order Constructs

PLS enables scholars to investigate models at a higher level of abstraction (Lohmöller, 1989) which is useful in estimating complex models (Chin, 1998). For this purpose, Wold (1982)

Table 1. Construct correlations and descriptive statistics of measures (n=103)

Construct	Mean	SD	ρ_v	ρ_c	α	1	2	3	4	5	6	7	8	9
1. Personalization	2.53	.91	.87	.93	.85	(.93)								
2. Context	3.15	.97	.89	.94	.88	.45	(.94)							
3. Informativeness	2.42	.91	.77	.87	.70	.40	.62	(.88)						
4. Entertainment	1.86	.83	.90	.94	.88	.36	.48	.65	(.95)					
5. Attitude	2.61	1.06	.84	.91	.81	.32	.58	.72	.70	(.92)				
6. Subjective norm	3.22	.97	.87	.93	.86	.12	.33	.27	.37	.40	(.94)			
7. Credibility	2.49	.92	.88	.93	.86	.19	.55	.40	.37	.35	.21	(.94)		
8. Privacy	2.87	1.20	.90	.94	.88	.26	.24	.25	.38	.28	.10	.30	(.95)	
9. BI_{Acc}	2.26	.93	.82	.93	.89	.30	.60	.66	.67	.79	.33	.37	.25	(.90)
10. CTN (1-4)	-	-	.62	.87	.87	-	-	-	-	-	-	-	-	-
11. TRU (5-8)	-	-	.46	.77	.78	-	-	-	-	-	-	-	-	-

Note: SD = standard deviation; ρ_v = average variance extracted; ρ_c = composite reliability; Cronbach's alpha (α) = $((\Sigma_{h \neq h'} \text{cov}(x_h, x_{h'}))/\text{var}(\Sigma_h x_h))$ $(p/(p-1))$ (Tenenhaus et al., 2005); square root of ρ_v on diagonal (in parentheses).

suggests the use of repeated indicators (i.e., the hierarchical component model) for measuring second-order constructs. That is, all indicators of the first-order constructs are reassigned to the second-order construct, as second-order models are a special type of PLS path modeling that uses manifest variables twice for model estimation. According to Hulland (1999), the researcher needs to decide whether it is more correct to think of the underlying construct as causing the observed measures (i.e., a reflective relationship) or of the measures as causing or defining the construct (i.e., a formative relationship). However, a prerequisite for the repeated indicators approach is that all indicators of the first-order and the second-order factors should be reflective. According to Jarvis et al. (2003), such a model is called a total disaggregation second-order factor model. It has a series of first-order latent factors with reflective indicators. These first-order factors are themselves reflective indicators of an underlying second-order construct. Moreover, the second order latent variable should be used as exogenous variable, because its variance is explained by its indicators and, otherwise, the specification of an additional source of variation (i.e., an antecedent construct) would be conceptually questionable (Diamantopoulos et al., 2008). Thus, all items included in our PLS analysis were configured as reflective indicators (Fornell and Larcker, 1981; Haenlein and Kaplan, 2004) and the second-order constructs *CTN* and *TRU* are considered exogenous variables. Reflective indicators are expressed as a function of their associated latent variables by using matrix algebra as suggested by Haenlein and Kaplan (2004):

$$y = \Lambda_y \eta + \varepsilon, \tag{7}$$

where Λ_y = loadings of indicators of endogenous variables, η = latent endogenous variable, and ε = measurement errors for indicators of endogenous variable.

Discriminant validity was assessed by examining the correlation matrix of the constructs. According to Fornell and Larcker (1981), satisfactory discriminant validity among constructs is obtained when the square root of the average variance extracted is greater than corresponding construct correlations. This implies that the variance shared between any two constructs is less than the variance shared between a construct and its indicators. For each pair of constructs, the square root of the average variance extracted exceeded their correlations, thus supporting the discriminant validity of the constructs.

EMPIRICAL ANALYSIS AND RESULTS

Hypotheses were tested using the SmartPLS 2.0 developed by Ringle et al. (2005). PLS path modeling is a component-based SEM approach that does not require multivariate normal data and places minimum requirements on measurement levels (Hulland, 1999; Tenenhaus et al., 2005). Moreover, the use of the PLS method is typically recommended in situations in which the sample size is small (Haenlein & Kaplan, 2004). The sample size in the present study is 103, thus supporting the use of PLS. In addition, PLS is viable for analyzing predictive research models that are in the early stages of theory development (Barclay et al., 1995), as is the model in the present study. Because PLS considers all path coefficients simultaneously and estimates multiple individual item loadings in the context of a theoretically specified model rather than in isolation, it helps to avoid biased and inconsistent parameter estimates for equations.

Both hypotheses were examined with full-sample using t-tests (df=188). First, PLS generates estimates of standardized regression coefficients for the paths in a structural equation model. Then, the bootstrap procedure approximates the sampling distribution of an estimator by resampling with replacement from the original sample, which is necessary to derive valid t-values. The analysis was conveyed using 500 bootstrap replications as suggested by Chin et al. (2003). Structural equation model and the results of the analysis are shown in Figure 1.

The explanatory power of the model for the dependent construct was measured by using the squared multiple correlations value (R^2). As PLS does not provide overall fit indexes for the model, Hulland (1999) suggests that researchers report R^2 values for all endogenous constructs included in their models. In the present study, the independent constructs explained 57% of the variance in intention to accept mobile advertisements, which is considered good for this kind of analysis.

PLS path modeling includes no proper single goodness-of-fit measure. However, to conclude our structural analysis, we calculate the goodness of fit (GoF) of the model using Tenenhaus et al. (2005) global fit measure for PLS. By taking the square root of the product of the variance extracted of all constructs with multiple indicators and the average R^2 value of the endogenous constructs, we can calculate a fit measure ranging between 0 and 1. The measure was calculated using the second-order constructs and the dependent construct. According to the categorization by Cohen (1988) and using.50 as a cut-off value for communality (Fornell & Larcker, 1981), the GoF criteria for small, medium, and large effect sizes are .10, .25, and .36. In our model, the GoF is.59, which indicates a good fit of the model to the data.

Table 2 lists the results for the hypotheses. As hypothesized, the perceived advertisement content *(CTN)* has significant positive effect on consumers' behavioral intention to accept (BI_{Acc}) mobile advertisements (β=.48, p<.001), thus supporting Hypothesis 1. Moreover, perceived trust in advertisers *(TRU)* has significant positive effect on consumers' behavioral intention to accept (BI_{Acc}) mobile advertisements (β=.32, p<.05), thus supporting Hypothesis 2. In general, these findings support the arguments that both the content and trust are important predictors of consumers' acceptance of mobile advertising.

DISCUSSION AND CONCLUSION

Our study contributes to the literatures on social psychology and consumer marketing by extending

Figure 1. Structural equation model of the study

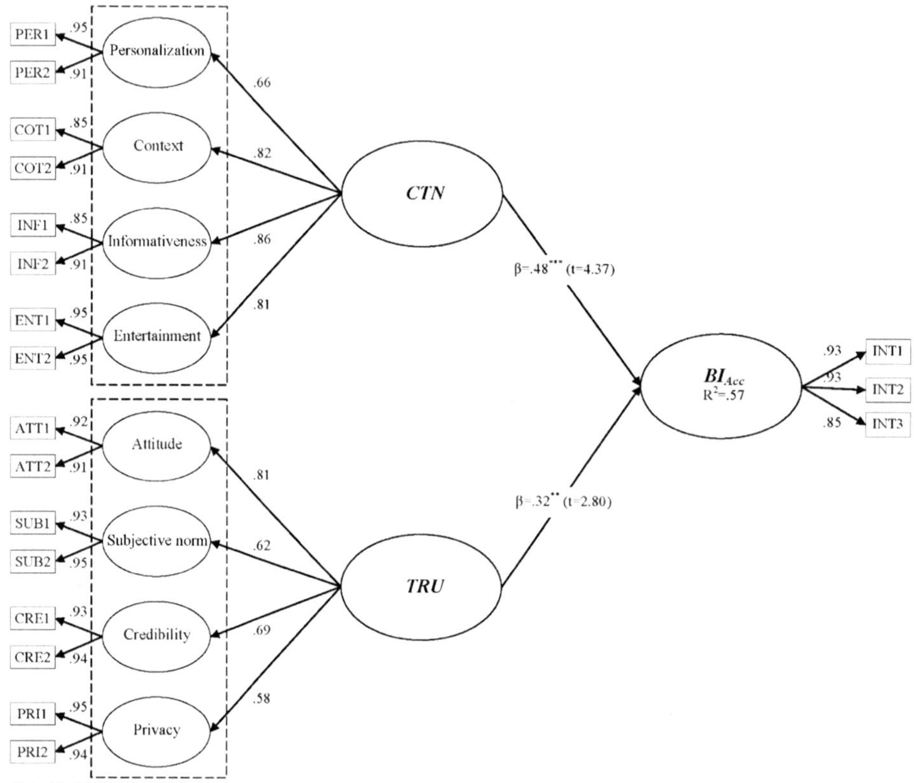

*Note: **p<0.01, ***p<0.001*

Table 2. Results of hypotheses testing (n=103, bootstrap samples=500, df=188)

H#	Relationship	Coefficient	*t*-value	*p*-value	Support
H_1	CTN → BI_{Acc}	.48	4.37	<.001	Yes
H_2	TRU → BI_{Acc}	.32	2.80	.006	Yes

the scope of application of the Theory of Reasoned Action (TRA) to the field of mobile advertising. The findings suggest that TRA is applicable in predicting consumers' behavioral intention to accept mobile advertisements. The results underline the importance of appealing content and trust in the advertisers in the consumers' behavioral intention to accept marketing messages. According to Bauer et al. (2005), the communication of advertising content over mobile media can only be effective if consumers permit the continuous reception of advertising messages on their mobile devices.

The findings are congruent with Heinonen and Strandvik (2007) and Carroll et al. (2005), who suggest that the acceptance of a mobile marketing message is likely to be influenced by the consumer's acceptance of the mobile medium, the relevance of the content and the context of the marketing communication. According to their findings, messages that are concise, funny, interactive, entertaining and relevant to the target group usually achieve higher levels of success. The present study improves the understanding of marketers' persuasive options. It provides evidence of the underlying dimensions that are relevant in designing content of mobile advertisements, and, in increasing consumers' trust in mobile advertisers.

These results are relevant for both scholars and business practitioners interested in how mobile advertising works. For scholars, the study provides sheds light on the dimensions of consumers' behavioral intention by showing that content and trust are multifaceted constructs that are products of a number of underlying attributes. The present study provides theoretical and methodological grounds for classifying these underpinnings. For mobile advertisers, our findings highlight the importance of focusing the type of content in terms of personalization and context-specificity and the function of the message's information and entertainment value. In addition, the study underlines the significance of encouraging trust in terms of personal and social dimensions, and consumers' perception of the advertiser's credibility and the protection or safety of their private information.

Although the present study provides solid evidence of the roles of content and trust in consumer behavior, it is not free from limitations. As our sample consisted of only 103 respondents, caution needs to be exercised in generalizing the results. Moreover, the established higher-order constructs should be validated though other empirical methods and larger sample sizes. In addition, our data includes demographic control variables, but the empirical analysis did not consider possible differences among consumer groups in terms of age, gender, cultural background, and other characteristics.

REFERENCES

Ahokas, I., & Kaivo-oja, J. (2003). Benchmarking European information society developments. *Foresight*, *5*(1), 44–54.. doi:10.1108/14636680310471280

Ajzen, I. (1991). The Theory of Planned Behavior. *Organizational Behavior and Human Decision Processes*, *50*, 179–211..doi:10.1016/0749-5978(91)90020-T

Ajzen, I., & Fishbein, M. (1980). *Understanding Attitudes and Predicting Social Behavior.* New Jersey: Englewood Cliffs.

Barclay, D., Higgins, C., & Thompson, R. (1995). The partial least squares (PLS) approach to causal modeling: Personal computer adoption and use as an illustration. *Technology Studies*, *2*, 285–309.

Barwise, P., & Strong, C. (2002). Permission-Based Mobile Advertising. *Journal of Interactive Marketing*, *16*(1), 14–24..doi:10.1002/dir.10000

Bauer, H. H., Reichardt, T., Barnes, S. J., & Neumann, M. M. (2005). Driving Consumer Acceptance of Mobile Marketing: a Theoretical Framework and Empirical Study. *Journal of Electronic Commerce Research*, *6*(3), 181–192.

Carroll, A., Barnes, S. J., & Scornavacca, E. (2005). Consumer Perceptions and Attitudes Towards Mobile Marketing. In Barnes, S. (Ed.), *Unwired Business: Cases in Mobile Business* (p. 109). Hershey, PA, USA: IRM Press.

Chin, W. W. (1998). The Partial Least Squares Approach to Structural Equation Modeling. In Marcoulides, G. A. (Ed.), *Modern Methods for Business Research* (pp. 295–336). Mahwah, NJ: Lawrence Erlbaum.

Chin, W. W., Marcolin, B. L., & Newsted, P. R. (2003). A partial least squares latent variable modeling approach for measuring interaction effects: Results from a Monte Carlo simulation study and an electronic-mail emotion/adoption study. *Information Systems Research*, *14*, 189–217.. doi:10.1287/isre.14.2.189.16018

Choi, S. M., & Rifon, N. J. (2002). Antecedents and Consequences of Web Advertising Credibility: A Study of Consumer Response to Banner Advertisements. *Journal of Interactive Advertising, 3*(1).

Cohen, J. (1988). *Statistical Power Analysis for the Behavioral Sciences* (2nd ed.). Hillsdale, NJ: Lawrence Erlbaum Associates.

Davis, F. D., Bagozzi, R. P., & Warshaw, P. R. (1989). User Acceptance of Computer Technology: A Comparison of Two Theoretical Models. *Management Science, 35*(8), 982–1002.. doi:10.1287/mnsc.35.8.982

Diamantopoulos, A., Riefler, P., & Roth, K. P. (2008). Advancing Formative Measurement Models. *Journal of Business Research, 61*(12), 1203–1218..doi:10.1016/j.jbusres.2008.01.009

Diamantopoulos, A., & Siguaw, J. (2000). *Introducing Lisrel: A Guide for the Uninitiated.* London: Sage.

Drossos, D., Giaglis, G. M., Lekakos, G., Kokkinaki, F., & Stavraki, M. G. (2007). Determinants of Effective SMS Advertising: An Experimental Study. *Journal of Interactive Advertising, 7*(2).

Ducoffe, R. H. (1996). Advertising value and advertising on the Web. *Journal of Advertising Research, 36*(5), 21–35.

Eastin, M. S. (2001). Credibility Assessments of Online Health Information: The Effects of Source Expertise and Knowledge of Content. *Journal of Computer-Mediated Communication, 6*(4).

European Commission. (2008a). *Towards a Single European Telecoms Market: Focus on Finland, Information Society and Media report.* Retrieved March 27, 2009, from http://ec.europa.eu/information_society/tl/research/

European Commission. (2008b). *Towards a Single European Telecoms Market: Focus on the Netherlands, Information Society and Media report.* Retrieved March 27, 2009, from http://ec.europa.eu/information_society/tl/research/

Ferris, M. (2007). Insights on Mobile Advertising, Promotion and Research. *Journal of Advertising Research, 47*(1), 28–37..doi:10.2501/S0021849907070043

Fishbein, M., & Azjen, I. (1975). *Belief, attitude, intention and behavior: An introduction to theory and research.* Reading, MA: Addison-Wesley.

Fornell, C., & Larcker, D. F. (1981). Evaluating structural equation models with unobservable variables and measurement error. *JMR, Journal of Marketing Research, 18*(1), 39–50.. doi:10.2307/3151312

Gopal, R., & Tripathi, A. (2006). Advertising via wireless networks. *International Journal of Mobile Communications, 4*(1), 1–16..doi:10.1504/IJMC.2006.008194

Haenlein, M., & Kaplan, A. M. (2004). A beginner's guide to partial least squares analysis. *Understanding Statistics, 3*(4), 283–297..doi:10.1207/s15328031us0304_4

Heinonen, K., & Strandvik, T. (2007). Consumer responsiveness to mobile marketing. *International Journal of Mobile Communications, 5*(6), 603–617..doi:10.1504/IJMC.2007.014177

Homer, P. M. (1990). The Mediating Role of Attitude Toward the Ad: Some Additional Evidence. *JMR, Journal of Marketing Research, XXVII*, 78–86..doi:10.2307/3172553

Hulland, J. (1999). Use of Partial Least Squares (PLS) in strategic management research: A review of four recent studies. *Strategic Management Journal, 20*(2), 195–204..doi:10.1002/(SICI)1097-0266(199902)20:2<195::AID-SMJ13>3.0.CO;2-7

Jarvis, C. B., Mackenzie, S. B., & Podsakoff, P. M. (2003). A Critical Review of Construct Indicators and Measurement Model Misspecification in Marketing and Consumer Research. *The Journal of Consumer Research*, *30*, 199–218.. doi:10.1086/376806

Jingjun Xu, D. (2006-2007). The Influence of Personalization in Affecting Consumer Attitudes Toward Mobile Advertising in China. *Journal of Computer Information Systems*, ▪▪▪, 9–19.

Jun, J. W., & Lee, S. (2007). Mobile Media use and its Impact on Consumer Attitudes Toward Mobile Advertising. *International Journal of Mobile Marketing*, *2*(1), 50–58.

Kaas, K. P. (1990). Nutzen und Kosten der Werbung. *ZfbF*, *42*(6), 492–504.

Karjaluoto, H., & Alatalo, T. (2007). Consumers' attitudes towards and intention to participate in mobile marketing. *International Journal of Services Technology and Management*, *8*(2/3), 155–173.. doi:10.1504/IJSTM.2007.012866

Kroeber-Riel, W., & Weinberg, P. (2003). *Konsumentenverhalten* (8th ed.). München: Vahlen.

Lee, S., Tsai, Y., & Jih, W. (2006). An Empirical Examination of Customer Perceptions of Mobile Advertising. *Information Resources Management Journal*, *19*(4), 39. doi:10.4018/irmj.2006100103

Leppäniemi, M., & Karjaluoto, H. (2005). Factors influencing consumers' willingness to accept mobile advertising; a conceptual model. *International Journal of Mobile Communications*, *3*(3), 197–213..doi:10.1504/IJMC.2005.006580

Leppäniemi, M., & Karjaluoto, H. (2008). Exploring the effects of gender, age, income and employment status on consumer response to mobile advertising campaigns. *Journal of Systems and Information Technology*, *10*(3), 251–265.. doi:10.1108/13287260810916943

Lohmöller, J.-B. (1989). *Latent Variable Path Modeling with Partial Least Squares*. Heidelberg: Physica-Verlag.

MacKenzie, S. B., & Lutz, R. J. (1989). An Empirical Examination of the Structural Antecedents of Attitude Toward the Ad in an Advertising Pretesting Context. *Journal of Marketing*, *53*(2), 48–65..doi:10.2307/1251413

Mitchell, A. A., & Olson, J. C. (1981). Are product attribute beliefs the only mediator of advertising effects on brand attitude? *JMR, Journal of Marketing Research*, *18*(3), 318–332..doi:10.2307/3150973

Muk, A. (2007). Consumer's intentions to opt in to SMS advertising. *International Journal of Advertising*, *26*(2), 177–198.

Nantel, J., & Sekhavat, Y. (2008). The impact of SMS advertising on members of a virtual community. *Journal of Advertising Research*, *48*(3), 363–374..doi:10.2501/S002184990808032X

Okazaki, S., Katsukura, A., & Nishiyama, M. (2007). How Mobile Advertising Works: The Role of Trust in Improving Attitudes and Recall. *Journal of Advertising Research*, *47*(2), 165–178.. doi:10.2501/S0021849907070195

Rappaport, S. D. (2007). Lessons from Online Practice: New Advertising Models. *Journal of Advertising Research*, *47*(2), 135–141..doi:10.2501/S0021849907070158

Rettie, R., Grandcolas, U., & Deakins, B. (2005). Text Message Advertising: Response Rates and Branding Effects. *Journal of Targeting. Measurement and Analysis for Marketing*, *13*(4), 304–312.. doi:10.1057/palgrave.jt.5740158

Ringle, C. M., Wende, S., & Will, S. (2005). *SmartPLS 2.0 (M3) Beta*. Hamburg, http://www.smartpls.de

Schlosser, A. E., Shavitt, S., & Kanfer, A. (1999). Survey of Internet users' attitudes toward Internet advertising. *Journal of Interactive Marketing, 13*(3), 34–54..doi:10.1002/(SICI)1520-6653(199922)13:3<34::AID-DIR3>3.0.CO;2-R

Sternthal, B., Dholakia, R., & Leavitt, C. (1978). The persuasive effect of source credibility: Tests of cognitive response. *The Journal of Consumer Research, 4*(4), 252–260..doi:10.1086/208704

Tenenhaus, M., Vinzi, V. E., Chatelin, Y.-M., & Lauro, C. (2005). PLS path modeling. *Computational Statistics & Data Analysis, 48*(1), 159–205..doi:10.1016/j.csda.2004.03.005

Tsang, M. M., Ho, S.-C., & Liang, T.-P. (2004). Consumer Attitudes Toward Mobile Advertising: An Empirical Study. *International Journal of Electronic Commerce, 8*(3), 65–78.

Venkatesh, V., Morris, M. G., Davis, G. B., & Davis, F. D. (2003). User Acceptance Of Information Technology: Toward A Unified View. *Management Information Systems Quarterly, 27*(3), 425–478.

Wold, H. (1982). Soft Modeling: the basic design and some extensions. In Jöreskog, K. G., & Wold, H. (Eds.), *Systems under indirect observation: causality, structure and prediction. Part II.* Netherlands: North-Holland Publishing company.

Zanot, E. J. (1984). Public attitudes toward advertising: The American experience. *International Journal of Advertising, 13*, 3–15.

APPENDIX

List of Original Items

Construct	Item	Loading	Weight	
Context[a]	COT1	.95	.55	I consider mobile advertising related to a specific time or date useful
	COT2	.94	.51	I consider mobile advertising related to me being in a specific location useful
Entertainment[a]	ENT1	.95	.52	I feel that receiving mobile advertisements is pleasant
	ENT2	.95	.53	I feel that receiving mobile advertisements is enjoyable and entertaining
Personalization[a]	PER1	.95	.61	I feel that mobile advertising is personalized for my usage
	PER2	.91	.47	Contents in mobile advertisements are personalized
Intention[a]	INT1	.93	.41	I am willing to receive mobile advertising in the future
	INT2	.93	.38	I intend to opt-in to receiving mobile advertising within the next six months if I have the opportunity to do so
	INT3	.85	.32	I intend to opt-in to receiving mobile advertising whenever I have a chance
Attitude[a]	ATT1	.92	.56	Overall, I like the idea of using mobile advertising
	ATT2	.91	.53	I believe that mobile advertising can benefit both the marketing company and me
Credibility[a]	CRE1	.93	.53	I feel that mobile advertisements are credible
	CRE2	.94	.54	The content provided by mobile advertising is credible
Privacy[a]	PRI1	.95	.56	I believe that a marketing company uses my data only for a purpose that I have approved
	PRI2	.94	.49	I believe that my mobile operator uses my data only for a purpose that I have approved
Informativeness[a]	INF1	.85	.50	Mobile advertisements provide the information I need
	INF2	.91	.63	I feel that mobile advertising is a good source for timely information
Subjective norm[a]	SUB1	.93	.49	I am more likely to participate in mobile marketing if the message is forwarded to me by someone I know
	SUB2	.95	.58	I am more likely to participate in mobile advertising if it is recommended by someone I know

Notes:[a]The response options ranged from 1 = "totally disagree" to 5 = "totally agree"

This work was previously published in International Journal of E-Services and Mobile Applications, Volume 1, Issue 4, edited by Ada Scupola, pp. 1-15, copyright 2009 by IGI Publishing (an imprint of IGI Global).

Chapter 11
Exploring the Factors Behind the Resistance to Mobile Banking in Portugal

Pedro Cruz
ISG- Business School, Portugal

Tommi Laukkanen
University of Joensuu, Finland

Pablo Muñoz
Universidad de Salamanca, Spain

ABSTRACT

M-banking has been highlighted as one of the most promising e-commerce services. However, the adoption of financial mobile services is still far behind market expectations. This study seeks to provide academics and bank managers alike with a better understanding of resistance to m-banking. Based on Ram and Sheth's (1989) Theory of Innovation Resistance, a SEM model was built to describe resistance barriers. A total number of 3,852 observations, of which 2,344 were effective for SEM analysis, were obtained from an Internet survey at a Portuguese bank. The "functional barriers" revealed more severe barriers than the "psychological" ones. Latent scores were used to compare consumers' perceptions and behaviour. Results indicate a significantly higher resistance among non-users, and demographic and behavioural profiles were established, promoting wider knowledge and possible enhancement of m-banking adoption. Conclusions and managerial implications are provided.

INTRODUCTION

The technological development of mobile devices (e.g. mobile phones, PDAs, smartphones), along with increasing wireless Internet penetration, facilitates a convergence of *all in the pocket, anytime, anywhere services*: games, music, photography, television, Internet, publicity, information and localization, among others.

Consumers find mobile services to be value-added, convenient and compatible with their lifestyle (Fife, 2007). However, reports show that there is a marginal use of mobile devices for transactional purposes, the rate being as low as

DOI: 10.4018/978-1-60960-607-7.ch011

6% (Kleijnen et al., 2007). In Europe, the difference between consumer interest and real usage reaches 50% in some m-commerce services, such as financial services (Sybase, 2008). This finding informs both the state of incipient development and the potential of this distribution channel.

According to Eurostat (2007), the average penetration rate of mobile phones in Europe in 2006 was higher than 100%. Portugal is no exception. During 2007, mobile phone penetration achieved an impressive record of 125% (Anacom, 2008). Simultaneously, the data/voice ratio is also increasing at a remarkable pace (OfCom Statistics, 2008). The relatively low number of computers connected to the Internet, as well as the large number of mobile phones, makes Portugal a high-potential market for mobile services, such as banking.

Third generation mobile services are framed within a context of caution, where fewer risk investments and marketing actions are carried out (Bouwman et al., 2007). Therefore, insight into why consumers are not adopting mobile services is needed.

The aim of this study is to investigate the inhibitors of mobile banking in Portugal, in terms of barriers to adopting the service. According to several studies and marketers, a wider knowledge of the phenomena is needed to develop strategies aimed at enhancing m-banking adoption.

The article includes a revision of relevant literature on technological innovation adoption and some succeeding research hypotheses are formulated. Based on Ram and Sheth's (1989) Theory of Innovation Resistance, a higher-order SEM model was built to describe resistance barriers to mobile banking. Hypotheses with implicit difference tests were contrasted through the use of structural equations' latent scores. Online banking users' perceptions and behaviour were compared and results indicate a significantly higher resistance among non-users. Furthermore, demographic and behavioural profiles were established. Results are followed by discussion and a reflection on the limitations of the study.

CONCEPTUAL FRAMEWORK AND DEVELOPMENT OF HYPOTHESES

Current literature is largely concentrated on the adoption of innovation and the factors influencing adoption and dissemination of ICT technological innovations. This adoption process has been systematised in holistic models (Lin, 2003), quantitative structural models like the Technology Acceptance Model (T.A.M., Davis et al., 1989), the Extended T.A.M. or the Unified Theory of Acceptance and Use of Technology (Venkatesh et al., 2003). Other models have focused on the motives lying beneath resistance to innovation adoption (Ellen et al., 1991; Bowman et al., 2007). However, relatively little attention has been paid to the latter approach (the theory of innovation resistance). Some authors, like Sheth (1981), have highlighted the importance of studying the process of resistance to innovation, instead of its adoption.

According to Sheth (1981) and Ellen et al. (1991), consumer decisions on innovation are based on two main aspects: on the one hand, perception of the service's benefits, and, on the other, the perception of risk. Ram and Sheth (1989) provide a more comprehensive analysis on resistance to innovation, focusing on psychological and functional barriers. This view argues that psychological barriers are reflected in tradition and image barriers; while functional barriers derive from usage barrier, value barrier and risk barrier.

Ram and Sheth (1989) argue that psychological barriers are created by conflicts related to customers' prior beliefs and values rather than by the actual usage of innovation. Therefore, it seems that these barriers are related to a broader discussion about technology readiness (Parasuraman, 2000), referring to customers' mental willingness to accept new technologies. The tradition barrier is the extent of incompatibility between an established routine and a certain innovation (Ram & Sheth, 1989). Earlier literature has also identified resistance to change as an important inhibitor in the adoption of Internet banking (Bradley & Stewart, 2002; Kuisma et al., 2007). For example, consumers with high social interaction needs may prefer going to a physical branch instead of adopting electronic services (Marr & Prendergast, 1993).

Image barrier is related to adverse associations to the uniqueness of a certain product or service (Ram & Sheth, 1989). In the context of technological innovations, this relates to the discussion about different types of anxiety towards computers (Kay, 1993) or technology itself, referring to consumers' negative state of mind about technology tools (Meuter et al., 2003). A possible image barrier to mobile banking services could be a result of a negative hard-to-use image of mobile technology, or just of technology in general.

As one of the three functional barriers, the usage barrier is related to the complexity and incompatibility of consumer workflows, habits or practices, and refers to the degree to which an individual considers an innovation to be relatively difficult to understand and use (Ram & Seth, 1989). In the case of technological innovation, it is clear that the usage barrier is closely related to the concept of ease-of-use, defined, by the Technology Acceptance Model (Davis et al., 1989), as the extent to which an individual considers an innovation to be free of effort. Furthermore, perceived ease-of-use is closely connected to the concept of complexity (Davis, 1989; Teo & Pok, 2003), defined as the extent to which an innovation is perceived as difficult to understand and use (Rogers, 2003). In the mobile services context, the small size of the devices (screens and keypads) might inhibit the progress of the service process (Laukkanen & Lauronen, 2005; Laukkanen, 2007b). In addition, the simplicity of PIN handling is also stated as an important aspect in the use of electronic banking services (Kuisma et al., 2007).

The value barrier is related to the performance and monetary value of an innovation in comparison to its alternatives (Ram & Sheth, 1989). It converges to Rogers' (2003) concept of relative advantage, which resembles the perceived usefulness construct (Wu & Wang, 2005) derived from the T.A.M. literature. For example, Pikkarainen et al. (2004) state that consumers use internet banking because of the benefits it provides comparatively to other banking delivery channels.

Therefore, and leaning on earlier literature, we argue that the greater the perceived advantage (Brown et al., 2003) in terms of convenience (Fife, 2007) and from a financial perspective (Luarn & Lin, 2005; Kuisma et al., 2007), the more mobile banking is likely to be adopted (De Marez et al., 2007).

According to Im et al. (2008), if there is uncertainty about the technology's outcome, a risk barrier will be created. Ram and Sheth (1989) note that, to some extent, all innovations represent uncertainty, and therefore, risk is inherent to them. The risk barrier in the mobile banking context is the extent to which a person perceives that the mobile device will perform a job inconsistently or inaccurately (Lee et al., 2003). Online transactions in general, and mobile transactions in particular, may raise security and privacy concerns among consumers (e.g. Barnes & Corbitt, 2003; Siau & Shen, 2003). Moreover, there are some uncertainties related to the actual usage of the service, as some consumers appear to fear that they might make mistakes when conducting their bank affairs via a computer or a mobile phone (Kuisma et al., 2007; Laukkanen & Lauronen, 2005; Laukkanen, 2007b). In fact, some customers may be more worried about their own errors made during the service usage process than about its security issues in general (Laukkanen & Lauronen, 2005). In mobile banking, data input and output mechanisms are especially considered to impede the banking process, creating feelings of insecurity (Laukkanen & Lauronen, 2005; Laukkanen, 2007b). Finally, regarding the mobile service, some consumers are worried that the memory capacity or the battery of

Figure 1. Conceptual model

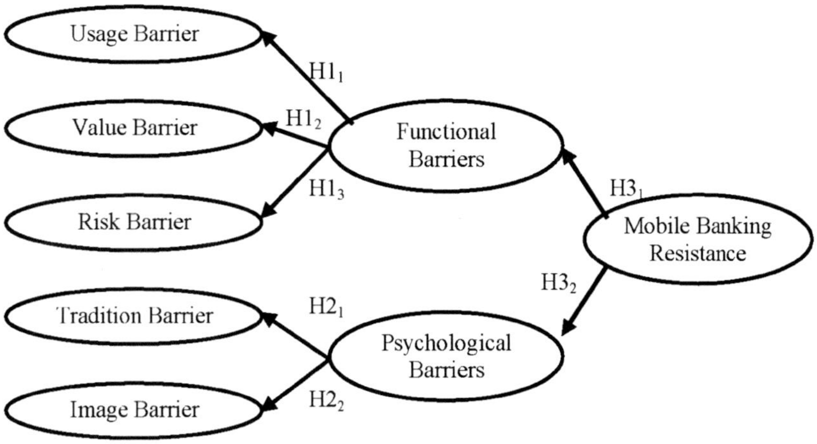

the mobile phone may run out posing uncertainty regarding completion (Siau & Shen, 2003).

Following the earlier literature on resistance to innovation, we expect a model where functional barriers reflect "usage", "value" and "risk" first order constructs; while "tradition" and "image" are the first order constructs determining the "psychological barriers". Based on former theoretical relations, we present the following hypotheses (Figure 1):

$H1_{1-3}$: Functional barriers are reflected positively on usage barrier ($H1_1$), value barrier ($H1_2$) and risk barrier ($H1_3$).

$H2_{1-2}$: Psychological barriers are reflected positively on tradition barrier ($H2_1$) and image barrier ($H2_2$).

$H3_{1-2}$: Overall resistance to mobile banking is reflected positively on functional barriers ($H3_1$) and psychological barriers ($H3_2$).

Previous studies have shown that increased exposure to technology increases the probability of developing favourable attitudes towards new channels (e.g. Bellman et al., 1999; Modahl, 2000). Consumers with more exposure to technologies (such as Internet) are those who are more willing to adopt direct channels (Dholakia & Uusitalo, 2002; Lohse et al., 2000). Bigne et al. (2005) found that there is a greater likelihood of being an m-shopper when the buyer has more experience as an Internet purchaser. The same pattern seems to occur between knowledge and use of the Internet and electronic banking. Karjaluoto et al. (2002) state that previous experience in computer and Internet use influences consumer attitudes and behaviour towards Internet banking. Teo and Pok (2003) also argue that frequent access to Internet banking enhances the predisposition to adopt further innovations, such as mobile banking. Similarly, in the study on bank customers' channel attribute preferences, Laukkanen (2007a) argues that the higher the usage frequency of internet banking, the higher the potential for mobile banking adoption. Accordingly, we established a set of hypotheses based on the notion that the use of Internet and mobile devices (both previous and current) influence consumer resistance to, and frequency of use of, mobile banking:

H4: Resistance to mobile banking services is significantly lower among consumers who have been using Internet banking for a longer time.

H5: Resistance to mobile banking services is significantly lower among the service (mobile) users.

H6: Resistance to mobile banking is significantly lower among consumers who use mobile banking more frequently.

Research has been interested in examining the demographics of consumers in the adoption of new technologies. Earlier literature argues that age is related to attitudes and the adoption of technology. The elderly have traditionally been considered resistant to change and rather belong to the late majority or laggards in terms of adoption rates of new technology-based services (Gilly & Zeithaml, 1985; Oumlil & Williams, 2000). This appears to be the case also in mobile banking, as the mature segment (55+) is evinced to have higher resistance to the service than the younger segment (Laukkanen et al., 2007). In addition, several studies on Internet banking (Howcroft et al., 2002; Karjaluoto et al., 2002; Muñoz & Cruz, 2004; Polatoglu & Ekin, 2001) have confirmed this and found that a typical Internet banking user is a relatively young male with a high level of education. However, Laukkanen and Pasanen (2008) argue that a typical mobile banking user is more likely to be middle-aged (30-49 years old) than a young man. The same pattern was found for Japan (Flinders, 2008) and China (Laforet & Xiaoyan, 2005). Taking into account the findings in the earlier literature, we suggest the following hypothesis:

H7: Comparing mobile service users and non-users, the sample distribution is not uniform in terms of age.

When compared to men, women perceive more risk in online purchases (Garbarino & Strahilevitz, 2004). Males seem to place greater value on the benefits of non-store shopping (Citrin et al., 2003; Marcus & Kitayama, 1991) and tend to perceive m-commerce favourably (Yang, 2005). Gender seems to have a significant impact on mobile services usage (Li et al., 2008; Nysveen, Pedersen & Thorbjørnsen, 2005; Nysveen et al., 2005; Venkatesh et al., 2003) and a few studies highlight a male preponderance among users of mobile banking services (Flinders, 2008; Laforet & Xiaoyan, 2005; Laukkanen & Pasanen, 2008; Suoranta & Mattila, 2004). Based on the above findings on consumer acceptance of e-commerce and m-commerce, the following hypothesis is formulated:

H8: Among Internet banking users, the proportion of men using mobile banking is significantly higher than women.

RESEARCH METHODOLOGY AND SAMPLE DESCRIPTION

Based on the theory of resistance to innovation and on literature on banking technologies, particularly on mobile banking, a survey questionnaire was designed. The five first-order adoption barriers were measured through the 17 statements included in Appendix 1. A seven-point *Likert* scale ranging from "totally disagree" (1) to "totally agree" (7) was used. The questionnaire was posted on the web page of a large Portuguese bank, between June 28th and July 13th 2007. Information on sex, age, type of mobile device, length of time using Internet banking, total duration and frequency of mobile service use was also gathered. From the initial set of 3,857 responses, 2,344 were considered valid for the empirical analysis. No repeated responses from the same user were allowed and, in order to minimize bias (pattern or random responses), cases were filtered when matching (at least) one of the following scenarios:

- If a respondent completed (the whole survey) in less than 4 minutes (554 cases were excluded). Five minutes was considered the "minimum" time for a comprehensive reading of the survey.
- If the responses followed some pattern: more than 95% with the same answer (358 cases)

at the *Likert*-type questions (reversed scale variables included).

We also settled the exclusion of cases (597 cases) that presented missing values at relevant variables, in particular those used by SEM. After evaluating the trade-off between additional uncertainty (due to a possible imputation method) and having a smaller dataset, the later situation was selected.

Out of the valid responses, approximately one third (776) belonged to mobile banking users and two thirds (1,568) to non-users.

The sample is largely male dominated (62.9%) and relatively young (45.4 percent of the respondents being 30-39 years old) (see Appendix 2).

The respondents are relatively experienced users regarding Internet banking: 66.6% have used the Internet service for more than 3 years. As to mobile banking services, the majority of the users can be considered as early adopters, given that 49.3% of them have less than 3 months usage experience. The frequency of use is low, in view of the fact that the majority (52.3%) of users resort to mobile banking services less than once a week (see Appendix 2).

The main goal of this research is to develop a higher-order model to explain resistance to mobile banking, including original items embedded in a multiple-indicator model defined by first-order constructs (conceptualised as "barriers").

A first-order construct can be conceptualised as a unidimensional set of observed items. The relationship between indicators and correspondent unobserved constructs (first-order constructs) typically assumes a reflective scale: factors "driving" the indicators (i.e., these indicators are assumed to be observable instances of their unobserved construct, and thus changes in the latent variable are "indicated" by observable changes in the items in its measure). Given that ours is considered a higher-order measurement model, constructs will have other latent variables acting as its "indicators".

Each first and second-order construct was assumed to be a function of two components (Anderson & Gerbing, 1988; Rindskopf & Rose, 1988): a component that is shared with the other factors ([Xi]) and corresponds to the higher-level construct of interest, and a component unique to that factor ([Zeta]). The first-order factors may be expressed in terms of the second-order factors (the same reasoning can be adopted when interpreting relationships between second and third-order constructs):

[Eta] = [Gamma][Xi] + [Zeta]

where [Xi] is a second-order factor that is a component of [Eta] shared with the other first-order factors, [Zeta] is the unique component of each first-order factor, and [Gamma] is the pattern coefficient.

Following Anderson and Gerbing's (1988) approach, a structural equation model (with AMOS software) was estimated and its fit and construct reliability, convergent and discriminate validity were checked.

The overall guiding principle of the modelling approach is to compartmentalize the tasks into sections (Anderson & Gerbing, 1988; Gerbing et al., 1994): a first-order confirmatory factor model was estimated before proceeding to the construct representation at the second-order level. Latent scores were obtained at the first-order confirmatory level and operated as "indicators" of higher-order constructs during the subsequently step. The final model assumed latent scores (from first-order constructs) as observed variables.

When interpreting the meaning of "Mobile Banking Resistance" higher-order construct (see Figure 2), we should keep in mind that the majority (70%) of respondents do not use mobile banking service, so their resistance is "materialised" into several barriers, according to their perceptions.

In order to contrast all hypotheses, some differences tests implicit in H4, H5, H6 and H7 were assessed through the use of non-parametric

Figure 2. Standardised parameter estimates

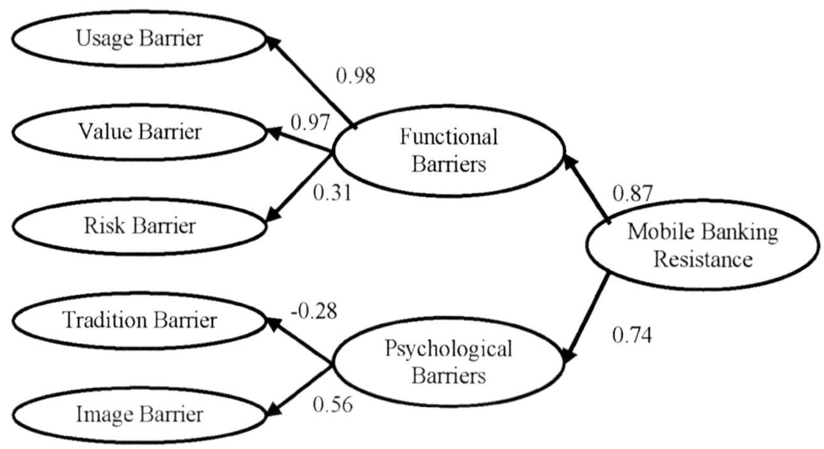

Kolmogorov-Smirnov Z and *Mann-Whitney U* tests (original variables and latent scores did not have a normal distribution).

RESULTS

Our Mobile Resistance Model was defined by 15 items (B8 and B9 were excluded due to high cross-saturation products) that identified 5 distinct facets [Eta] underlying the 3 constructs associated with Functional Barriers [X1] and 2 constructs with Psychological Barriers [X2]. The [Eta] constructs of Usage, Value, Risk, Tradition and Image were operationalised as first-order latent variables.

Additionally, the Functional and Psychological barriers constructs were considered as "manifestations" [Eta] of a higher-order construct: Mobile Banking Resistance [X3].

Validation of the constructs was rigorously analysed within the framework of structural equation modelling of latent variables. Reliability was ensured due to acceptable *Cronbach's Alphas* and composite reliability coefficients (Table 1). Given the low *Cronbach's Alphas* values, some explanations should be addressed. It is well known that this statistic depends partially on the research context and on the number of items measuring a latent variable. According to Nunnally (1967), *Cronbach's Alphas* values around 0.5 to 0.6 could be considered in early research stages. Our theory and measure scale is relatively exploratory in this context, a fact that could explain the low values obtained for constructs like Tradition and Image Barriers (0.57 and 0.6, respectively). Traub (1994) and DeVellis (1991) stress that the *Cronbach's Alpha* statistic is affected by a reduced number of items, a fact confirmed by Diamantopoulos (1996), arguing that when having reflective constructs with less than 3 indicators, values around 0.5-0.6 should not imply the construct's exclusion.

Convergent validity is ensured by acceptable *t* tests (sig.<0.05) for all loadings and causal relations (average loadings' size is 0.75). The discriminant validity was stringently assessed using the Fornell and Larcker (1981) test: each construct's extracted variance surpasses the recommended level of 0.5 (Hair et al., 1998) and exceeds the quadratic correlations between any pair of constructs (see Table 2). This suggests that each construct has more internal (extracted) variance than shared variance.

We assessed multicollinearity by examining *Tolerance* and the *Variance Inflation Factor* (VIF), two of the most common collinearity diagnoses. A VIF no greater than 3.52, and a minimum *Tol-*

Table 1. Confirmatory factorial analysis and reliability

Constructs	Variables	Standardised Coefficients	*t*	α	ave	ρ	R^2
Usage Barrier	B1	0.79	--				0.96
	B2	0.66	34.28	0.89	0.60	0.88	
	B3	0.80	43.83				
	B4	0.89	50.13				
	B5	0.71	36.61				
Risk Barrier	B10	0.78	--	0.84	0.59	0.85	0.10
	B11	0.79	38.72				
	B12	0.87	41.18				
	B13	0.60	28.27				
Tradition Barrier	B14	0.49	--	0.57	0.52	0.70	0.08
	B15	0.93	20.63				
Image Barrier	B16	0.76	--	0.60	0.51	0.71	0.31
	B17	0.71	21.95				
Value	B6	0.74	--	0.67	0.50	0.70	0.94
	B7	0.69	22.60				

erance value of 0.86 among observed variables, lead us to conclude that the model is free from multicollinearity problems (Myers, 1984). The individual R-squares (for endogenous constructs) are all greater than the recommended 0.10 cut-off (Falk & Miller, 1992), except for the Tradition Barrier (functional=0.75, psychological=0.56, usage=0.96; value=0.94; risk=0.10; tradition=0.08; image=0.31).

The estimated structural model presented an acceptable fit ($\chi 2$=2036.36; d.f.=88; p=0.00; CFI=0.925; RMSEA=0.078; GFI=0.926; NFI=0.922). The value for RMSEA is relatively high but still within an acceptable range (lower than 0.08), according to Novak et al. (1999).

It is therefore appropriate to examine the meaning of simultaneous relations among constructs.

The third-order structural model showed that Functional Resistance to the adoption of mobile services is mainly due to usage barrier (standardised structural impact=0.98) and value (0.97) barrier. Risk barrier also positively influenced the Functional Barriers, and its structural effect reached 0.31 (Figure 2). On the other hand, tradition had a negative influence over Psychological Barriers, the standardised structural coefficient being –0.28, while the image barrier's standardised impact was 0.56. Hence, hypotheses H1 and H2 were corroborated, except for H2$_1$ (where tradition had a significant negative impact).

As stated, overall resistance to Mobile Banking leads to Functional and Psychological Barriers, the effects being 0.87 and 0.74 respectively, and therefore corroborating H3.

Using latent scores from the model, some non-parametric tests were operated (due to the non-normality of data) to contrast H4, H5 and H6. Overall resistance to mobile banking services was not affected by variation in the duration of Internet banking use (sig.>0.05 for all categories). Accordingly, H$_4$ is not supported.

Both *Kolmogorov-Smirnov Z* and *Mann-Whitney U* tests showed significant statistical differences between users and non-users in the eight constructs (Table 3). Consequently, the data supports H5.

Table 2. Discriminant validity

	RISK	IMAGE	TRADIT	VALUE	USAGE
RISK	*0.59*				
IMAGE	0.17	*0.51*			
TRADIT	0.01	0.01	*0.52*		
VALUE	0.04	0.11	0.07	*0.50*	
USAGE	0.04	0.13	0.07	0.43	*0.60*

Diagonal: AVE; off diagonal: quadratic correlation between constructs

To test H6, we analysed the resistance barriers' variance across the frequency of mobile service use categories and concluded that there are significant differences (sig.=0.000). This result shows, for all barriers, except for Tradition, that resistance barriers decrease considerably with the rise of frequency of mobile service use (see Figure 3), partially corroborating H6.

The profile of respondents concerning age confirms a predominance of middle-aged online banking users: 62.6% are 30-49 years old (see Appendix B).

However, a comparison of the m-banker profile with that of the e-banker (non m-banker) shows that, generally, sample distribution is fairly uniform in terms of age, as confirmed by the values obtained in the Chi-Square analysis (χ^2= 4.599, p = 0.467, d.f.=5). Hence, H7 is not supported by the data.

An analysis of gender proportion shows that there is a male prevalence among mobile service users (68.4%), which is significantly higher when compared to the male proportion among non-users (60.2%). A Chi-square test rejects the independence between usage and gender (sig.=0.000), confirming the proportion imbalance (and thus corroborating H8).

In order to explore the existence of a possible impact of the technology, e.g. type of mobile device, over overall resistance, some non-parametric tests were applied. Results show that those respondents with a "basic" mobile phone have a significantly higher (sig.=0.000) level of overall resistance (3.793) when compared to those with "advanced" mobile devices (3.274)- 3G mobile

Table 3. Resistance levels (latent scores)

	All N=2334	Users n=776	Non-Users n=1568	Sig.*
Overall Resistance	3.3905	2.7217	3.9958	0.000
Psychological Resistance	1.9717	1.9417	2.3344	0.000
Functional Resistance	2.4710	2.3807	3.3129	0.000
Usage Barrier	3.4292	2.5686	3.7813	0.000
Value Barrier	2.9797	2.2617	3.1596	0.000
Risk Barrier	3.3467	3.2035	3.6571	0.000
Tradition Barrier	4.0211	4.5009	3.8323	0.000
Image Barrier	2.0123	1.9833	2.3708	0.000

Note: 2334 valid cases used in SEM (without missing values for the relevant variables)
* *Kolmogorov-Smirnov* Z and *Mann-Whitney U* tests employed

Figure 3. Perceived barriers (1st order) across frequency of mobile services use

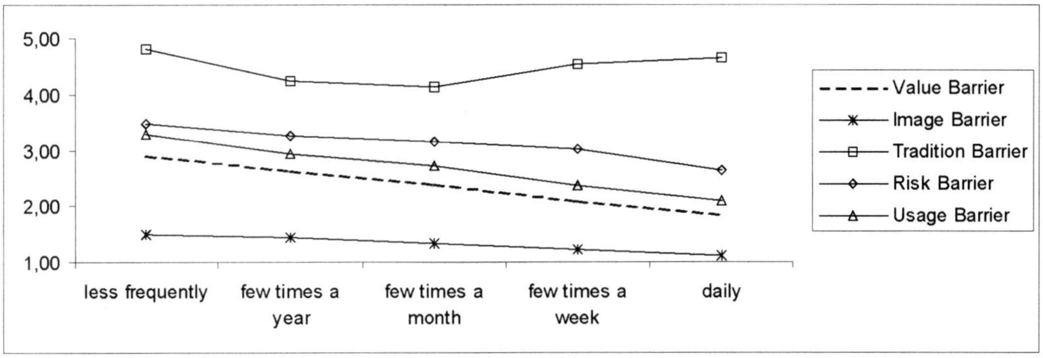

phones, PDAs or Smartphones. A Chi-Square test excludes the idea of independence between "type of mobile" and "usage" of mobile services (sig.=0.000), simultaneously showing that the majority of mobile banking users have a "3G handset" (67%), while 55% of the non-users have a basic mobile device (see Appendix C).

Appendix D provides a comprehensive list of accepted and rejected hypotheses.

CONCLUSION AND MANAGERIAL IMPLICATIONS

The current mobile banking user is predominantly male and handles a 3G handset, a fact that provides valuable information about target-groups to be considered as a priority in the implementation of communication methods and strategies, namely, women and consumers with "basic devices" (typically non-users of mobile banking services). Banks should provide adequate conditions not only to allow for device upgrading, but also at the commercial level (virtual demonstration, price comparison engines) and financial level (credit simulation). For banks, this could mean a potential twofold source of revenue: credit concession and enhancement of mobile banking services.

Results show that resistance to mobile banking leads to psychological and functional barriers, the latter being the main contributor to overall resistance. Campaigns highlighting the functional usability and relative advantage of this service (when compared to other types of banking services) should be a priority of banks' strategic endeavours (since they are the most intense determinants of functional resistance). Consumers should be convinced that mobile banking services have a high value-for-money and are easy to handle. To overcome the aforementioned barriers, service trials and incentives for a regular use of this service, as well as demonstration videos explaining the simplicity of how to handle it, are highly recommended (Ram & Jung 1991; Rogers 2003). Some banks offer free access to trial Internet services with simulated accounts. Similar actions should be implemented with mobile services. Results confirm the effectiveness of these recommendations: when consumers try the services at least once, their usage, value and risk barriers drop significantly. At the same time, all resistance barriers (except tradition) decrease as consumers use mobile banking more frequently, indicating that the increase in use and knowledge of this channel will undoubtedly modify their behaviour and response to marketing actions (Dahlen, 2002). On the other hand, offering a greater value-for-money for mobile services, when compared to alternatives, can reduce the value barrier. Banks can either raise the price of alternatives (branch,

Internet banking or ATM) or lower the price of transactions involving a mobile device. As far as the risk barrier is concerned, the second higher mean registered, banks should concentrate on effective communication strategies and provide regular information about mobile applications concerning security, and privacy policy and PIN code use. The recommended trialability of mobile services could simultaneously reduce usage barrier and risk barrier (Ram & Sheth, 1989).

Tradition stands out as a negative determinant of resistance, underlying the compatibility of mobile banking services with established routines (use of in-person branch services, ATMs, Internet banking). The high number of branches, POS terminals and ATMs in Portugal can explain this reality. In fact, Portugal is the European (EU27) country with the highest density of ATMs (European Central Bank, 2008). In 2006, the Portuguese population per ATM ratio was around half the EU27 average (721 versus 1,362) and the density of branches in Portugal easily undercuts the EU27 average (1,759 vs. 2,123 inhabitants/branch). It seems clear that online users find other channels as complementary or even necessary to mobile service use. The physical branch is quite important in this context, and some evidence points to an increasing shift towards new functions, such as advisory, consultancy and support for electronic channels (European Union Study Report, 2008). Branch staff also plays a crucial role as change agents (Ram, 1989) given that they provide face-to-face information and influence the innovation process (mobile banking services).

Image is an important determinant of psychological barriers, in spite of presenting the lowest barrier score. This indicates that Portuguese consumers seem to have a positive image of technology in general and of mobile services in particular. ATM penetration and the "contactless toll payment systems" (invented and widespread in Portugal) are good examples of how technology applied to financial services is perceived by the Portuguese.

The relatively high penetration rate (33.1%) and low scores in resistance to mobile banking among on-line users (only tradition exceeds 3.5) may reflect the fact that Portugal has, simultaneously, a low number of Internet-connected computers and a high mobile penetration rate, which makes for a very favourable context for mobile services (Narinder, 2007). Another explanation for such results might be related to a possible sampling bias, as only online banking users participated in the study.

Limitations and Future Research

This research is limited for several reasons. First, the collection of responses was restricted to a single survey and the resulting single-method bias restricts the inferences that can be drawn from this study (Podsakoff & Organ, 1986). Second, a self-selected online survey should be kept in mind when interpreting and generalizing findings: the inclusion of each customer (and their correspondent opinions and perceptions) is potentially subject to self-desirability bias. Finally, the less than optimal statistics for reliability: a possible explanation could lie on the random order of the questions and to the minimum item-per-construct strategy adopted (Image and Tradition constructs are measured with 2 items each).

Future research is needed, specifically on the role of personal innovativeness and on cultural variables in mobile banking adoption.

REFERENCES

Anacom (2008). Serviço Nacional Móvel: relatório do 4° trimestre de 2007, Autoridade Nacional de Comunicações. Retrieved May 18, 2008, from http://www.anacom.pt /template12.jsp? categoryId=250442

Anderson, J., & Gerbing, D. (1988). Structural equation modeling in practice: A review and recommended two-step approach. *Psychological Bulletin, 103*(3), 411–423. doi:10.1037/0033-2909.103.3.411

Barnes, S., & Corbitt, B. (2003). Mobile banking: concept and potential. *International Journal of Mobile Communications, 1*(3), 273–288. doi:10.1504/IJMC.2003.003494

Bellman, S., Lohse, G., & Johnson, E. (1999). Predictors of online buying behaviour. *Communications of the ACM, 42*(12), 32–38. doi:10.1145/322796.322805

Bigne, E., Ruiz, C., & Sanz, S. (2005). The Impact of Internet User Shopping Patterns and Demographics on Consumer Mobile Buying Behaviour. *Journal of Electronic Commerce Research, 6*(3), 193–209.

Bouwman, H., Carlsson, C., Molina-Castillo, F., & Walden, P. (2007). Barriers and drivers in the adoption of current and future mobile services in Finland. *Telematics and Informatics, 24*(2), 145–160. doi:10.1016/j.tele.2006.08.001

Bradley, L., & Stewart, K. (2002). A Delphi study of the drivers and inhibitors of Internet banking. *International Journal of Bank Marketing, 20*(6), 250–260. doi:10.1108/02652320210446715

Brown, I., Cajee, Z., Davies, D., & Stroebel, S. (2003). Cell phone banking: predictors of adoption in South Africa - an exploratory study. *International Journal of Information Management, 23*(5), 381–394. doi:10.1016/S0268-4012(03)00065-3

Citrin, A., Stern, D., Spangenberg, E., & Clark, M. (2003). Consumer need for tactile input. An Internet retailing challenge. *Journal of Business Research, 56*(11), 915–922. doi:10.1016/S0148-2963(01)00278-8

Dahlen, M. (2002). Learning the web: Internet User Experience and Response to Web Marketing in Sweden. *Journal of Interactive Advertising (online), 3*(1), Retrieved October 16, 2008, from http://www.jiad.org/article27

Davis, F. D. (1989). Perceived usefulness, perceived ease of use, and user acceptance of information technology. *MIS Quarterly, 13*(3), 319–340. doi:10.2307/249008

Davis, F. D., Bagozzi, R. P., & Warshaw, P. R. (1989). User acceptance of computer technology: A comparison of two theoretical models. *Management Science, 35*, 982–1003. doi:10.1287/mnsc.35.8.982

DeMarez, L., Vyncke, P., Berte, K., Schuurman, D., & De Moor, K. (2007). Adopter segments, adoption determinants and mobile marketing. *Journal of Targeting. Measurement and Analysis for Marketing, 16*(1), 78–95. doi:10.1057/palgrave.jt.5750057

DeVellis, R. (1991). *Scale development: theory and applications.* London: Sage Publications.

Dholakia, R., & Uusitalo, O. (2002). Switching to Electronic Stores: Consumer Characteristics and the Perception of Shopping Benefits. *International Journal of Retail and Distribution Management, 30*(10), 459–469. doi:10.1108/09590550210445335

Diamantopoulos, A. (1996). A model of publication performance of marketing academics. *International Journal of Research in Marketing, 13*, 163–180. doi:10.1016/0167-8116(95)00039-9

Ellen, P., Bearden, W., & Sharma, S. (1991). Resistance to technological innovations: an examination of the role of self-efficacy and performance satisfaction. *Journal of the Academy of Marketing Science, 19*(4), 297–307. doi:10.1007/BF02726504

European Central Bank. (2008, October). *EU Banking Structures*. Retrieved November 5, 2008, from http://www.ecb.int/pub/pdf/other/eubanking structures2008en.pdf.

European Union Study Report. (2008). ICT and e-Business Impact in the Banking Industry. *Study Report No. 06/2008*. Retrieved October 21, 2008, from Eurostat (2007). *European business: Facts and figures, Eurostat, Luxembourg*. Retrieved May 24, 2008, from http://epp.eurostat.ec.europa.eu

Falk, R., & Miller, N. (1992). *A Primer for Soft Modelling*, Akron, OH: University of Akron Press.

Fife, E. (2007). *Worldwide Mobile Data Services Study 2007*. Retrieved June 5, 2008, from http://www.marshall.usc.edu /assets/041/ 9099.pdf

Flinders, K. (2008). E-bank Japan sets mobile banking example. *Computer Weekly*. Retrieved October 14, 2008, from http://www.computerweekly.com /Articles/2008/10/13/232634/e-bank-japan-sets-mobile-banking-example.htm

Fornell, C., & Larcker, D. (1981, February). Evaluating structural equation models with unobservable variables and measurement error. *JMR, Journal of Marketing Research*, *28*, 39–50. doi:10.2307/3151312

Garbarino, E., & Strahilevitz, M. (2004). Gender differences in the perceived risk of buying online and the effects of receiving a site recommendation. *Journal of Business Research*, *57*, 768–775. doi:10.1016/S0148-2963(02)00363-6

Gerbing, D. W., Hamilton, J. G., & Freeman, E. B. (1994). A large-scale second-order structural equation model of the influence of management participation on organizational planning benefits. *Journal of Management*, *20*(4), 859–885. doi:10.1016/0149-2063(94)90033-7

Gilly, M., & Zeithaml, V. (1985, December). The Elderly Consumer and Adoption of Technologies. *The Journal of Consumer Research*, *12*, 353–357. doi:10.1086/208521

Hair, J., Anderson, R., Tatham, R., & Black, W. (1998). *Multivariate Data Analysis* (5th ed.). Englewood Cliffs, NJ: Prentice-Hall.

Howcroft, B., Hamilton, R., & Hewer, P. (2002). Consumer Attitude and the Usage and Adoption of Home-banking in the United Kingdom. *International Journal of Bank Marketing*, *20*(3), 111–121. doi:10.1108/02652320210424205

http://www.ebusiness-watch.org/studies/sectors/banking /banking.htm

Im, I., Kim, Y., & Han, H.-J. (2008). The effects of perceived risk and technology type on users' acceptance of technologies. *Information & Management*, *45*, 1–9.

Karjaluoto, H., Mattila, M., & Pento, T. (2002). Factors underlying attitude formation towards online banking in Finland. *International Journal of Bank Marketing*, *20*(6), 261–272. doi:10.1108/02652320210446724

Kay, R. H. (1993). An exploration of theoretical and practical foundations for assessing attitudes toward computers: the computer attitude measure (CAM). *Computers in Human Behavior*, (9): 371–386. doi:10.1016/0747-5632(93)90029-R

Kleijnen, M., Ruyter, K., & Wetzels, M. (2007). An assessment of value creation in mobile service delivery and the moderating role of time consciousness. *Journal of Retailing*, *83*(1), 33–46. doi:10.1016/j.jretai.2006.10.004

Kuisma, T., Laukkanen, T., & Hiltunen, M. (2007). Mapping the reasons for resistance to Internet banking: a means-end approach. *International Journal of Information Management*, *27*(2), 75–85. doi:10.1016/j.ijinfomgt.2006.08.006

Laforet, S., & Xiaoyan, L. (2005). Consumers' attitudes towards online and mobile banking in China. *International Journal of Bank Marketing, 23*(5), 362–380. doi:10.1108/02652320510629250

Laukkanen, T. (2007a). Customer preferred channel attributes in multi-channel electronic banking. *International Journal of Retail & Distribution Management, 35*(5), 393–412. doi:10.1108/09590550710743744

Laukkanen, T. (2007b). Internet vs. mobile banking: comparing customer value perceptions. *Business Process Management Journal, 13*(6), 788–797. doi:10.1108/14637150710834550

Laukkanen, T., & Lauronen, J. (2005). Consumer value creation in mobile banking services. *International Journal of Mobile Communications, 3*(4), 325–338. doi:10.1504/IJMC.2005.007021

Laukkanen, T., & Pasanen, M. (2008). Mobile banking innovators and early adopters: How they differ from other online users? *Journal of Financial Services Marketing, 13*(2), 86–94. doi:10.1057/palgrave.fsm.4760077

Laukkanen, T., Sinkkonen, S., Kivijärvi, M., & Laukkanen, P. (2007). Innovation resistance among mature consumers. *Journal of Consumer Marketing, 24*(7), 419–427. doi:10.1108/07363760710834834

Lee, E.-J., Lee, J., & Eastwood, D. (2003). A two-step estimation of consumer adoption of technology-based service innovations. *The Journal of Consumer Affairs, 37*(2), 256–282.

Li, S., Glass, R., & Records, H. (2008). The Influence of Gender on New Technology Adoption and Use–Mobile Commerce. *Journal of Internet Commerce, 7*(2), 270–289. doi:10.1080/15332860802067748

Lin, C. (2003). An interactive communication technology adoption model. *Communication Theory, 13*(4), 345–365. doi:10.1111/j.1468-2885.2003.tb00296.x

Lohse, G., Bellman, S., & Johnson, E. (2000). Consumer buying behaviour on the Internet: findings from panel data. *Journal of Interactive Marketing, 14*(1), 15–29. doi:10.1002/(SICI)1520-6653(200024)14:1<15::AID-DIR2>3.0.CO;2-C

Luarn, P., & Lin, H.-H. (2005). Toward an understanding of the behavioral intention to use mobile banking. *Computers in Human Behavior, 21*(6), 873–891. doi:10.1016/j.chb.2004.03.003

Markus, H., & Kitayama, S. (1991, April). Culture and the self: implications for cognition, emotion and motivation. *Psychological Review, 98*, 224–252. doi:10.1037/0033-295X.98.2.224

Marr, N., & Prendergast, G. (1993). Consumer adoption of self-service technologies in retail banking: is expert opinion supported by consumer research. *International Journal of Bank Marketing, 11*(1), 3–10. doi:10.1108/02652329310023381

Marsh, H. W., & Hocevar, D. (1985). Application of confirmatory factor analysis to the study of self-concept: First- and higher-order factor models and their invariance across groups. *Psychological Bulletin, 97*, 562–582. doi:10.1037/0033-2909.97.3.562

Meuter, M. L., Ostrom, A. L., Bitner, M. J., & Roundtree, R. (2003). The influence of technology anxiety on consumer use and experiences with self-service technologies. *Journal of Business Research, 56*(11), 899–906. doi:10.1016/S0148-2963(01)00276-4

Modahl, M. (2000). *Now or never*. Harper Collins: New York.

Muñoz, P., & Cruz, P. (2004). Flow in eBank Satisfaction Measure. *33rd European Marketing Academy Conference Proceedings*, Book of Abstracts, Murcia, Spain.

Myers, R. (1986). *Classical and modern regression with applications*. Boston, Duxbury Press.

Narinder, C. (2007). Fast growth of Mobile Communications in India: Lessons for emerging markets. *6th Annual Global Mobility Roundtable*, Los Angeles, USA.

Novak, T., Hoffman, D., & Duhachek, A. (2003). The influence of Goal-Directed and experiential activities on online flow experiences. *Journal of Consumer Psychology*, *13*(1-2), 3–17. doi:10.1207/S15327663JCP13-1&2_01

Nunnally, J. (1967). *Psychometric Theory*. NY: McGraw Hill.

Nysveen, H., Pedersen, P., & Thorbjørnsen, H. (2005). Explaining intention to use mobile chat services: Moderating effects of gender. *Journal of Consumer Marketing*, *22*(5), 247–256. doi:10.1108/07363760510611671

Nysveen, H., Pedersen, P., Thorbjørnsen, H., & Berthon, P. (2005). Mobilizing the Brand: The Effects of Mobile Services on Brand Relationships and Main Channel Use. *Journal of Service Research*, *7*(3), 257–276. doi:10.1177/1094670504271151

OfCom Statistics. (2008). *Communications - The next decade*. Retrieved May 13, 2008, from http://www.ofcom.org.uk /research/

Oumlil, A., & Williams, A. (2000). Consumer Education Programs for Mature Consumers. *Journal of Services Marketing*, *14*(3), 232–243. doi:10.1108/08876040010327239

Parasuraman, A. (2000). Technology Readiness Index (TRI): a multiple-item scale to measure readiness to embrace new technologies. *Journal of Service Research*, *2*(4), 307–320. doi:10.1177/109467050024001

Pikkarainen, T., Pikkarainen, K., Karjaluoto, H., & Pahnila, S. (2004). Consumer acceptance of online banking: an extension of the technology acceptance model. *Internet Research*, *14*(3), 224–235. doi:10.1108/10662240410542652

Podsakoff, P. M., & Organ, D. W. (1986). Self-reports in organizational research: Problems and prospects. *Journal of Management*, *12*, 531–544. doi:10.1177/014920638601200408

Polatoglu, V., & Ekin, S. (2001). An Empirical Investigation of the Turkish Consumers Acceptance of Internet Banking Services. *International Journal of Bank Marketing*, *19*(4), 156–165. doi:10.1108/02652320110392527

Ram, S. (1989). Successful innovation using strategies to reduce consumer resistance: an empirical test. *Journal of Product Innovation Management*, *6*(1), 20–34. doi:10.1016/0737-6782(89)90011-8

Ram, S., & Jung, H. (1991). Forced Adoption of Innovations in Organizations: Consequences and Implications. *Journal of Product Innovation Management*, *8*(2), 117–126. doi:10.1016/0737-6782(91)90005-J

Ram, S., & Sheth, J. (1989). Consumer resistance to innovations: the marketing problem and its solutions. *Journal of Consumer Marketing*, *6*(2), 5–14. doi:10.1108/EUM0000000002542

Rindskopf, D., & Rose, T. (1988). Some theory and applications of confirmatory second-order factor analysis. *Multivariate Behavioral Research*, *23*, 51–67. doi:10.1207/s15327906mbr2301_3

Rogers, E. (2003). *Diffusion of Innovations* (5th ed.). New York: Free Press.

Sheth, J. (1981). Psychology of innovation resistance: the less developed concept (LDC) in diffusion research. *Research in Marketing*, *4*(3), 273–282.

Siau, K., & Shen, Z. (2003). Mobile communications and mobile services. *International Journal of Mobile Communications*, *1*(1/2), 3–14. doi:10.1504/IJMC.2003.002457

Suoranta, M., & Mattila, M. (2004). Mobile Banking and Consumer Behaviour – New Insights into the Diffusion Pattern. *Journal of Financial Services Marketing*, *8*(4), 354–366. doi:10.1057/palgrave.fsm.4770132

Sybase (2008). *Research Reveals Global Appetite for Mobile Banking Services.* Retrieved May 21, 2008, from http://www.sybase.com / detail?id=1054292

Teo, T., & Pok, S. (2003). Adoption of WAP-enabled mobile phones among Internet users. *Omega*, *31*, 483–498. doi:10.1016/j.omega.2003.08.005

Traub, R. (1994). *Reliability for the social sciences: theory and applications.* London: Sage Publications.

Venkatesh, V., Morris, M., Davis, G., & Davis, F. (2003). User acceptance of Information Technology: towards a unified view. *MIS Quarterly*, *27*(3), 425–477.

Venkatesh, V., Ramesh, V., & Massey, A. (2003). Understanding usability: In mobile commerce. *Communications of the ACM*, *46*(12), 53–56. doi:10.1145/953460.953488

Wu, J.-H., & Wang, S.-C. (2005). What drives mobile commerce? An empirical evaluation of the revised technology acceptance model. *Information & Management*, *42*(5), 719–729. doi:10.1016/j.im.2004.07.001

Yang, K. (2005). Exploring factors affecting the adoption of mobile commerce in Singapore. *Telematics and Informatics*, *22*(3), 257–277. doi:10.1016/j.tele.2004.11.003

APPENDIX A: STATEMENTS MEASURING THE BARRIERS

Usage barrier
B1. In my opinion, mobile banking services are easy to use *
B2. In my opinion, the use of mobile banking services is convenient *
B3. In my opinion, mobile banking services are fast to use *
B4. In my opinion, progress in mobile banking services is clear *
B5. The use of changing PIN codes in mobile banking services is convenient *
Value barrier
B6. The use of mobile banking services is economical *
B7. In my opinion, mobile banking does not offer any advantage compared to handling my financial matters in other ways
B8. In my opinion, the use of mobile banking services increases my ability to control my financial matters by myself *
Risk barrier
B9. I fear that while I am paying a bill by mobile phone, I might make mistakes since the correctness of the inputted information. is difficult to check from the screen
B10. I fear that while I am using mobile banking services, the battery of the mobile phone will run out or the connection will otherwise be lost
B11. I fear that while I am using a mobile banking service, I might tap out the information of the bill wrongly
B12. I fear that the list of PIN codes may be lost and end up in the wrong hands
B13. I trust that while I am using mobile banking services, third parties are not able to use my account or see my account information *
Tradition barrier
B14. Patronizing in the banking office and chatting with the teller is a nice occasion on a weekday
B15. I find self-service alternatives more pleasant than personal customer service *
Image barrier
B16. In my opinion, new technology is often too complicated to be useful
B17. I have such an image that mobile banking services are difficult to use

Note: B8 and B9 (in italics) were excluded due to high cross-saturation products
* Reversed Scale

APPENDIX B: DESCRIPTIVE STATISTICS OF RESPONDENT'S CHARACTERISTICS (USAGE AND DEMOGRAPHICS)

		Frequency				
Variable	**Category**	**N**	**%**	**Users (%)**	**Non-user (%)**	**Sig. (χ^2)**
Sex	Male	1473	62.9	36	64	0.000
	Female	869	37.1	28.2	71.8	
	Total (valid)	2342				
Age	<25	162	6.9	36.4	63.6	0.899
	25-29	448	19.1	31.7	68.3	
	30-39	1063	45.4	33.1	66.9	

continued on following page

Appendix B. continued

	40-49	403	17.2	32.3	67.7	
	50-59	172	7.3	34.9.	65.1	
	>59	94	4.0	34.0	66	
	Total (valid)	2342				
Time using Internet	<3 months	111	5.8	36	64	0.764
	3-6 months	107	5.6	37.4	62.6	
Banking	6-12 months	229	12.1	29.7	70.3	
	12-18 months	187	9.9	35.8	64.2	
	18 months- 3 years	514	27.1	32.5	67.5	
	3-5 years	705	37.1	32.5	67.5	
	> 5 years	45	2.4	33.3	66.7	
	Total (valid)	1898				
Time using Mobile	<3 months	71	49.3	33.8	66.2	0.612
	3-6 months	5	3.5	20	80	
Banking	6-12 months	6	4.2	50	50	
	12-18 months	7	4.9	14.3	85.7	
	18 months- 3 years	15	10.4	46.7	53.3	
	3-5 years	39	27.1	28.2	71.8	
	> 5 years	1	0.7	0	100	
	Total (valid)	144				
Mobile Device	Basic	927	47.6	23.6	76.4	0.000
	Advanced (3G, PDA, Smart-phone)	1020	52.4	42.7	57.3	
	Total (valid)	1947				
Mobile Services	Some	776	33.1			
Usage	None	1568	66.9			
		2344				
Frequency of	Daily	99	12.8			
	Few times a week	271	34.9			
Mobile Banking use	Few times a month	217	28			
	Few times a year	84	10.8			
	Less frequently	105	13.5			
	Total (valid)	776				

APPENDIX C: CROSSTABULATION ON MOBILE DEVICE VS USE OF MOBILE SERVICES

			Use of Mobile Services		
			None	Some	Total
Mobile Device	Basic	N	708	219	927
		% of Use of Mobile Services	54.8%	33.4%	47.6%
	Advanced	N	584	436	1020
		% of Use of Mobile Services	45.2%	66.6%	52.4%
Total		N	1292	655	1947

Note: Sig. (χ^2)=0.000

APPENDIX D: HYPOTHESES CONTRASTS (SUMMARY)

Hypothesis	Relation/Statement	Expected Sign/Difference	Contrast
$H1_1$	Usage Barrier (SEM model)	+	Corroborated
$H1_2$	Value Barrier (SEM model)	+	Corroborated
$H1_3$	Risk Barrier (SEM model)	+	Corroborated
$H2_1$	Tradition Barrier (SEM model)	+	Not Supported (-)
$H2_2$	Image Barrier (SEM model)	+	Corroborated
$H3_1$	Functional Barriers (SEM model)	+	Corroborated
$H3_2$	Psychological Barriers (SEM model)	+	Corroborated
H4	Lower Resist. among experienced internet users	-	Not Supported (no sig. dif.)
H5	Lower Resist. among mobile users	-	Corroborated
H6	Lower Resist. among more frequent mobile banking users	-	Partially Corroborated (not corroborated for Tradition)
H7	Resist. different among age groups	+	Not Supported
H8	Male proportion	+	Corroborated

This work was previously published in International Journal of E-Services and Mobile Applications, Volume 1, Issue 4, edited by Ada Scupola, pp. 16-35, copyright 2009 by IGI Publishing (an imprint of IGI Global).

Chapter 12
Consumer Value of Context Aware and Location Based Mobile Services

Henny de Vos
Novay, The Netherlands

Timber Haaker
Novay, The Netherlands

Marije Teerling
Novay, The Netherlands

Mirella Kleijnen
Free University, The Netherlands

ABSTRACT

Context aware services have the ability to utilize information about the user's context and adapt services to a user's current situation and needs. In this article the authors consider users' perceptions of the added value of location awareness and presence information in mobile services. The authors use an experimental design, where stimuli comprising specific bundles of mobile services were presented to groups of respondents. The stimuli showed increasing, manipulated, levels of context-awareness, including location of the user and location and availability of buddies as distinct levels. Their results indicate that simply adding context aware features to mobile services does not necessarily provide added value to users, rather the contrary. The potential added value of insight in buddies' location and availability is offset by people's reluctance to share location information with others. Although the average perceived value overall is rather low there exists a substantial minority that does appreciate the added context aware features. High scores on constructs like product involvement, social influence and self-expressiveness characterize this group. The results also show that context aware service bundles with utilitarian elements have a higher perceived value than bundles with hedonic elements. On the basis of the different results some guidelines for designing context aware mobile services are formulated.

DOI: 10.4018/978-1-60960-607-7.ch012

INTRODUCTION

In contrast to the tremendous success of mobile voice communication services, the uptake of mobile data services has been slow (Carlsson, 2006). With 'mobile services' we refer to all kinds of innovative services that combine technologies and concepts from the domains of telecommunication, information technology, and consumer electronics. While voice communication has become a commodity and mobile phone penetration has risen close to or even higher than 100% in Europe, the adoption of mobile services is hampered by an apparent lack of added value (Ojala et al., 2003), a mismatch between launched applications and the everyday needs of target users (Carlsson, 2006; Steinfeld, 2004), and ineffective business models (Carlsson, 2006; Steinfield, 2004; De Reuver and Haaker, 2009). However, the introduction of new phones such as the iPhone in combination with flat fee data subscriptions has led to a clear increase in data traffic and use of mobile services and mobile internet. Especially news and video sites, e.g. Youtube, and social media services are increasingly used on the newest mobile phones.

A core value element in mobile services is that they can be used anytime and anywhere, which enables people to communicate or to access information at any location, any time and in any situation (De Reuver, 2008). The intention to use mobile services is however found to depend on the situational context (Bouwman et al., 2008). Therefore, mobile services and applications that adapt to the context may provide greater added value (Bae et al., 2006; Klemettinen, 2007). Context awareness deals with the ability to utilize information about the user's environment (context) in order to adapt services to the user's current situation and needs (Dockhorn Costa, 2005). The most well known form of context awareness is location awareness, which is used to adapt services to the current location of the user (Raper et al., 2007). Navigation services, e.g. in-car personal navigation, are among the most popular location based services (Berg Insight, 2007). Other examples of context aware mobile services deal with social context. For example presence services like MSN Mobile Messenger allow users to share information on social context like their online or offline status, or availability for communication. Twitter provides a social messaging utility for users to share what they are currently doing or thinking (Twitter, 2009). Context characteristics can also be derived from sensors, e.g. bio-sensors to measure heart rates, and used to adapt or trigger specific service behavior (Klemettinen, 2007; Koolwaaij et al., 2006).

Lately a significant amount of research has been conducted with regard to technological opportunities and possibilities of context aware mobile services (Hegering, 2004). These services often combine multiple forms of context awareness with personal preferences stored in profiles to create tailored solutions to meet users' needs (Klemettinen, 2007). Research focuses particularly on mobile services architectures, context modeling (e.g. Sinderen et al, 2006), context management (e.g. Kranenburg et al., 2006), context reasoning (e.g. Tao Gu et al., 2005), and on building reference applications. Less attention has been given to user evaluations of context aware services and applications. Relatively little is known about the actual added value of different context aware features in mobile services, and about the factors that contribute to user adoption of these services. For example the degree of control with respect to location disclosure is important (Consolvo et al., 2005). However, research in customer needs and behavior concerning context aware mobile services is hindered by the fact that respondents have difficulty understanding the specific characteristics of services that are not yet available. Decisions regarding adoption of context aware services revolve around consumers' perceptions of their potential use and expected benefits and costs, and depend on technology and service characteristics (Bouwman et al., 2007a), personal characteristics (Teerling et al., 2007; Kwon et al.

2007) and context characteristics (Laukkanen & Lauronen, 2005).

In this article we therefore address the following question: what are users' perceptions of the added value of location awareness and presence information in mobile services? To address this question we used an experimental design, where stimuli, in our case specific bundles of mobile services, were presented to groups of respondents.

The mobile services that we use in our study comprised increasing, manipulated, levels of context awareness. We included location of the user and location and availability of buddies as distinct levels of context awareness. Two experiments were conducted. The first experiment focused on a utilitarian setting, i.e. the included services provided predominantly pragmatic or goal-directed value such as local searches (yellow pages) and directions from the current location. The second experiment focused on a hedonic setting, i.e. the services provided mostly hedonic or entertaining value such as taking pictures and automatic options for geo-tagging, annotating, storing and sharing these pictures. The aim of the article is to contribute to the understanding of consumer value and adoption of context aware features in mobile services. Such understanding supports the design of value driven mobile service bundles.

The article is based on earlier work (Vos et al., 2008) and is organized as follows. First we elaborate on (customer value of) context aware mobile services and provide our main hypothesis. This is followed by a description of our research approach and experimental design. Subsequently we present our main findings and end with some conclusions regarding our research and implications for the design of context aware mobile services.

LOCATION BASED AND CONTEXT AWARE SERVICES

A service is called context aware if in its operation it uses context information other than explicit application logic related input (Abowd et al., 1998). The most well known form of context awareness is location awareness, which is used to adapt services to the current location of the user or users (Raper et al., 2007). Giaglis et al. (2003) provide a taxonomy of location based services including emergency services (e.g., automotive assistance), navigation services (e.g., personal navigation), information services (e.g., mobile yellow pages), marketing services (e.g., mobile advertising), tracking services (e.g., vehicle tracking) and even billing services (e.g., location-sensitive billing).

In-car personal navigation that is based on GPS location and advanced digital map technology is among the most popular applications. Personal navigation is moving from dedicated devices to advanced mobile devices that are equipped with GPS functionality and sufficient storage and computational capacity. The expected increase in penetration levels of GPS equipped devices is considered an important driver for the adoption of location based services (Berg Insight, 2007).

Location-based services (LBS) are growing in the U.S., largely thanks to the FCC's E-911 legislation, that required location information to be included in emergency calls made from mobile phones (totaltelecom, 2007). Raper et al. (2007), in their comprehensive overview of the field of location based services, label user acceptance for spatial cognition and interfaces of LBS as a key issue. Kallio (2005) and Kangas & Kinnunen (2005) find that user acceptance and actual use is strongly influenced by the context or situation of use. Kaasinen (2005) did a study about attitudes towards LBS. She found that users specifically requested instant on/off, dynamic scale adjustment, well designed labeling (of landmarks), push services that are under user control, and can be refined or cancelled easily, reliable services that are only used occasionally and support for content creation.

Zipf (2004) surveyed potential LBS users about their concepts of 'nearness', the map presentation and the type of content likely to be of interest. Raper (2006) reviewed the operational constraints on successful LBS deployment based on experience from 'live' LBS implementations, and highlighted context systems, positional fusion and application development platforms for LBS as key challenges for LBS.

Other forms of context awareness include social awareness and group awareness. Well-known examples of social aware services are presence services, for instance MSN Messenger. A more elaborate example of a context sharing mobile application is iyouit (iyouit, 2009; Böhm et al. 2008). Iyouit is an application that allows users to share there whereabouts with buddies, or any other piece of context information they might want to share, ranging from body data to pictures, local weather information, and even subjective data such as moods and experiences.

Information gathered from multiple context sources may also be used to adapt service behavior, e.g., providing recommendations about who to reach in a given situation or via which communication channel a person might be best reached (Ter Hofte et al., 2006). Sharing of social context information has the potential of supporting (dynamic) groups of people within social networks to more efficiently and effectively perform their activities (Ter Hofte et al., 2006). However, users' willingness to disclose their location is dependent on who can see their location, for what purpose and with what accuracy (Consolvo et al., 2005).

Hegering et al. (2004) advance several management challenges that are specifically hard when provisioning context aware services, such as configuration of the context value chain, fault management, accounting, performance (quality of context) and security. Regarding the adoption of context aware services the tedious balancing of costs and benefits need mentioning. Financial costs may be one obstacle (KPMG, 2005) but other user sacrifices and uncertainties, e.g. related to privacy, trust in the service, or required cognitive effort (Teerling et al., 2007) are important here as well.

Our study aims at gaining insight into the added value of context aware aspects in mobile services. In our experimental design we consider increasing, manipulated, levels of context awareness, including location of the user and location and presence of buddies as distinct levels. User context in this respect is automatically derived without any effort of the user. A straightforward hypothesis appears to be

Hypothesis 1: increasing levels of location and presence awareness in mobile services are accompanied by increasing levels of customer value.

This hypothesis is addressed in two experiments, one in a utilitarian setting and one in a hedonic setting (Kleijnen et al., 2007). Pragmatic context aware services have been around for some time as business to employee services that focus on providing location dependent information in transport or workforce management and hence aim at providing utilitarian value. As a first step towards consumer adoption, those context aware services that carry a predominantly pragmatic value, e.g. navigation services, may meet with higher valuation than services with a high 'fun factor'. We therefore hypothesize:

Hypothesis 2: location and presence aware service bundles with utilitarian elements have a higher perceived value than bundles with hedonic elements.

RESEARCH METHOD

To be able to investigate the role of context awareness we used an experimental design, where stimuli, in our case specific thematic sets of mobile services, are presented to groups of respondent. The sets of mobile services comprise increasing levels of context awareness and focus on specific themes. We emphasize that user context is automatically derived without any effort of the user.

Figure 1. Overview of levels of (automatically derived) context awareness considered in the experiments

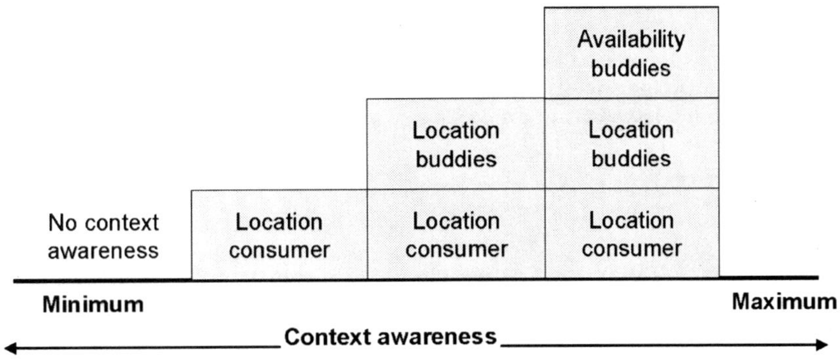

We distinguish four levels of context awareness, i.e. no context awareness, location of the user, location of the user and selected buddies, and location of the user and location and availability of selected buddies, see Figure 1. We added these four levels of context awareness to two specific mobile service bundles, i.e. a hedonic and a utilitarian service bundle. Hedonic services are characterized by elements of fun and entertainment, while utilitarian services are typically more focused on efficient and pragmatic time management (Kleijnen et al., 2007). These differences were operationalized by service descriptions that focused on these key concepts. Utilitarian services were presented by a service bundle that focused on effective time management; allowing consumers to plan their traveling in a more effective way. Hedonic services were represented by entertainment services that are rapidly gaining popularity in the market, e.g. moblog (mobile blogging) and mobile picture sharing. Taking into account these two different manipulations and the four context awareness levels, a 4x2 design is obtained, containing eight different scenario descriptions

The scenarios were developed in accordance with recent trends and service introductions in the market. Given the lack of a single killer application for mobile data services, and given that the mobile device and services should fulfill personal user needs in varying circumstances, a bundle of services is more capable of providing the right 'bundle of benefits' (Kottler, 1999; Stremersch & Tellis, 2002). A core compact bundle of services was therefore developed for the hedonic and the utilitarian setting. Scenarios were developed by adding the possibilities of specific levels of context awareness to this core set. To maintain the highest possible level of realism, a large mobile service provider developing context aware services was involved in the development of the scenarios. Additionally, the scenarios were subjected to a qualitative pretest in which 20 respondents (10 consumers and 10 mobile service experts) were asked to critically evaluate the scenarios in terms of clarity and realism. Second, we subjected the scenarios to a large scale quantitative pre-test using 191 respondents. The results of both the interviews and the quantitative pretest resulted in minor adaptations of the scenarios, related to fine-tuning sentence formulations rather than radical changes in content. A summary of the scenarios with the highest levels of context awareness are given in Figure 2.

VARIABLES AND SCALES

Respondents were asked to give their opinion with respect to different value scales, i.e. intention to use and overall value. All scales met the required standards with respect to their unidimensionality

Figure 2. Summary of two mobile service scenario's, i.e. location of user and location and availability of buddies

Context aware features and service aspects			
Recognition of user's current location; Recognition of location of buddies; Sharing location information with buddies; Sharing availability information with buddies; Visualizing location and availability information on map			
Utilitarian mobile service description		**Hedonic mobile service description**	
Search Yellow Pages for addresses. Provide directions from user or buddy location to specific addresses and/or buddies automatically. Contact buddies that are available. Visualize locations and directions on map.	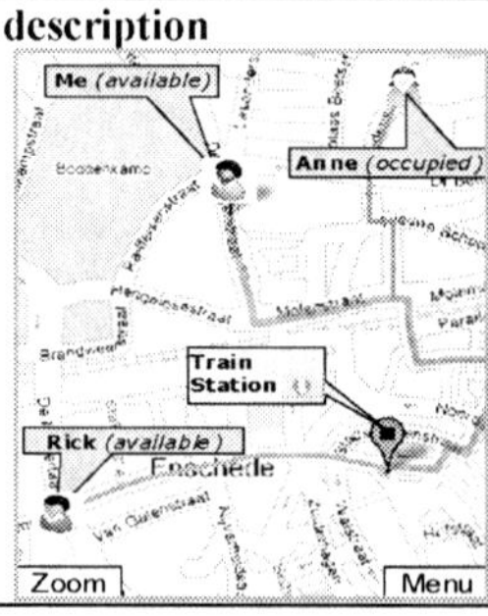	Pictures (made with mobile phone) and messages are automatically stored in a moblog (mobile diary) and labelled with date, time and location. Moblog entries and current locations of different buddies can be shared and visualized on a map.	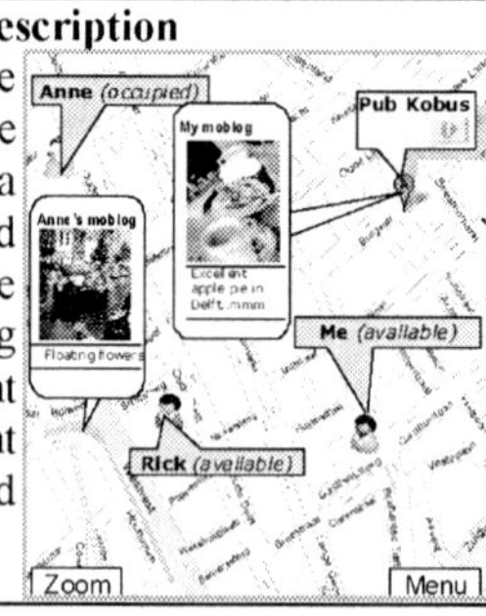

and reliability. Table 1 presents an overview of the scales used to measure the various constructs.

Next to relating value to levels of context awareness, we also want to explore the effects of other factors on consumer value perception. Besides common background characteristics, such as age, gender, income, we employed a range of semantic differential scales to measure the independent variables in our study. With alpha's ranging from 0.74 to 0.93 these scales are:

- Cognitive effort, i.e. the perceived effort of learning to use the service (Kleijnen et al., 2007),
- Control, i.e. the usefulness of mobile services for making decisions (Kleijnen et al., 2007),
- product involvement, i.e. the familiarity with and interest in mobile services (Zinkhan & Locander, 1988),
- impulsiveness (Puri, 1996),
- innovativeness with respect to mobile services (Kleijnen et al., 2004),

Table 1. Scales used to measure the various constructs

Construct	*Based on*	*Items*	*Cronbach's alpha*	*AVE*
Intention to use	Cronin et al. (2000)	It is likely that I will take a yearly subscription to these mobile services. I will definitely take a subscription to these mobile services. I would be willing to pay 5 Euros per month for unlimited use of these mobile services.	0.88	80.5%
Total value	Cronin et al. (2000)	Overall, the value of these mobile services is… The extent to which these mobile services fulfill my needs is… (scaling from "very low" to "very high" on a 7-point scale)	0.96	95.8%
Hedonic value	Voss et al. (2003)	Fun; Delightful; Thrilling; Enjoyable	0.91	66.4%
Utilitarian value	Voss et al. (2003)	Effective; Helpful; Functional; Practical	0.94	71.2%

- privacy, i.e. attitude towards privacy issues concerning mobile technology (Kleijnen et al., 2007; Malhotra et al., 2004),
- self expressiveness, i.e. to what extend does using the mobile service help you to express to others who you are (Nysveen et al., 2005) and,
- social influence, i.e. in how far others influence ones attitude toward mobile services (Lu et al., 2005).

Since the study focuses on context awareness, we also included characteristics with respect to situational context of respondents. For example questions about how many days each week are respondents at work, alone, or in a hurry. These questions are referred to as 'life-style' characteristics. All scales were considered in the aforementioned qualitative and quantitative pre-tests.

The questionnaire contained subsequent parts, starting with general questions on attitude towards mobile technology and general characteristics. Following this, a particular mobile service scenario was presented. The respondents were then asked to imagine themselves in that situation and answer questions concerning the particular situation and the use of these mobile services, e.g. value, intention to use, cognitive effort. The final part of the questionnaire concerned background characteristics, e.g., gender, life-style.

SAMPLE

We conducted a between-subjects experiment, whereby we approached consumers from a consumer panel to participate in our research. Participants were randomly assigned to one of the scenarios. After screening the questionnaires for incompleteness and abnormalities (Tabachnik and Fidell 1996), we were left with a total sample of 403 respondents that was almost evenly divided over the various scenarios, i.e., between 49 and 53 respondents for each scenario. The sample contained 56% women and 44% men and a broad range of age categories: younger than 25 years: 6%, 25 to 35 years: 19%, 36 to 45 years: 25%, 45 to 55 years: 25% and older than 55 years: 25%.

RESULTS

The manipulations, i.e., level of context awareness and utilitarian versus hedonic, were checked in control questions. The answers to these questions proved that respondents overall recognized and understood the manipulations.

Table 2. Assessment of total value and intention to use for the different utilitarian (util) and hedonic (hed) scenarios

Context awareness		**Total value**			**Intention to use**		
		util	hed	sig	util	hed	sig
Basic	Average	3.34	2.48	*	2.16	1.94	
	Stdev	1.61	1.43		1.46	1.47	
LBS	Average	3.33	2.64	*	2.45	1.80	*
	Stdev	1.71	1.54		1.52	1.19	
LBS&CA	Average	2.85	2.39		1.82	1.59	
	Stdev	1.63	1.38		1.13	0.98	
LBS&CA+	Average	2.79	2.19	*	2.18	1.59	*
	Stdev	1.43	1.32		1.53	0.97	

* sig <.05, i.e. significant difference between hedonic and utilitarian scenario.

Table 3. Attitude towards sharing of location information

	N	Average	Standard deviation	Percentage rating ≥ 4
Information on my location should be very accurate	301	2.38	1.66	27.2%
Explicit consent to service providers on use of location information	301	5.51	1.92	85.4%
Availability of exact location of buddies	201	2.15	1.48	23.9%
Explicit consent to buddies on use of location information	201	5.66	1.80	87.6%

Note: measurement on a 7-point scale, where 1=totally disagree to 7=totally agree. Number of respondents differs as some questions were only relevant for certain scenarios.

Table 3 gives an overview of respondents' rating for different aspects. The ratings should be related to the scale on which they were measured, i.e. from 1=very low to 7=very high, with 4 being the neutral point. The ratings are presented for both the utilitarian and hedonic scenarios, with different levels of context awareness, i.e. Basic=no context awareness, LBS=location of user, LBS&CA=location of user and buddies, LBS&CA+=location of user and location and availability of buddies.

In general a decreasing trend can be observed in the ratings when context awareness is added to the services, where the utilitarian services are valued more than the hedonic ones. However the differences in rating between the context awareness levels are not significant, except for the difference in intention to use between LBS and LBS&CA in the utilitarian scenario. The difference in rating for total value between the utilitarian and the hedonic scenarios are significant for all levels of context awareness, except for LBS&CA. This confirms our second hypothesis.

Average ratings for total value and intention to use are rather low, ranging from 1.59 for intention to use in the hedonic LBS&CA scenario to 3.34 for total value in the basic utilitarian scenario. All ratings are far below the mean of the scale. The averages are that low, because of the high percentage of respondents who perceive no value with respect to the presented mobile services (29% for total value and 50% for intention to use). Figure 3 gives an overview of the distribution of ratings among respondents, i.e. the percentage of respondents with a specific rating. The graphs combine the ratings of all four utilitarian and hedonic scenarios, respectively. On average only 31% of the respondents rate the presented scenario with a 4 or higher for total value. For intention to use it is even worse; 7% give a rating equal to or above 4. Respondents appear to be more positive about utilitarian service bundles

Figure 3. Distribution of total value and intention to use ratings for utilitarian and hedonic scenarios

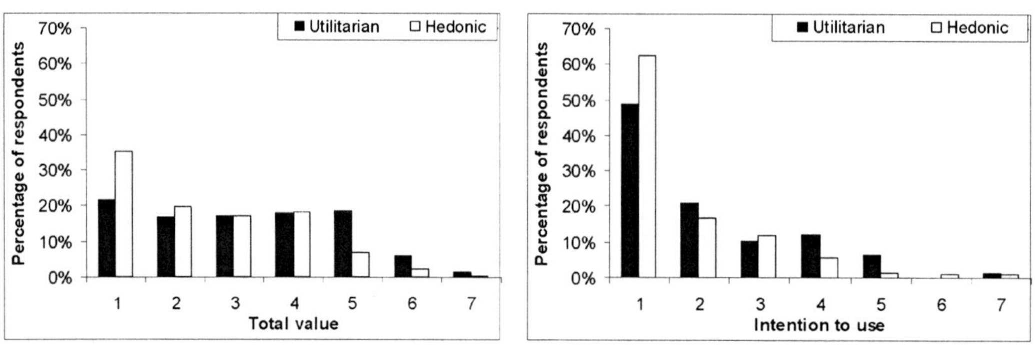

in comparison with the hedonic ones, i.e. lower ratings apply more often to the hedonic scenarios. The results also show a high variation between respondents. Such average low valuation with a high variation between individuals is also found in other studies on the value of mobile services (Haaker & de Vos, 2007; Bouwman et al., 2007b). They also conclude that such a low valuation does not imply that such services are not attractive for consumers or that providers cannot successfully exploit such services. Substantial minorities exist that do value the services. In fact providers would likely not be unhappy if all positive raters (31% of the mobile phone users) would become their customers. In section 4.2 we will discuss in more detail the characteristics of the respondents that score high on total value and intention to use.

Our hypothesis was that adding increasing levels of context awareness to a service bundle would increase the perceived value. We have to reject this hypothesis. No significant increases in total value or intention to use were observed, rather the contrary. This may be understood from users' attitude towards sharing of context information as we will discuss in section 4.1. When assessing the explanatory value of the service characteristics (i.e. hedonic versus utilitarian, level of context awareness) by means of a regression analysis, we have to conclude that such characteristics only account for 6.2% of the total variance in total value. Hedonic services as compared to utilitarian services have a significant negative effect on total value. Effects of context aware aspects are not significant. Personal characteristics explain 24% of the variance in total value, with significant effects between 0.166 and 0.229 on total value for product involvement, control, social influence and self-expressiveness. Life style characteristics account for 4% of the variance, with a significant positive effect for people that are more than 2 days a week 'at ease'.

ATTITUDE TOWARD SHARING OF CONTEXT INFORMATION

One issue with respect to context aware services is that the context of the user has to be available for service providers in order to be able to deliver the services. Literature shows that consumers want control over the use of their context information. Their concerns and opinions regarding this issue may have an effect on their value perceptions. Table 3 summarizes the attitude of respondents toward use of context information. Location information of the user is not only shared with the service provider, but in some scenarios location information is also shared with selected buddies. Most respondents (>80%) would like to give explicit consent to both service providers and buddies for the use of location information. Respondents do not care very much about the accuracy of location information. The number of respondents differs as some questions were only relevant for certain scenarios.

Table 4 presents respondents' willingness to share location information with specific people. Respondents dislike sharing information with unknown people. Colleagues and managers are sometimes allowed to see location information. Most respondents (> 80%) indicate that their partner should be given access to information on location and availability. This is particularly high, considering the fact that some of the respondents might not have a partner. The willingness to disclose location information to family (around 50%) and friends (around 25%) is much lower.

Both Table 3 and Table 4 show that respondents are reluctant to share location information. Underlying reasons are likely that they are concerned about their privacy or think that sharing information is not useful. It is clear that such opinions have a negative effect on the perceived value of services that share such information.

Table 4. Attitude toward sharing of location information with specific persons

Type of information	Share with ...					
	Unknown person	Partner	Family	Friends	Colleague	Manager
Location information:						
- at home	2%	87%	56%	33%	10%	7%
- at work	4%	81%	49%	25%	23%	21%
- travelling	2%	87%	51%	27%	12%	11%
- away	2%	82%	41%	25%	5%	2%
Availability information	18%	83%	78%	70%	50%	44%

CHARACTERISTICS OF THE 'HIGH RATERS'

The results showed that on average respondents perceive a low value with respect to context aware service bundles. However, the variation between respondents in perceived total value is high. In fact, the average is dominated by people that perceive a (very) low value as was shown by the distribution of ratings (see Figure 3). A substantial minority is however rather positive about the services. This section deals with the respondents indicating that they value the presented set of mobile services with a four or higher. We denote these respondents with the 'high raters'. Note that '4' is the mean rating on the 7-point scale.

A total of 126 respondents indicated that they valued the presented set of mobile services with a 4 or higher. This is 31%. Figure 4 visualizes the percentage of people with a positive rating for each scenario. The basic utilitarian scenario is evenly distributed: about 50% rate the services as positive and 50% as negative. The percentage of high raters decreases in the utilitarian scenarios when context aware aspects are added. For the highest level of context awareness 23% value the services with a 4 or higher. The hedonic scenarios give a similar result although the differences are smaller than in the utilitarian scenario. In fact, the percentage of high raters for the highest context aware level is almost the same for the utilitarian and the hedonic scenarios.

Figure 4. Percentage of respondents scoring '4' or higher on total value within the scenarios

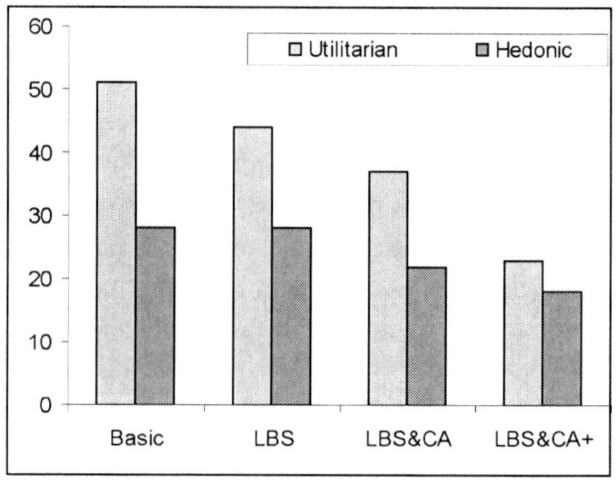

Table 5 gives an overview of the specific characteristics of the high raters as compared to the others. Chi-square and t-tests were used to find significant differences on a 5% significance level. The high raters score significantly higher in most scenarios on social influence, i.e. the extent to which they are influenced by others to use mobile services such as presented to them. A high score on self-expressiveness characterizes especially the high raters of the utilitarian scenarios. High scores on product involvement and innovativeness characterize the high raters for the

Table 5. Characteristics of the high raters, i.e. significant differences on personal characteristics as

	Utilitarian (5% sig level)	Hedonic (5% sig level)
Basic	Control, self expressiveness, daily travelling of 1 hour or more, alone	Product involvement, cognitive effort
LBS	Social influence, self expressiveness	Product involvement, innovativeness, self expressiveness
LBS&CA	Social influence, Self expressiveness	Innovativeness, cognitive effort (-), social influence
LBS&CA+	Social influence, self expressiveness	

hedonic scenarios. Cognitive effort, i.e., whether it takes a lot of effort to learn how to use the service, is significantly lower for the high raters of the LBS&CA hedonic scenario. Unexpectedly, for the basic scenario cognitive effort for the high raters is higher than that of low raters. For the basic utilitarian scenario a few life style characteristics differ significantly, i.e. the respondents who are often alone and travel a lot value the service higher. For the hedonic scenario with LBS&CA+ no significant difference between high and low raters could be found for specific factors.

Next to the characteristics listed in Table 5 the high raters also differ significantly in their attitude toward sharing of location information. They appreciate more that their location information can be used by service providers, they think their locations information should be very accurate and they like information of buddies to be available to them. Regarding consent the high raters do not differ significantly. High raters are significantly more willing to share availability information with unknown people, partners and colleagues (sig. < 0.1). High raters do not significantly differ with respect to gender, age, income and level of education.

DISCUSSION

This study considers the changes in perceived user value from adding different, manipulated, levels of context aware features to mobile services. The location of the user, and the location and availability of buddies are included as distinct levels. Our results indicate that simply adding context aware features to mobile services does not automatically result in added value for users. Rather the contrary is true. Bundles of services that automatically use the location and availability of buddies were found to have a lower value than bundles without these features. There is an apparent trade off between the usefulness and added value of location information of buddies on the one hand and the privacy one needs to give up when sharing this information with buddies on the other hand. On average the potential added value does not outweigh the loss of privacy. This is exemplified by users' general reluctance to disclose their location to other people. The willingness to share location information with one's partner is above 80%, but this drops sharply to around 50% and 25% for sharing with family and friends, respectively. This result confirms the finding in (Consolvo et al., 2005) that people's willingness to share location information depends on whom it is to be shared with. For most respondents the group of likely buddies for location sharing is limited to family members. A second finding of our study is that context aware service bundles with predominantly utilitarian elements have a higher perceived value than bundles with mostly hedonic elements. Hedonic mobile service bundles are considered of lower value for all levels of context awareness.

Although average value ratings are low, we still find that about 31% of the respondents perceived a higher rating, i.e., 4 or more on a 7-point scale.

High raters differ on specific background characteristics when comparing them to low raters. The high raters are in general people that are familiar with and interested in mobile services. They think that using such mobile services improves their status among friends and they use such mobile services to express their personality.

We are aware that one key problem when studying value and intention to use of services is that when these services are assessed in a laboratory (controlled, safe, reproducible) the results may have limited external validity as compared to studies 'in the wild' (semi-controlled, unpredictable, not reproducible) (Kallio et al., 2005). Still, as we carefully checked and iteratively improved our manipulations in the pre-test and the actual experiment, we feel that our results provide new insights into users' value perceptions of context aware features in mobile services. Another point to mention here is the age distribution of our respondents, which is relatively 'at age' given the domain of mobile services and may therefore bias our results. Over 50% of the respondents have an age above 45, and less than 25% below 36. However, the observed valuations did not differ significantly with age.

The results of this study provide some guidelines for providers of context aware services. These relate to the observed reluctance to share location information outside the family circle, the higher perceived value of utilitarian service bundles and the characteristics of the respondents with high ratings. With respect to marketing of such services, the emphasis should be on highlighting aspects that refer to characteristics of the respondents with high ratings, such as product involvement and status aspects related to social influence. As people prefer utilitarian (pragmatic) services and basically just want to share location information with their partner, providers could first target pragmatic service elements in a family setting. Providers should also reflect on the privacy issues of consumers and how these are addressed to decrease the concerns of users and hence increase perceived value.

Although we covered only a limited range of context aware aspects, i.e. user and buddy location and availability of buddies, it is likely that our guidelines are also applicable to other types of context information.

Further research will focus on gaining more insight into effects of personal characteristics on value and intention to use. Other value measures, such as utilitarian value or hedonic value might give rise to further guidelines for service providers. Since context aware services imply use in different user contexts, another line of research will be to explore the conditional value and the relation between user context and perceived value in more detail.

ACKNOWLEDGMENT

This research has been supported by the Dutch Freeband Communication Research Program (FrUX project and AWARENESS project) under contract BSIK 03025.

REFERENCES

Abowd, D., Dey, A. K., Orr, R., & Brotherton, J. (1998). Context-awareness in wearable and ubiquitous computing. *Journal Virtual Reality*, *3*(3), 200–211. doi:10.1007/BF01408562

Bae, J. S., Lee, J. Y., Kim, B. C., & Rye, S. (2006, August 1-4). Next Generation Mobile Service Environment and Evolution of Context Aware Services. *International Conference, EUC 2006*, Seoul, Korea.

Böhm, S. Koolwaaij, J., Luther, M., Souville, B., Wagner, M., & Wibbels, M. (2008, October 27–29). Introducing IYOUIT. *Proceedings of the International Semantic Web Conference (ISWC'08),* Germany, *vol. 5318 of LNCS*, (pp. 804–817). Springer Verlag.

Bouwman, H., Carlsson, C., Molina-Castillo, F., & Walden, P. (2007a). Barriers and drivers in the adoption of current and future mobile services in Finland. *Telematics and Informatics*, *24*, 145–160. doi:10.1016/j.tele.2006.08.001

Bouwman, H., de Vos, H., & van der Wijngaert, L. (2008, July). Context-sensitive mobile services for police officers. *7th International Conference on Mobile Business*, Barcelona.

Bouwman, H., Haaker, T., & de Vos, H. (2007b). Mobile Service Bundles: The example of navigation services. *Electronic Markets*, *17*(1), 20–28. doi:10.1080/10196780601136757

Carlsson, C. (2006). Special issue on mobile technology and services. *Electronic Commerce Research and Applications*, *5*(3), 189–191. doi:10.1016/j.elerap.2006.07.001

Consolvo, S., Smith, I., Matthews, T., LaMarca, A., Tabert, J., & Powledge, P. (2005, April 2-7). Location Disclosure to Social Relations: Why, When, & What People Want to Share. Portland, Oregon, USA.

Cronin, J., Brady, M. K., & Hult, G. T. (2000). Assessing the effects of quality, value, and customer satisfaction on consumer behavioral intentions in service environments. *Journal of Retailing*, *76*(2), 193–218. doi:10.1016/S0022-4359(00)00028-2

De Reuver, M. (2008). The mobile context explored. Bouwman, H., De Vos, H., & Haaker, T. (Eds.), *Mobile service innovation and business models*. Heidelberg-Berlin: Springer.

De Reuver, M., & Haaker, T. (2009, August). Designing viable business models for context-aware mobile Services. *Telematics and Informatics*, *26*(3), 240–248. doi:10.1016/j.tele.2008.11.002

Dockhorn Costa, P., Ferreira Pires, L., & van Sinderen, M. (2005). Designing a Configurable Services Platform for Mobile Context-Aware Applications. [JPCC]. *International Journal of Pervasive Computing and Communications*, *1*(1), 13–25.

Giaglis, G. M., Kourouthanassis, P., & Tsamakos, A. (2002). Towards a Classification Framework for Mobile Location Services. In B.E.Mennecke & T.J. Strader (Eds.), *Mobile Commerce: Technology, Theory, and Applications*. Hershey, PA: Idea Group Publishing.

Haaker, T. I., & de Vos, H. (2007, September 2-5). Customer preferences for bundled content services. *18th European Regional ITS Conference*, Istanbul, Turkey.

Hegering, G.-G., Küpper, A., Linnhoff-Popien, C., & Reiser, H. (2004). Management Challenges of Context-Aware Services in Ubiquitous Environments. *Self-Managing Distributed Systems, Book Series Lecture Notes in Computer Science, Volume 2867/2004* (pp. 246-259). Berlin/Heidelberg: Springer.

Insight, B. (2007). *Connected Navigation Devices*. LBS Research Series, www.berginsight.com

Iyouit (2009). www.iyouit.eu. Accessed 31 March 2009.

Kaasinen, E. (2005). User acceptance of location-aware mobile guides based on seven field studies. *Behaviour & Information Technology*, *24*, 37–49. doi:10.1080/0144929051233 1319049

Kallio, T., Kankainen, A., Kaikkonen, A., Kekalainen, A., & Cankar, M. (2005). Usability Testing of Mobile Applications: A Comparison between Laboratory and Field Testing. *Journal of Usability Studies*, *1*, 4–16.

Kangas, E., & Kinnunen, T. (2005). Applying user-centered design to mobile application development. *Communications of the ACM*, *48*(7), 55–59. doi:10.1145/1070838.1070866

Kleijnen, M. H. P., De Ruyter, K., & Wetzels, M. G. M. (2007). An assessment of value creation in mobile service delivery and the moderating role of time consciousness. *Journal of Retailing*, *83*(1), 33–46. doi:10.1016/j.jretai.2006.10.004

Klemettinen, M. (Ed.). (2007). *Enabling Technologies for Mobile Services: The MobiLife Book*. Chichester, England: John Wiley & Sons Ltd.

Koolwaaij, J., Tarlano, A., Luther, M., Morhs, B., Battestini, A., & Vaidya, R. (2006, July). *ContextWatcher – Sharing context information in everyday life. International Conference on Web Technologies, Applications, and Services*, Calgary, Canada.

Kotler, P. (1999). *Principles of Marketing*. London: Prentice Hall Europe.

KPMG. (2006). *Consumers and Convergence, challenges and opportunities in meeting next generation customer needs*. KPMG International.

Kranenburg, H, & van,., Bargh, M.S., Iacob, S., & Peddemors, A. (2006). A Context Management Framework for Supporting Context Aware Distributed Applications. *IEEE Communications Magazine*, *44*(8), 67–74. doi:10.1109/MCOM.2006.1678112

Kwon, O., Choi, K., & Kim, M. (2007). User acceptance of context-aware services: self-efficacy, user innovativeness and perceived sensitivity on contextual pressure. *Behaviour & Information Technology*, *26*(6), 483–498. doi:10.1080/01449290600709111

Laukkanen, T., & Lauronen, J. (2005). Consumer value creation in mobile banking services. *International Journal of Mobile Communications*, *3*(4), 325–338. doi:10.1504/IJMC.2005.007021

Lu, J., Yao, J. E., & Yu, C. S. (2005). Personal Innovativeness, Social Influences and Adoption of Wireless Internet Services via Mobile Technology. *The Journal of Strategic Information Systems*, *14*(3), 245–268. doi:10.1016/j.jsis.2005.07.003

Malhotra, Naresh K., Sung S. K., & Agarwal, J. (2004). Internet Users' Information Privacy Concerns (IUIPC). *The Construct, the Scale, and a Causal Model. Information Systems Research*, *15*(4), 336–355. doi:10.1287/isre.1040.0032

Meuter, M. L., Bitner, M. J., Ostrom, A. L., & Brown, S. W. (2005, April). Choosing among alternative service delivery modes: an investigation of customer trial of self-service technologies. *Journal of Marketing*, *26*, 61–83. doi:10.1509/jmkg.69.2.61.60759

mITF (mobile IT Forum) (2004). *Flying carpet: Towards the 4th Generation Mobile Communications System, version 2.0*. mITF retrieved from: www.mitf.org/public_e/archives/Flying_Carpet_Ver200.pdf

Nysveen, H., Pedersen, P. E., & Thorbjørnsen, H. (2005). Intentions to Use Mobile Services: Antecedents and Cross-Service Comparisons. *Journal of the Academy of Marketing Science*, *33*(3), 330–346. doi:10.1177/0092070305276149

Ojala, T., Korhonen, J., Aittola, M., Ollila, M., Koivumäki, T., Tähtinen, J., & Karjaluoto, H. (2003). SmartRotuaari - Context-aware mobile multimedia services. *2nd International Conference on Mobile and Ubiquitous Multimedia* (pp. 9-18), Norrköping, Sweden.

Olllo, www.olllo.nl. Accessed 12 November 2007.

Puri & Radhika. (1996). Measuring and Modifying Consumer Impulsiveness: A Cost-Benefit Accessibility Framework. *Journal of Consumer Psychology*, *5*(2), 87–113. doi:10.1207/s15327663jcp0502_01

Raper, J. (2006). Design constraints on operational LBS. In G. Gartner, W. Cartwright, & M. Peterson (Eds), *Location Based Services and TeleCartography. Lecture Notes in Geoinformation and Cartography*. Berlin Heidelberg: Springer.

Raper, J., Gartner, G., Karimi, H., & Rizos, C. (2007). A critical evaluation of location based services and their potential. *Journal of Location Based Services*, *1*(1), 5–45. doi:10.1080/17489720701584069

Steinfield, C. (2004). The development of Location Based Services in Mobile Commerce. In B. Preissl, H. Bouwman, & C. Steinfield (Eds), *Elife after the Dot.com bust*. Berlin; Springer Verlag.

Stremersch, S., & Tellis, G. (2002). Strategic bundling of Products and Prices: A new synthesis for marketing. *Journal of Marketing*, *66*(1), 55–72. doi:10.1509/jmkg.66.1.55.18455

Tabachnik, B. G., & Fidell, L. S. (1996). *Using Multivariate Statistics*. New York: Harper Collins.

Tao, G., Pung, H. K., & Zhang, D. Q. (2005). A service-oriented middleware for building context-aware services. *Journal of Network and Computer Applications*, *28*(1), 1–18. doi:10.1016/j.jnca.2004.06.002

Teerling, M., Haaker, T., & Vos, H. de (2007, October 18-21). *Understanding Consumer Value of Mobile Service Bundles, NAEC*, Lake Garda, Italy.

Ter Hofte, G. H., Mulder, I., & Verwijs, C. (2006). Close Encounters of the Virtual Kind: A Study on Place-based Presence. AI & Society. *The Journal of Human-Centred Systems*, *20*(2).

Totaltelecom (2007). *Location based services*. Retrieved from: www.totaltele.com/View.aspx?ID=91162&t=4, last accessed 12 November 2007.

Twitter (2009). *Twitter.com*. Accessed 31 March 2009.

van Sinderen, M. J., Halteren, A. T., Wegdam, M., Meeuwissen, B., & Eertink, H. E. (2006, September). Supporting Context-Aware Mobile Applications: An Infrastructure Approach. *IEEE Communications Magazine*, 96–104. doi:10.1109/MCOM.2006.1705985

Vos, H. de, Haaker, T., Teerling, M., & Kleijnen, M. (2008, June). Consumer Value of Context Aware and Location Based Mobile Services. *21st Bled eConference*, Bled, Slovenia.

Voss, K. E., Spangenberg, E. R., & Grohmann, B. (2003, August). Measuring the Hedonic and Utilitarian Dimensions of Consumer Attitude. *JMR, Journal of Marketing Research*, *40*, 310–320. doi:10.1509/jmkr.40.3.310.19238

Zinkhan, G. M., & Locander, W. B. (1988). ESSCA: A Multi-dimensional Analysis Tool for Marketing Research. *Journal of the Academy of Marketing Science*, *16*, 36–46. doi:10.1007/BF02723324

Zipf, A., & Joest, M. (2004). User Expectations and Preferences regarding Location-based Services - results of a survey. *2nd Symposium on Location Based Services and TeleCartography*. Vienna, Austria.

This work was previously published in International Journal of E-Services and Mobile Applications, Volume 1, Issue 4, edited by Ada Scupola, pp. 36-50, copyright 2009 by IGI Publishing (an imprint of IGI Global).

Chapter 13
Location-Based Services (LBS) in Micro-Scale Navigation:
Shortcomings and Recommendations

Suleiman Almasri
Anglia Ruskin University, UK

Muhammad Alnabhan
Brunel University, UK

Ziad Hunaiti
Anglia Ruskin University, UK

Eliamani Sedoyeka
Anglia Ruskin University, UK

ABSTRACT

Pedestrians LBS are accessible by hand-held devices and become a large field of energetic research since the recent developments in wireless communication, mobile technologies and positioning techniques. LBS applications provide services like finding the neighboring facility within a certain area such as the closest restaurants, hospital, or public telephone. With the increased demand for richer mobile services, LBS propose a promising add-on to the current services offered by network operators and third-party service providers such as multimedia contents. The performance of LBS systems is directly affected by each component forming its architecture. Firstly, the end-user mobile device is still experiencing a lack of enough storage, limitations in CPU capabilities and short battery lifetime. Secondly, the mobile wireless network is still having problems with the size of bandwidth, packet loss, congestions and delay. Additionally, in spite of the fact that GPS is the most accurate navigation system, there are still some issues in micro scale navigation, mainly availability and accuracy. Finally, LBS server which hosts geographical and users information is experiencing difficulties in managing the huge size of data which causes a long query processing time. This paper presents a technical investigation and analysis of the performance of each component of LBS system for pedestrian navigation, through conducting several experimental tests in different locations. The results of this investigation have pinpointed the weaknesses of the system in micro-scale environments. In addition, this paper proposes a group of solutions and recommendations for most of these shortcomings.

DOI: 10.4018/978-1-60960-607-7.ch013

INTRODUCTION

As a result of the rapid progress in wireless technologies and the development of portable and smart mobile devices, new range of location based services have been emerged (Rainio, 2001). The integration of wireless mobile networks, Global Positioning System (GPS), geographic information systems (GIS) and the Mobile device technologies allow the users to receive new range of services. For example, locating and tracking objects with dynamic distribution and also information regarding the nearest business or Point of Interest (PoI) such as ATM, restaurant or hotel (Junglas & Spitzmuller, 2005). In addition, new applications in different aspects of life are utilizing the LBS concept such as Mobile Learning (Hunaiti et al., 2008), Mobile Marketing, and Visually Impaired Guidance (Hunaiti et al., 2005).

As LBS system encompasses a group of technologies: mobile device with location sensing capabilities connected to a LBS database server over a mobile wireless network and the Internet (see Figure 1). Therefore, LBS performance is subject to the overall performance of these subsystems forming its architecture (Lee et al. 2005). The performance of LBS system can also be affected by the type of application implemented within the system. For instance, if a desktop application has been made over LBS system, it might not be successful due to the capabilities of the LBS system components (Benford, 2005).

Since LBS systems are services based on the location of the end user, one of the most important components is the location-sensing method. There are a number of available positioning methods and each one could be used based on the objective of the application being designed. For example, Cell of Origin (CoO) which depends on the distance between the end user and the radio base station of a cellular network (Giaglis et al., 2002). Short Range Positioning which is widely used by in-door LBS applications based on infrared transmitters and receivers (Bahl et al., 2000). Global Navigation Satellite Systems (GNSS) which are a satellite-based location sensing method such as GPS which is operated by the United States, GLONASS which is operated by the Russian Federation and GALILEO which is being built and funded by the European Union. In this paper, GPS is presented and investigated as it is the most

Figure 1. The main components of a typical lbs system

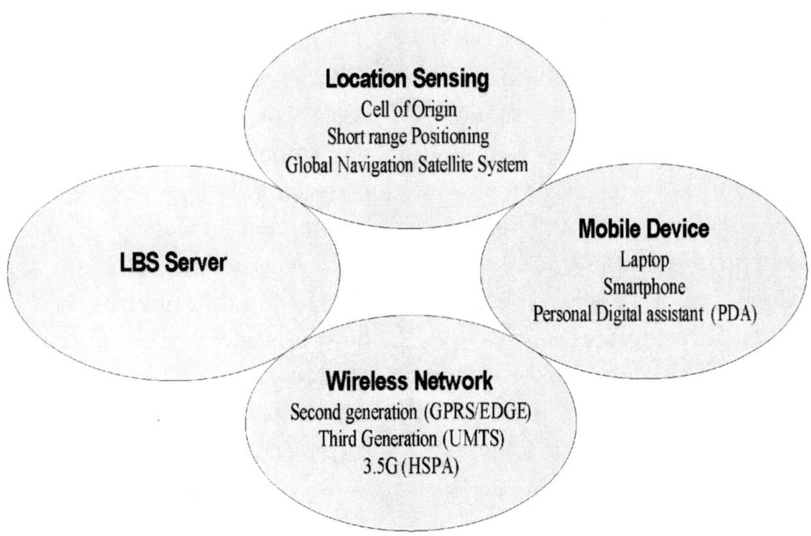

commonly used in real-time out doors LBS applications.

An essential step towards addressing the shortcomings of LBS systems is an all-inclusive investigation enabling a full understanding of the issues and allows the successful solution for LBS system. Therefore, this study has been conducted to evaluate the overall performance of LBS for the use in micro environments. This has been achieved by carrying out studies on GPS performance, wireless networks, and mobile devices capability.

This paper consists of four main sections, the first of which is the GPS performance issues. This section presents and discusses several experimental results in terms of GPS availability and Accuracy. The second section covers the main issues with the mobile devices for example, Personal Digital Assistant (PDA) or mobile smart-phones. The third section of this paper provides technical tests and information on selected mobile wireless networks. At the end of this paper a presentation of some recommendations of possible solution for each LBS component in order to enhance the whole LBS system performance.

GPS PERFORMANCE ISSUES

GPS is being widely utilized in many localizing applications such as mobile tracking and guidance, health assistive and learning applications. GPS is considered as an essential component in such applications providing the required information responsible for identifying where the LBS service should be delivered (Almasri et al., 2008).

GPS constellation is made up of a 24 satellites (RIZOS, 2007) enabling the end user to receive signal any time, anywhere on the globe (Wittea & Wilson, 2004). GPS receiver calculates its position through measuring the distance between those satellites and the receiver itself. This distance is determined by calculating how long the signal has taken to reach the receiver (Wittea & Wilson, 2004).

In metropolitan environments, GPS signal could be blocked by urban canyons, foliage or skyscrapers (Boulos et al., 2007). The satellite signals are also extensively affected by multi-path due to high obstructions; hence this degrades the satellites availability and dramatically reduces the positioning accuracy. In order to demonstrate the performance of GPS in such environments, both GPS availability and accuracy were evaluated in London-UK, which represents one of the largest and busiest environments in the world.

GPS Availability

GPS satellite availability is the capability of finding a clear direct view between the receiver and at least four GPS satellites (Wittea & Wilson, 2004). Failure to receive signals from four satellites will result in inaccurate position estimations that degrade the LBS system performance dramatically by sending inaccurate services to wrongly identified user's location.

In order to investigate the availability of the GPS service, several experimental scenarios measuring the number of satellites being tracked at three different sites, Canary Wharf, London Bridge and Stratford were conducted. Those three sites have been selected according to how urban the location is. Canary Wharf is the most urban site, whereas the London Bridge site is semi urban, and finally, Stratford is the open sky site. Throughout the experiments, several measurement trials took place during different periods of the day (morning, afternoon and night) taking into consideration satellites constellation changes and environment effects.

The experimental setup consists of an off-the-shelf mobile device (HTC P3300 PDA) with built-in standard GPS receiver which has been used to gather GPS signals as an example of the current available devices which might be utilised by the end users. This receiver uses SiRF star III GPS protocol (SiRF Technology, Inc., 2008), via GSC3e/LP, 50MHz ARM7TDMI processor GPS

Figure 2. Measurement locations at Canary Wharf (London) [Images from: MS Virtual Earth]

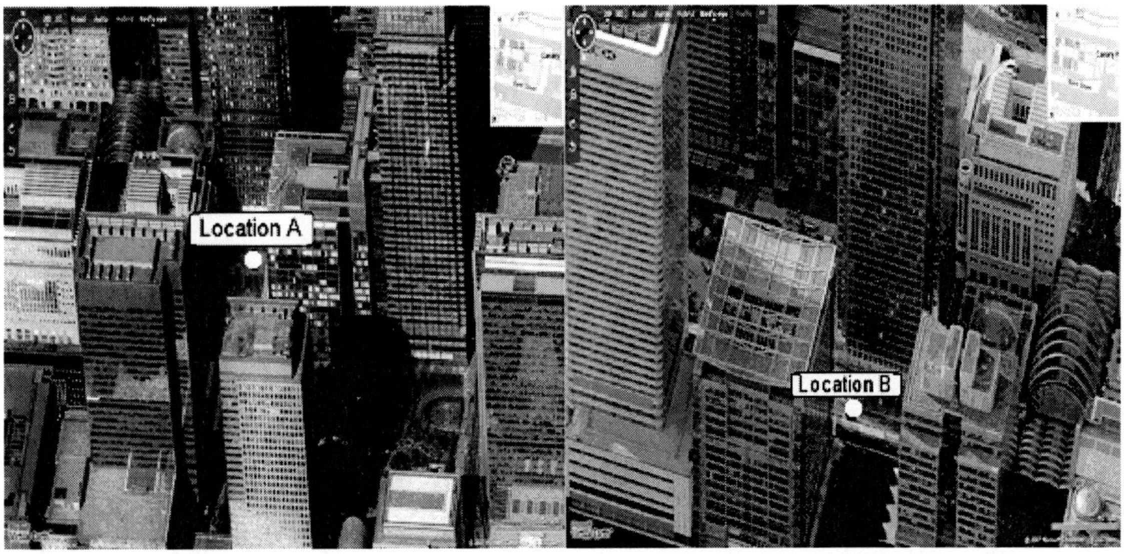

chip, with a total of 20 parallel satellite channels, and uses an internal GPS antenna (HTC P3300, 2008).

First Experimental Scenario

The first site was Canary Wharf which can be seen in Figure 2. It is the busiest and most urban area in London. This area has been chosen by the researchers to measure the availability of the satellite signal being blocked by surrounding tall buildings.

Two measurement locations have been chosen to demonstrate different degrees of signal blocking, the first one was at location A and the second was at location B, see Figure 2.

At Canary Wharf site (Location A) eight static measurement trials took place. As summarized in Table 1, the maximum number of satellites being tracked at this location was 6, and the minimum was 3, where the average was around 5 Satellites. The complete list of satellites signal received in this test can be seen in Table 1.

Similarly, at Location B another set of eight static measurement trials were conducted. The maximum number of satellites could be received was 5, and the minimum was 3, whereas the average was around 4 Satellites. The list of satellites can be seen in Table 2. It can be noticed by comparing Table 1 with Table 2 that satellite number 13, for example, was available at (location A) but has been blocked sometimes at this location.

Table 1. Summary of satellites signals available at Canary Wharf (Location A)

Trial	Num of Sats	Sat ID
1	4	25,27,02,04
2	5	13,25,27,02,04
3	6	13,25,27,02,04,23
4	5	13,25,27,02,04
5	3	25,27,023,04
6	5	13,25,27,02,04
7	5	13,25,27,02,04
8	5	13,25,27,02,04

Second Experimental Scenario

The second site where static measurements took place was London Bridge site, which is shown in Figure 3. This site has been chosen and classified

Table 2. Summary of satellites signal available at Canary Wharf (Location B)

Trial	Num of Sats	Sat ID
1	5	13,25,27,04,08
2	3	27,02,04
3	5	25,27,02,04,23
4	5	25,13,27,02,04
5	3	25,27,04
6	4	25,27,02,04
7	3	25,27,04
8	3	25,27,04

as a semi-urban environment since no skyscrapers are available in this area. Two measurement locations at this site have been selected: the first one was at location A (open area), and the second was at location B (near to a 10-story building), as shown in Figure 3.

In the view of measuring satellite availability, a set of measurement trials were conducted at both locations with respect to the time periods as described in the first scenario. As presented in Table 3, the maximum number of satellites was 8, whereas the minimum was 5, and the average was around 7 satellites.

In addition, at the second location which is exactly near the 10-story building another set of GPS signals were received and recorded. The maximum number of satellites was 5, while the minimum was 3, and the average was around 4 satellites. A summary of this test results can be seen in Table 4.

Third Experimental Scenario

The third site in this field work was at Cedars Road in Stratford which is shown in Figure 4. This area has been selected demonstrating an open sky area with no obstacles blocking the signal. The same location has been also selected by Ordnance Survey (The UK Navigational and Geographical body) to install a GPS reference station which detects GPS errors and therefore provides the end users with correction information accordingly.

At this location, five different static measurement trials were conducted. The maximum number of satellites was 7, whereas the minimum was 4, and the average was around 6 satellites. A summary of this test results is listed in Table 5.

Figure 3. Measurement locations at London Bridge area [Image From: MS Virtual Earth]

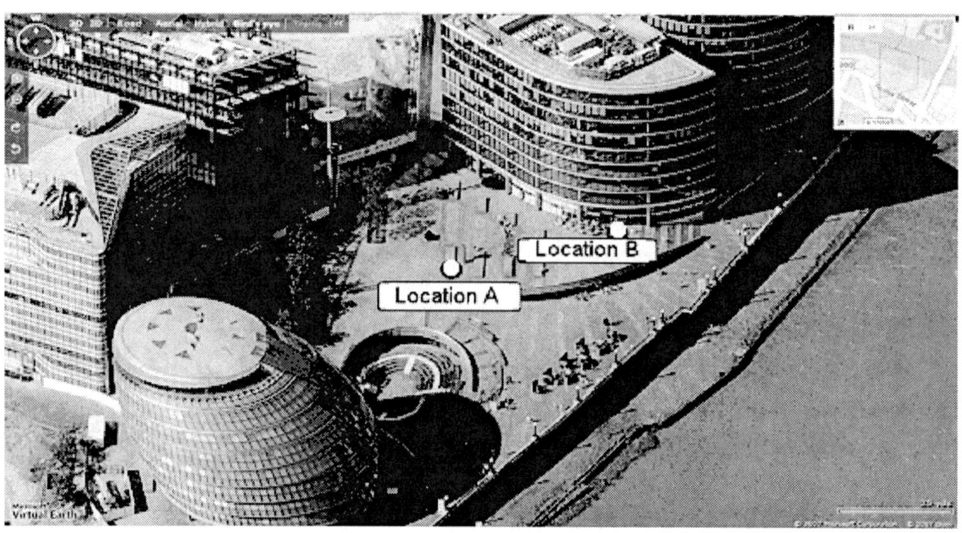

Table 3. Summary of satellites signal available at London Bridge (Location A)

Trial	Num of Sats	Sat ID
1	7	13,04,20,25,17,31,23
2	5	13,04,20,25,31
3	7	13,04,23,20,02,25
4	8	13,04,23,20,02,25,17,31
5	7	13,04,23,20,17,27,31
6	6	13,04,23,20,25,17
7	7	13,04,23,20,02,25,27
8	7	13,04,23,20,20,02,25,27

Table 4. Satellites availability at London Bridge (Location B)

Trial	Num of Sats	Sat ID
1	5	13,23,20,25,17
2	5	13,23,20,25,27
3	5	13,23,20,25,17
4	4	23,20,25,17
5	3	20,02,25
6	3	20,25,27
7	4	04,20,25,27
8	5	13,04,20,02,25

Figure 4. Measurements location at Cedars Road (Stratford)[Image from MS Virtual Earth]

Discussion

Figure 5 shows the average of satellites being tracked during previously mentioned scenarios. It is clear that the number-of-satellites in the urban environment (Canary Wharf) is the least within the group. While it is not stable in the semi urban environment (London Bridge), i.e. in some locations it is around 7 satellites whereas in other locations it is just 4. The best location found was the open environment (Stratford) where the number of satellites was stable around 6 satellites.

As mentioned previously, GPS receivers need to have a clear view to at least four satellites to deduce its location. In addition, the more satellites available the better accurate positioning could be achieved. Therefore the outcome of this experimental test has confirmed that GPS availability

Table 5. Satellites availability at Stratford

Trial	Num of Sats	Sat PRN
1	4	10,08,27,24
2	6	10,08,27,26,13,15
3	6	10,08,27,26,24,15
4	6	10,08,27,26,15,24
5	7	10,08,26,28,15,24,28

Figure 5. GPS availability in urban, semi-urban and open-sky locations

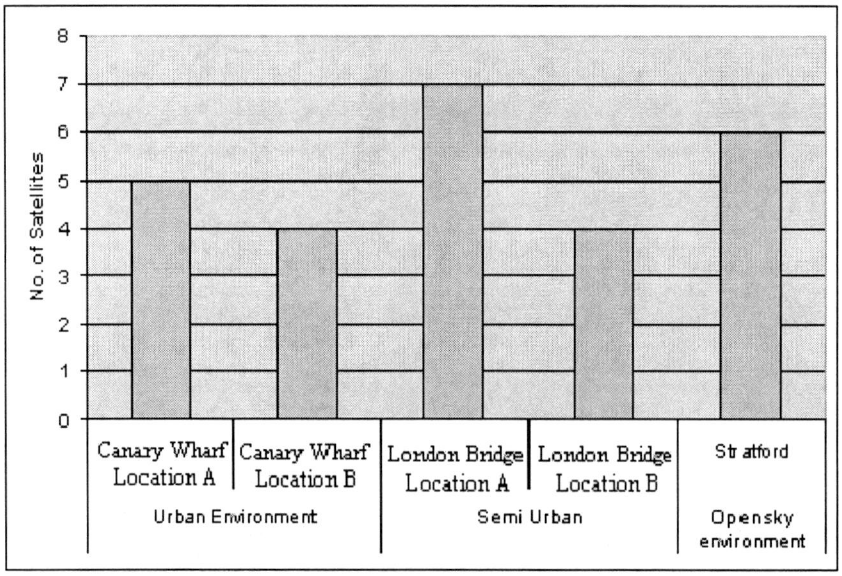

is subject to the environmental condition of the end user's location.

GPS Accuracy

GPS receivers always try to obtain signals from a number of satellites, the healthier the signals received the more accurate the positioning fixes (Steiniger et al., 2007). This requires a line of sight to the satellites but most of the time, tall buildings or foliage objects block the GPS signals from being received by GPS receivers. Even though, if a location does have a clear view of the sky, the accuracy of GPS positioning can be affected by tropospheric delays, clock and ephemeris errors, satellite geometry and receiver's internal noise. Therefore, stand alone GPS receivers provide accuracy levels between 5 to 10 metres; this is not suitable for some types of LBS applications that deal with safety and emergency situations (Theiss et al., 2005).

In this phase of the test trial, data collected by previously explained experimental scenarios are used to measure the accuracy of GPS positioning at the same measurement locations.

At location A in Canary Wharf, several position fixes were obtained and plotted on a map (see Figure 6). Few of the measured position samples were close to the exact location (the star), whereas the majority were far away from it. The measurements showed that the position samples had an error range between 1 to 25 metres in a 95% confidence level, with an average error of 18 metres from the exact point.

At the second location in Canary Wharf which seems to be more urban because it is located between 3 tower buildings (location B in Figure 2), the position samples which have been measured were plotted on a map in Figure 7. The results showed that the best position fix had an average of 2 metres error, whereas the worst fix had an error of around 50 metres. The calculated average of all measured positions at this location was 25 metres with a 95% confidence level.

Obviously, high tower buildings blocked the satellite signal. Hence, three satellites are not enough to provide an accurate position which will lead to receive information related to wrong location. This affects the performance of the LBS

Figure 6. GPS accuracy at Canary Wharf (Location A)

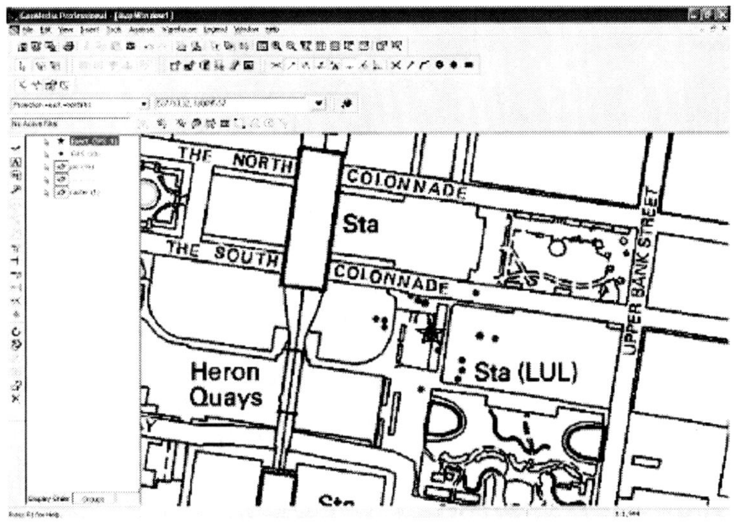

Figure 7. GPS accuracy at Canary Wharf (Location B)

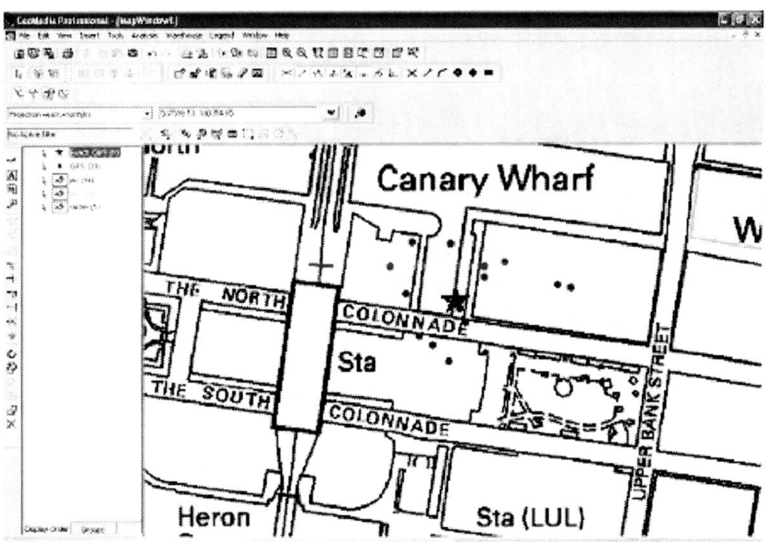

system badly because users need to have an accurate poisoning as well as helpful service.

At the second selected site which is considered to be a semi-urban environment (London Bridge), position samples were collected at the two previously identified measurement locations. For the first location (location A in Figure 3), position fixes were collected and plotted on the map shown in Figure 8. In this location, the best measured position sample had an error of 1 metre from the exact point, the worst position samples were about 18 metres from the exact fixed point, and the calculated average distance was 11 metre from the exact point.

At location B in London Bridge area, position samples were also collected and plotted on a map

Figure 8. GPS accuracy at London Bridge (Location A)

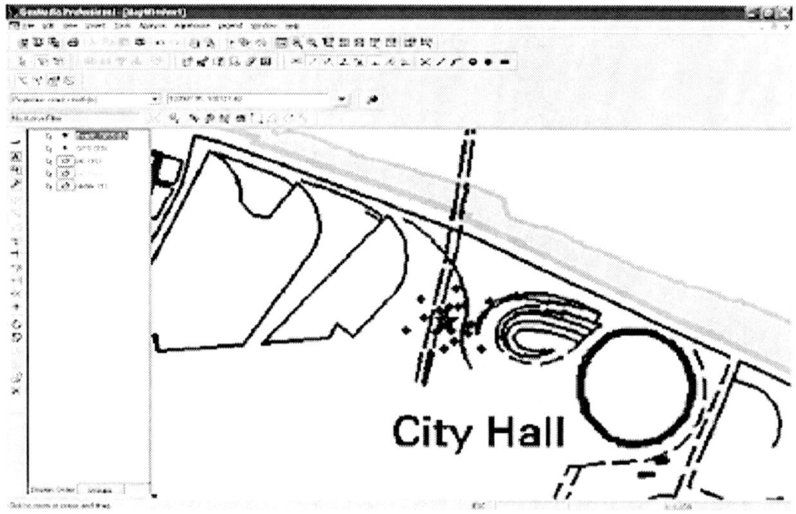

Figure 9. GPS accuracy at London Bridge (Location B)

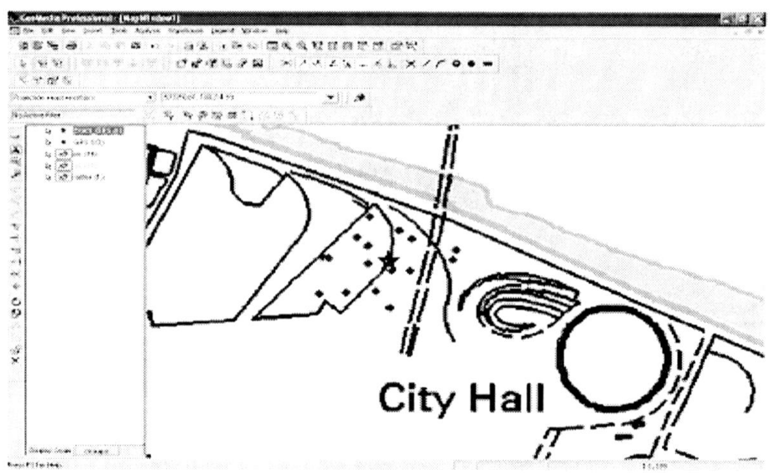

as shown in Figure 9. The results demonstrated that the best position fixes had an error of 2 metres from the exact point, and the worst samples had an error of around 25 metres. The calculated error average of all measured positions at this location was about 16 metres in a 95% confidence level. It is clear that the difference between the two averages in those two locations (A and B) has come out because of the impact of the 10-story building. It has blocked three satellites signal which reduced the availability and accuracy of the GPS system. Evidently, the user should try to be away from high buildings so as to get healthier signals and to have a good direct sight. This will cause the GPS receiver to fix less noisy position samples, which will enhance the efficiency of the whole LBS system.

At the third site, Stratford, it is very important to know that there are no high buildings around this area. Therefore, the GPS receiver is supposed to fix a very accurate position. The collected position samples at this location are plotted on a

Figure 10. GPS accuracy at Water Lane (Stratford, London)

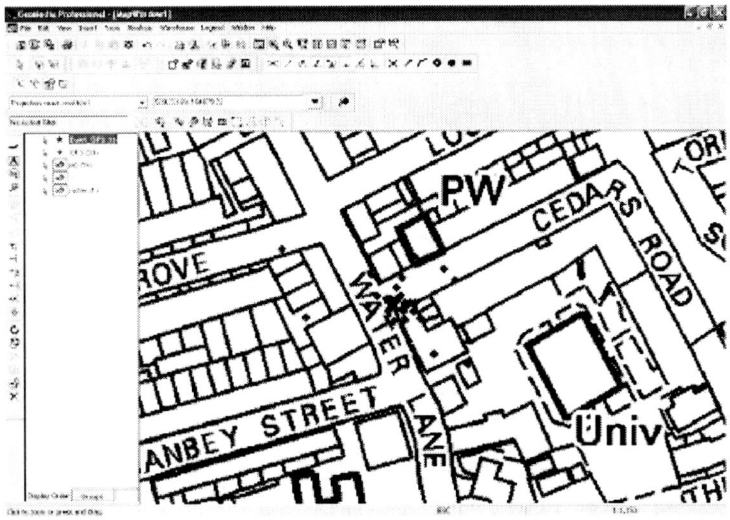

map as shown in Figure 10. The majority of the measurements are very close to the exact point with errors closer to 1 metre. The average error at this location was 8 metres in a 95% confidence level.

Discussion

When a clear satellite view was available (such as Stratford area), the availability of satellite signal was excellent. Hence, the accuracy of positioning process was very good as well. The LBS server in this case will send information to the users according to their accurately reported location; this will enhance the efficiency and the performance of the server and the whole LBS system.

The results obtained from these conducted test trials showed that the GPS service performance can be dramatically affected by the surrounding land infrastructure such as high buildings and difficult terrains. Hence, GPS accuracy is still a problem in urban environments. The best accuracy results were found in Stratford where the GPS receiver was able to track a number of healthy satellites (more than 4) with good geometry. Accordingly, standalone GPS positioning service accuracy is still not suitable for accuracy demanding LBS applications providing emergency and health assistive services, which is usually implemented in micro-scale applications.

MOBILE DEVICE PERFORMANCE ISSUES

Another important component of LBS system is the user Mobile Unit. Mobile devices are still experiencing many problems in terms of memory, battery and performance (Lee et al., 2005). The size of the mobile device is one of the most important issues in this technology. Manufactures always try to produce the smallest mobile device with better performance. But, minimising the device size has resulted in reducing the memory space and battery lifetime as can be seen in Table 6 which provides a list of mobile devices along with their capabilities. It is obvious that the size of internal memory (RAM) is either 64 MB or 128 MB.

According to the experiments carried out by the researchers on the HTC p3300 PDA, loading the operating system and all the necessary files consumes around 49% of the memory space, and

Table 6. Technical specifications of A group of PDAs and Smartphones

Manufacturer	PDA	OS	RAM	Processor (MHz)
Hewlett Packard	HP hw6915	WM5	64MB	416
Hewlett Packard	HP 614	WM6	128MB	520
Fujitsu-Siemens	Loox T830	WM5	64MB	416
Asus	P526	WM6	64MB	200
Asus	P750	WM6	128MB	520
Acer	C530	WM5	64MB	300
HTC	P3300	WM5	64MB	201
HTC	Advantage	WM5	128MB	624
i-mate	Ultimate 9502	WM6	128MB	400
Eten	Glofiish X600	WM6	64MB	400
Toshiba	Portege G710	WM6	64MB	.
Toshiba	Portege G810	WM6	128MB	400
Samsung	i780	WM6	128MB	.
Nokia	N95	Symbian S60	100MB	

the rest is available for the use by all applications installed.

In order to pinpoint the memory space and battery performance, two test scenarios have been conducted using a HTC P3300 PDA and a laptop to simulate mobile devices. The aim of the first task was monitoring the free and used spaces of the device internal memory. Before starting the test, the free memory space of the mobile device was 18.93 Mbytes, whereas the already occupied space was around 18.7 Mbytes. Next, the PDA has been connected to the internet to simulate a LBS server. Two different-size maps have been downloaded from the server to the PDA device. The first of which is a map of 5 Miles radius and the second is a map of 1 Mile radius.

When trying to download the large map, the PDA has run out of memory space and couldn't save it due to the lack of sufficient space. But in the other case it managed to receive and save the small map successfully.

The Pie chart in Figure 11 describes the results which show that almost half (49%) the PDA's memory is used just to run the device. The test shows that 13% of the PDA's memory is required

Figure 11. The memory usage in a HTC P3300 mobile device

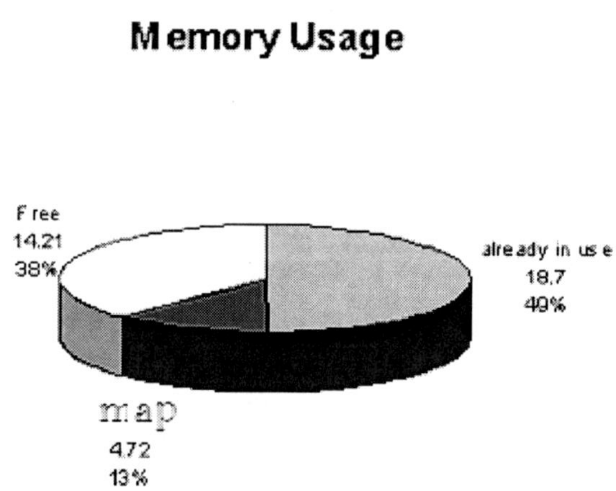

to successfully download the small map, whereas the large map requires 25.2 MB of memory space which is more than the 18.93MB free memory space. This means that it is not possible to use large maps according to the available memory resources.

The main issue for users is that they could have downloading nearly completed and then have it canceled due to the insufficient memory space. This would waste the user's time and the users would also still have to pay for the service as well as the internet connection time even if they could not use it. This would lead users to avoid utilising LBS applications due to them being impractical.

The second task of the mobile device investigations was monitoring the battery discharge rate. A file was uploaded to an FTP server on the internet for testing purposes and a battery monitoring software was installed on a laptop to measure the battery discharge rate while downloading this file.

The results of the battery discharge rate showed that there is a notable difference between the discharge rates while downloading a file from the server, and while the laptop is idle. As can be seen in Figure 12, the discharge rate when the laptop was idle was -28442.25, whereas it was -35266.9048 while it was in downloading mode. The difference between the discharge rates is 6824.66. This is a factor when considering that most mobile device users do not want to be constantly charging their devices.

Discussion

The result of the two tests conducted showed that mobile devices are still facing problems in terms of lack of space and short battery life time. Therefore, the users still need to use additional equipments (e.g. Extra storage) in order to install necessary applications. Consequently, LBS providers should mind such problems while sending huge amount of information to the end user's mobile device.

Since most LBS systems are designed to be used on the move, most users will not be able to charge their mobile devices battery if they are away from a ready source of power. Most mobile devices have a very limited power supply which means that when downloading they are using more of that limited power.

The laptop which has been used in this test is not like a normal mobile device (PDA or smartphone) in terms of power saving features that most mobile devices have. Even though there is still a notable difference in the battery discharge rate. The difference would be even more pronounced on a PDA or mobile phone.

NETWORK PERFORMANCE ISSUES

The efficiency of the wireless communication link between the end user mobile device and LBS server is an important issue in LBS as it has a direct impact on the overall performance (Chakravorty & Pratt, 2002). The mobile device should have the ability to be connected to a wireless network

Figure 12. Mobile battery discharge rates

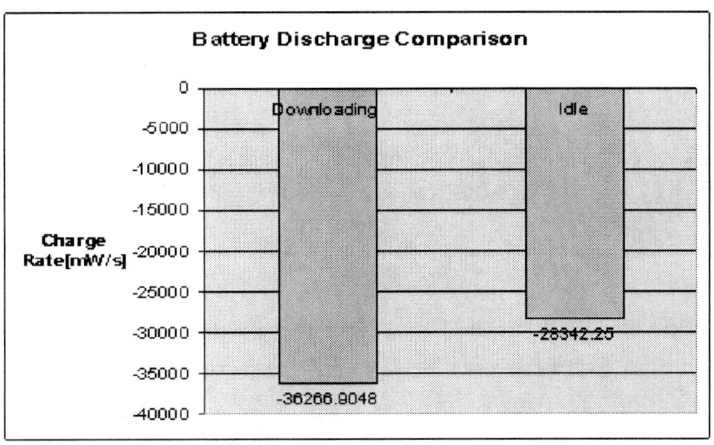

in order to access the internet and the server. In this paper two types of wireless networks have been studied, the 3.5G or what is known as High Speed Downlink Protocol Access HSDPA (Kim et al., 2005; Necker, 2006), and Wireless Local Area Network (WLAN) (O'Hara & Petrick, 2004). These types of communication channels still suffer from lack in bandwidth, delay and packet loss particularly when many users are accessing the mobile network (Benford et al., 2003). Sharing the bandwidth by big number of users reduces the data transfer rates and degrades the quality of service. In this section, three main issues of wireless network were tested: delay, packet loss and throughput.

Delay

It is the time taken for sending information from source to destination over a network link. In order to evaluate this delay, the mobile device was used to send Internet Control Message Protocol (ICMPs) to measure the Round Trip Time (RTT) (Andersson, 2001). In this test, around 20 ICMPs have been sent to www.rightclicktz.com, which simulates the LBS server, in idle mode using both WLAN and HSDPA links. The results were recorded and summarised in Figure 13.

Both technologies were stable, however the average RTT of the HSDPA was around 170 ms, but it was around 75 ms while using WLAN network.

Another detailed test has been conducted through sending ICMPs again to the same web site (www.rightclicktz.com) but in different scenarios. During this test, three tasks were completed: measuring the delay with a clear link, measuring the delay while in downloading mode (RTT+DL) and measuring the delay while in uploading mode (RTT+UL). While running this test, the downlink quality started to fluctuate particularly during peak time unlike the uplink channel, as can be seen in Figure 14.

The idea of testing both uplink and downlink channels simultaneously was to simulate congested networks in urban environments. This test has been conducted by sending ICMPs to the LBS server while both channels were in use (i.e. sending and receiving files from the server). The results were similar to the above test, as can be seen in Figure 15. The wireless network performance in urban environments was worse than suburban ones; hence the use of LBS in urban environments, where the users are most likely to use it, is still a big challenging issue.

Figure 13. A comparison between WLAN and HSDPA Links

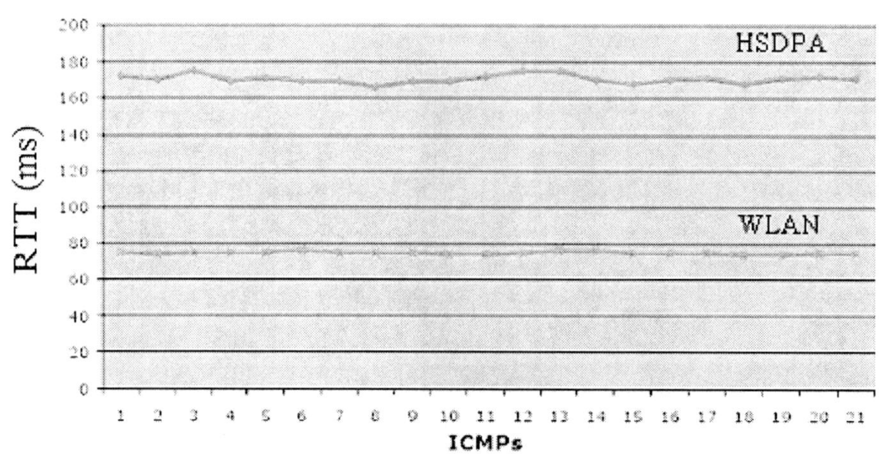

LBS users are a key concern in the system. Hence, when a large number of users access the wireless network all together, the transfer rates become very low. Therefore, in order to test the user's impact on the network, the same test has been conducted in two different areas: Chelmsford-UK, a suburban area with fewer users, and London-UK, an urban area with many users.

The Number of users per area showed a significant effect on the delay experienced by users. The results, which can be seen in Figure 16, showed that as the number of users increases the delay also increases.

In LBS systems, downlink channel performance is much more important than uplink channel simply because the users send few packets (enquiries) and receives loads of data and services (Almasri et al., 2008). It is obvious in Figure 16 that the delay (RTT) for both downlink and uplink channels in an urban environment (London) is much higher than the same delay for both channels in suburban environment (Chelmsford). This confirms the idea that the number of users is a very important factor in the efficiency of any LBS system.

In summary, the results showed clearly that the performance of wireless network is lower in urban environments where many users are accessing the network. It is essential to take into consideration these shortcomings, which can be minimised by establishing LBS applications which doesn't require high speed or long accessing time in order not to contribute in congesting the network channel.

Figure 14. A fluctuated DL channel and a stable UL channel.

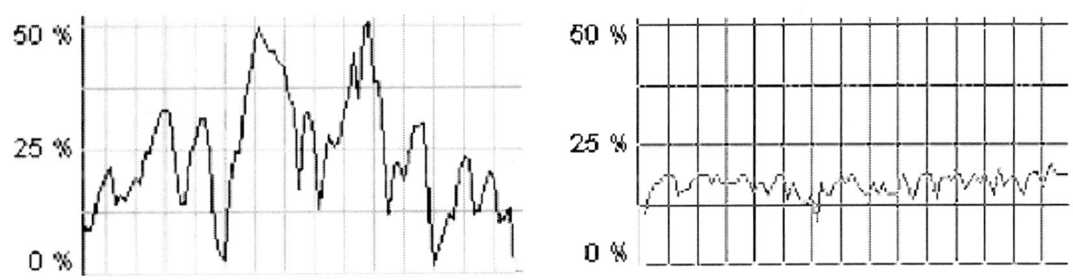

Figure 15. Testing the performance of DL and UL simultaneously.

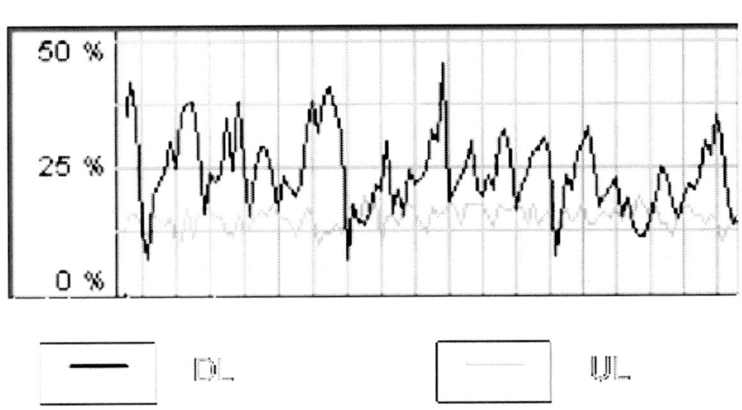

Figure 16. Delay in suburban environment (Chelmsford) VS urban environment (London) using HSDPA Link

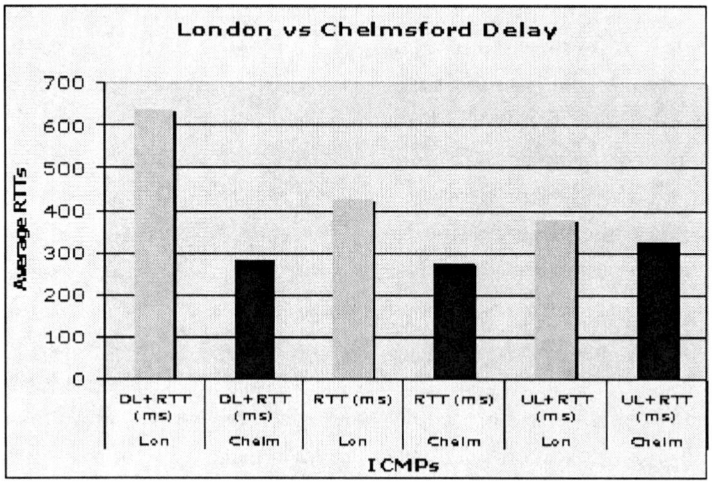

Moreover, some consideration should be taken into account when establishing LBS application to be designed in a way to work over different types of wireless network technologies (Heterogeneous networks) since each technology has different capabilities as seen by comparing WLAN and HSDPA technologies.

Packet Loss

It is very important to measure the number of packet losses as it is the main reason for data retransmission. Moreover, data retransmission causes more delay in the network which diminishes the efficiency of LBS systems. The following test was conducted to monitor the packet loss during different times of a day and in different locations. Around 100 packets, in-groups of 4, were sent to the private domain (www.rightclictz.com), and then all the packets that did not return were recorded. The test was conducted in two different scenarios: the first of which was sending them while in busy mode (downloading/uploading a file) and the second was while in idle mode.

The results showed no difference on packet loss for both scenarios. However, the situation was different when tested in different locations. While using the HSDPA network, the results did not show any significant packet loss in suburban area, where an average of 1% loss during less-peak times up to 7% loss during high-peak times were experienced (see Figure 17). The situation was different in the urban environment (London), at different times as well. The losses were from 2% during less-peak time up to 25% loss during high-peak times. It can also be noted that DL experiences more packet loss than UL.

However, for regular usage, the effect of packet losses was not that much. For a light internet uses like e-mailing and browsing, packet loss would not cause any big problems since undelivered packets will be retransmitted. The situation would be different if a user was utilising real time applications like internet calls or other applications like LBS systems that rely on real time response (Hunaiti et al., 2005).

Bandwidth and Throughput

Throughput is the maximum connection speed that the user could achieve while communicating with LBS server. In order to test this throughput, a file

Figure 17. Packet loss (DL) – Suburban environment (Chelmsford) Vs urban environment (London)

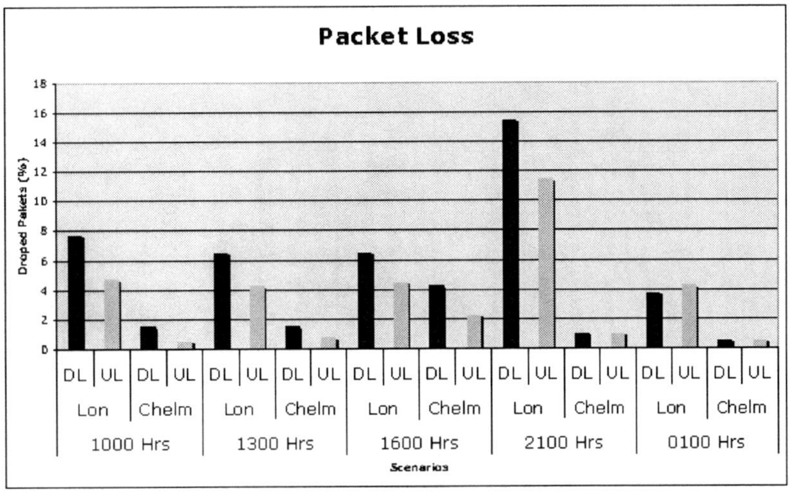

has been used to measure the time taken to upload/download it to/from the server over the internet. The test was also performed at different times in different locations (Chelmsford and London). HSDPA supports 5, 10 and 15 different codes in order to provide users with suitable bandwidth allocation depending on their requirements. This would offer a theoretical connection speed (throughput) of up to 14 Mbps (960 Kbps per code) to a user assuming that this user has been allocated with 15 codes (QUALCOMM, 2008). However this would not be achieved unless there was a single user, situated very close to the base station.

The test showed significant variations on throughput depending on number of users at a time (see Figure 18). In this situation, the better throughput was experienced in Chelmsford, the suburban area with few users.

The same applied when the test was conducted in London. Due to the huge number of users, the bandwidth allocated per user was extremely balanced. The 3.5G data card (HSDPA) supports a theoretical 1.8 Mbps speed. However, the tests results showed that this throughput could never be achieved in any environment or at any scenario.

Theoretically, downlink channel is supposed to provide higher speed than the downlink one, but this was not always true. At times during high-peak hours, uploading the same-size file took less time compared to downloading it. The reason might be related to the fact that more users are utilising the DL channel than the UL one, which leaves UL channel less engaged.

The test results have shown also that the throughput was very much related to the stability of the link, where at high-peak hours, the UL channel appears to be more stable than DL. However, the experience from a normal user's point of view did not show such a big significant throughput problems. Surfing the Internet, however slower than wired networks, was smooth and with fair speed.

To sum up, the major challenge experienced during the network analysis was the number of users per cell which was not stable especially in the DL channel. This is associated to the delay, which enlarges the number of dropped packets and results to low connection speeds (throughput). However, delay, packet loss and throughput, are still in an acceptable level unless the wireless network is used for real time applications such as LBS.

Figure 18. Throughput results using HSDPA in different environments

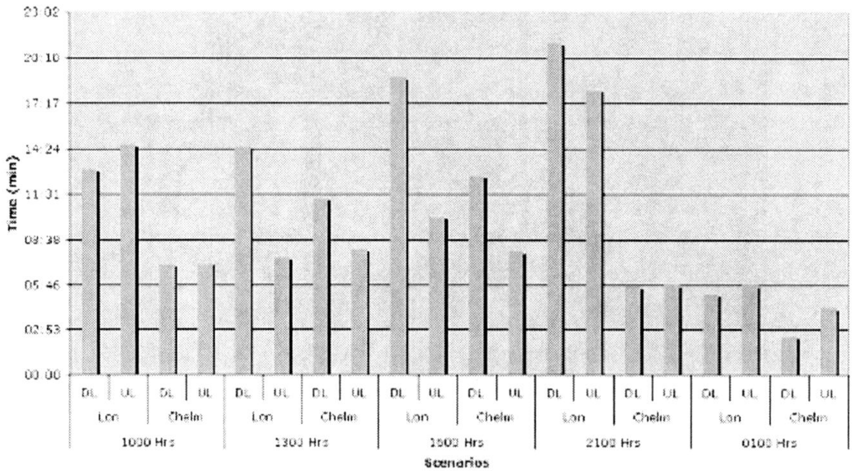

CONCLUSION AND RECOMMENDATIONS

In this paper, an all-inclusive analysis of LBS performance has been presented. This has been accomplished through evaluating each component including GPS, Mobile devices and Wireless Networks.

The outcome of this study did enable a full understanding of the issues that might hinder the deployment of LBS applications in a wider scale mainly within micro-environment. These issues along with some recommendations can be summarised as follows:

- **GPS availability and accuracy:** The test results showed that the satellite signal has been blocked sometimes by obstacles such as buildings. This causes the GPS receiver to fix inaccurate location which drives the LBS server to send incorrect information accordingly. It is recommended to use accuracy enhancement methods of GPS by means of: Differential GPS, or through sending correction information over the internet (e.g. OSNet, SISNET, etc.). The integration of such accuracy improvement methods with the stand alone GPS receivers currently utilized in LBS requires several software and hardware adaptation.
- **Mobile Devices:** The test results showed that mobile devices such as PDA and smartphones are still experiencing a lack of enough storage and short battery life time. It is recommended that the size of information, services and maps which is normally sent by the LBS system to the end user to be reduced as much as possible. There is a need also to develop new methods that will effectively utilise extra storage (e.g. Mini SD memory) to be used for LBS purposes.
- **Wireless Networks:** The test results showed that the network is still experiencing problems with the latency, packet loss, and bandwidth. It is recommended that LBS system utilise new coming technologies such as High-Speed Uplink Packet Access (HSUPA) (Peterson & Tarallo, 2007), Long Term Evolution (LTE) (Rao et al., 2009) and Worldwide Interoperability for Microwave Access (WiMAX) (Westwick, 2008). Therefore, when designing a LBS application, the wireless network performance and capabilities should be taken

into consideration. For example, applying new strategies of data-flow management can help in maintaining good performance.

Finally, more research work is being conducted to find a new method that could be applied on LBS system in order to contribute in overcoming the investigated problems. The design of the new method takes into consideration the findings and results presented in this paper.

REFERENCES

Almasri, S., Hunaiti, Z., Sedoyeka, E., & Balachandrab, W. (2008). Zone-Based Update Strategy for Location based Services (LBS). *IEEE International Symposium on Industrial Electronics,* Cambridge, UK.

Andersson, C. (2001). *GRPS and 3G Wireless Applications*. Wiley.

Bahl, P., Padmanabhan, V., & Balachandran, A. (2000). Enhancements to the RADAR user location and tracking system. *Microsoft Co. Technical Report MSR-TR-2000 12.*

Benford, S., Anastasi, R., Flintham, M., Drozd, A., Crabtree, A., & Greenhalgh, C. (2003, September). Coping with Uncertainty in a Location-Based Game. *IEEE Pervasive Computing / IEEE Computer Society [and] IEEE Communications Society,* 34–41. doi:10.1109/MPRV.2003.1228525

Benford, S. (2005). Future Location-Based Experiences. *JISC: Technology & Standards Watch.* Retrieved October 30, 2008, from http://www.jisc.ac.uk/ uploaded_documents /jisctsw_05_01.pdf

Boulos, M., Rocha, A., Martins, A., Vicente, M., Bolz, A., Feld, R., Tchoudovski, I., & Braecklein, M. Nelson5, J., Laighin, G., Sdogati, C., Cesaroni, F., Antomarini, M., Jobes A., & Kinirons, M. (2007). CAALYX: a new generation of location-based services in healthcare. *International Journal of Health Geographics, 12,* 6–9.

Chakravorty, R., & Pratt, I. (2002). Performance Issues with GPRS. *Journal of Communication and Networks, 4*(4), 266–281.

Giaglis, G., Kourouthanasis, P., & Tsamakos, A. (2002). Towards a classification network for mobile location services. In B.E. Mennecke & T.J. Strader (Eds.), *Mobile Commerce: Technology, Theory and Applications.* Hershey, PA: Idea Group Publishing.

HTC P3300 PDA. (2008). Retrieved from http://www.htc.com/www /product.aspx?id=392

Hunaiti, Z., Almasri, S., Matter, N., Sedoyeka, E., & Fenton, A. (2008, April). Extending Location-Based Services into M-learning. *The 3rd International Conference on Interactive Mobile and Computer Aided Learning,* Jordan.

Hunaiti, Z., Garaj, V., Balachandran, W., & Cecelja, F. (2005). Mobile Link Assessment for Visually Impaired Navigation System. *Proceedings of the IEEE IMTC, 2,* 883–887.

Junglas, I. A., & Spitzmuller, C. (2005). A Research Model for Studying Privacy Concerns Pertaining to Location-Based Services. *Proceedings of the 38th Annual Hawaii International Conference* (pp. 180a-180b).

Kim, S., Kim, H., & Lee, K. (2005). Unified Optimal Power Allocation Strategy for MIMO Candidates in 3GPP HSDPA. *ETRI Journal, 27*(6), 769–776.

Lee, D., Zhu, M., & Hu, H. (2005). When Location-Based Services Meet Databases. *Mobile Information Systems, 1*(2), 81–90.

Necker, M. (2006). A comparison of scheduling mechanisms for service class differentiation in HSDPA networks. *International Journal of Electronics and Communications, 60,* 136–141. doi:10.1016/j.aeue.2005.11.014

O'Hara, B., & Petrick, A. (2004). *IEEE 802.11 handbook: a designer's companion*. IEEE Standards Association. ISBN 0738144495, 9780738144498

Peterson, J., & Tarallo, J. (2007). Wireless technology issue overview. Bell Labs Technical Journal (Special Issue: Wireless Network Technology), *11*(4), 1-4

QUALCOMM. (2008). *HSDPA for Improved Downlink DataTransfer.* Retreived on April 2008, from http://www.cdmatech.com/ download_library/pdf/ hsdpa_downlink_wp_12-04.pdf

Rao, A., Weber, A., Gollamudi, S., & Soni, R. (2009). LTE and HSPA+: Revolutionary and evolutionary solutions for global mobile broadband. *Bell Labs Technical Journal (Special Issue: 4G Wireless Technologies, 13*(4), 7–34

Rainio, A. (2001). Location-Based Services and Personal Navigation in Mobile Information Society. *International Conference FIG Working Week.*

RIZOS, C. (2007). The Future of Global Navigation Satellite Systems. *Technical report of The Institute of Electronics. Information and Communication Engineers, 107*(2), 25–29.

SiRF Technology, Inc. (2008). Retrieved from http://www.sirf.com/products /gps_chip.html.

Steiniger, S., Neun, M., & Edwardes, A. (2007). Foundations of Location Based Services. *Lecture Notes on LBS*. Retrieved in October 2008, from http://www.geo.unizh.ch /publications/cartouche/ lbs_lecturenotes_steinigeretal2006.pdf

Theiss, A., David, C., & Yuan, C. (2005). Global Positioning Systems: an analysis of applications, current development and future implementations. *Computer Standards & Interfaces, 27*(2), 89–100. doi:10.1016/j.csi.2004.06.003

Westwick, R. (2008). Mobile WiMAX versus LTE: A comparison of next generation mobile broadband technologies. *Journal of Telecommunications Management, 1*(1), 79–85.

Wittea, T., & Wilson, A. (2004). Accuracy of non-differential GPS for the determination of speed over ground. *Journal of Biomechanics, 37*(12), 1891–1898. doi:10.1016/j.jbiomech.2004.02.031

This work was previously published in International Journal of E-Services and Mobile Applications, Volume 1, Issue 4, edited by Ada Scupola, pp. 51-71, copyright 2009 by IGI Publishing (an imprint of IGI Global).

Section 4
E-Services:
Systems, Architecture and Methodologies

Chapter 14
Understanding Telehealth:
Constructing Meaning to Promote Assimilation

Joachim Jean-Jules
Université de Sherbrooke, Canada

Alain O. Villeneuve
Université de Sherbrooke, Canada

ABSTRACT

During the past few decades, many healthcare authorities sought to integrate new methods of delivering care to patients. Among the priorities faced by these organizations, a major issue arose of how to provide healthcare to patients who live in rural or remote regions suffering from a lack of accessible professional resources and services that comply with WHO's call for providing fair access to healthcare. Many attempts were made to integrate new technologies such as telehealth into the healthcare system, but in many cases, telehealth was not successful due in part to limited assimilation into healthcare organizations and work practices. Telehealth addresses operational issues such as a shortage of professionals in rural or underserved geographical regions. Using a breadth of reference theories such as institutional theory, structuration theory, and organizational learning theory, we propose a conceptual model that integrates the determinants of telehealth assimilation and identifies factors that impinge upon the process of assimilation. We posit that telehealth assimilation can only be understood by taking a multilevel approach to the phenomenon, whereby assimilation starts at the individual level, permeates through other organizational levels such as groups, and finally ends at the organizational and inter-organizational level. Further, assimilation of technological innovations must be considered within their institutional context. Derived from our conceptual model, we make several propositions and hope that our work will significantly guide future research and managerial actions geared toward integrating healthcare in the workplace.

DOI: 10.4018/978-1-60960-607-7.ch014

INTRODUCTION

Ensuring equal access to health care is one of the challenges currently faced by many western countries (WHO, 2006). Countries have to ensure that the level of healthcare services provided is not only adequate but that those services are fairly distributed (Bourgueil et al., 2006). The importance of this issue is even greater for countries like Canada with vast territories. For example, the province of Quebec alone has an area equivalent to three times that of France and 20% of Quebec's population lives in rural areas that account for 80% of the inhabited territory.

Many studies (Gravelle and Sutton, 2001; Roos et al. 1997; Politzer et al., 1991) have found geographic variations in the delivery of health services in western countries despite the existence of universal healthcare systems. These geographical variations in the provision of health care are linked to numerous factors including staff shortages, local policy, terms of physicians' practices, transportation, and other variables (Field and Briggs, 2001). Inadequate delivery of health services translates into healthcare inequalities (Veugelers et al., 2004). Several studies show that low doctor-to-patient ratios in rural areas compared to those in urban areas are associated with poorer population health in the former (Gulliford, 2002; Vogel, 1998).

Some western countries have adopted a number of policies to address the issue of geographical distribution of health professionals and access to healthcare. However, most measures proved unsuccessful. For example, policies - such as providing financial support for training and living - aimed at increasing the number of professionals in remote areas did not work as expected (Bourgeuil et al, 2006). The United States first implemented financial aid in the 1970s and more lately, financial aid measures have been implemented in Canada (Ward, 2004). Financial aid often takes the form of grants or loans with preferential terms coupled with a requirement of settling in deficit areas for a certain period. These measures proved successful only in the short term since physicians often bought back their loan or grant (Bourgeuil et al., 2006). Despite financial support for settling in remote areas, physicians still continue to settle in areas with high population density even though competition is stronger in those areas (Barer and Stoddart, 1999).

Some measures targeted the physician training through adaptable curricula and internships taking place in rural or remote areas. It was believed that early experience in remote areas would positively influence a student's motivation to practice in such areas. In addition, other measures were developed aimed at limiting the installation of doctors in well-deserved areas and at requiring foreign doctors, through immigration policies, to settle in remote regions. In Canada, foreign doctors receive loans for settlement but must service remote areas for five years; otherwise, they have to pay back their loans. Some of these measures, as was the case in British Columbia, were abandoned due to violation of personal freedom of establishment (Ward, 2004).

Aid for the settlement or maintenance of exercise in deficit areas also took other forms such as financial incentives (higher wages or aid in kind). Higher wages for doctors in rural areas is the most common type of financial aid in Quebec and Ontario (Bilodeau et al. 2006; Bilodeau and Leduc, 2003; Barer and Stoddart, 1999). This measure, too, only proved satisfactory in the short term (Bourgueil et al., 2006). An alternative to financial aid or support is aid in kind that primarily breaks the isolation of health professionals in terms of daily practice conditions and lifestyle (Bilodeau et al., 2006). It includes full or partial reimbursement of expenses related to continuing education and travel due to geographical location or even the financing of investments in collaborative projects between health professionals such as telemedicine, professional networks, group practice, and others (Bourgueil et al., 2006). Nevertheless, aid in kind has not yet proven successful.

Due to the absence of long-lasting benefits of past policies and measures, other solutions must be explored. Many new solutions include information and communication technologies (ICT). ICT growth is phenomenal and its use in all forms of organizations is increasing. Not only do private or public organizations use ICT, but healthcare organizations do as well. Many driving forces are identified in the literature that promote ever increasing use of ICT in healthcare such as government pressures to reduce costs, chronic and ever-increasing health workforce shortages, consumer expectations for higher and safer healthcare, and new or changing models of healthcare delivery (Wooton et al., 2009). Thus, ICT is seen as an ideal solution to improve healthcare delivery and to alleviate the inequalities of traditional face-to-face healthcare delivery.

Telehealth is a broad term describing all possible variations of healthcare services using ICT including tele-education, teleconsultation, and teletraining, among others. As an ICT-based innovation, implementation of telehealth has emerged as a key strategy for providing healthcare services to underserved or difficult-to-serve populations and low-cost specialty services to areas where full-time staffing is not economical. It is expected to provide many other benefits such as shortening the timeframe for decision-making related to diagnosis and treatment, cutting emergency transfer costs, reducing expenses for patient travel from remote regions to healthcare service points, reducing delays in providing healthcare, promoting continuous healthcare, and attracting and retaining clinicians in remote regions.

In this respect, Telehealth is a technological response devised to facilitate medical education, improve access to healthcare, enhance the quality of service delivery, improve the effectiveness of public health and primary care interventions, and provide a solution to the global shortage of health professionals. Telehealth is made possible by the rapid development of medical technologies and ICT. Technological advances allow the introduction of new equipment that significantly change healthcare practices. Used in the context of telehealth, ICTs are complex. Telehealth relies on a bundle of information and communication technologies including data capturing devices, data processors, data storage, communication devices, and display devices. Data captured, stored, communicated, or otherwise processed are of different forms such as sounds, images, waveforms, and alphanumeric data.

ÉCOUTER

Lire Phonétiquement

Given the centrality of information technology in telehealth, many information systems (IS) studies are devoted to telehealth systems. Their examination reveal at least three salient research streams including 1) user adoption/acceptation of telehealth systems, 2) efficacy of telehealth systems, and 3) effectiveness of telehealth systems.

Numerous, heterogeneous research concerns characterize the user adoption/acceptation research stream (Mitchell, Mitchell and Disney, 1996; Hu and Chau, 1999; Cohn and Goodenough, 2002). Studies of management practices aimed at promoting the use of telehealth systems (Dansky et al., 2005) suggest that results differ when the focus is on strategic objectives such as clinical excellence or technological leadership versus cost control; those focusing on clinical excellence achieve higher levels of system acceptance. Using a sample of 32 home health care agencies in the Commonwealth of Pennsylvania, Rumberger (2005) made a business case for telehealth in home health agencies. Using an information technology adoption model, he studied the strategic and financial impact of telehealth. Although overall results were inconclusive due to data limitations, a break-even analysis supported that telehealth is financially beneficial in the context of home care delivery. Yant (2005) investigated the impact of telehealth adoption on nurses' job satisfaction and organizational performance and found that there was no difference in job satisfaction between telehealth adopters and non-adopters. Predictors of job satisfaction included prior job satisfaction levels, intensity of telehealth use, organizational support for telehealth, and perceptions of telehealth's usefulness. While agencies adopting telehealth showed higher levels of organizational performance than non-adopters, the same predictors could not explain performance as for job satisfaction. Complex interactions between key stakeholders and industry infrastructure underpin the telecardiology innovation process (Cho and Mathiassen, 2007). Using an adaptation from de Ven and de Ven' framework, the process primarily reveals a push-type with a small group of medical innovators with limited financial and technological resources and little infrastructure support sustaining the process. In the same area of research, telecardiology innovation was examined through an Actor Network Theory lens and considered as a negotiation process between multiple stakeholders involving structure, culture, power, and politics (Constantidines and Barrett, 2006).

Telehealth innovations are believed to have great potential to support appropriate healthcare. However, proponents of the technology are not necessarily those who use. As technology evolves, often the case when early versions are prototypes, new users play a larger role in the innovation's diffusion (Cho, Mathiassen and Gallivan 2009). As it is deployed further across organizations - notwithstanding the resilience of stakeholders - the innovation process remains fragile due to constraints such as lack of proper financing, misalignment of medical interest with business interests, and incompatibility of the innovation with actual IT infrastructures (Cho, Mathiassen and Robey, 2007). In the end, there is no assurance of full deployment and success of the innovation.

Efficacy is another concern found in the literature on telehealth. Most research on efficacy focuses on patient outcomes from the perspective of several medical specialties such as dermatology, cardiology, and home-based care (Krupinski an al., 2004; Winters and Winters, 2004; Schofield et al., 2005). Results demonstrate that although telehealth benefits a patients' well-being, it is challenging to achieve such satisfactory outcomes (Balas and Iakovidis, 1999; Finkelstein et al., 2003).

Other research examines the effectiveness of telehealth systems in varying contexts when compared to conventional face-to-face healthcare delivery (Picolo et al., 2000; Nordal et al., 2001; Bishop et al., 2002). Most commonly, effectiveness is measured in terms of system costs. Results are inconclusive with some studies suggesting that telemedicine is more cost-effective (Loane, 2001) and others concluding that telemedicine is more expensive than traditional care delivery (Doolittle, 1998; Nobel, 2005). Contradictions may be explained by the level of system use; Doolittle et al. (2003) suggests that the more frequently a system is used, the more cost-effective it is. Undoubtedly, all this research contributes to our understanding of telehealth. However, in our view, the literature does not tell the whole story; many gaps exist that need to be addressed.

First, since telehealth programs demonstrate clinical value and technical feasibility (Shore, Brooks and al., 2007; Duplantie, Gagnon and al., 2007; Putnam, 2007), they must move from experimental settings to the real world. This calls for adjustments to healthcare organizations' administrative and clinical routines (Saga and Zmud, 1994; Zucker, 1977) as well as adjustments to their work systems and technological configuration to account for assimilation as a technological innovation (Kwon, 1987; Cooper and Zmud, 1990; Chatterjee and Segars, 2001; Keen and McDonald, 2000).

Second, widespread deployment does not always accompany innovation adoption (Fichman and Kemerer, 1999). For example, Eveland and Tornatzky (1990) and Cooper and Zmud (1990) report that new technology with widespread adoption can fail to deploy fully.

To be truly valuable, a technological innovation must be routinized and infused in the adopting organization's operational or managerial work systems (Zmud and Apple, 1992). Consequently, to materialize the benefits of telehealth systems, we need to understand better the mechanisms through which assimilation into administrative and clinical practices occurs. However, little is known about the process of telehealth systems assimilation and about assimilation enabling and impeding factors since most studies focus on user acceptance and adoption; little is said about what happens after the initial adoption decision.

This chapter adds to the knowledge of assimilation of large-scale IS by developing a mixed determinants model of telehealth systems. First, we explain the nature of telehealth systems and the concerns raised by their deployment. On this basis, we elicit the underlying mechanisms of their assimilation. Next, we develop a conceptual model by integrating insights from three influential theories: social cognition theory of organizational learning, institutional theory, and structuration theory. Then, we proceed with a contribution to both theory and practice. Finally, we conclude by offering future research steps.

THEORETICAL BACKGROUND

In this section, we first propose a conceptualization of telehealth system to identify its characteristics. We then describe the role and nature of the technological artifact and their articulation with the social and institutional contexts in which the systems and the adopting organization are embedded (Orlikowski and Iacono, 2001). Such a theorizing strategy ensures that both the technological artifact and the context are considered.

Understanding Telehealth Systems

Generally, Telehealth systems are large-scale ICT systems linking two or more organizations and many categories of actors. Social and organizational context considerations are essential to ensure systems deployment success. An understanding of social relations, work structures, cultural factors, and the organization's experience with IT are additional imperatives. As the scope of a telehealth project is large, several decisions must be made at the organization's operational and tactical levels while it unfolds. Nevertheless, decisions are too numerous and technology is too diffuse and complex to be grasped by a single person's cognitive ability. Moreover, acquisition and deployment decisions of such systems are not typically situated within the discretionary power of a single member of the organization since other organizations also participate in the project (Eveland and Tornatzky, 1990 p. 124). When deployment of an information system requires complex organizational arrangements regarding the location of individual decisions, its adoption and application are the product of numerous decisions dictated by economic and social forces that go beyond managerial logic. An organization's institutional properties, its work politics, and certain environmental characteristics (Orlikowski, 1992) are included in the conceptual model through relevant social theories to illuminate organizational ICT assimilation. Most telehealth initiatives aim at extending conventional health systems by overcoming their limitations such as the inability to provide proper and accessible care to large geographical areas. Combining primary as well as specialized healthcare services, Telehealth projects not only enable healthcare coverage beyond geographical boundaries, they offer organizational integration of entities that have disparate vocations. The rationale for this type of integration goes beyond the logic specific to each organization. All of these reasons suggest that institutional factors influence assimilation of telehealth systems, if only to mobilize the necessary resources required for system integration.

Since telehealth systems comprise a variety of technologies such as home healthcare monitoring, telemedicine, and tele-education, any conceptual model must account for the nature of the technology. Telehealth applications are diverse and deployed based on numerous technologies such as teleconferencing, digital imaging, data storage and retrieval (sound, images, and alphanumeric data), robotics, two-way interactive television, and others. An appropriate infrastructure enabling data transmission and data entry, creation of multimedia databases, and communication between different partner organizations must be in place. Accordingly, telehealth information systems are comprised of different classes of technology that vary from those that interact with users to those that require higher interaction with professionals as the technologies intrude into healthcare practices. Some technologies not only support clinical activities, they also provide reasoning behind medical evaluation and training (Grémy and Bonnin, 1995). Finally, they include infrastructure-type technologies, the purpose of which consists of managing large flows of information coming from numerous sources and going to numerous places, services, or individuals within the organization. These technologies have no purpose in isolation; they are an integral part of the healthcare organization (Grémy and Bonnin, 1995). When faced with such a broad range of technologies, individuals often feel threatened, partly about losing control over their work schedule and professional autonomy. Moreover, given that the systems impose their own representations of the world, individuals' choices become limited. Feelings can range from fascination for some users to frustration for others.

Telehealth systems combine several heterogeneous parts made up of numerous intrinsically complex components (Paré and Sicotte, 2004). Therefore, it is appropriate to conceptualize them as focal technological systems linked by a web of connections to its social, political, and institutional contexts (Kling and Scacchi, 1982). Despite the reality that the technological artifact is central to telehealth systems, it is one element of an ensemble that also comprises the necessary components to apply the technical artifact to a given socio-economic activity (Kling and Dutton, 1982; Illich, 1973). Among these components, we find commitment, additional resources such as training, qualified personnel, organizational agreements, and the political and reward systems. In short, we find all that is necessary to promote the effective management of systems (Kling and Scacchi, 1982). The ensemble view of telehealth systems entails at least three components. First, the necessity to consider these systems as embedded in dynamic and complex social and organizational contexts and the examination of the manner in which different social and organizational influences contribute to model its deployment with how different user groups integrate them (Orlikowski and Iacono, 2001). Second, the structural potential of information technology (IT) on the rules and resources and the spirit of these systems must be accounted for (Giddens, 1984; DeScantis and Poole, 1994). Finally, it is necessary to consider uncertainties and, consequently, the interpretation of problems due to the diversity of the technologies involved, the organizational agreements they require, and changes to organizational schemas and procedures for everyday action (Barley, 1986).

The previous developments lead us to believe that the assimilation of telehealth systems not only entails that we proceed with shared social and organizational representations of these systems, but that they are susceptible to influences related to these systems' attributes, to characteristics of the institutional context, and to interactions between technological innovation and the organizational milieu.

Understanding Assimilation

Assimilation is often conceptualized as a learning process (Attewell, 1992; Fichman and Kemerer,

1997, 1999). Our conceptualization differs slightly as we model assimilation as two processes: technological routinization and infusion. Routinization refers to the fact that as time passes, the system ceases to be perceived as a novelty and becomes commonplace (Saga and Zmud, 1994; Ritti and Silver, 1986; Zucker, 1977). Infusion refers to the system's embeddedness in organizational procedures and work architectures as it links different organizational elements such as roles, formal procedures, and emerging routines (Cooper and Zmud, 1990; Kwon, 1987). As telehealth systems combine diverse technologies, their development and implementation - like that of complex systems - become a continual process (Weick, 1990).

Underlying Mechanisms of Assimilation

Making sense of telehealth systems is necessary due to the nature of the system and because telehealth constitutes a new form of health services delivery. Telehealth systems are complex, a reality that is due not only to their technological constituents, but also to the institutional agreements that deployment requires. For example, the Sores Care Tele Assistance Project (SCTAP) implemented by the Sherbrooke's University Integrated Health Network (Réseau Universitaire Intégré de Santé – RUIS) deploys seventy-three information technologies within sixty-five points of care. The deployment is comprised of eight phases, each including the organization of new clinical services and the elaboration of a deployment strategy of new virtual clinics and other entities. (RUIS de Sherbrooke, 2007).

Healthcare systems create problems never before experienced by managers or by health care professionals (Weick, 1990). Systems and technologies evolve and because of their complexity, users have difficulties grasping their inner workings, which leads to misunderstandings and misinterpretations (Weick, 1990b). Since they are exogenous to the organizational context, their introduction creates a certain hiatus among existing meaning, legitimization, and domination systems and the new exigencies related to the organization's daily actions (Barley, 1986). Therefore, these new technologies affect organizational members' capacity to reason on telehealth systems structures because technologies, in general, and new technologies, in particular, are interpretively flexible (Weick, 1990, 1990b), allow for different possible and plausible interpretations by diverse social groups, and can be misunderstood, uncertain, and complex (Pinch and Bijker, 1987; Weick, 1990, 1990b; Orlikowski and Gash, 1994).

This interpretive flexibility results from the presence of information technology artifacts as well as from cognitive schemas (Orlikowski and Iacono, 2001; Chae, 2002). While artifacts are made up of material resources like equipment, applications (Kling, 1987), non-material resources, networking capacity, programming languages (Chae, 2002), and structural elements built-into technology (DeSanctis and Poole, 1994), schemas point to generalizable procedures that take root in the context or pre-existing institutions (Chae, 2002). Within the context of implementation and utilization, procedures can be associated with the social organization of computerization (Kling and Iacono, 1989). Further, since telehealth systems often involve several organizations and individuals, this may lead to equivocal interpretations. For example, the SCTAP project groups thirteen Health and Social Services Centers (Centres de Santé et des Services Sociaux -- CSSS), spreading through three of Quebec's geographical regions and involving ninety nurses. Consequently, schemas are complex, numerous, and embedded in multiple structures. There is, thus, a strong need to build shared, unequivocal meanings; divergent meanings may lead to different courses of action impinging on the integration of the system into work routines.

Precedent arguments stress both participants' perception of ambiguity regarding the nature of telehealth systems and users' need to make sense of these technologies to use and integrate them into daily practice. Following Orlikowski and Gash (1994), participants develop hypotheses and expectations to build comprehension about what the technology is and what it is good for; their interpretation of the technology often dictates how he/she will use it. Thus, sense making is essential for assimilating the technology into work practice. Further, sense making clarifies the system's rationale and the development of its underlying philosophy or, in other words, its spirit (De Sanctis and Poole, 1994). It helps with understanding the nature of the technology and its consequences within a given context (Orlikowski and Gash, 1994).

Sense making is also a process that modifies the user's cognitive structure (mental models) that can be described as "a built-up repertoire of tacit knowledge that is used to impose structure upon, and impart meaning to otherwise ambiguous social and structural information to facilitate understanding" (Gioia, 1986, p.56). Sense making, however, is not a straightforward process. For example, it has been suggested that personal schemas interact with organizational contexts during the sense making process (Elsbach, Barr and Hargadon, 2005). Due to the variety of organizational contexts, the fact that several organizations are involved in telehealth projects (e.g. primary care, secondary care, etc.), and the broad range of disciplines often involved (e.g., doctors, nurses, radiologists, etc.), interpretations differ among actors to the point that it may be difficult to come to a common understanding; hence, system assimilation slows. The artifact context (e.g. telehealth systems) also combines with personal schemas during the process of sense making. More precisely, schemas about the self interact with the artifact context as one sees himself/herself performing his/her job with the technology, thereby creating expectations of performance and behaviour (Elsbach et al., 2005). To this extent, it becomes difficult to share understanding with others since participants have their own role perceptions of how they will interact with and through the technology. As participants make use of the technology and interactions with others and the technology emerge as it is deployed, roles are redefined and better understood.

Technology Routinization and Infusion

In early research, IS implementation was viewed as a final step in which a new system is configured and installed. Since the work of Kwon and Zmud (1987), the process research stream in IS innovation implementation has instead focused attention on post-installation organizational behaviours (Saga and Zmud, 1994). Expanding the work of Kwon and Zmud (1987), Cooper and Zmud (1990) introduced a model of implementation that is comprised of the following stages: initiation, adoption, adaptation, acceptance, routinization, and infusion. Implementation is conceptualized as a change effort to ensure that the new system will successfully be put in use (Lucas, 1981). Implementation success is key to the sustainability of the system thereafter. For example, Cooper and Zmud's model draws upon Lewin's (1952) planned change theory by considering the "changing state" while acceptance, routinization, and infusion mirror the "refreezing state" (Saga and Zmud, 1994). Other scholars moved away from the planned change perspective focusing either on issues arising from implementation process or on how to manage the group along with the interpersonal process to stimulate change. The IS literature on implementation later proposed a complementary view to the planned change perspective whereby IS implementation is better conceptualized as a dynamic process of mutual adaptation between technology and organization (Leonard-Barton, 1988). Further contributing to this line of reasoning, Orlikowski (1992/1993) as well as Orlikowski and Gash (1994) propose a situated change perspective. Central to this view is the subtle, smooth, and significant way through which organizational transformation is enacted. Such an organizational transformation is grounded in the organization's routines or work systems (Alter, 2003) and subsequently in the ongoing work practices of organizational actors. These organizational changes are enacted as users accommodate to contingencies, exceptions opportunities, breakdowns, and even unintended consequences they face in their task (Barley, 1990). Scholars widely acknowledge that new technologies trigger changes in organizational routines (Barley, 1986; Tyre and Orlikowski, 1994) and this disruption in routines is largely associated with cognitive and interpersonal changes (Barley, 1986; Orlikowski, 1993).

Adopting this line of reasoning, we contend that when a new technology is implemented, the organizational actors engage in a sense making process and develop new cognitive frames about the technology by making cognitive, interpersonal, and organizational adjustments that allow the new technology and the corresponding new routines to become ongoing practice (Edmonson et al., 2001). Accordingly, routinization follows Hall and Loucks' (1977) theory of cognitive adaptations and Elsbach, Barr, and Hargadon's (2005) theory of situated cognition. Individuals go through several understandings (meanings) of the technology as they first make sense of it before using it. From the interaction between their own event schemas that relate to how the process will unfold and the physical context (geographical location of participants), they start developing a collective mindset (Elsbach et al., 2005). Through use of the technology, their sharing of experiences, and coordination of activities with others, they refine their understandings (meanings) of it. At this stage, the artifact context (technology put to use) interacts with the individual's own schemas to augment the solution that subsequently becomes more attractive and desirable. Users then confront their meanings with those of others and refine their understanding of the nature of the technology and what it can be used for; at times, conflicts arise that lead to the new system's social representation (Lauriol, 1998).

Coordinating their activities with other users is an important means of assimilating the technology because it is the very nature of organizational learning, the primarily outcomes of which are routinization and infusion. Coordination helps users build new cognitive coordinations, memorize them, repeat them, and transpose them to new situations. Coordination also implies norms and rules and promotes the incorporation of technology into organizational routines. The institutional/cultural context has been proposed as a determinant of option attractiveness (Elsbach et al., 2005). Therefore, normative pressures and the salience of the solution (telehealth) help in considering the new ways of doing as attractive thereby sustaining the technology in the workplace. As long as the organization evolves in its understanding of the system and its possibilities, it modifies the workplace architecture to increase the extent of system use to integrate the systems' use and to accomplish activities not previously considered feasible (Saga and Zmud, 1994).

In summary, assimilation starts as an individual learning process through which participants later develop new schemas to represent the system. Subsequently, through interaction and coordination, an organizational learning process occurs by which individuals' schemas evolve, leading to a shared understanding that initially blends into routines but ends up integrated into daily practices and organizational beliefs, transcending the individuals who were at the beginning of the process.

DETERMINANTS OF ASSIMILATION

In this section, we explore the factors that influence the assimilation of telehealth systems. We contend that system innovation attributes, institutional environment properties, and interaction between technology and organizations are important determinants of telehealth system assimilation.

Assimilation begins at the individual level before developing at the group level and then it finally permeates to the organizational level. We assert that for our model to hold, individuals and groups do not need to belong to the same organizations; they can be dispersed throughout the participating organizations or even geographically dispersed. Although organizational assimilation takes root at the individual level and many factors influence its process - including personal factors, group factors etc. - in this chapter, we only discuss organizational factors that significantly affect assimilation.

Organizational assimilation is influenced by specific organizational capacities and by the interaction between telehealth systems, the organization, and a network of actors in the organizational field. Organizational capacities most relevant here are those associated with the technological and sociocognitive environments. The former primarily refers to the technological infrastructure's IT capacities and the latter to organizational climate. Regarding the interaction between the system and the organization, we refer to compatibility between the technological innovation and the institutional system on one hand, and the organization's existing technological systems on the other. In reference to the network of actors, we refer to the organizing vision of telehealth and some institutional influences through policymaking and the support of mediating institutions.

IT Capacities

Organizational assimilation implies that changes must be tailored to the technological architecture to facilitate how organizational elements such as roles, formal procedures, and emergent routines connect (Cooper and Zmud, 1990; Kwon, 1987). The organization must possess IT capacities that cover both technological and organizational dimensions (Bharadwaj, Sambamurthy and Zmud, 1999) to ease the required changes. In particular, the organization needs the capacity to maintain a close, ongoing partnership between the heads of business and IT processes. It also needs the capacity to adjust operating and technological processes to maintain their efficiency and effectiveness and exploit the capacities of emerging IT (Bharadwaj et al., 1999). A flexible, integrated IT infrastructure is critical since it ensures continuous compatibility and interoperability between telehealth systems and the existing organizational systems (Broadbent and Weill, 1997; Kayworth, Chatterjee and Sambamurthy, 2001) and avoids fragmentation and lack of integration of the various systems. By IT infrastructure, we mean a shared organizational resource comprised of physical elements (such as technological artifacts) and intellectual elements (such as knowledge and know-how) (Broadbent, Weill, Brien and Neo, 1996), all of which are regulated by standards (Kayworth et al., 2001). The physical elements determine the infrastructure's IT capacities, namely its ability to combine the organization's resources to support effectiveness (Amit and Shoemaker, 1993). In fact, an integrated IT infrastructure must constitute a platform on which the organization's shared IT capacities are articulated (Weill, Subramani and Broadbent, 2002). Intellectual elements materialize in the relational and integration architectures required to exploit IT capacities. Relational architecture is required to configure the operating modalities for IT capacities while the function of integration architecture is to coordinate IT capacities in relation to the business capacities they are designed to support (Sambamurthy and Zmud, 2000). IT infrastructure sophistication has been found to have a significant influence on the assimilation of information technologies (Armstrong and Sambamurthy, 1999).

Ultimately, an organization that has the necessary capacities a) to maintain a close interrelationship between IT professionals and IT users, b) to continually adjust the interface of its operating and technological processes, and c) whose technological infrastructure is flexible enough not only to allow these adjustments but to permit the exploitation of emerging technologies, provides favorable conditions for the routinization and infusion of a technological innovation. This leads to the following proposition:

Proposition 1: The presence of a flexible infrastructure that can integrate existing and emerging operating and technological processes has a positive impact on the organizational assimilation of telehealth systems.

Organizational Climate

For the same reasons mentioned in the previous section, we believe that the work environment - in particular the organizational climate - constrains or promotes the organizational assimilation of telehealth systems. Organizational climate is defined as the individual perceptions concerning the most salient characteristics of the organizational context (Schneider, 1990). Unlike psychological climate, it corresponds to patterns of meaning shared by the individual members of the organization with regard to certain characteristics of the organizational context (Tracey, Tannenbaum and Kavanagh, 1995). It results from the interaction among objective elements that are observable in the organizational context and the perceptual processes of individual members of the organization (Schneider, 1983). Therefore, it is possible to influence perception of the climate by appropriately configuring the elements and processes of the organizational context (Kozlowski and Hults, 1987). Organizational climate is useful for understanding the normative response trends of organizational members (Kozlowski and Hults, 1987). An overall conceptualization of organizational climate may prove irrelevant for studying a specific phenomenon (Kozlowski and Hults, 1987). Rather, the concept of organizational climate should be considered as a broad, multidimensional perceptual domain for which construct definition depends on the variable of interest (Schneider, 1985). Moreover, an organizational environment may have several climates depending on whether individuals attach specific meanings to separate sets of organizational events or factors (Schneider and Reichers, 1983). From this point of view, a work environment is characterized by a climate of service, a climate of workplace safety, or a climate of self-fulfillment (Mikkelsen and Gronhaug, 1999). Bearing this in mind, we consider only the dimensions of climate that appear to be most relevant to routinization (or institutionalization) and infusion of telehealth systems.

Telehealth relies on information and communication technologies. Those technologies are in constant evolution calling for frequent adjustments in usage techniques and ways of using them. To ensure adequacy of use and to improve patient care through the technology, updates in skills are warranted if users are to benefit from technological advances. For example, medical, administrative, and IT staff members must constantly upgrade their skills since the institutionalization of a technological innovation entails that staff competencies be developed and updated (Kozlowski and Hults, 1987). Since healthcare systems are integrated with other organizational information systems, the latter must be updated to accommodate changes in the former. The structure of healthcare organizations is such that individuals often belong to one or more groups dedicated to specific activities, but they have to work with members of other groups to provide patient care (Dawson et al., 2008). When this reality is reinforced by appropriate integration strategies - as was the case with Quebec's health and social services networks - it becomes a climate of integration that fosters professional interdisciplinarity and cohesion among case managers, teams, and departments, which then become institutionalized (Kozlowski and Hults, 1987). Such a climate is favorable both to the routinization of telehealth systems and to their use in extended and integrated ways leading to infusion (Zmud and Apple, 1992). Therefore, it is reasonable that if healthcare organizations implement strategies that reinforce learning, continuous upgrading, and integration of their staff members' competencies, they could induce normative responses by their members that promote continuous upgrading of their skill level. An organizational climate that unites these three dimensions (learning, upgrading, and integration of skills) could prove favorable to the assimilation of telehealth systems. This leads to the following conjecture:

Proposition 2: An organizational climate that promotes upgrading skills has a positive impact on the organizational assimilation of telehealth systems.

Compatibility

Telehealth systems are not deployed in a vacuum; they are deployed in organizational contexts with well-established social structures such as practices, professional culture, technologies, and other socio-technical elements (Gosain, 2004). It is essential to consider the impact of these structural elements on the assimilation of telehealth IS because studies suggest that they either constitute barriers to IS implementation or facilitate it by providing the necessary infrastructure or by strengthening the organization's absorptive capacity (Kling and Iacono, 1989; Chae and Poole, 2005). Similarly, because of the organizational arrangements they require, telehealth IS have the capacity to structure the behaviours of the organizations involved. The deployment of a new and complex system not only triggers a process of mutual structuring among the host organizations and the system, but also raise the possibility of a mismatch (institutional misalignment) between the institutional regime of the organizations involved and the institutional logics conveyed by the technology (Gosain, 2004). It is possible that conflicting structural components may come into contact. We must acknowledge that the healthcare community is characterized by strong disciplinary heritages, themselves characterized by well-established values, norms, and cultural and cognitive schemas that shape and guide the behaviours of members of the medical or other professions. Disciplinary heritages are developed and maintained by institutional agents such as the educational system, supervisory agencies, professional associations, and others. Heritages are comprised of rules, work procedures, protocols, and codes of ethics. Technologies used in daily practice are adopted as long as they stabilize professional practice. Indeed, disciplines as bodies of knowledge preserve concepts, practices, and values that are employed in action (Pickering, 1995). Moreover, information systems are "complex social objects" (Kling and Scacchi, 1982) laden with intentionality (Chae and Poole, 2005). Because of the artifacts that compose them, IS form a unique combination of three kinds of agency: material, human, and disciplinary (Chae and Poole, 2005; Pickering, 1995). Agency denotes the capacity of a person or an object to produce a particular result. Human agency refers primarily to individuals' reflexivity and ability to adjust actions with the aim of achieving a defined goal, which may be a plan or intention (Chae and Poole, 2005). Material agency is physical or biological in nature; it is revealed in the actions of the specific powerful forces of certain generative mechanisms (Harre and Madden, 1975). Finally, disciplinary agency corresponds to the shaping and orientation of human action by cultural and conceptual systems (Pickering, 1995). In fact, in the development phase of a system, it goes through a "registration process" during which the dominant interests and the developers' responses to institutional pressures and their vision of the world are reflected in the system's functioning (Latour, 1992). In this sense, telehealth systems developers have the opportunity to incorporate rules, functionalities, and resources into the systems that are capable of structuring the users' interaction with the system.

The discussion above helps point out that the interaction between telehealth systems and the organization raises the problem of compatibility between the systems and the organization's operating infrastructure on one hand and the organization's technological infrastructure in the sense of software, hardware, and IT management procedures on the other. In the first case, we refer to operational compatibility, whereas the second corresponds to technological compatibility (Jones and Beatty, 1998). If telehealth systems are compatible with existing technological infrastructure and operating procedures, this compatibility facilitates the assimilation of these systems into the organizations involved in the telehealth project:

Proposition 3: Telehealth systems' compatibility with the organization's operating infrastructure has a positive impact on their assimilation.

Proposition 4: Telehealth systems' compatibility with the organization's existing technological infrastructure has a positive impact on their assimilation.

Organizing Vision

In the case of technological innovations like telehealth systems, institutional processes come into play from the beginning of the diffusion process. A heterogeneous network of parties creates and employs an organizing vision that reduces uncertainties surrounding the systems and their applications (Swanson and Ramiller, 1997). This organizing vision allows members of the network to make sense of the innovation. An organizing vision aims at making the spirit of the technology explicit, namely the underlying philosophies of the IT artifacts as well as the motives for its development (Chae, 2002).

Users do not rely exclusively on their own interpretations of the technology during the early phase of sensemaking since they use the organizing vision as a starting point. From their original understanding, they search, probe, and confirm meanings with others (Swanson and Ramiller, 1997). They assess interpretations that exist within their own organization, yet are consistent with those of partner organizations due to the presence of an organizing vision. However, some pitfalls and limitations arise during the process of sensemaking.

Telehealth projects vary in nature and objectives; they sometimes vary in configuration (same technologies but configured differently). Some of the system components are immature technologies or prototypes; those system components are not well defined and their applications are not well understood (Swanson and Ramiller, 1997). Meanings assigned to such technologies reflect incompleteness and instability (Rosemberg, 1994). An organizing vision is warranted to provide a structure that helps users understand the nature and role of telehealth systems given social, technical, and economic contexts (Klecun-Dabrowska and Cornford, 2002). It also reduces equivocalities and creates shared understandings of the system.

The sociocognitive theory of organizational learning suggests that for a given technology, each group develops a set of shared understandings, thus forming a technological frame. This frame emerges from interactions among group members through coordination of interdependent activities (Schein, 1985; Strauss, 1978) and through the way the group's specific norms influence its members (Porac et al., 1989; Grégory, 1983; Van Maanen and Schein, 1979). By formulating expectancies, assumptions, and knowledge about key aspects of telehealth systems, the organizing vision contributes to make congruent the groups' technological frames. Consequently, organizations are less likely to experience difficulties or conflicts linked to a systems' implementation and use (Orlikowski and Gash, 1994 p.180). Therefore, we propose the following conjecture:

Proposition 5: A compelling organizing vision positively influences the assimilation of telehealth systems.

The organizing vision also provides a structure of legitimization that complements and reinforces the structure of interpretation by including in its discourse aspects directed toward justifying the technological innovation. Pointing to the systems' legitimacy, discourse deals with technical and functional arguments with political, organizational, and business arguments. This discourse is not simply formulated in terms of the low-cost associated with electronically delivered services; it also includes dominant social values and principles like improving the quality of citizens' lives (Klecun-Dabrowska and Cornford, 2002). This dimension of the organizing vision is dedicated to communicating not only the expected benefits of the innovation, but also its spirit. To illustrate, by promoting telehealth as a means to bring specialized care services to remote regions and to recruit and maintain physicians in those regions, the organizing vision capitalizes on current norms and societal values. In doing so, it legitimates telehealth and mobilizes resources needed for successful deployment (Orlikowski and Gash, 1994, Swanson and Ramiller, 1997). The preceding development suggests the following conjectures:

Proposition 6: An organizing vision that promotes telehealth based on the health system's current values and social norms positively influences the assimilation of telehealth systems.

Proposition 7: Perceived social benefits positively influence the assimilation of telehealth systems.

Policymaking

Many barriers have been identified with respect to the diffusion of innovations, two of which are particularly relevant here: lack of reimbursement and legal liability. In the domain of reimbursement, few public or private payers reimbursed healthcare services provided through advanced ICT in the USA during the early years of ICT. Some improvements have been made but are still too restrictive. The Balance Budget Act expands coverage to include telemedicine services, but at the same time introduces new requirements that keep people from using telemedicine under current Medicare conditions (DHSS, 2001). Likewise, Gagnon et al. (2001) report that the reimbursement system made some services less available and less accessible in the telehealth project implemented in Iles-de-la Madeleine, a remote region of Québec, Canada.

Legal liability is an important issue in the diffusion of telehealth. Medical errors account for some 440,00 to 980,00 deaths every year in the U.S.; these errors are attributable in great part to the decentralized, fragmented, complex nature of the U.S. healthcare delivery system (IOM, 1999). According to IOM (1999), the relationship between the complexity of the healthcare system and medical errors has serious, adverse effects on the diffusion of telehealth. Who is accountable for a medical error, the referent or the referring physician? There is no clear answer. Yet, specialists may be reluctant to give medical advice through ICT. We hence propose:

Proposition 8: Policymaking directed toward facilitating the reimbursement of telehealth services positively influences the assimilation of telehealth systems.

Proposition 9: Policymaking directed toward establishing the liabilities of telehealth professionals when an error is reported positively influences the assimilation of telehealth systems.

Mediating Institutions

Telehealth shares the properties of both technological and administrative innovations (Robinson et al., 2003). Consequently, deployment issues stem from incorporation in the organizational and clinical routines (Paré and Sicotte, 2004). Partnerships with mediating institutions like knowledge generating organizations (consulting, universities, etc.) specialized in advanced technologies know-how can help lower knowledge barriers associated with technology integration (Attewell, 1992).

Proposition 10: Collaboration with some intermediating institutions positively influences the assimilation of telehealth systems.

DISCUSSION

This chapter develops a theoretical perspective for understanding the assimilation of telehealth systems on the grounds of their nature and the issues raised by their deployment. In so doing, it contributes to research by pointing out the value of grounding the theorizing of assimilation of telehealth systems in the fundamental characteristics of the technological artifact such as its interpretive flexibility and the complex organizational arrangements necessary for its deployment. Unlike in prior studies of assimilation of IS innovations (Chatterjee et al., 2002; Purvis et al., 2001; Gallivan, 2001; Meyer and Goes, 1988), assimilation is conceptualized as an organizational phenomenon that has its theoretical origins in individual interpretation. Second, the model uncovers the interplay between the institutional context and cognitive regulations.It makes more salient the fact that assimilation of telehealth systems is, essentially, a multilevel phenomenon.

Our study also carries implications for practice. Particularly, it points to the importance of examining the assimilation process within the continuum of IS phenomena surrounding the deployment of telehealth systems. It helps understand why managerial actions directed toward facilitating the assimilation process should be employed as early as the adoption phase. Issues related to compatibility between the systems and the organization's work infrastructure should be managed during the development phase. Even though routinization and infusion are post-implementation behaviours, factors that influence them should be taken under consideration before systems acquisition.

Our model can be challenged on a few grounds. For example, telehealth, as a large, complex system, can be compared to large-scale systems such as enterprise resources planning (ERP). We address such challenges here. First, we recognize that lessons learned from other types of systems help guide the modeling of telehealth assimilation and that throughout this chapter, we made use of relevant literature. We believe that telehealth departs from many other types of information systems since in addition to encompassing several levels within the organization as in the case with ERP, telehealth crosses organizational boundaries since several organizations often partner in developing and implementing the system. This contrasts with large systems like ERP where suppliers are located upstream on the value chain and clients are downstream. Unlike ERP, telehealth is not only vertical along the value chain but also lateral across a range of business partners. To that extent, the arrangement resembles a network organizational form (Powell, 1991), far removed from traditional hierarchical forms.

Telehealth requires complex arrangements both within the organization and across partners with all partners looking at mutual benefits and shared interests. It also differs since its inception, implementation, and infusion rely on users as opposed to ERP where the technology is institutionalized from the top down and imposed upon users; in the latter case, use of the system is prescribed and mandatory. This is not to say that use of telehealth is voluntary; use depends more highly on users' perceptions and interpretation (sense making) since, unlike organizational and societal interests, personal interests play an important role in adopting the technology. Sense making is far more complex since it not only requires users from one organization to build common understanding, it also requires that users from the participating organizations share mental models. This is a demanding undertaking. Once users build an understanding of the system, such understanding is shared with other groups and permeates through other levels in the organization and across partners. Understanding and meaning evolve as users gain experience with the system and integrate it into their work through a spiral form of adaptation and adjustment. This is in contrast with ERP systems where users are instructed to do their work in a certain way and where there is less provision for adapting and adjusting the technology to one's needs. ERP systems are rigid and impose work routines upon the users while telehealth is malleable; not only are the work routines changed and adjusted, the system itself is crafted to fit with organizational needs. Our stance, then, has been to recognize that telehealth is a unique phenomenon that needs to be studied through different lenses than traditional business information systems and technology. We are confident that our proposed model is a valuable contribution to the body of knowledge about telehealth.

FUTURE WORK

Future work focuses on extending the current model of assimilation of IS and on addressing related open research issues. Despite the fact that we take individual sense making as the starting point of routinization, we focus on examining the assimilation of telehealth systems at the organizational level. Like prior studies, our exclusive focus on the organizational level ignores micro and meso explanations of the phenomenon (House et al., 1995). A more detailed account is needed to make explicit both the structure (Kozlowski and Klein, 2000) and the functional relationships of the phenomenon (Morgesson and Hofmann, 1999). It is necessary to account for how individual, group, and organizational characteristics interact to structure the process of assimilation. Approaching this phenomenon from a sociocognitive viewpoint leads us to a multilevel perspective that results in an integrated understanding of how the assimilation process unfolds across levels in organizations and across organizations; it adds depth and richness to the theoretical model. Once complete, we intend to use this a priori theory as a starting point to build a substantive theory of telehealth assimilation in the context of healthcare in Quebec.

This new research direction enhances the managerial significance of this work. A more comprehensive and integrative multilevel model of assimilation, by making explicit the functional relationships at individual, group, and organizational levels, highlights the range of managerial interventions required to secure the IS assimilation and, consequently, telehealth systems' effectiveness. In addition, the model provides, beyond policy, a better understanding of the locus of authority for each specific managerial intervention. In so doing, it is helpful in enhancing the effectiveness of managerial actions and in smoothing the IS governance aspects regarding telehealth systems.

At the time of this writing, we have already collected data from a wide variety of participants by means of semi-structured interviews. Using a case study approach (Yin, 2003), we proceeded with testing our research model. Participants include a broad range of disciplines and our sample covers three of the four University Integrated Health Networks in Quebec. Preliminary results are promising and are globally in support of our research hypotheses. In addition, some additional contributing factors seem to emerge from the data helping us refine our theory. We look forward to completing this part of our research agenda and to sharing results with the research community.

REFERENCES

Amit, R., & Schoemaker, P. J. H. (1993). Strategic assets and organizational rent. *Strategic Management Journal*, *14*(1), 33. doi:10.1002/smj.4250140105

Armstrong, C. P., & Sambamurthy, V. (1999). Information Technology assimilation in firms: the influence of senior leadership and IT infrastructures. *Information Systems Research, 10*(4), 304–327. doi:10.1287/isre.10.4.304

Attewell, P. (1992). Technological diffusion and organizational learning: The case of business computing. *Organization Science, 3*(1), 1. doi:10.1287/orsc.3.1.1

Balas, E. A., & Iakovidis, I. (1999). Distance technologies for patient monitoring. *British Medical Journal, 310*, 1309.

Barer, M. L., & Stoddart, G. L. (1999). *Improving access to needed medical services in rural and remote Canadian communities: Recruitment and retention revisited.*

Barley, S. R. (1986). Technology as an occasion for structuring: Evidence from observations of CT scanners and the social order of radiology departments. *Administrative Science Quarterly, 31*(1), 78. doi:10.2307/2392767

Barley, S. R. (1990). The alignment of technology and structure through roles and networks. *Administrative Science Quarterly, 35*(1), 61. doi:10.2307/2393551

Bharadwaj, A., Sambamurthy, V., & Zmud, R. (1999). *IT capabilities: Theoretical perspectives and empirical operationalization.*

Bilodeau, H., & Leduc, N. (2003). Inventory of main factors determining the attraction, installation and retention of physicians in remote areas. *Cahiers de Sociologie et de Demographie Medicales, 43*(3).

Bilodeau, H., Leduc, N., & Schendel, N. (2006). *Analyse des facteurs d'attraction, d'installation et de maintien de la pratique médicale dans les régions éloignées du Québec.* Groupe De Recherche Interdisciplinaire En Santé. Université De Montréal.

Bishop, J. E., O'Reilly, R. L., Daddox, K., & Hutchinson, K. (2002). Client satisfaction in a feasibility study comparing face-to-face interviews with telepsychiatry. *Journal of Telemedicine and Telecare, 8*(4). doi:10.1258/135763302320272185

Bourgueil, Y., Mousques, J., & Tajahmadi, A. (2006). *Comment améliorer la répartition géographique des professionnels de santé: Les enseignements de la littérature internationale et les mesures adoptées en France.* Institut De Recherche Et Documentation En Économie De La Santé. Broadbent, M., & Weill, P. (1997). Management by maxim: How business and IT managers can create IT infrastructures. *Sloan Management Review, 38*(3), 77.

Chae, B. (2002). *Understanding Information Systems as social institutions: Dynamic institutional theory.* Unpublished doctoral dissertation, Texas A&M University, United States -- Texas.

Chae, B., & Poole, M. S. (2005). The surface of emergence in systems development: Agency, institutions, and large-scale Information Systems. *European Journal of Information Systems, 14*, 19–36. doi:10.1057/palgrave.ejis.3000519

Chatterjee, D., Grewal, R., & Sambamurthy, V. (2002). Shaping up for e-commerce: Institutional enablers of the organizational assimilation of Web technologies. *Management Information Systems Quarterly, 26*(2), 65. doi:10.2307/4132321

Chatterjee, D., & Segars, A. H. (2001). *Transformation of the enterprise through e-business: An overview of contemporary practices and trends.* Report to the Advanced Practices Council of the Society for Information Management, (July 2001).

Cho, S., & Mathiassen, L. (2007). The role of industry infrastructure in telehealth innovations: A multi-level analysis of a telestroke program. *European Journal of Information Systems, 16*(6), 738–750. doi:10.1057/palgrave.ejis.3000718

Cho, S., Mathiassen, L., & Gallivan, M. (2009). Crossing the diffusion chasm: From invention to penetration of a telehealth innovation. *Information Technology & People*, *22*(4), 351. doi:10.1108/09593840911002450

Cho, S., Mathiassen, L., & Robey, D. (2007). Dialectics of resilience: A multi-level analysis of a telehealth innovation. *Journal of Information Technology*, *22*(1), 24–35. doi:10.1057/palgrave.jit.2000088

Cohn, J. R., & Goodenough, B. (2002). Health professionals attitudes to videoconferencing in pediatric heathcare. *Journal of Telemedicine and Telecare*, *8*(5). doi:10.1258/135763302760314243

Cooper, R. B., & Zmud, R. W. (1990). Information Technology implementation research: A technological diffusion approach. *Management Science*, *36*(2), 123. doi:10.1287/mnsc.36.2.123

Dansky, K. H., & Ajello, J. (2005). Marketing telehealth to align with strategy. *Journal of Healthcare Management*, *50*(1), 19–30.

Dawson, J. F., Gonzalez-Roma, V., Davis, A., & West, M. A. (2008). Organizational climate and climate strength in UK hospitals. *European Journal of Work and Organizational Psychology*, *17*(1), 89. doi:10.1080/13594320601046664

RUIS de Sherbrooke. (2007). La téléassistance en soins de plaies, orientée vers une amélioration continue de la qualité des soins. *Manuel D'Organisation De Projet,* 1-63.

DeSanctis, G., & Poole, M. S. (1994). Capturing the complexity in advanced technology use: Adaptive structuration theory. *Organization Science*, *5*(2), 121–147. doi:10.1287/orsc.5.2.121

DHSS. (2001). *Report to Congress on telemedicine*. Secretary of Health and Human Services.

Doolittle, G., Williams, A., & Cook, D. J. (2003). An estimation of costs of a pediatric telemedicine practice in public schools. *Medical Care*, *41*, 100–109. doi:10.1097/00005650-200301000-00012

Doolittle, G., Williams, A., & Harmon, A. (1998). A cost measurement study for a tele-oncology practice. *Journal of Telemedicine and Telecare*, *4*, 84–88. doi:10.1258/1357633981932000

Duplantie, J., & Gagnon, M.-P. (2007). Telehealth and the recruitment and retention of physicians in rural and remote regions: A Delphi study. *Canadian Journal of Rural Medicine*, *12*(1), 30–36.

Edmondson, A. C., Bohmer, M. R., & Pisano, G. P. (2001). Disrupted routines: Team learning and new technology implementation in hospitals. *Administrative Science Quarterly*, *46*(4), 685. doi:10.2307/3094828

Elsbach, K. D., Barr, P. S., & Hargadon, A. B. (2005). Identifying situated cognition in organization. *Organization Science*, *16*(4), 422–435. doi:10.1287/orsc.1050.0138

Eveland, J. D., & Tornatzky, L. (1990). The deployment of technology. In Tornatzky, L., & Fleischer, M. (Eds.), *The processes of technological innovation*. Lanham, MD: Lexington Books.

Fichman, R. G., & Kemerer, C. F. (1997). The assimilation of software process innovations: An organizational learning perspective. *Management Science*, *43*(10), 1345. doi:10.1287/mnsc.43.10.1345

Fichman, R. G., & Kemerer, C. F. (1999). The illusory diffusion of innovation: An examination of assimilation gaps. *Information Systems Research*, *10*(3), 255. doi:10.1287/isre.10.3.255

Field, K. S., & Briggs, D. J. (2001). Socio-economic and locational determinants of accessibility and utilization of primary healthcare. *Health & Social Care in the Community,* 9.

Finkelstein, J., Khare, R., & Vora, D. (2003). Home automated telemanagement system to facilitate self-care of patients with chronic diseases. *Journal of Systemics. Cybernetics and Informatics, 44*, 35–56.

Gagnon, M.-P., Fortin, J. P., & Cloutier, A. (2001). La télémédecine au service d'une région: Recherche évaluative sur le projet des Iles-de-la-Madeleine. *Ruptures, 8*, 53–70.

Gallivan, M. J. (2001). Organizational adoption and assimilation of complex technological innovations: Development and application of a new framework. *The Data Base for Advances in Information Systems, 32*(3), 51.

Giddens, A. (2005). *La Constitution de la Société: éléments de la théorie de la structuration (traduit de l'anglais par Michel Audet)*. Paris, France: Quadridge PUF.

Gosain, S. (2004). Enterprise Information Systems as objects and carriers of institutional forces: The new iron cage. *Journal of the Association for Information Systems, 5*(4), 151–182.

Gravelle, H., & Sutton, M. (2001). Inequality in the geographical distribution of general practitioners in England and Wales 1974-1995. *Journal of Health Service Resources Policy, 6*.

Gregory, K. L. (1983). Native-view paradigms: Multiple cultures and culture conflicts in organization. *Administrative Science Quarterly, 28*(3), 359–376. doi:10.2307/2392247

Grémy, F., & Bonnin, M. (1995). Evaluation of automatic health Information Systems. What and how? E. In Van Gennip, & J. Talmon (Eds.), *Assessment and evaluation of Information Technologies in medicine*. Amsterdam, The Netherlands: IOS Press.

Gulliford, M. C. (2002). Avaibility of primary care doctors and population in England: Is there an Association? *Journal of Public Health Medicine*, 24.

Hall, G., & Loucks, S. (1977). A developmental model for determining whether the treatment is actually implemented. *American Educational Research Journal, 14*(3), 263–276.

Harre, R., & Madden, E. H. (1975). *Causal powers: A theory of natural necessity*. Oxford, UK: Blackwell.

Hu, P., & Chau, P. (1999). Physician acceptance of telemedicine technology: An empirical investigation. *Topics in Health Information Management, 19*(4), 20–35.

Illich, I. (1973). *Tools for conviviality*. New York, NY: Harper and Row.

Institute of Medicine (IOM). (1999). *To err is human: Building a safer health system*. Washington, DC: National Academic Press.

Jones, M. C., & Beatty, R. C. (1998). Towards the development of measures of perceived benefits and compatibility of EDI: A comparative assessment of competing first order factor models. *European Journal of Information Systems, 7*(3), 210–220. doi:10.1057/palgrave.ejis.3000299

Kayworth, T. R., Chatterjee, D., & Sambamurthy, V. (2001). Theoretical justification for IT infrastructure investments. *Information Resources Management Journal, 14*(3), 5. doi:10.4018/irmj.2001070101

Klecun-Dabrowska, E., & Cornford, T. (2002). *The organizing vision of telehealth*. Gdansk, Poland: ECIS.

Kling, R., & Dutton, W. H. (1982). The computer package, dynamic complexity. In Danziger, J. N., Dutton, W., Kling, R., & Kraemer, K. L. e. (Eds.), *Computers and politics: High technology in American local governments* (pp. 22–50). New York, NY: Columbia University Press.

Kling, R., & Iacono, C. (1989). The institutional character of computerized Information System. *Office Technology & People, 1*(1), 24–43.

Kling, R., & Scacchi, W. (1982). The Web of computing: Computer technology as social organization. *Advances in Computers, 21*, 1–90. doi:10.1016/S0065-2458(08)60567-7

Kozlowski, S. W. J., & Hults, B. M. (1987). An exploration of climates for technical updating and performance. *Personnel Psychology, 40*, 539–563. doi:10.1111/j.1744-6570.1987.tb00614.x

Kozlowski, S. W. J., & Klein, J. K. (2000). A multilevel approach to theory and research in organizations: Contextual, temporal, and emergent process. In Klein, J. K., & Kozlowski, S. W. J. (Eds.), *Multilevel theory, research, and methods in organizations* (pp. 3–90). San Francisco, CA: Jossey-Bass.

Krupinski, E. A., Engstrom, M., Baker, G., Levine, N., & Weinstein, R. S. (2004). The challenges of following patients and assessing outcomes in dermatology. *Journal of Telemedicine and Telecare, 10*, 21–24. doi:10.1258/1357633043227764149

Kwon, T. H. (1987). *A study of the influence of communication network on MIS institutionalization.* Unpublished dissertation, The University of North Carolina at Chapel Hill, United States -- North Carolina.

Kwon, T. H., & Zmud, R. W. (1987). Unifying the fragmented models of Information Systems implementation. In Boland, R. J., & Hirscheim, R. A. (Eds.), *Critical issues in Information Systems*. New York, NY: John Wiley.

Latour, B. (1992). The sociology of a few mundane artifacts. B. In Bijker, and J. Law (eds.), *Shaping Technology/Building Society Studies in Sociotechnological Change* (135-150). Cambridge, Massachussets: MIT Press.

Lauriol, J. (1998). Les représentations sociales dans la décision. In Laroche, H., & Nioche, J.-P. (Eds.), *Repenser la stratégie*. France: Vuibert.

Leonard-Barton, D. (1988). Implementation as mutual adaptation of technology and organization. *Research Policy, 17*(5), 251. doi:10.1016/0048-7333(88)90006-6

Loane, M. A., Oakley, A., Rademaker, M., Bradford, N., Fleischl, P., & Kerr, P. (2001). A cost-minimization analysis of the societal cost of real-time tele-dermatology compared with conventional care. *Journal of Telemedicine and Telecare, 7*, 233–238. doi:10.1258/1357633011936453

Lucas, H. C. (1981). An experimental investigation of the use of computerized–based graphics in decision making. *Management Science (Pre-1986), 27*(7), 757.

McKee, J. J., Evans, N. E., & Owens, F. J. (1996). Digital transmission of 12-lead electrocardiograms and Duplex speech in the telephone bandwidth. *Journal of Telemedicine and Telecare, 2*(1). doi:10.1258/1357633961929150

Meyer, A. D., & Goes, J. B. (1988). Organizational assimilation of innovations: A multilevel contextual analysis. *Academy of Management Journal, 31*(4), 897–923. doi:10.2307/256344

Mikkelsen, A., & Gronhaug, K. (1999). Measuring organizational learning climate. *Review of Public Personnel Administration, 19*(4), 31. doi:10.1177/0734371X9901900404

Mitchell, B. R., Mitchell, J. G., & Disney, A. P. (1996). User adoption issues in renal telemedicine. *Journal of Telemedicine and Telecare, 2*(2). doi:10.1258/1357633961929835

Morgeson, F. P., & Hofmann, D. A. (1999). The structure and function of collective constructs: Implications for multilevel research and theory development. *Academy of Management Review, 24*(2), 249.

Nobel, S. M., Coast, J., & Benger, J. R. (2005). A cost-consequences of minor injuries telemedicine. *Journal of Telemedicine and Telecare*, *11*, 15–19. doi:10.1258/1357633053430421

Nordal, E. J., Moseng, D., Kvammen, B., & Lochen, M. (2001). A comparative study of teleconsultation versus face-to-face consultations. *Journal of Telemedicine and Telecare*, *7*(5). doi:10.1258/1357633011936507

Organisation Mondiale de la Santé. (2006). *Rapport sur la santé dans le monde: Travailler ensemble pour la santé*. Genève, Switzerland: OMS.

Orlikowski, W. J. (1992). The duality of technology: Rethinking the concept of technology in organizations. *Organization Science*, *3*(3), 398–427. doi:10.1287/orsc.3.3.398

Orlikowski, W. J. (1993). Learning from notes: Organizational issues in groupware implementation. *The Information Society*, *9*(3), 237–250. do i:10.1080/01972243.1993.9960143

Orlikowski, W. J., & Gash, D. C. (1994). Technological frames: Making sense of Information Technology in organizations. *ACM Transactions on Information Systems*, *12*(2), 174–207. doi:10.1145/196734.196745

Orlikowski, W. J., & Iacono, C. S. (2001). Research commentary: Desperately seeking "it" in IT research - A call to theorizing the IT artifact. *Information Systems Research*, *12*(2), 121. doi:10.1287/isre.12.2.121.9700

Paré, G., & Sicotte, C. (2004). Les technologies de l'information et la transformation de l'offre de soins. *Cahier Du GReSI, 4*(4).

Pickering, A. (1995). *The mangle of practice*. Chicago, IL: The University of Chicago Press.

Picolo, D., & Smolle, D. (2000). Tele-dermoscopy: Results of a multicentre study on 43 pigmented skin lesions. *Journal of Telemedicine and Telecare*, *6*(3), 132–137. doi:10.1258/1357633001935202

Pinch, T., & Bijker, W. (1987). The social connstruction of facts and artifacts. In Bijker, W., Hughes, T., & Pinch, T. (Eds.), *The social construction of technological systems* (pp. 159–187). Cambridge, MA: MIT Press.

Politzer, R. M., Harris, D. L., Gaston, M. H., & Mullan, F. (1991). Primary care physician supply and the medically underserved. A status report and recommendations. *Journal of the American Medical Association*, *266*(1), 104–109. doi:10.1001/jama.266.1.104

Porac, J. F., Thomas, H., & Baden-Fuller, C. (1989). Competitive groups as cognitive communities: The case of Scottish knitwear manufacturers. *Journal of Management Studies*, *26*(4), 397–416. doi:10.1111/j.1467-6486.1989.tb00736.x

Purvis, R. L., Sambamurthy, V., & Zmud, R. (2001). The assimilation of knowledge platforms in organizations: An empirical investigation. *Organization Science*, *12*(2), 117. doi:10.1287/orsc.12.2.117.10115

Putnam, J. M. (2007). Combining telehealth and e-learning: A case study in smoking cessation programming. *American Journal of Health Studies*, *22*(3), 130.

Ritti, R., & Silver, J. (1986). Early processes of institutionalization: The dramaturgy of exchange in interoganizational relations. *Administrative Science Quarterly*, *31*, 25–42. doi:10.2307/2392764

Roos, N. P., Fransoo, R., Carriere, K. C., & Frohlic, N. (1997). Needs-based planning: The case of Manitoba. *Canadian Medical Association Journal*, *157*(9), 15–1221.

Rosemberg, N. (1994). *Exploring the black box: Technology, economics and history*. Cambridge, UK: Cambridge University Press. doi:10.1017/CBO9780511582554

Rumberger, J. S. (2005). *Is there a strategic and financial business case for telehealth in home health agencies?* Unpublished doctoral dissertation, The Pennsylvania State University, United States -- Pennsylvania.

Saga, V., & Zmud, R. (1994). The nature and determinants of Information Technology acceptance, routinization and infusion. In Levine, L. (Ed.), *Diffusion, transfer, and implementation of Information Technology* (pp. 67–68). Amsterdam, The Netherlands: North-Holland.

Sambamurthy, V., & Zmud, R. W. (2000). Research commentary: The organizing logic for an enterprise's IT activities in the digital era - A prognosis of practice and a call for research. *Information Systems Research, 11*(2), 105. doi:10.1287/isre.11.2.105.11780

Schein, E. (1985). *Organizational culture and leadership.* San Francisco, CA: Jossey-Bass.

Schneider, B. (1983). Work climates: An interactionist perspective. In Freimer, N. W., & Geller, E. S. (Eds.), *Environmental psychology: Directions and perspectives* (pp. 153–178). New York, NY: Praeger.

Schneider, B. (1985). Organizational behavior. *Annual Review of Psychology, 36*, 573–611. doi:10.1146/annurev.ps.36.020185.003041

Schneider, B. (1990). The climate for service: An application of the climate construct. In Schneider, B. (Ed.), *Organizational climate and culture.* San Francisco, CA: Jossey-Bass.

Schneider, B., & Reichers, A. E. (1983). On the etiology of climates. *Personnel Psychology, 36*(1), 19–39. doi:10.1111/j.1744-6570.1983.tb00500.x

Schofield, R. S., Kline, S. E., Schmalfuss, C. M., Carver, H. M., & Aranda, J. M. (2005). Early outcomes of a care coordination-enhanced telehome care program for elderly veterans with chronic heart failure. *Telemedicine and E-Health, 11*, 20–27. doi:10.1089/tmj.2005.11.20

Shore, J. H., & Brooks, E. (2007). An economic evaluation of telehealth data collection with rural populations. *Psychiatric Services (Washington, D.C.), 58*(6), 830–835. doi:10.1176/appi.ps.58.6.830

Strauss, A. (1978). A social world perspective. In Denzin, N. (Ed.), *Studies in symbolic interaction (Vol. 1)*. Greenwich, CT: JAI Press.

Swanson, B. E., & Ramiller, N. C. (1997). The organizing vision in Information Systems innovation. *Organization Science, 8*(5), 458. doi:10.1287/orsc.8.5.458

Tracey, J. B., Tannenbaum, S. I., & Kavanagh, M. J. (1995). Applying trained skills on the job: The importance of the work environment. *The Journal of Applied Psychology, 80*(2), 239–252. doi:10.1037/0021-9010.80.2.239

Van Maanen, J., & Schein, E. (1979). Toward a theory of organizational socialization. *Research in Organizational Behavior, 1*, 209–264.

Veugelers, P. J., Yip, A. M., & Burge, F. (2004). Les inégalités en santé et dans la prestation des services de santé. *Maladies Chroniques Du Canada, 25*(3/4).

Vogel, R. L., & Ackerman, R. J. (1998). Is primary care physician supply correlated with health outcomes? *International Journal of Health Services*, 28.

Ward, T. (2004a). *Is there a light at the end of the tunnel - Can we resolve the physician distribution challenge in Canada?* Washington, DC: International Health Workforce Conference October 7-9.

Weick, K. (1990). The vulnerable system: An analysis of the Tenerife Air disaster. *Journal of Management, 16*(3), 571–593. doi:10.1177/014920639001600304

Weill, P., Subramani, M., & Broadbent, M. (2002). Building IT infrastructure for strategic agility. *MIT Sloan Management Review, 44*(1), 57.

Winters, J. M., & Winters, J. M. (2004). A telehomecare model for optimizing rehabilitation outcomes. *Telemedicine Journal and e-Health*, *10*, 200–212. doi:10.1089/tmj.2004.10.200

Wooton, R., Patil, N. G., Scott, R. E., & Ho, K. (2009). *Telehealth in the developing world.* London, UK: Royal Society of Medicine Press LTD.

Yant, B. (2005). *Impact of telehealth on job satisfaction of nurses and organizational performance of home health agencies.* Unpublished doctoral dissertation, The Pennsylvania State University, United States -- Pennsylvania.

Zmud, W. R., & Apple, L. E. (1992). Measuring technology incorporation/infusion. *Journal of Product Innovation Management*, *9*, 148–155. doi:10.1016/0737-6782(92)90006-X

Zucker, L. (1977). The role of institutionalization in cultural persistence. *American Sociological Review*, *42*(5), 726–743. doi:10.2307/2094862

Chapter 15
Introducing a First Step towards a Holistic Talent Management System Architecture

Andreas Eckhardt
Goethe University, Germany

Sven Laumer
Otto-Friedrich University Bamberg, Germany

Christian Maier
Otto-Friedrich University Bamberg, Germany

ABSTRACT

As one of the top issues for executives nowadays the recruiting, retaining, and skilling of scarce talent has received a lot of attention both in research and practice. While companies are in dire need of new strategies and integrated approaches in human resources, research has predominately observed the skill profile and general attitude of professionals to their work, their individual incentives, and consequentially, their turnover intention. We aim to relieve these needs in practice by introducing a first step towards a holistic system architecture of recruiting, retention, and development processes. For this purpose we use a design science approach to develop a talent management system architecture containing all respective processes and their related subsystems based on the next-generation holistic e-recruiting system invented by Lee (2007). This architecture could increase knowledge transfer and thereby improve the adjustment and performance of the entire HR process. Furthermore the increased amount of data gained through new implemented performance measurement subsystems offers management a variety of new strategic options for continuing to battle in a global "War for Talent."

DOI: 10.4018/978-1-60960-607-7.ch015

INTRODUCTION

Threat pinpointed, threat averted – end of story? Not really. Already since the late seventies the potential threat of a *talent shortage* for specific profiles as IT was recognized by corporations as well as research (LaBelle et al., 1980). Today, more than three decades later, the threat not only still exists but gets even worse. In terms of IT talent in the early stages Igbaria and Siegel (1992) advised organizations to keep an eye on the growing shortages of IT personnel and the increasing demand on the labor market. But they also identified that there might be multiple forces within these organizations which delay this process and distract from its importance (Igbaria & Siegel, 1992). A study of the career anchors of IT personnel predicted that due to labor shortages staff turnover will become an even more critical issue in the near future (Crepeau et al., 1992). Despite being recognized by several researchers and a few corporations, public institutions in the States like the General Accounting Office (GAO) still raised questions about the validity of the shortage estimates and the methodologies applied in the market surveys that were conducted (Garner & Weldon 1998; Ferratt et al., 1999). Further organizations, labor market analysts and external experts showed similar concerns about the various estimates of the talent shortages as well, but after a while they started to identify the problem and deduce that a talent shortage really exists (Cappelli, 2000b; Brock, 2003; Sadin, 2003; Frank et al., 2004; Farrell & Grant, 2005; Acharya & Mahanty, 2008).

Given this situation, top management is compelled to develop new effective strategies to remain within this "War for Talent" (Chambers et al., 1998; Keim & Weitzel, 2006). The talent shortage demands that the companies compete effectively against competitors on the job market (Patel, 2002). One way is to use information technology to gain a competitive advantage regarding the recruiting of employees. A study with Germany's Top-1000 companies indicates that there is an increasing usage of IT to manage job postings (mainly on the Internet) and applications (mainly received by Internet channels such as e-mail or websites). At a level of 85.9 per cent, the majority of vacancies in 2006 were communicated to potential candidates by means of the company's website and 59.0 per cent were published on internet job boards such as monster.com. At a level of 26.6 per cent job postings, printed newspapers or magazines are far behind. Therefore it is not surprising that 64.1 per cent of the hires generated are due to job postings on the internet. Furthermore the increasing usage of IT can be shown in the use of application methods as well. At a level of 51.3 per cent, more than half of the incoming applications are received by electronic channels by the Top-1000 companies in Germany (Eckhardt et al., 2007b). Concerning the value of this increasing IT use, *e-recruiting* is known to lead to significant reductions of time and cost per hire as well as improvements concerning applicant data and applicants' overall quality (Eckhardt et al., 2009). These findings enable the thinking that these companies using e-recruiting perform better in the search for talent.

However, as several researchers have stated, managers who do not pay attention to all issues regarding scarce talent will face serious challenges in engaging and maintaining their workforce (Agarwal & Ferratt, 2002). The open contest for competitors' employees is no longer a rarity but is an accepted business method. Cappelli (2000a) described the central point in the following way: 'To poach is fine, to be poached is not" (Cappelli, 2000a). Not all recruiting is the same; *retention* and *personnel development* should be shifted from broad programs to highly targeted efforts aimed at individuals and groups of key employees (Cappelli, 2000a). However, it is not always the best way for organizations to recruit new staff and offer them important positions, because they do not know the peculiarities of the company and its culture. Organizations could also use other methods to find skilled employees. If they regard the *internal labor market*, they have a many people with a high knowledge of the organization. These people could be trained and developed, so that they can carry out both present and future tasks (Hentze & Kammel 2001).

Based on the existence of these three critical issues for organizations the research question of our research paper is as follows: *How can recruiting, retention and, development processes be integrated in holistic talent management system architecture?*

To answer our research question we took a *design science approach* (Hevner et al., 2004). The approach is structured as follows. After the introduction we present the status quo of our method design science in research. In the proximate section 'Research Motivation' we discuss the problem relevance of our topic, especially the issue of a global talent shortage in the engineering and Information Systems field. The underlying literature review about the basic recruiting process, e-recruiting as an e-service as well as retention and development practices and strategies is shown in the section 'Theoretical Background'. The design of our architecture is provided after the literature review. In the further course the summarizing section 'Key Findings' offers implications for theory and practice by providing a number of suggestions for further research and about the way responsible managers could benefit by our proposed architecture. Finally we complete our design science approach with a short conclusion of our research contributions. But first of all we introduce design science as a research method in the following section.

DESIGN SCIENCE

Design science is part of the theory of design and action. It is a type of theory which tells the researcher 'how to do something' (Gregor, 2006). The theory is about the justification of theoretical knowledge and the related form, function and methods (Gregor & Jones, 2004; Walls et al., 1992). There are many different types and parts of design theory. Corresponding research can be found as: software engineering in a constructive form of research (Iivari, 1991; Iivari et al, 1998), as prototyping (Baskerville & Wood-Harper, 1998), as systems development (Burstein & Gregor, 1999) and as design science (Hevner et al., 2004; March & Smith, 1995; Peffers et al., 2008). Design Science was introduced into IS research by the work of March and Smith (1995). The core of their work was a framework designed to demonstrate the relationship, activities and outputs of design and natural science research. In total they named four elements of design science research, on one hand the developed constructs and models and on the other hand the methods used and the implementations carried out (March and Smith, 1995). Our design approach is based on the seven guidelines of Hevner et al. (2004) with the exception that we left out the design evaluation of our architecture for further research. We start with our motivation and the need for more holistic research approaches in the fields of e-recruiting, retention and personnel development.

RESEARCH MOTIVATION

Talent shortage is not exactly a new problem for the global economy, but the process of recognizing the problem took a while. But after this cognition process was successfully completed, researchers first tried to define the threat and examine its source. Interestingly, the first approach to analyzing the whole situation of labor market shortage and increasing the importance of recruiting and retaining talent came from practice. The McKinsey consultant Elizabeth Chambers and her colleagues briefly described the state of affairs within their corporate quarterly publication and declared that there was for all corporations a "War for Talent" (Chambers et al., 1998). Management practices were identified as a major source of the blatant disparity in the sense that driven turnover from the industry results in misplaced recruiting which makes it difficult to accommodate differences in contributions (Cappelli, 2000b). Within the "War for Technical Talent" project Cappelli (2000b) denied a real shortage of talents in IT and engineering but he identified a temporarily imbalance on the entire labor market due to rapidly rising wages and a shortage in the supply of talent. Increasing globalization and industrialization were mentioned by Stokes (2000) as particular reasons for regarding the shortage not just as a national problem but more and more as a worldwide shortage of talent. In 2003 the U.S. Bureau of Labor Statistics prognosticated a labor shortage of 10 million workers in 2008 (U.S. Department of Labor, Bureau of Labor Statistics, 2003). This situation was seen as a reflection of the boom period of the New Economy in the late 90s (Brock, 2003). Demographic reasons like the decelerated growth of the labor force due to the retirement of the big baby-boom generation were identified as an important reason for the massive talent shortage in the States (Frank et al., 2004). The percentage of children under the age of 18 years in the U.S. has dropped massively in recent years while the section of the population aged 65 or older is predicted to increase from 12 percent in 2000 to about 20 percent in 2030 (U.S. Census, 2000; Frank et al., 2004; Trauth et al., 2008). Frank et al. (2004) summarized the fact in the following way: '*In summary, the demographic time bomb fueled by aging baby boomers is not a guess it is an actuarial fact.*'

But, as forecast, the talent shortage did not remain a national problem: Its impact is felt globally (Stokes, 2000; Hewitt, 2002). In Europe the talent shortage is also occurring in important markets. For example in France by 2006 more people will leave the working population and retire than young people join the working population and start working. The situation in the neighboring country Germany is even worse. It is expected that in about 20 years 7 out of 10 Germans will be pensioners (Sadin, 2003). Surprisingly even in Asia's boom countries China and India the talent shortage issue has become a serious issue for politicians, managers and researchers. China's '*looming talent shortage*' (Farrell & Grant, 2005) could be traced back to two different causes. Firstly, the birth rate dropped below the rate necessary to maintain the same population size (Kahn, 2004), and as a result the Chinese workforce will begin to scale down in just 10 years (Jackson & Howe, 2004; Frank et al., 2004). Secondly the country suffers from a '*skilled-labor gap*' (Farrell & Grant, 2005). The number of highly-qualified university graduates is far too small to supply the requirements of multinational companies. The need of elementary reforms is obvious against the background of a national young talent pool of engineers which is no larger than the United Kingdom's. Talent shortage threatens China's rapid economic growth as well as its way up the value chain (Farrell & Grant, 2005). The other boom country India faces a different situation. They do not have a demographic problem like declining birth rates or a shrinking workforce. Nevertheless, high turnover rates occur as individual workers sell their skills for incrementally higher payment and for that reason talent shortage is becoming a topic taken more and more into account in India (Frank et al., 2004; Acharya & Mahanty, 2008).

In the following section we will present one possible solution to the talent shortage problem by discussing an IT architecture which aligns recruiting and retention management processes. Therefore we will follow the design science approach by Hevner et al. (2004) and will present the theoretical background of our research first.

THEORETICAL BACKGROUND

To provide the research rigor of our design science approach we based our research artifacts on a comprehensive review of literature about the corporate recruiting and retention processes.

The results of this literature review are provided in the following subsections.

Recruiting Process

As it is one of the critical business processes, the design of the recruiting process is discussed by various researchers in papers which contain valuable approaches to defining the structure and standards of classic staff recruitment. The recruiting approaches differ a lot when comparing large multinationals and small and medium-sized companies as two extremes. In smaller enterprises recruitment is more related to the availability of a known individual than to the process itself. Atkinson and Meager (1994) found evidence of a correlation between business size and the adoption of formal recruiting procedures and confirmed this hypothesis. Within their recruiting process large-scale enterprises also focus on process quality determinants, such as time and costs per hire or the overall quality of the candidates hired (Eckhardt et al., 2007a). Furthermore they are interested to discover if it is useful to use information systems in applicant tracking. Strohmeier (2007) found that applicant management systems turn out to be a good investment if the company receives a mid high number of applications.

Different researchers provide evidence for a systematic procedure for starting a recruiting process based on 4 common stages; an assessment if vacancies need to be filled, a definition and broad analysis of the job profile, the production of a job description and a person specification (Carroll et al., 1999). The overall process was divided into three phases; generating applicants, maintaining applicant status and influencing job choice decisions (Barber, 1998). Several activities are identified which have to be fulfilled for these phases. For the generation of applications companies can publish the adverts for vacancies using paper based (such as newspaper) or digital (website, job board) media. Certain recruiting activities supported by information technology may increase the possibility of an applicant's accepting a job offer (Barber, 1998)

Breaugh and Starke (2000) provide a phase-model to portray the recruitment process as a combination of activities, variables and strategic measures to achieve a number of recruitment objectives whose outcomes are compared afterwards. The phase model of the organizational recruitment process outlines the importance to view the entire process from a more holistic HR perspective integrating retention and development aspects as well. So recruitment objectives also include besides the cost and time of filling jobs and applicants' total number aspects as the corporate retention rate or concerning development as an individual's job performance or job satisfaction. After a comprehensive process with several recruitment activities all primary objectives are compared with the actual outcomes (Breaugh & Starke, 2000).

In addition Faerber et al. (2003) suggest a recruiting process integrating different recruitment functions. In their model Faerber et al. (2003) demonstrate the relationship of each recruiting task, its activities and objectives. Starting with the first contact between candidate and potential employer till the signing of the contract Faerber et al. provide a different structure. Based on prior work (Albert, 1998; Schneider, 1995) the process design contained the recruiter's five main tasks; short- and long-term candidate, applicant management, pre-selection as well as the final selection of candidates. Thus the management and tracking of applications in organizations is a main focus as well as the long-term attraction of applicants to build up an employer brand.

Beside these examples of research dealing with the design of the recruiting process and its related tasks little is known about the supporting IT function. An initial valuable approach to structure an e-recruiting process was made by Lee (2005a). He suggests a business process diagram illustrating an intra-organizational information flow and internal process events to simulate the flow of a recruiting process. The main activities of the recruiting process are the submission of job postings and the management of applications. Alike the process example by Breaugh and Starke (2000) aspects and events outside recruitment as employee development do also affect the recruiting process. These figures illustrate that the use of IT to support recruiting is increasing. Within the process diagram hiring needs initiate the process. In a next step the job requisition will be prepared and send for approval via the system. When the job requisition is approved it will be posted online (e.g. in job boards) or offline (e.g. in newspapers). The job postings remain active till a target number of applications are reached. In the next process step the applications will be evaluated afterwards selected candidates will be invited for a job interview. After a further selection the candidates are pre-screened before employment to gain information regarding their further performance. Following this pre-screening contracts are offered to the successful final candidates.

Based on his prior work Lee (2007) suggested an architecture for a *holistic e-recruiting system* (see Figure 2 for our adapted version). This system was intended to align all the activities and IT tools that support the recruiting process of a company. Lee (2007) provided an idealistic architecture for an e-recruiting system supporting all relevant parts of the recruiting process with IT. For Lee (2007) the central systems of the architecture are systems to manage job requisitions, application tracking, prescreening, job agents, candidate relationship as well as a system to monitor the performance of the recruitment process. All these systems are linked with a workflow management system as well as a database management system to provide central workflow management of the recruiting process and central storage of all relevant data. For example the applicant tracking management subsystem is in charge for collecting job applications, recording candidate resumes or e-portfolios, managing the status of each candidate, providing information for decision makers and for distributing information to other HR systems via interfaces. The job requisition subsystem is to structure and to streamline the management, development and publishing of job requisitions and online postings. The system enables HR managers to monitor current job postings and to close job postings as they are filled. The purpose of the prescreening management system is to assess the matching degree between applicants' qualifications and job requirements. Prescreen supports recruiters as it sorts a pool of resumes in order to rank them and to provide a list of top candidates. The communication with candidates is managed by the candidate relationship management system to establish long-term relationships with potential candidates. Furthermore *"it is the purpose of the system to provide applicants with a feeling they have an ongoing relationship with the company through a virtual human touch"* as pointed out by Lee (2007). This architecture is a holistic approach to summarize all possibilities how IT can support the recruitment process and how this support can be designed. As the figures of the German study illustrate (see introduction) there is an increasing usage of IT in recruiting, and companies have to structure their processes and IT architectures. Furthermore companies try to use innovative IT solutions to gain the competitive edge in the "war for talent".

Based on this previous work it is our aim to align the existing architecture of an e-recruiting process (Lee 2007) with retention process and suggest a way to combine different recruiting and retention management systems.

Personnel Recruiting as E-Service

The currently returning talent shortage in Europe (Thompson, 2007) demands not only new ways to recruit employees and more IT support, but also a new view of staff recruitment. The whole recruiting process needs to be more applicant-focused and task-orientated. Companies need to focus on corporate image and employer branding to appear in applicants' minds in a much more transparent way (Jaeger, 2007; Eckhardt et al., 2007a). Therefore employers need to restructure not only their staff recruitment but also the whole human resources department. Attracting, recruiting, retaining and developing employees are no longer separate tasks for separate departments with different managers. They all need to be combined in one focused process also known as "talent management" (Frosch & Trost, 2008). A major demand for the holistic talent management approach is an underlying "talent management architecture" (Eckhardt et al., 2008).

Over the last years more and more of the vacant positions were published online and in addition the number of digital applications via email or a web-based form rose (Malinowski et al., 2005). The internet also brought a new phenomenon to the labor market, "the passive applicant". In this case the candidate does not contact the potential employer with an application: Instead of that the candidate expects the employer to contact him via a recruiting platform (Keim et al., 2005). Platforms for contacting these passive jobseekers are provided by business networking communities such as LinkedIn or internet job portals like Monster. Other newer opportunities to find high-qualified employees are virtual worlds like Second Life. The first approach was made here by IBM who opened a virtual recruitment center in Second Life to contact experienced software developers (Laumer et al., 2008). A further new possibility to attract potential candidates is offered by the combination of e-assessment tools (Buzzetto-More & Alade, 2006) and online gaming (Laumer et al., 2009). Within this self-assessment (George & Smith, 1990) the candidate could discover the potential employer from the point of view of a "virtual employee" (Caligiuri & Phillips, 2003). With the help of this process the companies guide the candidates through their company, increase their transparency and offer the candidates the service how they might fit in the corporate environment (Phillips, 1998). Innovative ways to recruit and attract with the help of diverse e-services are needed as the problem of talent shortage spirals downward.

For example, for the year 2011 Germany's Top-1,000 companies expect a talent shortage for different IT and engineering professionals in R&D (von Stetten et al., 2011), as illustrated by Figure 1. The lowest supply is expected for the group of engineers in the R&D field. Solely 19.0 per cent of the companies questioned expect enough talent for this group on external labor markets. For the group of IT professionals only 20.4% of the companies estimate a sufficient availability. For job profiles in the companies' production 31.0 per cent of the study participants expect enough new workers. Four of ten assume the same availability for sales talent and 43.3 per cent for talent in the controlling and finance field. The highest availability rates are expected for the fields of HR and marketing. 58.8 per cent and 62.1 per cent expect here a sufficient availability of new talent on external labor markets (von Stetten et al., 2011).

As discussed in the introduction companies in Germany are using e-recruiting as a preferred method to search for new employees and to address the talent shortage. For more than two third of the German Top-1,000 companies these e-recruiting systems provide improvements. For example, 89.8 per cent indicates that e-recruiting systems are useful for the company, 87.6 per cent pointed out that the use of e-recruiting systems increase the effectiveness of the recruiting process, 78.4 per cent declared hat e-recruiting simplifies and 77.6 per cent stated that e-recruiting speeds up staff recruitment.

Based on these reported benefits of e-recruiting for staff recruitment the support of retention and personnel development with e-services might have benefits as well for companies. Therefore the aim of our paper is to provide an architecture integrating recruiting, retention and development processes based on e-services. The general issues of retention and development are discussed in the next sub-sections.

Figure 1. Expected Labor Supply for 2011 (Author's diagram adapted by von Stetten et al. 2011)

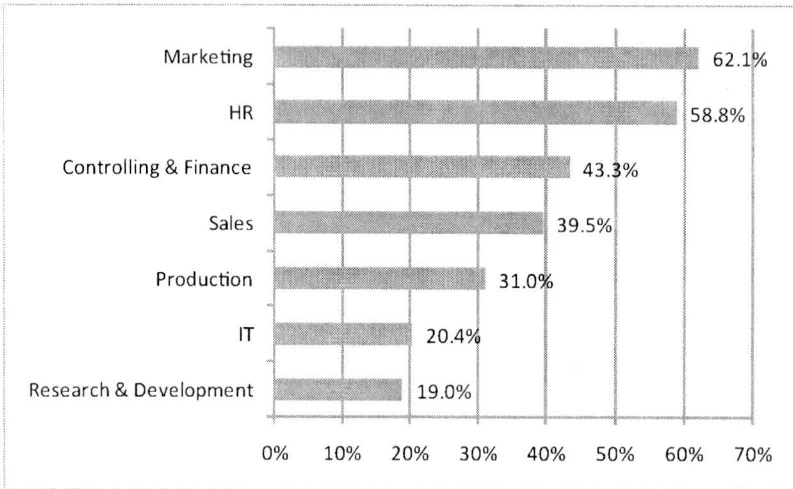

Personnel Retention

The necessary corollary to the recruiting of new employees is the retention of the existing workforce. Cougar (1984) was one of the first to analyze employees' motivation and wants. It turned out that individual measures and compensations had to be developed and implemented to retain the existing workforce. But Cappelli (2000a) also mentioned that these '*Golden Handcuffs*' are no longer a deterrent: Recruiters are forced to unlock these retaining handcuffs by offering a large signing bonus.

Several research approaches identified environmental and task related factors which influence an individual's turnover intention. For example Igbaria and Siegel (1992) developed a number of recommendations for executives to reduce the turnover intentions of IT employees significantly. The reduction of stress factors like role ambiguity and role conflict was mentioned by Igbaria and Siegel (1992) as a way to make employees stay with the corporation. They also encouraged companies to organize boundary spanning activities and continuous face-to-face meetings with other internal and external firm units. The individual talent should be assigned to more challenging and interesting projects with a high degree of autonomy and creativity. Furthermore, Igbaria and Siegel (1992) advised employers to show their high potentials individual recognition for their achievements and provide dual career paths which offer the opportunity for advancement within their respective field.

Other investigations found that individual career motives could also influence employees' retention. For that reason Agarwal and Ferratt (2000) provided two implications for further research. They took an extreme stand. On one hand they expected classic human resources strategies and practices not to influence an individual employee's decision to stay or leave a company. But they suggested that managers should understand the career motives of their staff as a first step and then build their retention measures around those motives in a second step. The career motives could be structured into three different subcategories; the preferred length of employment, the current career stage and career anchors in general. One strategy would be to recruit only those whose career motives fit in with the needs of the company (Agarwal & Ferratt, 2000). This strategy plays a particularly important role in the development of our integrated IT architecture in section 3. Agarwal and Ferratt (2000) examined the implementation of this strategy within their studies when one corporation hired only IT talents with an anchor of career development and security.

In a further study, Agarwal and Ferratt (2002) examined existing corporate practices for retaining IT professionals and summarized them in 11 different categories. To measure an IT professional's performance the companies often use an annual performance appraisal or 360 degree performance feedbacks. The factor of payment is implemented within compensation and benefits systems in the form of variable compensation or market anchor compensation (Agarwal & Ferratt, 2002). Another favored measure is the redesign of the individual's work arrangements by putting him in a job rotation process or letting him work in a team environment. Employability training in- or off-house could help to develop a talent's skills and his intention to stay respectively. Long-term career planning improves talent's business and leadership skills and lead to consideration of how they could achieve this career within the company (Agarwal & Ferratt, 2002). Opportunities for advancement or recognition in the form of non-monetary rewards (Laumer, 2009), monetary bonuses or extended vacations are also quite often used to retain a scarce talent as an IT professional within a corporation. Leadership training is often offered for both IT executives and regular employees. This could also be helpful to provide the IT talents with a feeling of belonging and being connected within a larger community. Female IT professionals in particular enjoy the opportunity of flexible work and time arrangements as well as child care services or telecommuting (Agarwal & Ferratt, 2002). Within their survey of key issues for IT executives Luftman et al. (2006) named the top six vehicles for retaining IT staff. They regarded the retention of IT professionals as an issue of top priority in order to avoid IT talent being poached by competitors (Cappelli, 2000). *'Organizations that do not have retention as a high priority will have a tough time'* (Luftman et al., 2006). The SIM members questioned said that open and honest communication is the key to retaining IT staff. An improved communication structure between IT professionals and their supervisors helps to create a 'family-like' working environment (Warner & Zhu, 2000). It was also revealed that a useful measure to retain IT professionals would be to provide them with challenging work experience (Luftman et al, 2006). Trust among colleagues and coworkers were mentioned as an important issue for retention by exactly half of the respondents. Surprisingly, opportunities for advancement identified by Agarwal and Ferratt (2002) as an important retention method four years earlier were only mentioned by a third of all participants. A good balance between work and outside life should be fundamental for retention as well, but came in as less important in this survey (Luftman et al., 2006).

Personnel Development

Nonetheless, recruiting external talent is essential for the further development of the company, turning the viewpoint to the internal resources might also represent an interesting opportunity, as the required personnel can also be acquired internally by selecting members of the company's current staff and preparing them suitably for their new responsibilities (Donahue, 2001). To do this, it is often necessary to introduce training measures which will enable staff to develop their capabilities (Cappelli, 2008; Donahue, 2001). The concept of personnel development is, however, characterized by "great heterogeneity and lack of clarity" (Becker, 2005 p. 2). As a result, we find a large number of definitions of personnel development in the literature. To provide a consistent understanding of the concept of personnel development in the sections that follow, we base ourselves on the definition of (Hentze & Kammel, 2001), who regard personnel development as a function of personnel management, whose aim it must be to provide all members of staff with training and qualifications enabling them to carry out both present and future tasks. In this process, those individual capabilities of the personnel which serve the achievement of the company's goals should be encouraged. This positive alteration in the qualifications of the staff is understood as a process consisting of a series of several linked activities (Hentze & Kammel, 2001).

Since a company may have several different reasons for developing the capabilities of its staff, we have chosen from the literature a personnel development process for our architecture which supports this aim. Such a process is defined by Scholz (2000). In this process the gaps in the capabilities of each employee are deduced from the difference between their 'capability profile' and the profile of requirements defined by the company. This reveals how suited an employee currently is for a given job. Building on this, those employees who are eligible to take part in development activities but have not yet been able to acquire competences that are available are given a deadline for completing the relevant training. Frequently, an employee's entire potential cannot be fully exploited because the company only has limited resources available. After successful completion of a training program the 'capability profile' is updated and checks are carried out to see whether the employee now satisfies the conditions of the required profile and/or whether the training program actually provides the desired competences (Scholz, 2000).

As shown in the last two subsections the retention and development of scarce talent is a significant challenge for management executives. While it is important to understand individual requirements, an integrated strategy could be the key to remain successful within a global talent shortage. In the following subsection we present our aligned architecture based on the work of Lee (2007).

THE TALENT MANAGEMENT ARCHITECTURE

"*Retention is king*" (Frank et al., 2004) and represents together with recruiting and developing the top issues for managing scarce talent as IT professionals (Luftman & Kempaiah, 2008). Several researchers in information systems, human resources and management are discussing different approaches to each of these processes and tasks. Based on this previous work we will in this subsection align the recruiting process with retention and developing activities and provide an IT architecture to align retention, developing and recruiting supported by an information system. For this purpose we will extend the holistic e-recruiting architecture discussed by Lee (2007) by means of important processes and systems which support retention as well as developing and highlight the interface between these three essential HR tasks.

For Lee (2007) the central subsystem of e-recruitment is a *work-flow management subsystem* connected with a database management subsystem managing the communication with the enterprise-wide database. In this database all information and data related to the recruiting process is stored. Therefore the retention and developing parts of an aligned architecture will use the workflow management system, database management system and the enterprise-wide database as well to manage the information flow and storage. For retention we added two important processes and developing employees needs one additional process.

First, the development of retention activities based on key employees, an analysis of their incentives must be conducted. As Brannick (2001) highlights, one of seven strategies for retaining top talent is to "use a great employee profile to drive retention" (p.30). Based on face-to-face interviews, discussion and internet based questionnaires; companies have to collect information on what motivates a top employee to work for, to join and to stay in the company. Brannick (2001) provides a list of questions that must be asked and discussed. Based on these findings companies should develop possible retention actions and provide a collection of incentives and activities that can be used to retain top talent. All findings must be stored in the enterprise-wide database.

Second, all retention approaches must be approved by top management. Therefore if there is a need for a retention activity an approval process must be started. Based on the findings made when developing retention activities, one must be selected or a new one must be developed based on the first process described. After the identification of retention needs and the suggestion of a retention action, the action must be approved and started. Finally its performance must be measured.

Third, the aim of the recruitment process is to hire the best candidate for a specific vacancy and to integrate them into the company's culture. Based on the findings of what the employee value proposition is (Brannick, 2001) and what the characteristics of retaining employees are, a subprocess must be designed to identify borderline turnover candidates while selecting applications for a vacancy.

Finally, to develop existing personal each position should be described by necessary skills, knowledge and competencies to be able to handle the tasks conscientiously. Moreover, each employee should also specify his strengths and weaknesses and store them in the database. If the organization identifies that one skill is missing particularly, they will be able ask their employees to take an active part in advanced educations to learn this special competency. Such a skill-oriented advanced education program could be particularly helpful if these skills are also missing on the external labor market.

Three explicit artifacts have been designed and developed in our research. Two of them result from retaining processes and one due to the developing one. The first research artifact produced is the developed system architecture for the corporate retention processes based on the research of Agarwal and Ferratt (2000). Our second artifact is the *alignment* of this retention architecture with the holistic e-recruiting system invented by Lee (2007). The third artifact based on prior research on personnel development, enables organizations to train their employees and identifies missing skills. Figure 2 illustrates the *process diagrams* for the processes discussed in this subsection and how subsystems can be used to support them. In the next section the design of the subsystems for supporting the retention process will be discussed.

Systems

The development of retention activities and the approval of a specific action are the two major retention processes. Both are managed using the "retain action management subsystem" and their performance is analyzed with the retention performance analysis subsystem. Both together are responsible for supporting retention processes with IT and manage the linkage to the recruiting process and subsystems. Thus the focus of the system is to streamline the developing of retention activities and to identify key employees as well as the need for retention for specific job profiles or tasks. Users are retention managers, top management and recruiting managers.

As Brannick (2001) highlights, it is important to "count what counts" (p. 30). Brannick (2001) explained that for improving a company's employee retention it is important to understand what the company is compared with what the company wants to be in terms of employee retention. Therefore turnover statistics must be provided by the human resources analysis system and the retention performance measurement system to measure on the one hand the actual retention or turnover rates and on the other how an implemented retention action is influencing these rates. Therefore it is important to break down turnover rates down by

Figure 2. An IT Architecture for aligning recruiting, developing retention processes (Author's diagram adapted by Lee 2007)

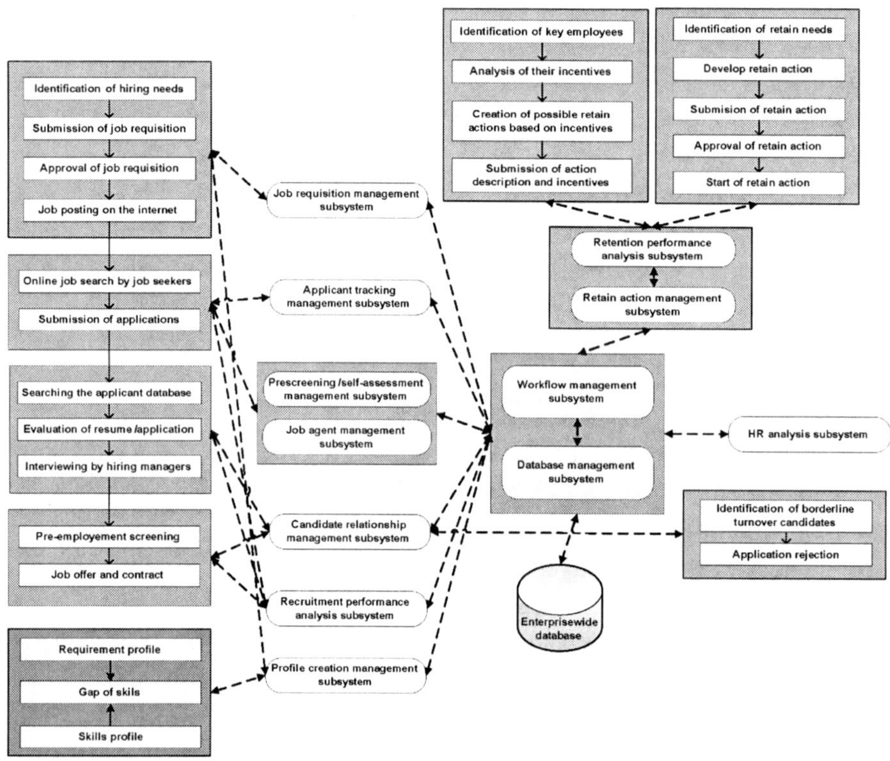

department, by job or by geographic region and to measure effectively. Brannick (2001) suggests, for example, measuring the "Major Employee Retention Rates" differentiated by voluntary and involuntary; overall, department and job; hourly and salaried; percentage of great employees who leave and come back. Based on the statistics, the need for retention activities can be monitored and key employees can be identified by the "retention management subsystem". The following paragraphs describe the subsystems in detail.

- ***Retention Action Management Subsystem***: The main objective of the RAMS is to streamline the development of retention activities. Based on the identification of retention needs or the initiation of the retention performance analysis subsystem, the RAMS is the basis for the development and approval of retention activities. It supports communication with all stakeholders and provides the data needed to develop effective activities. The descriptions of the activities developed are stored in the enterprise-wide database and can be accessed via the RAMS. The RAMS is responsible for every activity related to retention.
- ***Retention Performance Analysis Subsystem***: Together with the human resources analysis subsystem the RPAS is responsible for the measurement of the performance of implemented retention activities. The system can identify retention needs and therefore initiates the development of retention activities. Furthermore it can identify key employees of the company. Based on this analysis the system

can specify important target groups for retention activities and extract the key characteristics of key employees. These characteristics and the requirements of key employees can be the basis for the development of effective retention actions.

- ***Candidate Relationship Management Subsystem***: This system offers recruiters the possibility of establishing a long-term relationship with candidates as well as employees. Lee (2007) points out that the CRM-System "is designed to provide applicants [and employees] with a feeling they have an ongoing relationship with the company through a 'virtual human touch'". In Lee's architecture for an e-recruiting system, the CRM-system is designed to increase the chance of job acceptance by candidates. Lee explained the importance of a CRM-system by highlighting that "only companies with a positive long-term relationship with job applicants will gain a competitive edge". For an aligned architecture of recruiting and retention processes the CRM-system is responsible for communication with candidates and employees. However, its main objective is to support the selection of appropriate candidates. Using data analyzed to identify potential turnover candidates, the CRM-system can identify borderline turnover candidates and therefore advise the rejection of an application.
- ***Human resources analysis subsystem***: As Brannick (2001) highlights, it is important to "count what counts". The human resources analysis subsystem is responsible for monitoring the performance of the human resources processes. One aspect is analyzing hired employees and their turnover intention. One possible number is the turnover rate of employees in a company. The HR analysis subsystem is the basis for decisions regarding the rejection of applications, the identification of retention needs or the identification of key employees.
- ***Profile creation management subsystem***: In order to train the employees in their organization in an effective and efficient manner, an organization notes required skills for existing positions and matches these with existing skill profiles of their employees. Based on these analyses the subsystem gives an overview over essential skills which are not available to the necessary degree.

Interfaces

The retention management subsystem is the central system for managing retention activities. However, as outlined, retention cannot be discussed without recruiting. Thus, this section points out the interfaces between the recruiting management system and the retention management system.

First, as already discussed, there must be a sub-process for selecting borderline turnover candidates. If the company is looking for applicants for a job profile identified as critical for retention the "candidate relationship management subsystem" has to involve the additional process in the standard selection process of the recruiting function.

Second, if the company has implemented some retention activities which are major reasons for key employees to stay in the company, the company should use them to attract new employees while filling staff vacancies. Therefore the job requisition management subsystem has to search for implemented retention activities to include them in job ads etc. Thus the main arguments and motivations of key employees for working for the company, can be highlighted in the job ads and can attract candidates who were not sure whether to apply for the job or not.

Third, the recruiting performance analysis subsystem is aligned with the retention performance analysis subsystem to measure if the recruiting process selects candidates with low turnover rates or not, and how the staffing activities are influencing the retention rates of the company.

Fourth, if an employee is leaving the company, the candidate relationship management subsystem (in a long term focus) has to manage communication with the ex-employees in terms of initiating boomerang hires.

KEY FINDINGS

Our paper suggests an alignment of the architecture of a holistic e-recruiting system by Lee (2007) with a system for managing retention and personnel development. We aligned the model by including the concept of retention and modeled retention with two processes: developing retention activities based on incentives of key employees and approving retention actions. Therefore we connected the retention management subsystem and the retention performance analysis subsystem with the workflow management subsystem. In addition, for the need for statistics, and especially for turnover rates we introduced the HR analysis subsystem to measure key HR figures which can be used to identify key employees and the need for retention activities. Moreover, we introduced a deeper selection process to identify borderline turnover candidates and connected the process to the candidate relationship management subsystem. The profile creation management subsystem is responsible for identifying rare skills and to call attention to the need for training the staff to stop this development. The aligned architecture is a basis for developing an individual architecture to support recruiting, personnel development and retention with IT.

Limitations

Like any conceptual research our research is limited as well. We discussed a theory and literature based approach to align retention, developing and recruiting process in an integrated system architecture by reviewing Information Systems, HR and management literature. The concept we recommended must be validated by a case study or empirical research to highlight the benefits. Even so, as Lee (2007) discussed, the holistic e-recruiting system is expensive and complex and a lot of companies have to choose a part of the system to support their needs.

Implications for Research

Future research can benefit from this approach. The combination of different disciplines like human resources, recruiting, information systems, talent management, retention is needed to develop information systems as a discipline which highlights the impact of IS on business and on specific services in particular. Therefore a critical design evaluation of our architecture is needed to find out which subsystem is useful for different companies (like small and medium sized ones compared to large ones), and additionally if it could provide insights for the business value of IT.

Implications for Practice

Apart from our limitations, the suggested architecture could be a basis for the development of an information system supporting the recruiting, development and retention activities of an enterprise. This is indeed a major implication for practitioners. In the following paragraphs we discuss the managerial implications for every added component of the architecture.

- ***Retention Action Management Subsystem***: Although retention activities are used in HR management, they are often not systematically planned and are adopted in an undirected way. To streamline these activities the RAMS serves as kind of a steering tool for an appropriate retention program. The HR management benefits in two different ways from the RAMS, on the one hand by means of the tool for identifying key employees retention activities could be executed with a high degree of goal-orientation, and on the other hand by means of the alarm system for sudden retention needs the activities could be approved immediately. The identification of key employees and their individual retention needs constitutes an especially important factor for the recruiting process as well. With the help of the aligned recruiting and retention architecture the HR management is able to identify potential turnover candidates as early as the candidate selection phase in staff recruitment as they know which employee is easier to retain and which is not.

- ***Retention Performance Analysis Subsystem***: The RPAS is responsible for the performance measurement of the retention activities implemented. A more focused human resources management based on clearly defined quality criteria in terms of time, cost and process quality is the future for HR. Companies need to be competitive like some who have already managed to effectively reduce their time and costs per hire and to increase the overall quality of their employees as well their employee related data due to standardization (Muenstermann et al., 2010a; Muenstermann et al., 2010b) or long-term employer branding (Gatewood et al., 1993). The RPAS permits the HR management to analyze their retention incentives in terms of time, cost and quality and so to improve the overall process.
- ***Candidate Relationship Management Subsystem***: A shortage of highly qualified employees is not only an effect of demography and a shrinking workforce; it is also an effect of misleading recruiting. In boom periods companies tend to recruit nearly every acceptable candidate instead of analyzing their potential. In difficult economic times companies neglect the staff recruitment altogether. The key to this complex situation is a new form of countercyclical staff recruitment. For a successful form of countercyclical recruitment companies need to take advantage of candidate relationship management (Lee, 2005b). A CRMS allows the HR management to build up a long-term relationship with prospective candidates in difficult economic times in order to recruit them in boom periods (Lee, 2005b). Examples of this relationship between company and candidate are invitations to corporate events, lectures and seminars or birthday and Christmas cards. Empirical research has shown that candidates expect companies to offer a lot more in the recruiting and retention process than they have done so far (Weitzel et al., 2009).
- ***HR analysis subsystem***: As described in detail in the introduction, the future in HR is the combination of different tasks like attracting, staffing, retaining and developing "talents" in one overall function. In order to produce the success of the new "talent management" function (Frosch & Trost, 2008), all processes such as recruiting and retention need to be analyzed and compared in terms of efficiency and value adding to the overall HR. The HRAS supports the HR management by evaluating the quality criteria and producing management ratios for benchmarks with other internal processes or external HR processes in other companies.
- ***Profile creation management subsystem***: In difficult economic situations, if an organization has too many employees, they could train their personal to acquire rare skills and competencies, to be prepared for the economic boom. In boom-

ing stages as organizations plan to expand, the PCMS can help to prepare such steps early. Organizations which are willing to expand to China will benefit in such a situation from Chinese speaking highly skilled employees that are aware of the organizational culture of the parent group. Such a PCMS could be helpful to realize business strategies or to act foresighted. An example for the latter one could be the age distribution of an organization. If the system realizes that special skills are only allocated to elderly people, it will alert the HR department to train people to acquire these.

Our paper was motivated by the global rising shortage of talent and the problem of high turnover rates in the US, Europe and Asia. One can conclude that an aligned architecture for recruiting, retention and development is especially necessary for a competitive environment on the labor market. On the one hand the architecture suggested could protect a company from being poached (because the company knows why employees work for it and the architecture can support these incentives) and on the other hand it can support the poaching of employees for your company (because the company knows the reasons why it is a valuable employer).

CONCLUSION

"To poach is fine, to be poached is not" highlighted Peter Cappelli in 2000(a). Today retention and skilling is already a critical issue for managing scarce groups as IT staff (Luftman & Kempaiah, 2008) and the war for talent still exists. Therefore an integrated strategy of retention and development can prevent a company on the one hand from being poached and on the other hand can help them to develop poaching activities. We support a strategy like this by providing a system architecture based on a design science approach (Hevner et al., 2004).

REFERENCES

Acharya, P., & Mahanty, B. (2008). Manpower shortage crisis in Indian Information Technology industry. *International Journal of Technology Management*, *38*(3), 235–247. doi:10.1504/IJTM.2007.012712

Agarwal, R., & Ferratt, T. W. (2000). Retention and the career motives of IT professionals. *Proceedings of the 2000 ACM SIGCPR Conference on Computer Personnel Research*, (pp. 158-166). April 2000, Chicago, Illinois, United States.

Agarwal, R., & Ferratt, T. W. (2002). Enduring practices for managing IT professionals. *Communications of the ACM*, *45*(9). doi:10.1145/567498.567502

Albert, G. (1998). *Betriebliche Personalwirtschaft*. Kiehl: Ludwigshafen a.R.

Barber, A. E. (1998). *Recruiting employees*. Thousand Oaks, CA: Sage Publications.

Baskerville, R., & Wood-Harper, A. T. (1998). Diversity in Information Systems action research methods. *European Journal of Information Systems*, *7*(2), 90–107. doi:10.1057/palgrave.ejis.3000298

Becker, M. (2005). *Personalentwicklung. Bildung, Förderung und Organisationsentwicklung in Theorie und Praxis* (*Vol. 4*). Stuttgart, Germany: Schaeffer-Poeschel.

Brannick, J. (2001). Seven strategies for retaining top talent. *The Journal of Business Strategy*, *22*(4), 28–31. doi:10.1108/eb040183

Breaugh, J. A., & Starke, M. (2000). Research on employee recruitment: So many studies, so many remaining questions. *Journal of Management*, *26*(3), 405–434. doi:10.1177/014920630002600303

Brock, F. (2003, Oct. 12). Seniority: Who'll sit at the boomers' desks? *The New York Times*.

Burstein, F., & Gregor, S. (1999). The systems development or engineering approach to research in Information Systems: An action research perspective. In B. Hope & P. Yoong (Eds.), *Proceedings of the 10'' 'Australasian Conference on Information Systems*, Victoria University of Wellington. New Zealand, (pp. 122-134).

Buzzetto-More, N. A., & Alade, A. J. (2006). Best practices in e-assessment. *Journal of Information Technology Education*, *5*, 251–269.

Caligiuri, P. M., & Phillips, J. M. (2003). An application of self-assessment realistic job previews to expatriate assignments. *International Journal of Human Resource Management*, *14*(7), 1102–1116. doi:10.1080/0958519032000114228

Cappelli, P. (2000a). A market-driven approach to retaining talent. *Harvard Business Review*, *78*(1), 103–111.

Cappelli, P. (2000b). *Is there a shortage of Information Technology workers?* (pp. 1–25). McKinsey and Company Project Review.

Cappelli, P. (2008). Talent management for the twenty-first century. *Harvard Business Review*, *86*(3), 74–81.

Carroll, M., Marchington, M., Earnshaw, J., & Taylor, S. (1999). Recruitment in small firms: Processes, methods and problems. *Employee Relations*, *21*(3), 236–250. doi:10.1108/01425459910273080

Chambers, E. G., Foulon, F., Handfield-Jones, H., Hankin, S. M., & Michaels, E. G. III. (1998). The war for talent. *The McKinsey Quarterly*, 1.

Cougar, J. D. (1984). Research issues in Information Systems personnel management. In McFarlan, F. W. (Ed.), *Fifth Anniversary Research Colloquium: The Information Systems Research Challenge* (pp. 217–230). Boston, MA: Harvard Business School Press.

Crepeau, R. G., Crook, C. W., Goslar, M. D., & McMurtrey, M. E. (1992). Career anchors of Information Systems personnel. *Journal of Management Information Systems*, *9*(2), 145–160.

Donahue, K. B. (2001). Time to get serious about talent management. *Harvard Business Review*, *79*(7), 6–7.

Dychtwald, K., Erickson, T., & Morison, B. (2004). It's time to retire retirement. *Harvard Business Review*, *82*(3), 48–57.

Eckhardt, A., Laumer, S., & Weitzel, T. (2008). *Extending the architecture for a next-generation holistic e-recruiting system*. In International Conference on Information Resources Management (Conf-IRM), Niagara Falls, Ontario, Canada.

Eckhardt, A., von Stetten, A., & Laumer, S. (2009). Value contribution of IT in recruiting – A multinational causal analysis. In *Proceedings of the 2009 ACM SIGMIS CPR Conference* (SIGMIS 2009), Limerick.

Eckhardt, A., Weitzel, T., Koenig, W., & Buschbacher, J. (2007a). How to convince people who don't like IT to use IT - A case study on e-recruiting. In *Proceedings of the 13th Americas Conference on Information Systems* (AMCIS 2007), Keystone, Colorado.

Faerber, F., Keim, T., & Weitzel, T. (2003). An automated recommendation approach to personnel selection. In *Proceedings of the 2003 Americas Conference on Information Systems*, Tampa, 2003

Farrell, D., & Grant, A. J. (2005). China's looming talent shortage. *The McKinsey Quarterly*, 4.

Frank, F. D., Finnegan, R. P., & Taylor, C. R. (2004). The race for talent: Retaining and engaging workers in the 21st century. *Human Resource Planning*, *27*(3), 12–25.

Frosch, M., & Trost, A. (2008). Die Trends im Talentmanagement. *Personalmagazin*, *1*, 50–51.

Garner, R., & Weldon, D. (1998, January 26). The numbers game. *Computerworld*, *32*(4), 88–93.

Gatewood, R. D., Gowan, M. A., & Lautenshlager, G. J. (1993). Corporate image, recruitment image, and initial job choice decisions. *Academy of Management Journal*, *36*(2), 414–427. doi:10.2307/256530

George, D. I., & Smith, M. C. (1990). An empirical comparison of self-assessment and organizational assessment in personnel selection. *Public Personnel Management*, *19*(2), 175–190.

Gregor, S. (2006). The nature of theory in Information Systems. *Management Information Systems Quarterly*, *30*(3), 611–642.

Gregor, S., & Jones, D. (2004). The formulation of design theories. In Linger, H., Fisher, J., Wojtkowski, W., Zupaneie, J., Vigo, K., & Arold, J. (Eds.), *Constructing the infrastructure for the knowledge economy: Methods and tools, theory and practice* (pp. 83–93). Boston, MA: Kluwer Academic Publishers.

Hentze, J., & Kammel, A. (2001). *Personalwirtschaftslehre*, 7th ed. Stuttgart.

Hevner, A. R., March, S. T., Park, J., & Ram, S. (2004). Design science in Information Systems research. *Management Information Systems Quarterly*, *28*(1), 75–105.

Hewitt, P. (2002). *Retirement trends foster global stagnation*. CSIS, April 5.

Igbaria, M., & Siegel, S. R. (1992). The reasons for turnover of Information Systems personnel. *Information & Management*, *23*, 321–330. doi:10.1016/0378-7206(92)90014-7

Iivari, J. (1991). A paradigmatic analysis of contemporary schools of IS development. *European Journal of Information Systems*, *1*(4), 249–272. doi:10.1057/ejis.1991.47

Iivari, J., Hirschheim, R., & Klein, H. K. (1998). A paradigmatic analysis contrasting Information Systems development approaches and methodologies. *Information Systems Research*, *9*(2), 1–30. doi:10.1287/isre.9.2.164

Jackson, R., & Howe, N. (2004). *The graying of the middle kingdom*. Center for Strategic & International Studies and the Prudential Foundation. April.

Jaeger, W., Jaeger, M., & Frickenschmidt, S. (2007). Verlust der Informationshoheit. *Personal*, *2*, 8–11.

Kahn, J. (2004, May 30). The most populous nation paces a population crisis. *New York Times*.

Keim, T., Malinowski, J., & Weitzel, T. (2005). Bridging the assimilation gap: A user-centered approach to IT adoption in corporate HR processes. In *Proceedings of the 2005 Americas Conference on Information Systems*, Omaha.

Keim, T., & Weitzel, T. (2006). *Strategies for hiring IT professionals: An empirical analysis of employer and job seeker behavior on the IT labor market.* 12th Americas Conference on Information Systems, Acapulco, Mexico.

LaBelle, C. D., Shaw, K., & Hellenack, L. J. (1980). Solving the turnover problem. *Datamation*, April, 144-152.

Laumer, S. (2009). Non-monetary solutions for retaining the IT workforce. In *Proceedings of the 15th Americas Conference on Information Systems* (AMCIS 2009). San Francisco, USA.

Laumer, S., Eckhardt, A., & Weitzel, T. (2008). Recruiting IT professionals in a virtual world. In *Proceedings of the 12th Pacific Asia Conference on Information Systems* (PACIS 2008), Suzhou, China.

Laumer, S., von Stetten, A., & Eckhardt, A. (2009). Online gaming to apply for jobs –The impact of self-assessment and e-assessment on recruiting. In *Proceedings of the 41th Hawaii International Conference on System Sciences* (HICSS 2009), Big Island, USA.

Lee, I. (2005a). An integrated economic decision and simulation methodology for e-recruiting process redesign. *International Journal of Simulation and Process Modelling*, *1*(3/4), 179–188. doi:10.1504/IJSPM.2005.007648

Lee, I. (2005b). An analytical model of e-recruiting investment decision: An economic employment approach. *IEEE Transactions on Engineering Management*, *52*(4), 486. doi:10.1109/TEM.2005.857569

Lee, I. (2007). The architecture for a next-generation holistic e-recruiting system. *Communications of the ACM*, *50*(7), 81–85. doi:10.1145/1272516.1272518

Luftman, J., & Kempaiah, R. (2008). Key issues for IT executives 2007. *MIS Quarterly Executive*, *7*(2), 99–112.

Luftman, J., Kempaiah, R., & Nash, E. (2006). Key issues for IT executives 2005. *MIS Quarterly Executive*, *5*(2), 81–99.

Malinowski, J., Keim, T., & Weitzel, T. (2005). Analyzing the impact of IS support on recruitment processes: An e-recruitment phase model. In *Proceedings of The Ninth Pacific Asia Conference on Information Systems*, Bangkok, Thailand

March, S. T., & Smith, G. F. (1995). Design and natural science research on Information Technology. *Decision Support Systems*, *15*, 251–266. doi:10.1016/0167-9236(94)00041-2

Muenstermann, B., Eckhardt, A., & Weitzel, T. (2010a). The performance impact of business process standardization - An empirical evaluation of the recruitment process. *Business Process Management Journal*, *16*(1), 29–56. doi:10.1108/14637151011017930

Muenstermann, B., von Stetten, A., Laumer, S., & Eckhardt, A. (2010b). The performance impact of business process standardization – HR case study insights. *Management Research Review*, *33*(9), 924–939. doi:10.1108/01409171011070332

Patel, D. (2002). Managing talent. *HR Magazine*, March.

Peffers, K., Tuunanen, T., Rothenberger, M., & Chatterjee, S. (2008). A design science research methodology for Information Systems research. *Journal of Management Information Systems*, *24*(3), 45–77. doi:10.2753/MIS0742-1222240302

Phillips, J. M. (1998). Effects of realistic job previews on multiple organizational outcomes: A meta-analysis. *Academy of Management Journal*, *41*(6), 673–690. doi:10.2307/256964

Sadin, M. (2003, October 6). Computers that care. *Newsweek*, p. 6.

Schneider, B. (1995). *Personalbeschaffung*. Peter Lang Europäischer Verlag der Wissenschaften: Frankfurt a.M.

Scholz, C. (2000). *Personalmanagement: Informationsorientierte und verhaltenstheoretische Grundlagen*, 5th ed. München.

Stokes, S. L. (2000). Attracting and keeping IT talent. *Information Systems Management*, *17*, 8–16. doi:10.1201/1078/43192.17.3.20000601/31235.2

Strohmeier, S. (2007). Research in e-HRM: Review and implications. *Human Resource Management Review*, *17*, 19–37. doi:10.1016/j.hrmr.2006.11.002

Thomas, S., & Ray, K. (2000). Recruiting and the Web: High-tech hiring. *Business Horizons*, *43*(5/6), 43–52. doi:10.1016/S0007-6813(00)89200-9

Thompson, V. (2007). Talent shortage? How to win with what you've got. *Super Vision*, *68*(2), 15–17.

Trauth, E. M., Quesenberry, J. L., & Yeo, B. (2008). Environmental influences on gender in the IT workforce. *The Data Base for Advances in Information Systems*, *8*(39).

U.S. Department of Labor. (2004). *Bureau of Labor Statistics*.

Von Stetten, A., Laumer, S., Eckhardt, A., Weitzel, T., & von Westarp, F. (2011). (in press). *Recruiting trends 2011*. Eine empirische Untersuchung mit den Top-1.000-Unternehmen aus Deutschland sowie den [-Unternehmen aus den Branchen Finanzdienstleistung, IT und Oeffentlicher Dienst, Research Report, Bamberg and Frankfurt a. Main.]. *Top (Madrid)*, 300.

Walls, J. G., Widmeyer, G. R., & El Sawy, O. A. (1992). Building an Information System design theory for vigilant EIS. *Information Systems Research*, *3*(1), 36–59. doi:10.1287/isre.3.1.36

Ward, M., Gruppen, L., & Regehr, G. (2002). Measuring self-assessment: Current state of the art. *Advances in Health Sciences Education : Theory and Practice*, *7*, 63–80. doi:10.1023/A:1014585522084

Warner, M., & Zhu, Y. (2002). Human resource management with Chinese characteristics: A comparative study of the People's Republic of China and Taiwan. *Asia Pacific Business Review*, *9*(2), 21–43.

Weitzel, T., Eckhardt, A., & Laumer, S. (2009). A framework for recruiting IT talent: Lessons from Siemens. [MISQE]. *MIS Quarterly Executive*, *8*(4), 175–189.

Chapter 16
Aligning Six Sigma and ITIL to Improve IT Service Management

Peter C. Chan
Hewlett Packard, USA

Shauntell R. Durant
Hewlett Packard, USA

Verna Mae Gall
University of Maryland - University College, USA

Mahesh S. Raisinghani
TWU School of Management, USA

ABSTRACT

Today's executives are challenged to deliver value to their shareholders in a global marketplace. To compete, organizations are setting business strategies for the entire organization. This management change presents a significant change for Information Technology (IT) organizations that have historically dealt with individual business units or functional domains. Now IT must support the enterprise's business strategy and the entirety of the enterprise. Technology organizations must deliver interoperation of processes, people, and technology to the entire enterprise.

To design, deliver, and manage IT services to meet an agreed level of quality, organizations are implementing IT Service Management (ITSM) and creating quality standards, which use the best practices of IT Infrastructure Library (ITIL). ITIL tells IT management what needs to be done and how it will get done from the process perspective. When undertaking an ITSM a project to implement ITIL, ITIL does not provide a method for measuring quality or identifying and completing process improvement projects. By integrating Six Sigma, a quality methodology, IT management will have the methodology and tools for measuring quality and improving processes. Adopting Six Sigma principles also will help IT focus on their customers and the business strategy, manage proactively based on facts, and reinforce collaboration across the enterprise.

DOI: 10.4018/978-1-60960-607-7.ch016

The framework in our exploratory research has been built upon a deductive study which has been developed through a literature review and synthesis and an exploratory inductive research which has been developed using a qualitative case study. It makes the case for leveraging ITIL and Six Sigma with ITSM in practice and opportunities for future research.

OVERVIEW OF INFORMATION TECHNOLOGY SERVICE MANAGEMENT (ITSM)

The rise of services oriented architecture, client server computing, virtualization and distributed applications have created a plethora of moving targets in the information technology (IT) organization. IT organizations have had to deal with a business that have historically been siloed by function or department and have been separated from the business. However as organizations have adopted an enterprise approach running their businesses, IT managers can no longer run an IT organization as a technology-based organization. They must be able to migrate to being a value-based service provider and contributor to the enterprise strategy instead of an overhead cost. IT Service Management (ITSM) is a process-based practice intended to align the delivery of IT services with needs of the enterprise, emphasizing benefits to customers. ITSM involves a paradigm shift from managing IT as stacks of individual components to focusing on the delivery of end-to-end services using best practice process models (WhatIs.com, 2008). Information Technology Infrastructure Library (ITIL) is a globally recognized collection of best practices for information technology (IT) service management.

ITSM audits are based on analysis of four key performance indicators in specific ways:

- Growth and value, which involves tracking revenue growth against investment and utilization.
- Budget adherence, which involves optimizing the use of available funds and avoiding unnecessary expenditures.
- Risk impact, which involves identifying and evaluating the consequences of risks taken or avoided.
- Communication effectiveness.

IT managers need to adopt a service management approach consisting of well-defined IT processes and a continuous improvement program in order to meet their customer's expectations and contribute to the enterprise's goals.

ITSM Evolves IT to Achieve Higher Result with Business Focus

To meet the challenge, IT managers in USA are adopting ITSM, a practice that has been used by their European counterparts for over 24 years. According to Salle (2004), the evolution of IT organizations typically mature through three stages: technology provider, service provider and strategic partner.

Continual maturity occurs as the IT organization moves through each stage. In the initial stage the IT organization operates as a technology provider and it is using IT infrastructure management (ITIM). In this stage, the focus is to manage and provide a solid infrastructure to the enterprise by maximizing the return on technical assets and controlling the infrastructure including its hardware and data. As an organization moves to becoming a service provider it will use ITSM which leverages ITIL to identify the "services its customers need and focusing on planning and delivering

Figure 1. Evolution of the IT Function within organizations (Salle, 2004, p. 1)

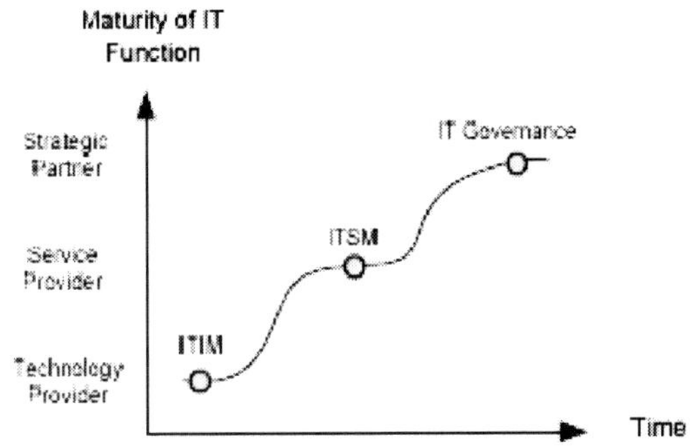

those services to meet availability, performance, and security requirements. In addition, IT is managing service-level agreements, both internally and externally, to meet agreed-upon quality and cost targets. Ultimately, when IT organizations evolve to IT business value management (IT Governance), they are transformed into true business partners enabling new business opportunities. In that stage, IT processes are fully integrated with the complete lifecycle of business processes improving service quality and business agility" (Salle, 2004, p. 1). Alternatively the Capability Maturity Model Integrative (CMMI) model from the Software Engineering Institute can also be used as the organization evolves in the maturity level of its processes.

ITIL

What is ITIL (ITIL Service Management Practices)?

The U.K. government's Central Computer and Telecommunication Agency created ITIL. It is a framework and does not require a license to practice and it is independent of any commercial solution or platform. In the last 24 years, ITIL has become a de-facto standard and most widely used accepted approach to define processes for IT service-oriented organizations. It aligns IT processes to overall business goals. ITIL is a guide for establishing common processes, roles and activities, with appropriate reference to each other and how communication lines should exist (ITIL, 2007). Although processes dictate how services are delivered, processes are of little interest to customers because they are not visible. ITIL is a public domain set of books that describes comprehensive and consistent best practice guidance in the area of organizing a coherent process for IT Service Management. In the late 1980's when it was originated it consisted of more than 30 books, but the release of ITIL V3 condensed the framework into 5 books.

The latest version is ITIL V3 and it is designed to address the significant shortcomings of ITIL V2. ITIL V3 has a broader scope and contains everything that was included in Version 2. Version 3 pays more attention to the areas of designing services suitable for the business and creating a strategy around this. Version 3 gives more advice on Continual Service Improvement (Damiano & McLaugin, 2005, p. 253). ITIL V3 addresses the life cycle of service management. It embraces a more holistic service management practice that included business and IT at strategic, tactical, and operational layer. While in comparison, ITIL V2 focused just on the operational layer of IT operations. The new version provides best practice guidance to implement a true life cycle of service

Table 1.

Book	Responsibility of Book
Service Strategy	Defines the policies and strategies to implement Service Management in line with the overall business strategy. It focuses on the planning aspect of the policies and strategies.
Service Design	Describes how to use the strategy to create design and specifications for the service. This book has more structure and explains a step-by-step approach to designing services. Much of what was in this book was already addressed by ITIL v2 with the exception of security management.
Service Transition	Covers configuration, change and release management. Change management focuses on how to assess and plan for changes. Configuration management has been extended to include service assets or IT assets, which are important for IT to be aware of. This book also Details how to get the specification into the live environment.
Service Transition	Defines how to best support the support the day- to-day running of the service throughout its life cycle. Provides guidance on running a live production environment.
Service Operation	Provides a framework for Continual Service Improvement. Service performance is measured at each stage ensuring that IT align and continually realign to the needs of the business. This book makes it clear that, for organizations to become more proactive, assessments must be a continual process, rather than one that only happens when a failure occurs.

management in five best practice guides. Version 3 of ITIL library consists of 5 five sets of books:

Benefits of ITIL for Service-Oriented Organizations

ITIL's framework of best practices can be used to assist organizations as their needs and technology evolve around processes. ITIL Service Management Practices offer benefits that demonstrate their value and ROI. Organizations can use ITIL to add more value to their business, be more agile with their responses, define standards, implement new technologies, adopt new trends, regulate compliance, and improve the quality of IT services. Additionally, companies gain higher productivity from both business and IT staff. These best practices for the support and delivery of IT services can help a company document IT processes as required for Sarbanes-Oxley (Worthen, 2005, n.p.). ITIL is part of the foundation of the COBIT model, which defines control objectives for IT in support of business processes. COBIT was explicitly chosen as the tool of choice for external auditors to use in IT audits for Sarbanes-Oxley. The ITIL process documentation and COBIT control objectives are a powerful combination that can accelerate Sarbox compliance. ITIL allows IT staff to become more innovative in their work. "The results of a recent survey by Axios Systems, show that over one-third of IT professionals have already adopted the ITIL framework, while another one-third plan to roll-out the framework within a year. In the same research, 70% of the Best-in-Class businesses had the ability to measure end-user satisfaction and 60% of the Best-in-Class businesses could compare service delivery standards against SLA provisions" (Brown, 2007, p.5).

What Does ITIL Not Do?

ITIL does not focus on a specific industry segment or restricted geographically. ITIL does not address *who* within the organization is going to be in charge of implementing each process. It does not address the *how* to use the tools that are needed to implement these processes. ITIL does not offer corporate or organization certificates but rather personal certification. Certification does not guarantee service quality and this is a common misconception within IT organizations. Many vendors and consultants offer services in restructuring ITIL, and many customers seek such guidance. However, ITIL does not promote an organizational

Figure 2. Hype Cycle for IT Operations Management, 2007 (Govekar et al., 2007)

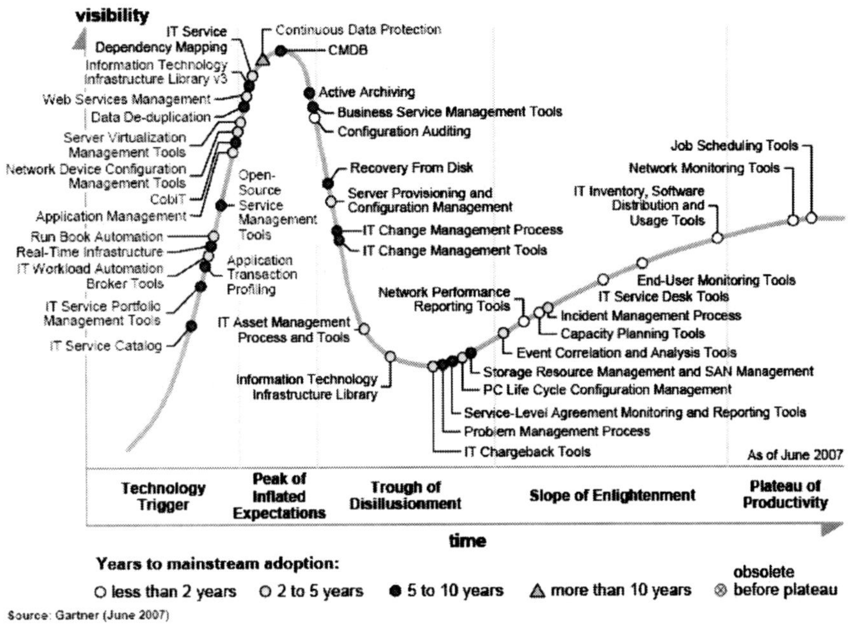

structure, or require any particular management. Nor does the ITIL mandate any particular workflow design or process. Here is a quote from ITIL: "For each of the processes described in this book, one or more roles are identified for carrying out the functions and activities required. It should be noted that organizations may allocate more than one role to an individual within the organization, or may allocate more than one individual role. The purpose of the role is to lace responsibility rather than to create an organizational structure" (ITIL Organization, 2006)

What You Need to Know Before Adopting ITIL?

ITIL's core practice guides are adaptable and applicable within a variety of organizational context. However, every organization is different and should have its own unique requirement. ITIL can be intimidating to understand and organizations should realize that you can not conform to, adopt, comply with or implement ITIL because it is not a *set* standard and you can not use what ITIL says verbatim.

IT is a service function providing effectiveness, value, and support to the enterprise and its strategy. Organizations should focus on service and business performance improvement first and not processes. An organization should put in place just enough process to answer the question for any given situation to avoid adding processes that don't add value or are not definitive.

Organizations have been adopting ITIL for approximately 25 years. As illustrated in figure 2, that is consistent with the diffusion theory organizations are moving towards becoming service providers as discussed by Salle in 2004.

SIX SIGMA

Six Sigma was introduced and created by the American electronics manufacturer Motorola, Inc.

In the mid-1980's CEO Bob Galvin established a goal to improve "all products – goods as well as services – by an order of magnitude (e.g. factor of 10) within 5 years" (Klefsjo & Wiklund & Edgeman, 2001, p. 31). During this time, Motorola's field sales force received and reported increased customer complaints about warranty claims. To address the issue, "Bill Smith, a senior engineer and scientist within the Communications Division" invented the new quality method, Six Sigma which standardized the counting defects; by reaching Six Sigma defects would be "near perfect," only 3.4 defects per million. Smith convinced Bob Galvin that this process would be a "key to addressing quality concerns" (McCarty, 2004, p. 1). "Six Sigma clearly focused resources at Motorola, including human effort, on reducing variation in all processes, that is to say manufacturing processes, administrative processes and all other processes. To set a clear measure on the improvement work, the program called Six Sigma was launched in 1988" (Klefsjo et al., 2001).

Antony, (2007) states, "There have been three recognizable generations of six sigma. The first generation of six sigma lasted for a period of 8 years (1987-1994) and the focus was on reduction of defects. Motorola was a great example of a successful first generation company. The second generation of six sigma spanned the period from 1994 to 2000 and the focus was on cost reduction" (p. 17). During this period the methodology was popularized by GE's, CEO Jack Welsh embracing the approach to quality. For example, "AlliedSignal attained savings of $US2 billion during a five-year period while General Electric saved $US1 billion over a two year window" (Klefsjo et al., 2001, Six Sigma origins section). The third generation, according to Antony (2007) started after 2000, and its focus "is on creating value to customers and the enterprise itself. The first companies to embrace third generation of six sigma are foreign. Examples of third generation of six sigma companies include Posco, Samsung, etc." (p. 17).

Six Sigma Themes

Although the beginning of Six Sigma is steeped in quantitative analysis, it is a misperception that Six Sigma is only a statistical measure with a goal of perfection. Pete Pande and Larry Holpp (2002) in the book "What is Six Sigma?" assert Six Sigma is more than a statistical and process methodology and they distribute Six Sigma into Six themes:

- Understand and focus on the customer.
- Be dependent on data- and fact-driven management.
- Master and improve your key processes.
- Manage proactively.
- Collaborate across boundaries within the organization.
- "Drive for perfection, but tolerate failure" (p. 14-16)

USING SIX SIGMA

Six Sigma Methodologies: DMAIC and DMADV

There are two key Six Sigma methodologies: DMAIC and DMADV. DMAIC is an acronym for Define, Measure, Analyze, Improve, and Control. DMADV is Define, Measure, Analyze, Design, and Verify (DMADV). According to Kerri Simon (2002) in the article, DMAIC versus DMADV, the DMAIC method is predominant in Six Sigma implementations and "should be used when a product or process is in existence at your company but is not meeting customer specification or is not performing adequately." Practitioners when employing DMAIC, seek to "root out and eliminate the causes of defects" (American Society for Quality). Pande and Holpp (2002) agree that organizations needing to fix broken processes can reap benefits of implementing DMAIC. By using DMAIC organizations will have proven and identified the extent of the problem. By assessing and measuring problems, an organization will be spending time and resources solving high value problems. When the problem is proven and selected for resolution, its root cause is identified using data and facts, not intuition and gut feel. As the DMAIC team works to create a solution, their purpose is to affect real change to the existing problematic process. The resulting solution must be well tested to manage risks to the organization. After the solution is implemented within the enterprise, the organization measures the results validates that the solution is sustaining change (2002, p. 30-31). In the table below, Chieh (2007), outlines in the chart below how DMAIC can be utilized to fix under performing processes....

When designing new processes, the approach used is Design for Sigma Six (DSFF). DSFF uses DMADV not DMAIC in order to design and move to market "new products or services measures by customer-based critical-to quality metrics" (Foster, 2007, p. 463). Simon expands the uses of DMADV beyond the creation of new products or services and advocates using DMADV approach after optimizing an "existing product or process" fails to meet "customer specification or Six Sigma level" (Simon, 2007, n.p.).

Six Sigma Key Roles

To carry out the work, Six Sigma identifies key roles and responsibilities. The Executives and Champions are responsible for selecting project that will be implemented by the organization. The associates responsible for implementing the projects and process improvements are Master Black Belts, Black Belts, Green Belts, Yellow Belts, and White Belts. It is important to note that successful Six Sigma implementations have a top-down management approach.

Tools of Six Sigma

The teams once formed will use a series of tools, which are employed with the Six Sigma methodology. The expansive toolset, which teams leverage to complete their work, is a primary benefit Six Sigma. Pande and Holpe (2002) recommend teams limit the use of tools to only those that help "get the job done" (p. 67) and categorize the tools as follows:

Criticism of Six Sigma

There is a recent debate of whether Six Sigma has a place in companies embracing the new innova-

Table 2. DMAIC in Mathematical Terms

Steps	Phase	Questions
1. Understand what process is to be improved and set a goal.	Define	> What is the Y or the outcome measure?
2. Measure the current state.	Measure	> What is Y's current performance?
3. Develop cause-and-effect theories of what may be causing the problem. 4. Search for the real causes of the problem and scientifically prove the cause-and-effect linkage.	Analyze	> What are the potential Xs or causes? > What may be causing the problem? > What are the real Xs or causes? > What is really causing the problem?
5. Take action.	Improve	> How can the understanding of the real causes of the problem be exploited to eliminate or reduce the size of the problem? > How can this Y = f(X) understanding be exploited?
6. Measure to verify improvement has taken place. 7. Take actions to sustain the gains.	Control	> Did Y really improve? > How can the Xs be controlled so the gains in Y remain?

(Chieh, 2007, n.p.)

Table 3.

Category	Tools
Gathering Ideas and Organizing Information	Brainstorming Affinity Diagramming Multi-voting Tree Diagram High-Level Process Map Flow Chart (Process Map) Cause and Effect (Fish Bone) Diagrams
Data Gathering	Sampling Operational Definitions Voice of the Customer Methods Check sheets and Spreadsheets
Process and Data Analysis	Process Flow Analysis Pareto Charts Histograms (Frequency Plot) Run (Trend Chart) Scatter Plot (Correlation) Diagram
Statistical Analysis	Test of statistical significance Correlation and regression Design experiments
Implementa-tion andProcess Management	Project Management Methods FEMA Stakeholder Analysis Force Field Diagram Process Documentation Balance Score Cards and Process Dash-boards

Adapted from Pande and Holpe (2002, p. 51- 67)

tion economy. The argument is that it is difficult for innovation and Six Sigma practices to coexist because by their very nature they are opposite. Six Sigma requires the organization to focus and direct its resources on operational excellence and perfection; it has a very low tolerance for risk because imperfection is increased when risk is introduced. On the other hand, innovation is by its very nature a high risk because it is dealing with new concepts (Rae, 2007, n.p.). Management professor Thomas Davenport contends that the time for Six Sigma has passed as companies are embracing innovation:

Process management is a good thing. But I think it always has to be leavened a bit with a focus on innovation and [customer relationships]." The discipline was developed as a systematic way to improve quality, but the reason it caught fire was its effectiveness in cutting costs and improving profitability. That makes it a powerful tool—if those are a company's goals. But as innovation becomes the cause du jour, companies are increasingly confronting the side effects of a Six Sigma culture (Grow & Hindo, 2007, n.p.).

Davenport's opinion is supported by research conducted by Benner and Tushman's research that suggests that Six Sigma will lead to more incremental innovation at the expense of more blue-sky work. The two professors analyzed the types of patents granted to paint and photography companies over a 20-year period, before and after a quality improvement drive. Their work shows that, after the quality push, patents issued based primarily on prior work made up a dramatically larger share of the total, while those not based on prior work plummeted (Hindo, 2007).

Examples of Six Sigma's fall from extreme favor can be seen at many fortune 100 companies including Home Depot, GE and 3M. The new CEO of Home Depot, Blake Frank has scaled back the strictness of the Six Sigma implementation enacted under the former CEO Robert Nardelli by giving more decision making to store managers. After the departure of James McNerney, 3M's leadership has changed its implementation of Six Sigma. Even GE, who under Jack Welsh popularized Six Sigma within the United States, is working to change how Six Sigma is used. Jack Immelt is attempting to move his team to innovate around a theme of "ecomangination" with mixed results (Grow & Hindo, 2007, n.p.).

This debate does not mean that Six Sigma cannot provide value to organizations. It does mean that Six Sigma cannot be implemented in a vacuum. Multiple methodologies are needed to support the enterprise strategy. For example, companies that are looking to innovate while supporting and improving processes and products are adopting the "Ambidextrous Organization" proposed by O'Rielly and Tushman's (2004) research that shows that successful companies pursue innovations because "they separate their new, exploratory units from their traditional, exploitative ones, allowing for different processes, structures, and cultures; at the same time" (p. 74). The organizations "maintain tight links across units at the senior executive level. In other words, they manage organizational separation through a tightly integrated senior team" O'Reilly & Tushman, 2004, p. 75). To describe these organizations, O'Reilly and Tushman (2004) have coined the term "ambidextrous organizations" and believe that the organizations have provided a "practical and proven model for forward-looking executives seeking to pioneer radical or disruptive innovations while pursuing incremental gains. A business does not have to escape its past; these cases show how to renew itself for the future. Their study showed that 90% of the ambidextrous organizations achieved their goals (O'Reilly & Tushman, 2004, p. 76).

IMPLICATIONS FOR RESEARCH

The framework in our exploratory research has been built upon a deductive study which has been developed through a literature review and synthesis and an exploratory inductive research which has been developed using a qualitative case study. It makes the case for leveraging ITIL and Six Sigma with ITSM in practice and opportunities for future research.

IT operations, as the production arm of IT departments, have been mostly ignored by IT research. There is a growing body of research on isolated aspects of operations services; but there is little research that is explicitly ITSM related. Despite the significant growth of ITSM practice in industry, there is no academic work or community of scholars that shares a common mission to understand how to advance it. Services are emerging in separate areas of academic, industry and government but few attempts to integrate them (Conger et al, 2007, p. 50). Since the early versions of ITIL lacked truly quantifiable business values, IT organizations are not interested in and supportive of IT Service Management processes. ITIL faces an uphill battle for acceptance and creditability that needs to be won across the entire organization.

Six Sigma is a measurement-driven approach to continuous process improvement that focuses on reduction of variation, consistency and high product quality. Therefore, in terms of IT service oriented industry, combining Six Sigma with ITIL can migrate current processes toward usable, measurable, ITIL-compatible processes as mentioned above with the case study on GE, however there is no research on version 3 because it is relatively recent, i.e., released in May 2007. Based on this we have shown the Six Sigma tools, which can be used to improve ITSM processes using the DMAIC model.

IMPLICATIONS FOR MANAGEMENT

At this point we have provided background on ITSM, ITIL and Six Sigma individually. But the question is can an organization's ITSM implementation benefit by integrating ITIL and Six Sigma? We conclude that ITIL and Six Sigma should be leveraged in tandem by IT organizations. Together with the ITIL best practices model and Six Sigma continual improvement and measurement, IT will be able to set boundaries and provide control elements for the senior IT management. ITIL is needed to provide the framework and best practices for ITSM. ITIL provide a set of guidelines to specify what an IT organization should do based on industry best practices. ITIL best practices process model is a key item to drive IT to meet the enterprises expectation.

However, ITIL does not define for an organization how it should be accomplished. According to Fry and Bolt (2004) "Six Sigma provides a process improvement approach that is based on statistical measurement, drives quality improvement, and reduces operational costs. It helps in developing detailed work instructions, and it defines a methodology for continually mapping, measuring, and improving the quality process. Six Sigma tells how, but it doesn't tell what to do nor does it specify any best practices specifically for ITSM...ITIL defines the "what" of service management and Six Sigma defines the "how" of quality improvement. Together, they make a great combination for improving the quality of IT service delivery and support" (p.2). ITSM, ITIL and Six Sigma combined can also assist IT govern itself and ensure that it is meeting and sustaining the enterprises strategies and objectives (Colbeck et al., 2005).

BUSINESS CASES

Integrating ITSM, ITIL and Six Sigma

Combining ITIL® and Six Sigma to Improve Information Technology Service Management at General Electric, Fry and Bolt (2004) outline the process that GE undertook in 2004 to improve internal practices for their IT service management. The IT Solutions Enterprise Planning & Strategy consulting group was engaged to develop a process improvement methodology that combined ITIL and Six Sigma to migrate from the current state to "measurable, ITIL-compatible processes" (p. 3). The group used Six Sigma to assess risk, compare current state and processes to the end goal, and then identified the solution which leveraged Remedy's "ITSM Suite to automate and improve ITSM processes" (Fry & Bolt, 2004, p. 3) The five objectives set by the team included 1) Achieve ISO Compliance for ITSM using ITIL for GE IT standards (ITIL is ISO 9000 compliant), 2) Use ITIL framework to define ITSM best practice standards, 3) Assess the current state of IT service by using the Capability Maturity Model (CMM) and Capability Assessment Tool against the ITIL standard, 4) Constantly improve ITSM processes using Six Sigma and Deming's Total Quality Management (TQM), and 5) Define measurement using Six Sigma to assure control and improvement (Fry & Bolt, 2004, p. 3)

To implement the methodology and meet the objectives set by the team, the team executed the Six Sigma DMAIC. The chart below outlines the project phase, goal or purpose and deliverable as stated by Fry and Bolt (2004) in the case study:

GE's project to improve their IT service management demonstrates that ITIL and Six Sigma can be leveraged together to implement and maintain ITSM within an organization. The benefits that GE found included 1) reducing cost by minimizing "potential downtime and the adverse effects of system, network, and application failures and install, move, adds, change and decommission implementations". 2) "Improve decision-making ability by facilitating access to information throughout the organization, and by enabling the enterprise wide use of outputs from an integrated framework of processes and tools through such devices as cross functional IT service dashboards" and 3) "Improve IT service levels by creating operational efficiencies and enabling a linked IT Service Management process loop for defining, measuring, analyzing, improving, and controlling service performance" (Fry & Bolt, 2004, p. 9).

Liberty Mutual, a diversified global insurer and sixth largest property and casualty insurer in the U.S., has introduced Lean Six Sigma and ITIL to dramatically reduce cycle times, improve quality and reduce cost. Although it is still in the early days of introducing Lean Six Sigma to the overall organization, preliminary results have been encouraging. Since 1994, the Boston-based company has almost tripled net revenue, to \$23.5 billion, while shedding noncore businesses like financial services and health care. For example, in one of the company's underwriting groups, process reengineering has helped reduce cycle time by as much as 70 percent. Within the IT group, Lean Six Sigma is the foundation for all process-improvement efforts. Stuart McGuigan, CIO and senior VP of Liberty Mutual Group, calls ITIL a "common-sense model" and believes

Table 4.

Phase	Goal or Purpose	Deliverable
Phase 1: Define Opportunities	Align ITSM strategy with the business, organizational, and technological strategies. The desired result is to set a definitive vision, scope, and strategic approach for ITSM operations.	>Team charter that describes the purpose, goals, and resources for the project. >Data collection plan that includes such items as interview schedules and questions. >Critical to Quality (CTQ) outline that identifies the critical success factors. >Current situation analysis that was created using the CAT tool. >Customers/Output—Process—Input/Suppliers (COPIS) "as-is" process map outline that provides an understanding of the current processes to which the solution
Phase 2: Measure performance	Create a current-state assessment of how well the current environment supports the ITSM strategy. The desired result is to determine current processes, issues, and the critical success factors—or Critical To Quality factors (CTQs)—of the desired future state environment.	>An accurate assessment of current process performance. >Detailed "as-is" process maps derived from the COPIS outline created in Phase 1 > Critical Success Factors (CTQ) summary chart
Phase 3: Analyze factors impacting performance	Examines the data collected in the Measure phase to generate a prioritized list of the sources of variation. The Analyze phase focuses improvement efforts by separating the "vital few" variables (those most likely responsible for the variation) from the "trivial many" (those least likely responsible for variation).	> Opportunities Table: Solutions mapped to process gap > Critical Success Factors (CTQs) benchmarked against ITIL best practices to identify opportunities for improvement > Cause and Effect Diagram > Pareto Chart of Opportunities
Phase 4: Improve Performance	Define and refine recommended tactical solutions based on information determined in phases 1-3. The desired result is a documented recommendation based on strategic fit, cost, and benefits.	> Risk Assessment/Failure Modes and Effect Analysis (FMEA)/Contingency Plan. > Prioritized potential solutions > Solution Pilot > "To Be" process maps
Phase 5: Control Performance	Propose a plan for designing and implementing the ITSM process improvement solution. The desired result is a true life cycle ITSM solution that allows for continuous improvement.	> Multi-Generational Product Plan (MGPP) > Process metrics defined > Full solution implemented > Control/Response Plan implemented > Risk mitigation actions complete

Source: Fry & Bolt, 2004, pp. 4-8.

it can empower a large organization like Liberty Mutual by helping the organization break work into meaningful components and achievable goals. He also says that ITIL gives him a common language that was missing during the days when each technology silo had its own terms and conventions. "ITIL," he says, "has given us an end-to-end vocabulary." (Brown, 2008).

In another case study, Integrated Mobile delivers end-to-end life cycle management of an enterprise's entire inventory of wireless assets through a single management interface. With capabilities built on the lean Six Sigma, ITIL and CMM principles, Integrated Mobile eliminates the obstacles, challenges and distractions of managing wireless services resulting in a highly efficient and effective mobile enterprise. The issue was managing mobile services and assets of more than 925 retail outlets across major metropolitan markets in 49 states with more than 76,000 employees. Adding more than 17,000 new employees annually and several new stores, including international outlets, could lead to a mobile environment that is overwhelming. The company needed to find an efficient way to handle the multitude of inquiries, tasks and transactions associated with their mobile communications since there was too much time and money being spent on maintaining wireless devices. In this state of chaos there was no time to identify ways to drive efficiencies and improve the end-user experience.

This specialty retailer turned to Integrated Mobile's iManage solution, a feature-rich, managed service that helped eliminate the chaos and excessive costs often associated with a mobile enterprise (Integrated Mobile, 2007).

Integrated Mobile's iManage solution includes:

- A Single Interface for All Wireless Management for All Carriers
- Supply Chain Automation
- Custom Kitting
- Provisioning and Fulfillment
- Standardized Handsets and Catalog Management that Supports Business Rules
- Approval Workflow
- Cost Allocation
- Credit Card Transaction Processing
- 24x7 Bilingual Administrative and Technical Tier 1 Support
- Asset Management
- Comprehensive Reporting, Transactions, Trends, Defects
- Validation of Transactions on the Carrier Bill
- MACD Processing
- Rate Plan Optimization
- Defect Elimination
- Optional iCare and iBill services for application and device support and invoice processing

This is how the specialty apparel retailer in this case study was able to improve satisfaction and productivity while driving down the cost of wireless with capabilities built on the lean Six Sigma, ITIL and CMM principles.

Lessons for IT Managers

It was to be expected that GE, an organization whose culture under the direction of Jack Welsh became entrenched in Six Sigma, leveraged Six Sigma when it implemented a project to use ITIL to improve its IT Service Management. But even organizations that are not Six Sigma implementations can benefit from using Six Sigma principles, methods, and tools. For example, the purpose of ITSM is to align the IT organization with the enterprise. This aligns to Pande and Holppe (2000) contention that within the Six Sigma implementation that there should be an emphasis and focus on the customer. Additionally, Six Sigma recommends collaborations across boundaries within the organization which is important to IT management because they need to determine service many divisions and functions and must provide enterprise level not silo level support and service. Using the DMAIC project management approach provides the IT organization a methodology to establish a vision and strategy for ITSM within the organization, assess the current state and its effectiveness in supporting the ITSM strategy, determine what is most likely for not meeting the defined strategy, identify a solution and propose a plan for implementation ITSM. Alternatively, if ITSM and ITIL are currently in place within the organization, the Six Sigma approach to organizing and getting work done including identifying and selecting problems to be solved could be used. Additionally, the Six Sigma tool kit provides tools that assist IT staff identify problems, determine root causes, and recommend ITIL solutions to fix existing problems. To provide sound IT service management, organizations should use ITIL and Six Sigma combined to ensure strategic alignment with their organization.

CONCLUSION

Over the next decade, executives will continue to be challenged to deliver value to share holders and other stakeholders. To support the organization, IT executives and managers will be challenged to transform their organization from delivering technology to providing service and ultimately becoming a business partner within the organization. To meet the challenge, we encourage the use of ITIL in combination with Six Sigma in IT Organizations with an ITSM implementation. IT managers need to leverage ITSM to develop, deliver and manage IT services to agreed upon quality standard. To have a sound IT management, managers should use ITIL v3 as best practices. Six Sigma DMAIC should be used to make certain that ITSM is aligned with the customer and to provide the mechanism to deliver and monitor all the IT service management processes. The case study at GE shows that ITIL and Six Sigma can be used in tandem to deliver ITSM and meet the organization's business objectives. Nevertheless, delivering ITSM using ITIL and Six Sigma is a one building block for building a world-class company. As Dr. Foster (2007), world-class competitors do not achieve the status because they adopted a "tool or technique." They achieve the status because they are good in "performing the fundamentals "including understanding their customers, products, employees, competitors' landscape, the market they are competing in and the enabling technologies. The world-class competitors succeed because they are creative and know how to make use of operations processes to their best advantages." (Foster, 2007, p. 509). ITIL and Six Sigma are just two tools that should be used for the IT organization to deliver world class IT Service Management and meet quality targets. As the IT organization successfully delivers ITSM, it will become a business partner within the organization

REFERENCES

Antony, J. (2007). Is Six Sigma a management fad or fact? *Assembly Automation, 27*(1), 17. doi:10.1108/01445150710724658

Brown, D. (2007). *ITIL V2 to V3 – Is it worth it?* Retrieved November 4, 2007, from http://www.axiossytems.com

Brown, P. (2008). *Case study: Satisfaction guaranteed.* Retrieved on August 6, 2008, from http://www.smartenterprisemag.com

Chieh, J. C. (2007, March 21). *Six Sigma basics: DMAIC like normal problem solving.* iSixSigma.com. Retrieved September 28, 2007, from http://europe.isixsigma.com/library/content/c070321b.asp

Chuck, E. (2002). *Success with Six Sigma -A mini-tutorial for the quality-minded.* Retrieved September 27, 2007, from http://www.freequality.org/sites/www_freequality_org/documents/Training/Classes Fall 2002/Six Sigma.doc

Colbeck, D., & Evans, I. (2005). IT governance. In Colbeck, D., & Evans, I. (Eds.), *The complete guide to IT service management* (1st ed., pp. 1–2). Ottawa, Canada: ColbeckEvans.

Conger, S. (2007). *Information Technology service management: An emerging area for academic research and pedagogical development.* New York, NY: ACM Digital Library.

Damiano, F., & McLauglin, K. (2007). *American ITIL.* New York, NY: ACM Digital Library.

Drucker, P. (2001). Management by objectives and self-control. In Drucker, P. (Ed.), *The essential Drucker* (1st ed., p. 112). New York, NY: HarperCollins.

Foster, S. (2007). *Managing quality: Integrating the supply chain* (3rd ed.). Upper Saddle River, NJ: Pearson.

Foster, S. (2007). Implementing and validating the quality system. In Foster, S. (Ed.), *Managing quality: Integrating the supply chain* (3rd ed., p. 509). Upper Saddle River, NJ: Pearson.

Fry, M., & Bolt, M. (2004). *Combining ITIL® and Six Sigma to improve Information Technology service management at General Electric.*

Govekar, M., Curtis, D., Brittain, K., Scott, D., Mingay, S., & Holub, E. … Phifer, G. (2007, June 22). *Hype cycle for IT operations management*, 2007 (ID Number: G00148631). Retrieved July 28, 2007 from http://www.gartner.com

Grow, B., & Hindo, B. (June 11, 2007). Six Sigma: So yesterday? In an innovation economy, it's no longer a cure-all. *Business Week.* Retrieved November 23, 2007, from http://www.businessweek.com/innovate/content/feb2007/id20070227_766365_page_2. html

Hammer, M., & Goding, J. (2001). Putting Six Sigma in perspective. Quality, *40*(10), 58-62. (Document ID: 84943086).

Hank, M. (2006). ITIL: What it is and what it isn't. *Business Communication Review, 36*(12). (WN: 063304081014).

Hindo, B. (2007, June 11). At 3M, a struggle between efficiency and creativity. *BusinessWeek.* Retrieved November 23, 2007, from http://www.businessweek.com/magazine/content/07_24/b4038406.htm

Integrated Mobile. (2007). *Specialty apparel retailer improves satisfaction and productivity while driving down the cost of wireless*. Retrieved on August 5, 2008, from http://www.integrated-mobile.com/case-studies

ITIL Organization. (2007). Retrieved November 15, 2007, from http://www.itil-officialsite.com/home/home.asp

Klefsjo, B., Wiklund, H., & Edgeman, R. L. (2001). Six Sigma seen as a methodology for total quality management. *Measuring Business Excellence, 5*(1), 31-35. (Document ID: 82396437).

O'Reilly, C., & Tushman, M. (2004, April). The ambidextrous organization. *Harvard Business Review, 82*(4), 74–81.

Pande, P., & Holp, L. (2002). *What is Six Sigma?*New York, NY: McGraw-Hill.

Rae, J. (2007, February 27). Debate: Six Sigma vs. innovation. *BusinessWeek.* Retrieved November 23, 2007, from http://www.businessweek.com/innovate/content/feb2007/id20070227_766365.htm

Salle, M. (2004). *IT service management and IT governance: Review, comparative analysis and their impact on utility computing*. Retrieved December 15, 2007 from http://www.hpl.hp.com/techreports/2004/HPL-2004-98.pdf

Simon, K. (2000, December 11). DMAIC versus DMADV. *iSixSigma.com.* Retrieved September 28, 2007, from http://europe.isixsigma.com/library/content/c001211a.asp

This manuscript was previously published in the International Journal of E-Services and Mobile Applications, Dr. Ada Scupola (Editor)

Weinberg, A. (2007). *Focusing on the customer: An interview with the head of Merrill Lynch's operations and IT.* Retrieved July 21, 2007 from http://www.mckinseyquarterly.com/article_abstract.aspx?ar=2021&L2=13

WhatIs.com. (2008). *IT service management definition.* Retrieved July 5, 2008, from http://www.bitpipe.com/tlist/IT-Service-Management.html

Worthen, B. (2005). ITIL and Sarbanes-Oxley (Sarbox). *CIO Magazine*. Retrieved on November 15, 2007, from http://www.cio.com/article/10519/_ITIL_and_Sarbanes_Oxley_Sarbox_

APPENDIX

Roles within Six Sigma

Table A1. The hierarchy and the roles within a Six Sigma organization are described by the American Society for Quality (http://www.asq.org) and Thomas Pyzdek as follows:

Roles	Description of Role
Executives/Leadership	Must be lead by the CEO because "Six Sigma involves changing major value streams that cut across organizational barriers (Pyzde, 2000). Provides overall alignment by establishing the strategic focus of the Six Sigma program within the context of the organization's culture (ASQ).
Champions	Champions are the next level within the organization. These are "high-level individuals who understand Six Sigma and are committed to its success." In a Six Sigma implementation there will also be "informal leaders" who are daily practitioners and are communicating the "Six Sigma message" within the organization (Pyzde, 2000). The role of a champion is to "translate the company's vision, mission, goals and metrics to create an organizational deployment plan and identify individual projects." They are also to "identify resources and remove roadblocks" (ASQ).
Sponsors	Sponsors own processes and systems that can be improved using the DAIMC process. They "help initiate and coordinate Six Sigma improvement activities in their areas of responsibilities" (Pyzde, 2000).
Master Black Belts	Within the project hierarchy, the Master Black belt is "highest level of technical and organizational proficiency." The Master Black Belt is responsible for providing technical leadership, including statistical training and assistance to Black Belts, developing key metrics, and strategy within the Six Sigma program. The Black Belt must have the same knowledge as the Black Belt; additionally, they must "understand the statistical mathematical theory on which the methods are based… There is usually about one Master Black Belts for every ten Black Belts, or about 1 Master Black Belt per 1,000 employees" (Pyzde, 2000). They are similar to a Program Manager within the Project Management hierarchy
Black Belts	Black Belts are respected leaders and technically inclined; they "receive 160 hours of classroom instruction, plus one-on-one project coaching from Master Black Belts or consultants." An organization will have on average "about one percent of their work force as Black Belt." A Black Belt, Project Manager, will complete between 5 to 7 projects each year. (Pyzde, 2000). Black Belts are responsible for leading "problem solving projects and the training of and coaching of project teams" (ASQ).
Green Belts	Green Belts are the equivalent of a project lead within the Six Sigma program. They are capable "of forming and facilitating Six Sigma teams and managing Six Sigma projects from concept to completion" (Pyzde, 2000). Green Belts are also responsible for assisting "with data collection and analysis for Black Belt projects" (ASQ).
Yellow Belts	Yellow Belts are project team members. They review "process improvements that support the project" (ASQ).
White Belts	White Belts "can work on local problem-solving teams that support." overall projects, but may not be part of a Six Sigma project team. They understand "basic Six Sigma concepts from an awareness perspective" (ASQ).
Adapted from American Society for Quality (http://www.asq.org) and Thomas Pyzdek (2000)	

Compilation of References

(2004). *BankData*. Beirut, Lebanon: Bilanbanques.

ABA Banking Journal (2001). E-banking 2001: where are we headed? 93(1)- 52-53.

ABL Association des Banques au Liban. (2004), Activity and Performance of the Lebanese banking sector. Available online at :http://www.abl.org.lb/ABL/Pictures/Part%20 IV-2004-2005%20eng.pdf

Abowd, D., Dey, A. K., Orr, R., & Brotherton, J. (1998). Context-awareness in wearable and ubiquitous computing. *Journal Virtual Reality*, *3*(3), 200–211. doi:10.1007/BF01408562

Acharya, P., & Mahanty, B. (2008). Manpower shortage crisis in Indian Information Technology industry. *International Journal of Technology Management*, *38*(3), 235–247. doi:10.1504/IJTM.2007.012712

Agarwal, R., & Ferratt, T. W. (2002). Enduring practices for managing IT professionals. *Communications of the ACM*, *45*(9). doi:10.1145/567498.567502

Agarwal, R., & Ferratt, T. W. (2000). Retention and the career motives of IT professionals. *Proceedings of the 2000 ACM SIGCPR Conference on Computer Personnel Research*, (pp. 158-166). April 2000, Chicago, Illinois, United States.

Ahokas, I., & Kaivo-oja, J. (2003). Benchmarking European information society developments. *Foresight*, *5*(1), 44–54..doi:10.1108/14636680310471280

Ajzen, I. (1991). The Theory of Planned Behavior. *Organizational Behavior and Human Decision Processes*, *50*, 179–211..doi:10.1016/0749-5978(91)90020-T

Ajzen, I. (1985). From intentions to actions: a theory of planned behaviour, in Kuhl, J. and Beckmann, J. (Eds), *Action control: From cognition to Behaviour*, Springer, Heilderberg. 11-39.

Ajzen, I., & Fishbein, M. (1980). *Understanding Attitudes and Predicting Social Behavior*. New Jersey: Englewood Cliffs.

Aladwani, A. M. (2001). Online banking: A field study of drivers, development challenges and expectations. *International Journal of Information Management*, *21*, 213–225. doi:10.1016/S0268-4012(01)00011-1

Albert, G. (1998). *Betriebliche Personalwirtschaft*. Kiehl: Ludwigshafen a.R.

Almasri, S., Hunaiti, Z., Sedoyeka, E., & Balachandrab, W. (2008). Zone-Based Update Strategy for Location based Services (LBS). *IEEE International Symposium on Industrial Electronics*, Cambridge, UK.

Alonso Mendo, F., & Fitzgerald, G. (2005). Theoretical approaches to study SMES e-business progression. *Journal of Computing and Information Technology*, *13*(2), 123–136. doi:10.2498/cit.2005.02.04

Amit, R., & Schoemaker, P. J. H. (1993). Strategic assets and organizational rent. *Strategic Management Journal*, *14*(1), 33. doi:10.1002/smj.4250140105

Anacom (2008). Serviço Nacional Móvel: relatório do 4° trimestre de 2007, Autoridade Nacional de Comunicações. Retrieved May 18, 2008, from http://www.anacom.pt / template12.jsp? categoryId=250442

Anckar, B. & DÍncau, D. (2002). Value-creation in Mobile Commerce: Findings from a Consumer Survey. *Journal of Information Technology Theory and Application*, *4*(1), 43–64.

Anderson, E., & Sullivan, M. (1993). The antecedents and consequences of customer satisfactions for firms. *Management Science, 12*(2), 125–143.

Anderson, J., & Gerbing, D. (1988). Structural equation modeling in practice: A review and recommended two-step approach. *Psychological Bulletin, 103*(3), 411–423. doi:10.1037/0033-2909.103.3.411

Andersson, C. (2001). *GRPS and 3G Wireless Applications*. Wiley.

Angehrn, A. (1997). Designing mature internet business strategies: The ICDT model. *European Management Journal, 15*(4), 361–369. doi:10.1016/S0263-2373(97)00016-9

Antony, J. (2007). Is Six Sigma a management fad or fact? *Assembly Automation, 27*(1), 17. doi:10.1108/01445150710724658

Ariely, D. (2000). Controlling information flow: on the importance of control in decision making and the value of information. *The Journal of Consumer Research, 27*, 233–248. doi:10.1086/314322

Armstrong, C. P., & Sambamurthy, V. (1999). Information Technology assimilation in firms: the influence of senior leadership and IT infrastructures. *Information Systems Research, 10*(4), 304–327. doi:10.1287/isre.10.4.304

Arndt, J. (1967). Role of product- related conservations in the diffusion of a new product. *JMR, Journal of Marketing Research, 4*, 291–295. doi:10.2307/3149462

Asociación Española de Comercio Electrónico y Marketing Relacional –AECEM-. (2008). Estudio sobre comercio electrónico b2c 2008. Retrieved November, 2008, from: http://www.red.es/media/2008-10/1224586013979.pdf

Asociación para la Investigación en Medios de Comunicación –AIMC-. (2008). Audiencia de Internet. abril-mayo 2008. Retrieved November, 2008, from: http://www.aimc.es/aimc.php

Attewell, P. (1992). Technological diffusion and organizational learning: The case of business computing. *Organization Science, 3*(1), 1. doi:10.1287/orsc.3.1.1

Badoc, M., Lavayssiere, B., & Copin, E. (1994). *E-marketing de la banque et de l'assurance*. Edition Organisation.

Bae, J. S., Lee, J. Y., Kim, B. C., & Rye, S. (2006, August 1-4). Next Generation Mobile Service Environment and Evolution of Context Aware Services. *International Conference, EUC 2006*, Seoul, Korea.

Bagozzi, R. P. (2007). The legacy of the technology acceptance model and a proposal for a paradigm shift. *Journal of the Association for Information Systems, 8*(4), 244–254.

Bahl, P., Padmanabhan, V., & Balachandran, A. (2000). Enhancements to the RADAR user location and tracking system. *Microsoft Co. Technical Report MSR-TR-2000 12*.

Bailey, J. P., & Bakos, J. Y. (1997). An exploratory study of the emerging role of electronic intermediaries. *International Journal of Electronic Commerce, 1*(3), 7–20.

Bakos, J. Y. (1991). A strategic analysis of electronic marketplaces. *Management Information Systems Quarterly, 15*(3), 294–310. doi:10.2307/249641

Balas, E. A., & Iakovidis, I. (1999). Distance technologies for patient monitoring. *British Medical Journal, 310*, 1309.

Banker, R., Chang, H., & Kao, Y. (2002). Impact of Information Technology on public accounting firm productivity. *Journal of Information Systems, 16*(2), 209–222. doi:10.2308/jis.2002.16.2.209

Barber, A. E. (1998). *Recruiting employees*. Thousand Oaks, CA: Sage Publications.

Barclay, D., Higgins, C., & Thompson, R. (1995). The partial least squares (PLS) approach to causal modeling: Personal computer adoption and use as an illustration. *Technology Studies, 2*, 285–309.

Barczac, G., Ellen, P. S., & Pilling, B. K. (1997). Developing typologies of consumer motives for use of technologically based banking services. *Journal of Business Research, 38*(2), 131–139. doi:10.1016/S0148-2963(96)00032-X

Barer, M. L., & Stoddart, G. L. (1999). *Improving access to needed medical services in rural and remote Canadian communities: Recruitment and retention revisited.*

Barley, S. R. (1986). Technology as an occasion for structuring: Evidence from observations of CT scanners and the social order of radiology departments. *Administrative Science Quarterly, 31*(1), 78. doi:10.2307/2392767

Barley, S. R. (1990). The alignment of technology and structure through roles and networks. *Administrative Science Quarterly*, *35*(1), 61. doi:10.2307/2393551

Barnes, S., & Corbitt, B. (2003). Mobile banking: concept and potential. *International Journal of Mobile Communications*, *1*(3), 273–288. doi:10.1504/IJMC.2003.003494

Barthelemy, J. (2008). The determinants of total IT outsourcing: An empirical investigation of French and German firms. *The Journal of Computer Information Systems*, April. Retrieved on November 29, 2010, from http://www.allbusiness.com/human-resources/workforce-management-hiring-consulting/1036323-1.html

Barua, A., Konana, P., Whinston, A. B., & Yin, F. (2001). Driving e-business excellence. *Sloan Management Review*, *43*(1), 36–44.

Barwise, P., & Strong, C. (2002). Permission-Based Mobile Advertising. *Journal of Interactive Marketing*, *16*(1), 14–24..doi:10.1002/dir.10000

Baskerville, R., & Wood-Harper, A. T. (1998). Diversity in Information Systems action research methods. *European Journal of Information Systems*, *7*(2), 90–107. doi:10.1057/palgrave.ejis.3000298

Bateson, J. (1985). Self-service consumer: An exploratory study. *Journal of Retailing*, *61*(3), 49–76.

Bauer, H. H., Reichardt, T., Barnes, S. J., & Neumann, M. M. (2005). Driving Consumer Acceptance of Mobile Marketing: a Theoretical Framework and Empirical Study. *Journal of Electronic Commerce Research*, *6*(3), 181–192.

Bauer, R. A. (1960). Consumer Behaviour as Risk Taking. In Hancock, R. S. (Ed.), *Dynamic Marketing for a Changing World*. Chicago: American Marketing Association.

Beatson, A., Coote, L. V., & Rudd, J. M. (2006). Determining consumer satisfaction and commitment through self-service technology and personal service usage. *Journal of Marketing Management*, *22*(7-8), 853–882. doi:10.1362/026725706778612121

Becker, M. (2005). *Personalentwicklung. Bildung, Förderung und Organisationsentwicklung in Theorie und Praxis* (*Vol. 4*). Stuttgart, Germany: Schaeffer-Poeschel.

Bellman, S., Lohse, G., & Johnson, E. (1999). Predictors of online buying behaviour. *Communications of the ACM*, *42*(12), 32–38. doi:10.1145/322796.322805

Benapudi, N., & Leone, R. P. (2003). Psychological implications of customer participation in co-production. *Journal of Marketing*, *67*(1), 14–28. doi:10.1509/jmkg.67.1.14.18592

Benbasat, I., Goldstein, D. K., & Mead, M. (1987). The case research strategy in studies of Information Systems. *Management Information Systems Quarterly*, *11*(3), 369–386. doi:10.2307/248684

Benford, S., Anastasi, R., Flintham, M., Drozd, A., Crabtree, A., & Greenhalgh, C. (2003, September). Coping with Uncertainty in a Location-Based Game. *IEEE Pervasive Computing / IEEE Computer Society [and] IEEE Communications Society*, 34–41. doi:10.1109/MPRV.2003.1228525

Benford, S. (2005). Future Location-Based Experiences. *JISC: Technology & Standards Watch*. Retrieved October 30, 2008, from http://www.jisc.ac.uk/ uploaded_documents /jisctsw_05_01.pdf

Bergeron, J. (2001). Les facteurs qui influencent la fidélité des clients qui achètent sur Internet [The factors influencing the loyalty of online purchasers]. *Recherche et Applications en Marketing*, *16*(3), 39–53.

Bhansali, C. (2005, June). Successful firms will be online on time. *Accounting Technology*.

Bhappu, A. D., & Schultze, U. (2006). The role of relational and operational performance in business-to-business customers' adoption of self-service technology. *Journal of Service Research*, *8*(3), 372–385. doi:10.1177/1094670506286571

Bharadwaj, A., Sambamurthy, V., & Zmud, R. (1999). *IT capabilities: Theoretical perspectives and empirical operationalization*.

Bhargava, H. K., & Choudhary, V. (2004). Economics of an information intermediary with aggregation benefits. *Information Systems Research*, *15*(1), 22–36. doi:10.1287/isre.1040.0014

Bhimani, A. (1996). Securing the commercial Internet. *Communications of the ACM*, *39*(6), 29–35. doi:10.1145/228503.228509

Bigne, E., Ruiz, C., & Sanz, S. (2005). The Impact of Internet User Shopping Patterns and Demographics on Consumer Mobile Buying Behaviour. *Journal of Electronic Commerce Research*, *6*(3), 193–209.

Bilodeau, H., & Leduc, N. (2003). Inventory of main factors determining the attraction, installation and retention of physicians in remote areas. *Cahiers de Sociologie et de Demographie Medicales*, *43*(3).

Bilodeau, H., Leduc, N., & Schendel, N. (2006). *Analyse des facteurs d'attraction, d'installation et de maintien de la pratique médicale dans les régions éloignées du Québec*. Groupe De Recherche Interdisciplinaire En Santé. Université De Montréal.

Bishop, J. E., O'Reilly, R. L., Daddox, K., & Hutchinson, K. (2002). Client satisfaction in a feasibility study comparing face-to-face interviews with telepsychiatry. *Journal of Telemedicine and Telecare*, *8*(4). doi:10.1258/135763302320272185

Black, N. J., Lockett, A., Wiklofer, H., & Ennew, C. (2001). The adoption of Internet financial services: A qualitative study. *International Journal of Retail & Distribution Management*, *29*(8), 390–398. doi:10.1108/09590550110397033

Blechar, J., Constantiou, I. D., & Damsgaard, J. (2006). Exploring the influence of reference situations and reference pricing on mobile service user behaviour. *European Journal of Information Systems*, *15*, 285–291. doi:10.1057/palgrave.ejis.3000618

Bobbitt, L. M., & Dabholkar, P. A. (2001). Integrating attitudinal theories to understand and predict use of technology-based self-service: The Internet as an illustration. *International Journal of Service Industry Management*, *12*(5), 423–450. doi:10.1108/EUM0000000006092

Böhm, S. Koolwaaij, J., Luther, M., Souville, B., Wagner, M., & Wibbels, M. (2008, October 27–29). Introducing IYOUIT. *Proceedings of the International Semantic Web Conference (ISWC '08)*, Germany, *vol. 5318 of LNCS*, (pp. 804–817). Springer Verlag.

Bolton, R. N. (2003). Marketing challenges of e-services. *Communications of the ACM*, *46*(6), 43–44. doi:10.1145/777313.777337

Boulos, M., Rocha, A., Martins, A., Vicente, M., Bolz, A., Feld, R., Tchoudovski, I., & Braecklein, M. Nelson5, J., Laighin, G., Sdogati, C., Cesaroni, F., Antomarini, M., Jobes A., & Kinirons, M. (2007). CAALYX: a new generation of location-based services in healthcare. *International Journal of Health Geographics*, *12*, 6–9.

Bourgueil, Y., Mousques, J., & Tajahmadi, A. (2006). *Comment améliorer la répartition géographique des professionnels de santé: Les enseignements de la littérature internationale et les mesures adoptées en France*. Institut De Recherche Et Documentation En Économie De La Santé. Broadbent, M., & Weill, P. (1997). Management by maxim: How business and IT managers can create IT infrastructures. *Sloan Management Review*, *38*(3), 77.

Bouwman, H., Carlsson, C., Molina-Castillo, F., & Walden, P. (2007). Barriers and drivers in the adoption of current and future mobile services in Finland. *Telematics and Informatics*, *24*(2), 145–160. doi:10.1016/j.tele.2006.08.001

Bouwman, H., Haaker, T., & de Vos, H. (2007b). Mobile Service Bundles: The example of navigation services. *Electronic Markets*, *17*(1), 20–28. doi:10.1080/10196780601136757

Bouwman, H., de Vos, H., & van der Wijngaert, L. (2008, July). Context-sensitive mobile services for police officers. *7th International Conference on Mobile Business*, Barcelona.

Boyer, K., Hallowell, R., & Roth, A. V. (2002). E-services: Operating strategy—A case study and a method for analyzing operational benefits. *Journal of Operations Management*, *20*, 175–188. doi:10.1016/S0272-6963(01)00093-6

Boztepe, S. (2007). User value: Competing theories and models. *International Journal of Design*, *1*(2), 55–63.

Bradley, L., & Stewart, K. (2002). A Delphi study of drivers and inhibitors of internet banking. *International Journal of Bank Marketing*, *20*(6), 250–260. doi:10.1108/02652320210446715

Bradley, L., & Stewart, K. (2002). A Delphi study of the drivers and inhibitors of Internet banking. *International Journal of Bank Marketing*, *20*(6), 250–260. doi:10.1108/02652320210446715

Brannick, J. (2001). Seven strategies for retaining top talent. *The Journal of Business Strategy*, *22*(4), 28–31. doi:10.1108/eb040183

Breaugh, J. A., & Starke, M. (2000). Research on employee recruitment: So many studies, so many remaining questions. *Journal of Management*, *26*(3), 405–434. doi:10.1177/014920630002600303

Brock, F. (2003, Oct. 12). Seniority: Who'll sit at the boomers' desks? *The New York Times*.

Brohman, M. K., Piccoli, G., & Martin, P. (2009). A design theory approach to building strategic network-based customer service systems. *Decision Sciences*, *30*(3), 403–430. doi:10.1111/j.1540-5915.2009.00242.x

Brohman, M. K., Watson, R. T., Piccoli, G., & Parasuraman, A. (2003). Data completeness: A key to effective net-based customer service systems. *Communications of the ACM*, *46*(6), 47–51. doi:10.1145/777313.777339

Brown, I., Cajee, Z., Davies, D., & Stroebel, S. (2003). Cell phone banking: predictors of adoption in South Africa - an exploratory study. *International Journal of Information Management*, *23*(5), 381–394. doi:10.1016/S0268-4012(03)00065-3

Brown, D. (2007). *ITIL V2 to V3 – Is it worth it?* Retrieved November 4, 2007, from http://www.axiossytems.com

Brown, P. (2008). *Case study: Satisfaction guaranteed.* Retrieved on August 6, 2008, from http://www.smartenterprisemag.com

Buchanan, T., & Smith, J. L. (1999). Using the Internet for psychological research: Personality testing on the World-Wide Web. *The British Journal of Psychology*, *90*, 125–144. doi:10.1348/000712699161189

Buell, R. W., Campbell, D., & Frei, F. X. (2010). Are self-service customers satisfied or stuck? *Production and Operations Management*, *19*(6), 679–697. doi:10.1111/j.1937-5956.2010.01151.x

Bughin, J. (2003). The diffusion of Internet banking in western Europe. *Electronic Markets*, *13*(3), 251–258. doi:10.1080/1019678032000108329

Bunduchi, R. (2005). Business relationships in Internet-based electronic markets: The role of goodwill trust and transaction costs. *Information Systems Journal*, *15*(4), 321–341. doi:10.1111/j.1365-2575.2005.00199.x

Burgess, S., Sellitto, C., & Wenn, A. (2005). Maturity in the websites of Australian wineries: A study of varying website content. *International Journal of Electronic Business*, *3*(5), 473–490. doi:10.1504/IJEB.2005.008521

Burgess, S. (2002). *Business-to-consumer interactions on the Internet: A model for small businesses*. PhD Thesis. Monash University, Melbourne Australia.

Burgess, S., & Zoppos, B. (2002). The features of selected Australian websites. *E-Conomy: From here to where? (CD Rom Proceedings)*, School of Management Information Systems, Edith Cowan University, Perth, December.

Burgess, S., Bingley, S., & Sellitto, C. (2005). A model for website development of micro tourism enterprises. *Illuminating Entrepreneurship: Proceedings of the Institute for Small Business and Entrepreneurship 28th National Conference*, Perlex Associates, UK.

Burke, R. R. (2002). Technology and the customer interface: what consumers want in the physical and virtual store. *Journal of the Academy of Marketing Science*, *30*, 411–432. doi:10.1177/009207002236914

Burstein, F., & Gregor, S. (1999). The systems development or engineering approach to research in Information Systems: An action research perspective. In B. Hope & P. Yoong (Eds.), *Proceedings of the 10'' Australasian Conference on Information Systems*, Victoria University of Wellington. New Zealand, (pp. 122-134).

Bury, L. (1999, August 28-29). On line and on time. *Accountancy*.

Buzzetto-More, N. A., & Alade, A. J. (2006). Best practices in e-assessment. *Journal of Information Technology Education*, *5*, 251–269.

Cagan, J., & Vogel, C. M. (2002). *Creating breakthrough products: Innovation from product planning to program approval*. Upper Saddle River, NJ: Prentice Hall.

Caligiuri, P. M., & Phillips, J. M. (2003). An application of self-assessment realistic job previews to expatriate assignments. *International Journal of Human Resource Management*, *14*(7), 1102–1116. doi:10.1080/0958519032000114228

Calof, J. L., & Beamish, P. W. (1995). Adapting to foreign markets: Explaining internationalization. *International Business Review*, *4*(2), 15–31. doi:10.1016/0969-5931(95)00001-G

Cameron, J. (2006). Governance structure, mechanisms and methods for managing collaborative e-business projects. *Journal of Chain and Network Sciences*, *6*(2), 155–174. doi:10.3920/JCNS2006.x072

Cameron, J. (2005). Ten concepts for an e-business collaborative project management framework. In *Proceedings of eIntegration in Action: 18th Bled eConference*, June 6-8 2005, Bled, Slovenia.

Cao, M., & Zhang, Q. (2005). Web site quality and usability in e-commerce. In Gao Y. (Eds) (2005): *Web System Design and Online Consumer Behavior* (pp. 107-124), Idea Group Publishing, Hershey.

Cappelli, P. (2000a). A market-driven approach to retaining talent. *Harvard Business Review*, *78*(1), 103–111.

Cappelli, P. (2000b). *Is there a shortage of Information Technology workers?* (pp. 1–25). McKinsey and Company Project Review.

Cappelli, P. (2008). Talent management for the twenty-first century. *Harvard Business Review*, *86*(3), 74–81.

Carlsson, C. (2006). Special issue on mobile technology and services. *Electronic Commerce Research and Applications*, *5*(3), 189–191. doi:10.1016/j.elerap.2006.07.001

Carlsson, C., Carlsson, J., Hyvönen, K., Puhakainen, J., & Walden, P. (2006). Adoption of Mobile Devices/ Services – Searching for Answers with the UTAUT. In *Proceedings of the 39th Hawaii International Conference on System Sciences, Hawaii*. (CD-ROM), Computer Society Press (10 pages).

Carlsson, C., Hyvönen, K., Repo, P., & Walden, P. (2005). Asynchronous Adoption Patterns of Mobile Services. In *Proceedings of the 38th Hawaii International Conference on System Sciences, Hawaii*. (CD-ROM), Computer Society Press (10 pages).

Carroll, M., Marchington, M., Earnshaw, J., & Taylor, S. (1999). Recruitment in small firms: Processes, methods and problems. *Employee Relations*, *21*(3), 236–250. doi:10.1108/01425459910273080

Carroll, A., Barnes, S. J., & Scornavacca, E. (2005). Consumer Perceptions and Attitudes Towards Mobile Marketing. In Barnes, S. (Ed.), *Unwired Business: Cases in Mobile Business* (p. 109). Hershey, PA, USA: IRM Press.

Cenfetelli, R., Benbasat, I., & Al-Natour, S. (2005). Information Technology mediated customer service: A functional perspective. In *Proceedings of International Conference of Information Systems*, Las Vegas, (pp. 725-738).

Chae, B., & Poole, M. S. (2005). The surface of emergence in systems development: Agency, institutions, and large-scale Information Systems. *European Journal of Information Systems*, *14*, 19–36. doi:10.1057/palgrave.ejis.3000519

Chae, B. (2002). *Understanding Information Systems as social institutions: Dynamic institutional theory.* Unpublished doctoral dissertation, Texas A&M University, United States -- Texas.

Chakravorty, R., & Pratt, I. (2002). Performance Issues with GPRS. *Journal of Communication and Networks*, *4*(4), 266–281.

Chambers, E. G., Foulon, F., Handfield-Jones, H., Hankin, S. M., & Michaels, E. G. III. (1998). The war for talent. *The McKinsey Quarterly*, 1.

Chaston, L., & Mangles, T. (2001). E-commerce and small UK accounting firms: Influence of marketing styles and orientation. *The Service Industries Journal*, *21*(4), 83–99. doi:10.1080/714005049

Chatterjee, D., Grewal, R., & Sambamurthy, V. (2002). Shaping up for e-commerce: Institutional enablers of the organizational assimilation of Web technologies. *Management Information Systems Quarterly*, *26*(2), 65. doi:10.2307/4132321

Chatterjee, D., & Segars, A. H. (2001). *Transformation of the enterprise through e-business: An overview of contemporary practices and trends*. Report to the Advanced Practices Council of the Society for Information Management, (July 2001).

Checkland, P. (1981). *Systems thinking, systems practice*. Chichester, UK: Wiley.

Chen, M., Tseng, C., & Chang, J. (2005). A survey investigation into the use of the Internet among accounting firms. *International Journal of Management, 22*(4), 649–660.

Chesbrough, H., & Spohrer, J. (2006). A research manifesto for services science. *Communications of the ACM, 49*(7), 35–40. doi:10.1145/1139922.1139945

Chi, L., & Holsapple, C. (2005). Understanding computer-mediated interorganizational collaboration: A model and framework. *Journal of Knowledge Management, 9*(1), 53–75. doi:10.1108/13673270510582965

Chieh, J. C. (2007, March 21). *Six Sigma basics: DMAIC like normal problem solving*. iSixSigma.com. Retrieved September 28, 2007, from http://europe.isixsigma.com/library/content/c070321b.asp

Chin, W. W., Marcolin, B. L., & Newsted, P. R. (2003). A partial least squares latent variable modeling approach for measuring interaction effects: Results from a Monte Carlo simulation study and an electronic-mail emotion/adoption study. *Information Systems Research, 14*, 189–217.. doi:10.1287/isre.14.2.189.16018

Chin, W. W. (1998). The Partial Least Squares Approach to Structural Equation Modeling. In Marcoulides, G. A. (Ed.), *Modern Methods for Business Research* (pp. 295–336). Mahwah, NJ: Lawrence Erlbaum.

Chircu, A. M., & Kauffman, R. J. (1999). Strategies of Internet middlemen in the intermediation/ disintermedation/ reintermediation cycle. *Electronic Markets, 9*(2), 109–117. doi:10.1080/101967899359337

Chiu, D., Kok, D., Lee, A., & Cheung, S. (2005). Integrating legacy sites into Web services with Webxcript. *International Journal of Cooperative Information Systems, 14*(1), 25–44. doi:10.1142/S0218843005001006

Cho, Y. K., & Menor, L. J. (2010). Toward a provider-based view on the design and delivery of quality e-service encounters. *Journal of Service Research, 13*(1), 83–95. doi:10.1177/1094670509350490

Cho, S., & Mathiassen, L. (2007). The role of industry infrastructure in telehealth innovations: A multi-level analysis of a telestroke program. *European Journal of Information Systems, 16*(6), 738–750. doi:10.1057/palgrave.ejis.3000718

Cho, S., Mathiassen, L., & Gallivan, M. (2009). Crossing the diffusion chasm: From invention to penetration of a telehealth innovation. *Information Technology & People, 22*(4), 351. doi:10.1108/09593840911002450

Cho, S., Mathiassen, L., & Robey, D. (2007). Dialectics of resilience: A multi-level analysis of a telehealth innovation. *Journal of Information Technology, 22*(1), 24–35. doi:10.1057/palgrave.jit.2000088

Cho, K.R. & Padmanabhan, P. (1995). Acquisition versus new venture: The choice of foreign establishment mode by Japanese firms. *Journal of International Management*, Fall, l(3), 255-85.

Choi, S. M., & Rifon, N. J. (2002). Antecedents and Consequences of Web Advertising Credibility: A Study of Consumer Response to Banner Advertisements. *Journal of Interactive Advertising, 3*(1).

Christensen, C. M., & Raynor, M. E. (2003). *Innovator's Solution: creating and sustaining successful growth* (p. 303). Harvard Business School Press.

Christensen, C. M., & Davis, A. B. (2006). The Future of Newspapers. Retrieved September 29th, 2008. Available at:http://www.forbes.com/home/leadership/2006/10/10/leadership-newspapers-media-lead-innovation-cx_cc_1011clayton.html

Chuck, E. (2002). *Success with Six Sigma - A mini-tutorial for the quality-minded*. Retrieved September 27, 2007, from http://www.freequality.org/sites/www_freequality_org/documents/Training/Classes Fall 2002/Six Sigma.doc

Chung, W., Chen, H., & Reid, E. (2009). Business stakeholder analyzer: An experiment of classifying stakeholders on the Web. *Journal of the American Society for Information Science and Technology, 60*(1), 59–74. doi:10.1002/asi.20948

Churchill, G. A. (1979). A paradigm for developing better measures of marketing constructs. *JMR, Journal of Marketing Research, 14*, 64–73. doi:10.2307/3150876

Citrin, A., Stern, D., Spangenberg, E., & Clark, M. (2003). Consumer need for tactile input. An Internet retailing challenge. *Journal of Business Research*, *56*(11), 915–922. doi:10.1016/S0148-2963(01)00278-8

Clemons, E. K., & Row, M. C. (1992). Information Technology and industrial cooperation: The changing economics of coordination and ownership. *Journal of Management Information Systems*, *9*(2), 9–28.

Cockburn, C., & Wilson, T. D. (1996). Business Use of the World-Wide Web. *International Journal of Information Management*, *16*(2), 83–102. doi:10.1016/0268-4012(95)00071-2

Cohen, J. (1988). *Statistical Power Analysis for the Behavioral Sciences* (2nd ed.). Hillsdale, NJ: Lawrence Erlbaum Associates.

Cohn, J. R., & Goodenough, B. (2002). Health professionals attitudes to videoconferencing in pediatric heathcare. *Journal of Telemedicine and Telecare*, *8*(5). doi:10.1258/135763302760314243

Colbeck, D., & Evans, I. (2005). IT governance. In Colbeck, D., & Evans, I. (Eds.), *The complete guide to IT service management* (1st ed., pp. 1–2). Ottawa, Canada: ColbeckEvans.

Collier, J. E., & Bienstock, C. C. (2006). Measuring service quality in e-retailing. *Journal of Service Research*, *8*(3), 260–275. doi:10.1177/1094670505278867

Conger, S. (2007). *Information Technology service management: An emerging area for academic research and pedagogical development*. New York, NY: ACM Digital Library.

Consolvo, S., Smith, I., Matthews, T., LaMarca, A., Tabert, J., & Powledge, P. (2005, April 2-7). Location Disclosure to Social Relations: Why, When, & What People Want to Share. Portland, Oregon, USA.

Consortium Service Innovation. (2002). *Getting started with KCS.* Official Consortium Service Innovation Site. Retrieved on November 30, 2010, from http://www.thinkhdi.com/files/pdfs/GettingStartedKCS.pdf

Cooper, V., Lichtenstein, S., & Smith, R. (2009). Successful Web-based IT support services: Service provider perceptions of stakeholder-oriented challenges. *International Journal of E-Services and Mobile Applications*, *1*(1), 1–20. doi:10.4018/jesma.2009092201

Cooper, V. A., Lichtenstein, S., & Smith, R. (2006a). Knowledge Transfer in Enterprise Information Technology Support using Web-based Self-service Systems. *International Journal of Technology Marketing*, *1*(2), 145–170.

Cooper, R. B., & Zmud, R. W. (1990). Information Technology implementation research: A technological diffusion approach. *Management Science*, *36*(2), 123. doi:10.1287/mnsc.36.2.123

Cooper, V. A. (2007). Knowledge Transfer in Enterprise IT Support Provisions using Web-based Self-Service, unpublished PhD Thesis, Deakin University, Melbourne, Australia.

Cooper, V. A., Lichtenstein, S., & Smith, R. (2006b). Enabling the Transfer of Information Technology Support Knowledge to Enterprise Customers Using Web-based Self-service Systems: Critical Success Factors from the Support Organisations Perspective, in Proceedings of the Seventeenth Australasian Conference on Information Systems (ACIS 2006), Adelaide, Australia.

Cougar, J. D. (1984). Research issues in Information Systems personnel management. In McFarlan, F. W. (Ed.), *Fifth Anniversary Research Colloquium: The Information Systems Research Challenge* (pp. 217–230). Boston, MA: Harvard Business School Press.

Cowan, R., David, P. A., & Foray, D. (2000). The explicit economics of knowledge codification and tacitness. *Industrial and Corporate Change*, *9*(2), 211–253. doi:10.1093/icc/9.2.211

Cox, D. F. (1967). *Risk taking and information handling in consumer behaviour*. Harvard University press.

Cranefield, J., & Yoong, P. (2005). Organisational factors affecting inter-organisational knowledge transfer in the New Zealand state sector: A case study. *The Electronic Journal for Virtual Organisations and Networks, 7*, December.

Crepeau, R. G., Crook, C. W., Goslar, M. D., & McMurtrey, M. E. (1992). Career anchors of Information Systems personnel. *Journal of Management Information Systems*, *9*(2), 145–160.

Cronin, J., Brady, M. K., & Hult, G. T. (2000). Assessing the effects of quality, value, and customer satisfaction on consumer behavioral intentions in service environments. *Journal of Retailing*, *76*(2), 193–218. doi:10.1016/S0022-4359(00)00028-2

Curran, J., & Meuter, M. (2005). Self-service technology adoption: Comparing three technologies. *Journal of Services Marketing*, *19*(2), 103–113. doi:10.1108/08876040510591411

Cyr, D., Bonnanni, C., Bowes, J., & Ilsever, J. (2005). Beyond Trust: Web site design preferences across cultures. *Journal of Global Information Management*, *13*(4), 24–52. doi:10.4018/jgim.2005100102

Dabholkar, P. A. (1996). Consumer evaluations of new technology-based self-service options: An investigation of alternative models of service quality. *International Journal of Research in Marketing*, *13*(1), 29–51. doi:10.1016/0167-8116(95)00027-5

Dabholkar, P. A. (1991). Using technology-based self-service options to improve perceived service quality. In M. C. Gilly, et al. (Eds.), *AMA Summer Educator's Conference Proceedings* (pp. 534-535). Chicago, IL: American Marketing Association.

Dahlberg, T., & Nyrhinen, M. (2006). A new instrument to measure the success of IT outsourcing. In *Proceedings of the 39th Hawaii International Conference on System Sciences.* Retrieved on November 30, 2010, from http://csdl2.computer.org/comp/proceedings/hicss/2006/2507/08/250780200a.pdf

Dahlen, M. (2002). Learning the web: Internet User Experience and Response to Web Marketing in Sweden. *Journal of Interactive Advertising (online)*, *3*(1), Retrieved October 16, 2008, from http://www.jiad.org/article27

Dailey, L. (2004). Navigational web atmospherics. Explaining the influence of restrictive navigation cues. *Journal of Business Research*, *57*(7), 1–9. doi:10.1016/S0148-2963(02)00364-8

Damanpour, F. (1987). The adoption of technological, administrative, and ancillary innovations: Impact of organizational factors. *Journal of Management*, *13*, 675–688. doi:10.1177/014920638701300408

Damiano, F., & McLauglin, K. (2007). *American ITIL*. New York, NY: ACM Digital Library.

Damsgaard, J., & Kelleher, C. (2007). *What drives the innovation, diffusion and adoption of mobile services? - An analysis of four alternative studies*. Los Angeles: Presented at LA Global Mobility Round Table.

Damsgaard, J., & Lyytinen, K. (1996). Government Strategies to Promote the Diffusion of Electronic Data Interchange (EDI): What we Know and What we don't Know. *Information Infrastructure and Policy*, *5*(3), 169–190.

Daniel, E. (1999). Provision of electronic banking in the UK and the Republic of Ireland. *International Journal of Bank Marketing*, *17*(2), 72–82. doi:10.1108/02652329910258934

Dansky, K. H., & Ajello, J. (2005). Marketing telehealth to align with strategy. *Journal of Healthcare Management*, *50*(1), 19–30.

Davis, F. D., Bagozzi, R. P., & Warshaw, P. R. (1989). User acceptance of computer technology: A comparison of two theoretical models. *Management Science*, *35*, 982–1003. doi:10.1287/mnsc.35.8.982

Davis, F. D. (1989). Perceived usefulness, perceived ease of use, and user acceptance of computer technology. *Management Information Systems Quarterly*, *13*(3), 319–340. doi:10.2307/249008

Davis, F. D., & Venkatesh, V. (2004). Toward Preprototype User Acceptance Testing of New Information Systems: Implications for Software Project Management. *IEEE Transactions on Engineering Management*, *51*(1), 31–46. doi:10.1109/TEM.2003.822468

Davis, F. D., Bagozzi, R. P., & Warshaw, P. R. (1989). User Acceptance of Computer Technology: A Comparison of Two Theoretical Models. *Management Science*, *35*(8), 982–1002..doi:10.1287/mnsc.35.8.982

Davis, F. D. (1989). Perceived usefulness, perceived ease of use, and user acceptance of information technology. *MIS Quarterly*, *13*(3), 319–340. doi:10.2307/249008

Davis, C. (2009). Plastic Logic announce content partners: USA Today, Zinio and more. Retrieved February 15th, 2009. Available at: http://www.slashgear.com/plastic-logic-announce-content-partners-usa-today-zinio-and-more-0933577/

Dawson, J. F., Gonzalez-Roma, V., Davis, A., & West, M. A. (2008). Organizational climate and climate strength in UK hospitals. *European Journal of Work and Organizational Psychology, 17*(1), 89. doi:10.1080/13594320601046664

De Reuver, M., & Haaker, T. (2009, August). Designing viable business models for context-aware mobile Services. *Telematics and Informatics, 26*(3), 240–248. doi:10.1016/j.tele.2008.11.002

De Reuver, M. (2008). The mobile context explored. Bouwman, H., De Vos, H., & Haaker, T. (Eds.), *Mobile service innovation and business models.* Heidelberg-Berlin: Springer.

Debreceny, R., & Gray, G. (2001). Financial reporting on the Internet and the Internet and the external audit. *European Accounting Review, 8*(2), 335–350. doi:10.1080/096381899336078

Deise, M. V., Nowikow, C., King, P., & Wright, A. (2000). *Exexutive's Guide to E-business: From Tactics to Strategy* (p. 272). Pricewaterhousecoopers L.L.P.

Del Aguila-Obra, A. R., Padilla-Melendex, A., & Serarols-Tarres, C. (2007). Value creation and new intermediaries on Internet. An exploratory analysis of the online news industry and the Web content aggregators. *International Journal of Information Management, 27*, 187–199. doi:10.1016/j.ijinfomgt.2006.12.003

Dell, M. (2000). *Direct from Dell: Strategies that revolutionized an industry*. New York, NY: Collins.

DeMarez, L., Vyncke, P., Berte, K., Schuurman, D., & De Moor, K. (2007). Adopter segments, adoption determinants and mobile marketing. *Journal of Targeting. Measurement and Analysis for Marketing, 16*(1), 78–95. doi:10.1057/palgrave.jt.5750057

DeSanctis, G., & Poole, M. S. (1994). Capturing the complexity in advanced technology use: Adaptive structuration theory. *Organization Science, 5*(2), 121–147. doi:10.1287/orsc.5.2.121

DeVellis, R. (1991). *Scale development: theory and applications.* London: Sage Publications.

Dholakia, R., & Uusitalo, O. (2002). Switching to Electronic Stores: Consumer Characteristics and the Perception of Shopping Benefits. *International Journal of Retail and Distribution Management, 30*(10), 459–469. doi:10.1108/09590550210445335

DHSS. (2001). *Report to Congress on telemedicine*. Secretary of Health and Human Services.

Diamantopoulos, A., Riefler, P., & Roth, K. P. (2008). Advancing Formative Measurement Models. *Journal of Business Research, 61*(12), 1203–1218..doi:10.1016/j.jbusres.2008.01.009

Diamantopoulos, A., & Siguaw, J. (2000). *Introducing Lisrel: A Guide for the Uninitiated.* London: Sage.

Diamantopoulos, A. (1996). A model of publication performance of marketing academics. *International Journal of Research in Marketing, 13*, 163–180. doi:10.1016/0167-8116(95)00039-9

Dicherson, M. D., & Gentry, J. W. (1983). Characteristics of adopters and non-adopters of home computers. *The Journal of Consumer Research, 10*(2), 225–235. doi:10.1086/208961

Ding, X., Verma, R., & Iqbal, Z. (2007). Self-service technology and online financial service choice. *International Journal of Service Industry Management, 18*(3), 246–268. doi:10.1108/09564230710751479

Diniz, E. (1998). Web banking in USA, *Journal of Internet Banking and E-commerce,*3(2), online at: http://www.arraydev.com/commerce/JIBC/9806-06.htm

Dixon, M., & Nixon, B. (2000). *E-banking: Managing your money and transactions online*. SAMS publishing, pp.244.

Dockhorn Costa, P., Ferreira Pires, L., & van Sinderen, M. (2005). Designing a Configurable Services Platform for Mobile Context-Aware Applications. [JPCC]. *International Journal of Pervasive Computing and Communications, 1*(1), 13–25.

DomainTools. (2008). Domain counts & Internet statistics. Retrieved May, 2008, from: http://www.domaintools.com/internet-statistics/.

Donahue, K. B. (2001). Time to get serious about talent management. *Harvard Business Review*, *79*(7), 6–7.

Donovan, R. J., Rossiter, J. R., Marcoolyn, G., & Nesdale, A. (1994). Store atmosphere and purchasing behaviour. *Journal of Retailing*, *70*(4), 283–294. doi:10.1016/0022-4359(94)90037-X

Doolittle, G., Williams, A., & Cook, D. J. (2003). An estimation of costs of a pediatric telemedicine practice in public schools. *Medical Care*, *41*, 100–109. doi:10.1097/00005650-200301000-00012

Doolittle, G., Williams, A., & Harmon, A. (1998). A cost measurement study for a tele-oncology practice. *Journal of Telemedicine and Telecare*, *4*, 84–88. doi:10.1258/1357633981932000

Douglas, A. K. (2004). *Marketing your firm on the Internet. Tax Practice Management*. January-February.

Driedonks, C., Gregor, S., & Wassenaar, A. (2005). Economic and social analysis of the adoption of B2B electronic marketplaces: A case study in the Australian beef industry. *International Journal of Electronic Commerce*, *9*(3), 49–72.

Drossos, D., Giaglis, G. M., Lekakos, G., Kokkinaki, F., & Stavraki, M. G. (2007). Determinants of Effective SMS Advertising: An Experimental Study. *Journal of Interactive Advertising*, *7*(2).

Drucker, P. (2001). Management by objectives and self-control. In Drucker, P. (Ed.), *The essential Drucker* (1st ed., p. 112). New York, NY: HarperCollins.

Ducoffe, R. H. (1996). Advertising value and advertising on the Web. *Journal of Advertising Research*, *36*(5), 21–35.

Dumas, J. S. (2003). User-Based Evaluations. In Jacko, J. A., & Sears, A. (Eds.), *The Human-Computer Interaction Handbook* (pp. 1093–1117). Lawrence Erlbaum Associates.

Dunn, W. N. (1981). *Public policy analysis*. Englewood Cliffs, NJ: Prentice-Hall.

Duplantie, J., & Gagnon, M.-P. (2007). Telehealth and the recruitment and retention of physicians in rural and remote regions: A Delphi study. *Canadian Journal of Rural Medicine*, *12*(1), 30–36.

Dychtwald, K., Erickson, T., & Morison, B. (2004). It's time to retire retirement. *Harvard Business Review*, *82*(3), 48–57.

Easingwood, C. J. (1986). New product development for service companies. *Journal of Product Innovation Management*, *3*(4), 264–275. doi:10.1016/0737-6782(86)90005-6

Eastin, M. S. (2001). Credibility Assessments of Online Health Information: The Effects of Source Expertise and Knowledge of Content. *Journal of Computer-Mediated Communication*, *6*(4).

Eaton, K. (2009). Forget Kindle 2: Fujitsu's E-Reader Screen is Bigger, and It's in Color. Retrieved February 15th, 2009. Available at: http://www.fastcompany.com/blog/kit-eaton/technomix/forget-kindle-2-fujitsus-e-reader-screen-bigger-color

Eckhardt, A., Laumer, S., & Weitzel, T. (2008). *Extending the architecture for a next-generation holistic e-recruiting system*. In International Conference on Information Resources Management (Conf-IRM), Niagara Falls, Ontario, Canada.

Eckhardt, A., von Stetten, A., & Laumer, S. (2009). Value contribution of IT in recruiting – A multi-national causal analysis. In *Proceedings of the 2009 ACM SIGMIS CPR Conference* (SIGMIS 2009), Limerick.

Eckhardt, A., Weitzel, T., Koenig, W., & Buschbacher, J. (2007a). How to convince people who don't like IT to use IT - A case study on e-recruiting. In *Proceedings of the 13th Americas Conference on Information Systems* (AMCIS 2007), Keystone, Colorado.

Edmondson, A. C., Bohmer, M. R., & Pisano, G. P. (2001). Disrupted routines: Team learning and new technology implementation in hospitals. *Administrative Science Quarterly*, *46*(4), 685. doi:10.2307/3094828

Ellen, P., Bearden, W., & Sharma, S. (1991). Resistance to technological innovations: an examination of the role of self-efficacy and performance satisfaction. *Journal of the Academy of Marketing Science*, *19*(4), 297–307. doi:10.1007/BF02726504

Elsbach, K. D., Barr, P. S., & Hargadon, A. B. (2005). Identifying situated cognition in organization. *Organization Science*, *16*(4), 422–435. doi:10.1287/orsc.1050.0138

Emarketer (2008). European B2C e-commerce: spotlight on the UK, Germany and France. Retrieved March, 2008, from: http://www.emarketer.com/Report.aspx?code=emarketer_2000426.

Ernst and Young. (2001), Global online retailing, *An Ernst and Young Special Report*, pp.142.

Eroglu, S. A., Machleit, K. A., & Davis, L. M. (2001). Atmospheric Qualities of Online Retailing: a Conceptual Model and Implications. *Journal of Business Research. Special Issue on Retail Strategy and Consumer Decision Research*, *54*(2), 177–184.

eScope (2007). Overview of the European Internet landscape. Retrieved November, 2008, from: http://blog.boom.ro/documente/prezentari/EscapeReports.ppt#1

Ettredge, M., Richardson, V., & Scolz, S. (2002). Dissemination of information for investors at corporate websites. *Journal of Accounting and Public Policy*, *21*, 357–369. doi:10.1016/S0278-4254(02)00066-2

European Central Bank. (2008, October). *EU Banking Structures.* Retrieved November 5, 2008, from http://www.ecb.int/pub /pdf/other/eubanking structures2008en.pdf.

European Commission. (2008a). *Towards a Single European Telecoms Market: Focus on Finland, Information Society and Media report.* Retrieved March 27, 2009, from http://ec.europa.eu/information_society/tl/research/

European Commission. (2008b). *Towards a Single European Telecoms Market: Focus on the Netherlands, Information Society and Media report.* Retrieved March 27, 2009, from http://ec.europa.eu/information_society/tl/research/

European Interactive Advertising Association. (2008). EIAA online shoppers 2008 – Report. Retrieved November, 2008, from http://www.eiaa.net/research/research-details.asp?SID=1&id=410&lang=6&origin=media-consumption.asp

European Union Study Report. (2008). ICT and e-Business Impact in the Banking Industry. *Study Report No. 06/2008*. Retrieved October 21, 2008, from Eurostat (2007). *European business: Facts and figures, Eurostat, Luxembourg.* Retrieved May 24, 2008, from http://epp.eurostat.ec.europa.eu

Eurostat (2008). Internet as a data source. Retrieved November, 2008, from: http://epp.eurostat.ec.europa.eu/portal/page?_pageid=2973,64549069,2973_64554228&_dad=portal&_schema=PORTAL

Evanschitzky, E., & Iyer, G. (Eds.). (2007). *E-services – Opportunities and threats. Journal of Value Chain Management, 1(1/2) (Special Issue)*. Wiesbaden, Germany: Deutscher Universitäts-Verlag.

Eveland, J. D., & Tornatzky, L. (1990). The deployment of technology. In Tornatzky, L., & Fleischer, M. (Eds.), *The processes of technological innovation*. Lanham, MD: Lexington Books.

Faerber, F., Keim, T., & Weitzel, T. (2003). An automated recommendation approach to personnel selection. In *Proceedings of the 2003 Americas Conference on Information Systems*, Tampa, 2003

Falk, R., & Miller, N. (1992). *A Primer for Soft Modelling*, Akron, OH: University of Akron Press.

Farrell, D., & Grant, A. J. (2005). China's looming talent shortage. *The McKinsey Quarterly*, 4.

Ferdinand, J., & Simm, D. (2007). Re-theorizing external learning: Insights from economic and industrial espionage. *Management Learning*, *38*(3), 297–317. doi:10.1177/1350507607079030

Ferris, M. (2007). Insights on Mobile Advertising, Promotion and Research. *Journal of Advertising Research*, *47*(1), 28–37..doi:10.2501/S0021849907070043

Fichman, R. G., & Kemerer, C. F. (1997). The assimilation of software process innovations: An organizational learning perspective. *Management Science*, *43*(10), 1345. doi:10.1287/mnsc.43.10.1345

Fichman, R. G., & Kemerer, C. F. (1999). The illusory diffusion of innovation: An examination of assimilation gaps. *Information Systems Research*, *10*(3), 255. doi:10.1287/isre.10.3.255

Field, J., Heim, G., & Sinha, K. (2004). Managing quality in the e-service system: Development and application of a process model. *Production and Operations Management*, *13*(4), 291–306. doi:10.1111/j.1937-5956.2004.tb00219.x

Field, K. S., & Briggs, D. J. (2001). Socio-economic and locational determinants of accessibility and utilization of primary healthcare. *Health & Social Care in the Community*, 9.

Fife, E. (2007). *Worldwide Mobile Data Services Study 2007*. Retrieved June 5, 2008, from http://www.marshall.usc.edu /assets/041/ 9099.pdf

Fink, D. (2007). Antecedents for building trust in professional e-services. *Integration and Innovation Orient to E-Society*, *2*, 41–49.

Finkelstein, J., Khare, R., & Vora, D. (2003). Home automated telemanagement system to facilitate self-care of patients with chronic diseases. *Journal of Systemics. Cybernetics and Informatics*, *44*, 35–56.

Fishbein, M., & Azjen, I. (1975). *Belief, attitude, intention and behavior: An introduction to theory and research*. Reading, MA: Addison-Wesley.

Flavián, C., Guinalíu, M., & Gurrea, R. (2006). The Role Played by Perceived Usability, Satisfaction and Consumer Trust on Website Loyalty. *Information & Management*, *43*, 1–14. doi:10.1016/j.im.2005.01.002

Flavián, C., Gurrea, R., & Orús, C. (2008, July). *The relevance of visual aspects for the website store*. Paper presented at the 15th International Conference on Retailing and Consumer Services Science, Zagreb, Croatia.

Flinders, K. (2008). E-bank Japan sets mobile banking example. *Computer Weekly*. Retrieved October 14, 2008, from http://www.computerweekly.com / Articles/2008/10/13/232634/ e-bank-japan-sets-mobile -banking-example.htm

Forbes, W., & Rothschild, M. (2000). *Toward an Understanding of Consumer Experience on the Internet: Implications for Website Design*. Paper presented at the 33rd Hawaii International Conference on Systems Sciences, Hawaii, Unitated States.

Fornell, C., & Larcker, D. F. (1981). Evaluating structural equation models with unobservable variables and measurement error. *JMR, Journal of Marketing Research*, *18*(1), 39–50..doi:10.2307/3151312

Foster, S. (2007). *Managing quality: Integrating the supply chain* (3rd ed.). Upper Saddle River, NJ: Pearson.

Foster, S. (2007). Implementing and validating the quality system. In Foster, S. (Ed.), *Managing quality: Integrating the supply chain* (3rd ed., p. 509). Upper Saddle River, NJ: Pearson.

Fowler, M. (2000). *Distilled: A brief guide to the standard object modeling language*. Addison-Wesley Professional Series.

Frambach, R. T., Roest, H. C. A., & Krishnan, T. (2007). The impact of consumer internet experience on channel preference and usage intentions across the different stages of the buying process. *Journal of Interactive Marketing*, *21*(2), 26–41. doi:10.1002/dir.20079

Frank, F. D., Finnegan, R. P., & Taylor, C. R. (2004). The race for talent: Retaining and engaging workers in the 21st century. *Human Resource Planning*, *27*(3), 12–25.

Frosch, M., & Trost, A. (2008). Die Trends im Talentmanagement. *Personalmagazin*, *1*, 50–51.

Fry, M., & Bolt, M. (2004). *Combining ITIL® and Six Sigma to improve Information Technology service management at General Electric.*

Furst, K., Lang, W.W. & Nolle, D.E. (2002). Internet Banking, *Journal of Financial services Research*, 22(1,2), 95-117.

Gagnon, M.-P., Fortin, J. P., & Cloutier, A. (2001). La télémédecine au service d'une région: Recherche évaluative sur le projet des Iles-de-la-Madeleine. *Ruptures*, *8*, 53–70.

Gallivan, M. J. (2001). Organizational adoption and assimilation of complex technological innovations: Development and application of a new framework. *The Data Base for Advances in Information Systems*, *32*(3), 51.

Garbade, K. (1982). *Securities markets*. New York, NY: McGraw-Hill.

Garbarino, E., & Strahilevitz, M. (2004). Gender differences in the perceived risk of buying online and the effects of receiving a site recommendation. *Journal of Business Research*, *57*, 768–775. doi:10.1016/S0148-2963(02)00363-6

Garner, R., & Weldon, D. (1998, January 26). The numbers game. *Computerworld*, *32*(4), 88–93.

Gassmann, O. (2006). Opening up the innovation process: towards an agenda. *R & D Management*, *36*(3), 223–228. doi:10.1111/j.1467-9310.2006.00437.x

Gatewood, R. D., Gowan, M. A., & Lautenshlager, G. J. (1993). Corporate image, recruitment image, and initial job choice decisions. *Academy of Management Journal*, *36*(2), 414–427. doi:10.2307/256530

Gatignon, H., & Robertson, T. S. (1985). A proposal inventory for new diffusion research. *The Journal of Consumer Research*, *1*(March), 249–267.

Gauvin, S., & Sinha, R. (1991). Innovativeness In industrial Organisations: A two-Stage Model of Adoption». Working paper, University of Laval. Canada.

Gebauer, J., & Shaw, M. (2002). Introduction to the special section: Business-to-business electronic commerce. *International Journal of Electronic Commerce*, *6*(4), 7–17.

Gefen, D. (2000). E-Commerce: The Role of Familiarity and Trust. *OMEGA: The International Journal of Management Science*, *28*, 725–737. doi:10.1016/S0305-0483(00)00021-9

Geib, M., Kolb, L., & Brenner, W. (2006). Collaborative customer management in financial services alliances. In Fjermestad, J., & Romano, N. C. (Eds.), *Electronic customer relationship management. New York, NY.* Armonk.

Gellman, R. (1996). Disintermediation and the Internet. *Government Information Quarterly*, *13*(1), 1–8. doi:10.1016/S0740-624X(96)90002-7

George, D. I., & Smith, M. C. (1990). An empirical comparison of self-assessment and organizational assessment in personnel selection. *Public Personnel Management*, *19*(2), 175–190.

Gerbing, D. W., Hamilton, J. G., & Freeman, E. B. (1994). A large-scale second-order structural equation model of the influence of management participation on organizational planning benefits. *Journal of Management*, *20*(4), 859–885. doi:10.1016/0149-2063(94)90033-7

Gerrad, P., & Cunningham, J. B. (2003). The diffusion of Internet banking among Singapore consumers. *International Journal of Bank Marketing*, *21*(1), 16–28. doi:10.1108/02652320310457776

Giaglis, G. M., Kourouthanassis, P., & Tsamakos, A. (2002). Towards a Classification Framework for Mobile Location Services. In B.E.Mennecke & T.J. Strader (Eds.), *Mobile Commerce: Technology, Theory, and Applications*. Hershey, PA: Idea Group Publishing.

Giddens, A. (2005). *La Constitution de la Société: éléments de la théorie de la structuration (traduit de l'anglais par Michel Audet)*. Paris, France: Quadridge PUF.

Gilly, M., & Zeithaml, V. (1985, December). The Elderly Consumer and Adoption of Technologies. *The Journal of Consumer Research*, *12*, 353–357. doi:10.1086/208521

Gogia, S. B. (2008). Providing tele-mental health services after disasters – A case based on the post tsunami experience. In Scupola, A. (Ed.), *Cases on managing e-services* (pp. 238–252). Hershey, PA: Idea Group Inc. doi:10.4018/9781605660646.ch016

Gong, H., Tate, M., & Alborz, S. (2007). Managing the outsourcing marriage to achieve success. In *Proceedings of Hawaii International Conference on the System Sciences* (HICSS 2007. IEEE Society Press. Gutek, B., Bhappu, A. D., Liao-Troth, M. A., & Cherry, B. (1999). Distinguishing between service relationships and encounters. *The Journal of Applied Psychology*, *84*(2), 218–233.

Gopal, R., & Tripathi, A. (2006). Advertising via wireless networks. *International Journal of Mobile Communications*, *4*(1), 1–16..doi:10.1504/IJMC.2006.008194

Gosain, S. (2004). Enterprise Information Systems as objects and carriers of institutional forces: The new iron cage. *Journal of the Association for Information Systems*, *5*(4), 151–182.

Gournaris, S., & Dimitriadis, S. (2003). Assessing service quality on the Web: Evidence from business-to-consumer portals. *Journal of Services Marketing*, *17*(5), 529–548. doi:10.1108/08876040310486302

Gourville, J. T. (2003). *Why consumers don't buy: the psychology of new product adoption.* Harvard Business School Publishing.

Govekar, M., Curtis, D., Brittain, K., Scott, D., Mingay, S., & Holub, E. ... Phifer, G. (2007, June 22). *Hype cycle for IT operations management*, 2007 (ID Number: G00148631). Retrieved July 28, 2007 from http://www.gartner.com

Gravelle, H., & Sutton, M. (2001). Inequality in the geographical distribution of general practitioners in England and Wales 1974-1995. *Journal of Health Service Resources Policy, 6*.

Gregor, S. (2006). The nature of theory in Information Systems. *Management Information Systems Quarterly, 30*(3), 611–642.

Gregor, S., & Jones, D. (2004). The formulation of design theories. In Linger, H., Fisher, J., Wojtkowski, W., Zupaneie, J., Vigo, K., & Arold, J. (Eds.), *Constructing the infrastructure for the knowledge economy: Methods and tools, theory and practice* (pp. 83–93). Boston, MA: Kluwer Academic Publishers.

Gregory, K. L. (1983). Native-view paradigms: Multiple cultures and culture conflicts in organization. *Administrative Science Quarterly, 28*(3), 359–376. doi:10.2307/2392247

Grémy, F., & Bonnin, M. (1995). Evaluation of automatic health Information Systems. What and how? E. In Van Gennip, & J. Talmon (Eds.), *Assessment and evaluation of Information Technologies in medicine*. Amsterdam, The Netherlands: IOS Press.

Grönroos, C. (2001). *Service management and marketing: A customer relationship management approach*. Chichester, UK: Wiley.

Grow, B., & Hindo, B. (June 11, 2007). Six Sigma: So yesterday? In an innovation economy, it's no longer a cure-all. *Business Week*. Retrieved November 23, 2007, from http://www.businessweek.com/innovate/content/feb2007/id20070227_766365_page_2. html

Gulliford, M. C. (2002). Avaibility of primary care doctors and population in England: Is there an Association? *Journal of Public Health Medicine*, 24.

Gullkvist, B. (2008). Emerging e-services in accounting: A longitudinal case study. In Scupola, A. (Ed.), *Cases on managing e-services*. Hershey, PA: Idea Group Inc. doi:10.4018/9781605660646.ch007

Gurrea, R. (2006). *La Prensa Digital: Un Nuevo Escenario para el Análisis del Comportamiento del Lector de Periódicos*. Unpublished doctoral dissertation, University of Zaragoza, Spain.

Gutierrez-Nieto, B., Fuertes-Callen, Y., & Serrano-Cinca, C. (2008). Internet reporting in microfinance institutions. *Online Information Review, 32*(3), 415–436. doi:10.1108/14684520810889709

Haaker, T. I., & de Vos, H. (2007, September 2-5). Customer preferences for bundled content services. *18th European Regional ITS Conference*, Istanbul, Turkey.

Haenlein, M., & Kaplan, A. M. (2004). A beginner's guide to partial least squares analysis. *Understanding Statistics, 3*(4), 283–297..doi:10.1207/s15328031us0304_4

Hagedorn-Rasmussen, P. (2006). *Making sense of 'e-knowation' – Exploring the relationship between emerging strategy, innovation and entrepreneurial nets of critical capabilities and resources*. Working paper nr. 8. Lyngby: E-service project.

Hagel-III. J., Durchslag, S., & Brown, J. S. (2002). *Orchestrating loosely coupled business processes: The secret to successful business collaboration*. Unpublished manuscript.

Hair, J., Anderson, R., Tatham, R., & Black, W. (1998). *Multivariate Data Analysis* (5th ed.). Englewood Cliffs, NJ: Prentice-Hall.

Hall, G., & Loucks, S. (1977). A developmental model for determining whether the treatment is actually implemented. *American Educational Research Journal, 14*(3), 263–276.

Hall, M. (2008, February 14). Recession-proof MSP readies new service. *Computerworld*.

Hammer, M., & Goding, J. (2001). Putting Six Sigma in perspective. Quality, *40*(10), 58-62. (Document ID: 84943086).

Hammond, K. (2001). B2C e-Commerce 2000-2010: What Experts Predict. *Business Strategy Review, 12*, 43–50. doi:10.1111/1467-8616.00165

Hank, M. (2006). ITIL: What it is and what it isn't. *Business Communication Review, 36*(12). (WN: 063304081014).

Hansen, M. T., Nohria, N., & Tierney, T. (1999). What's your strategy for managing knowledge? *Harvard Business Review, 77*(2), 106–116.

Harre, R., & Madden, E. H. (1975). *Causal powers: A theory of natural necessity*. Oxford, UK: Blackwell.

Hegering, G.-G., Küpper, A., Linnhoff-Popien, C., & Reiser, H. (2004). Management Challenges of Context-Aware Services in Ubiquitous Environments. *Self-Managing Distributed Systems, Book Series Lecture Notes in Computer Science, Volume 2867/2004* (pp. 246-259). Berlin/Heidelberg: Springer.

Heinonen, K., & Strandvik, T. (2007). Consumer responsiveness to mobile marketing. *International Journal of Mobile Communications*, *5*(6), 603–617..doi:10.1504/IJMC.2007.014177

Helander, M. G., & Khalid, H. M. (2000). Modelling the customer in electronic commerce. *Applied Ergonomics*, *31*, 609–619. doi:10.1016/S0003-6870(00)00035-1

Henderson, K., & Cowart, L. (2002). Bucking e-commerce trends: A content analysis comparing commercial real estate brokerage and residential real estate brokerage websites. *Journal of Corporate Real Estate*, *4*(4), 375–385. doi:10.1108/14630010210811967

Henten, A. (1994). *Impacts of information and communication technologies on trade in services*. Lyngby, Denmark: Technical University of Denmark.

Henten, A. (2005). *Internationalisation of knowledge services: The case of engineering consultancy services – Ramboll*. Presentation at Nordic Conference on Innovation and Value Creation in the Service Economy. Oslo, 25-26 October 2005.

Hentze, J., & Kammel, A. (2001). *Personalwirtschaftslehre*, 7th ed. Stuttgart.

Hernandez, J. M. C., & Mazzon, J. A. (2007). Adoption of Internet banking: proposition and implementation of an integrated methodology approach. *International Journal of Bank Marketing*, *25*(2), 72–88. doi:10.1108/02652320710728410

Hess, C. M., & Kemerer, C. F. (1994). Computerized loan originating systems: An industry case study of the electronic markets hypothesis. *Management Information Systems Quarterly*, *18*(3), 251–274. doi:10.2307/249618

Hevner, A., March, S., Park, J., & Ram, S. (2004). Design scienc in Information Systems research. *Management Information Systems Quarterly*, *28*(1), 75–105.

Hevner, A. R., March, S. T., Park, J., & Ram, S. (2004). Design science in Information Systems research. *Management Information Systems Quarterly*, *28*(1), 75–105.

Hewer, P., & Howcroft, B. (1999). Consumers distribution channel adoption and usage in the financial services industry: A review of existing approaches. *Journal of Financial Services Marketing*, *3*(4), 344–358.

Hewitt, P. (2002). *Retirement trends foster global stagnation*. CSIS, April 5.

Hill, T. P. (1977). On goods and services. *Review of Income and Wealth*, *23*(4), 314–339. doi:10.1111/j.1475-4991.1977.tb00021.x

Hiltunen, M., Heng, L., & Helgesen, L. (2004). Personalized Electronic Banking Services. In C.-M. Karat, J. Blom & J. Karat (Eds.), *Designing Personalized User Experiences in eCommerce* (5), 119-140. Netherlands: Kluwer Academic Publishers.

Hindo, B. (2007, June 11). At 3M, a struggle between efficiency and creativity. *BusinessWeek*. Retrieved November 23, 2007, from http://www.businessweek.com/magazine/content/07_24/b4038406.htm

Hirchman, E. C. (1980). Innovativeness, Novelty Seeking And consumer Creativity. *The Journal of Consumer Research*, *7*, 283–295. doi:10.1086/208816

Hofacker, C. F., Goldsmith, R. E., Bridges, E., & Swilley, E. (2007). E-services: A synthesis and research agenda. *Journal of Value Chain Management*, *1*(1/2), 13–44.

Hoffman, K. D. (2003). Marketing + MIS = E-service. *Communications of the ACM*, *46*(6), 53–55. doi:10.1145/777313.777340

Holak, S. L. (1988). Determinants of innovative durable adoption: An empirical study with implications for early product screening. *Journal of Product Innovation Management*, *5*(1), 50–69. doi:10.1016/0737-6782(88)90032-X

Holbrook, M. B., & Hirschman, E. C. (1982). The experimental aspects of consumption: consumer fantasies, feelings, and fun. *The Journal of Consumer Research*, *9*, 325–345. doi:10.1086/208906

Holland, C. P., & Lockett, A. G. (1997). Mixed mode network structures: The strategic use of electronic communication by organizations. *Organization Science*, *8*(5), 475–488. doi:10.1287/orsc.8.5.475

Holsapple, C. W., & Jin, H. (2007). Connecting some dots: E-commerce, supply chains and collaborative decision making. *Decision Line*, *38*(5), 14–21.

Homer, P. M. (1990). The Mediating Role of Attitude Toward the Ad: Some Additional Evidence. *JMR, Journal of Marketing Research*, *XXVII*, 78–86..doi:10.2307/3172553

Hong, W., Thong, J. Y. L., & Tam, K. Y. (2004). Designing Product Listing Pages on E-Commerce Websites: An Examination of Presentation Mode and Information Format. *International Journal of Human-Computer Studies*, *61*, 481–503. doi:10.1016/j.ijhcs.2004.01.006

Howcroft, B., Hamilton, R., & Hewer, P. (2002). Consumer Attitude and the Usage and Adoption of Home-banking in the United Kingdom. *International Journal of Bank Marketing*, *20*(3), 111–121. doi:10.1108/02652320210424205

Hsu, M., & Chiu, C. (2004). Predicting electronic service continuance with a decomposed theory of planned behavior. *Behaviour & Information Technology*, *23*(5), 359–373. doi:10.1080/01449290410001669969

HTC P3300 PDA. (2008). Retrieved from http://www.htc.com/www /product.aspx?id=392

http://www.ebusiness-watch.org/studies/sectors/banking /banking.htm

Hu, P., & Chau, P. (1999). Physician acceptance of telemedicine technology: An empirical investigation. *Topics in Health Information Management*, *19*(4), 20–35.

Hulland, J. (1999). Use of Partial Least Squares (PLS) in strategic management research: A review of four recent studies. *Strategic Management Journal*, *20*(2), 195–204.. doi:10.1002/(SICI)1097-0266(199902)20:2<195::AID-SMJ13>3.0.CO;2-7

Hunaiti, Z., Garaj, V., Balachandran, W., & Cecelja, F. (2005). Mobile Link Assessment for Visually Impaired Navigation System. *Proceedings of the IEEE IMTC*, *2*, 883–887.

Hunaiti, Z., Almasri, S., Matter, N., Sedoyeka, E., & Fenton, A. (2008, April). Extending Location-Based Services into M-learning. *The 3rd International Conference on Interactive Mobile and Computer Aided Learning*, Jordan.

IFRA eNews initiative (2006), retrieved August 14th, 2008. Available at:http://www.ifra.com/website/website.nsf/html/CONT_ENEWS?OpenDocument&ENW&

Igbaria, M., & Siegel, S. R. (1992). The reasons for turnover of Information Systems personnel. *Information & Management*, *23*, 321–330. doi:10.1016/0378-7206(92)90014-7

Ihlström Eriksson, C., Kalling, T., Åkesson, M., & Fredberg, T. (2008). Business models for m-service– Exploring the e-newspaper case from a consumer view. *Journal of Electronic Commerce in Organizations*, *6*(2), 29–57. doi:10.4018/jeco.2008040103

Ihlström Eriksson, C., Åkesson, M., Bergqvist, M., & Ljungberg, J. (2009). Forming a value network - analyzing the negotiations between actors in the e-newspaper case. In *Proceedings of the 42nd Hawaii International Conference on System Sciences, Hawaii*. (CD-ROM), Computer Society Press (10 pages).

Ihlström, C. (2004). *The Evolution of a New(s) Genre*. Doctoral Dissertation, Göteborg University.

Ihlström, C. (2005). The e-newspaper innovation - converging print and online. In *Proceedings of the International Workshop on Innovation and Media: Managing Changes in Technology, Products and Processes, Stockholm*.

Ihlström, C., Åkesson, M., & Nordqvist, S. (2004). From Print to Web to e-paper - the challenge of designing the e-newspaper. In *Proceedings of ICCC 8th International Conference on Electronic Publishing, ELPUB 2004, Brasilia*, pp. 249-260.

Ihlström, C., Svensson, J., & Åkesson, M. (2005). Participatory Design of Future Every Day IT Artifacts - Engaging readers and publishers in designing the e-newspaper. In *Proceedings of the 28th Information Systems Research Seminar in Scandinavia, Norway*. CD-ROM.

Iivari, J. (1991). A paradigmatic analysis of contemporary schools of IS development. *European Journal of Information Systems*, *1*(4), 249–272. doi:10.1057/ejis.1991.47

Iivari, J., Hirschheim, R., & Klein, H. K. (1998). A paradigmatic analysis contrasting Information Systems development approaches and methodologies. *Information Systems Research*, *9*(2), 1–30. doi:10.1287/isre.9.2.164

Illich, I. (1973). *Tools for conviviality*. New York, NY: Harper and Row.

Im, I., Kim, Y., & Han, H.-J. (2008). The effects of perceived risk and technology type on users' acceptance of technologies. *Information & Management*, *45*, 1–9.

Infopro (2003). *The Business Handbook*. Beirut. Lebanon.

Insight, B. (2007). *Connected Navigation Devices*. LBS Research Series, www.berginsight.com

Insley, R., Al-Abed, H., & Fleming, T. (2002), What's the definition of E-banking, Journal of Online Banking, available at: http://www.bankersonline.com/technology/gurustech081803d.html. Consulted in March 2006.

Institute of Medicine (IOM). (1999). *To err is human: Building a safer health system*. Washington, DC: National Academic Press.

Integrated Mobile. (2007). *Specialty apparel retailer improves satisfaction and productivity while driving down the cost of wireless*. Retrieved on August 5, 2008, from http://www.integratedmobile.com/case-studies

Internet World Stats. (2008). Home page. Retrieved May, 2008, from: www.internetworldstats.com.

Iqbal, Z., Verma, R., & Baran, R. (2003). Understanding consumer choices and preferences in transaction-based e-services. *Journal of Service Research*, *6*(1), 51–65. doi:10.1177/1094670503254276

ITIL Organization. (2007). Retrieved November 15, 2007, from http://www.itil-officialsite.com/home/home.asp

Iyer, K. N. S., Germain, R., & Frankwick, G. L. (2004). Supply chain B2B e-commerce and time-based delivery performance. *International Journal of Physical Distribution & Logistics Management*, *34*(8), 645–661. doi:10.1108/09600030410557776

Iyouit (2009). www.iyouit.eu. Accessed 31 March 2009.

Jackson, R., & Howe, N. (2004). *The graying of the middle kingdom*. Center for Strategic & International Studies and the Prudential Foundation. April.

Jacoby, J., & Kaplan, L. (1972), The components of perceived risk. *Advances in Consumer Research,* ed. Venkatesan M., Association for Consumer research, 287-291.

Jaeger, W., Jaeger, M., & Frickenschmidt, S. (2007). Verlust der Informationshoheit. *Personal*, *2*, 8–11.

Janda, S., Trocchia, P., & Gwinner, K. (2002). Consumer perceptions of Internet retail service quality. *International Journal of Service Industry Management*, *13*(5), 412–431. doi:10.1108/09564230210447913

Janssen, M., & Sol, H. G. (2000). Evaluating the role of intermediaries in the electronic value chain. *Internet Research: Electronic Networking Applications and Policy*, *19*(5), 406–417.

Janssen, M., & Verbraeck, A. (2005). Evaluating the information architecture of an electronic intermediary. *Journal of Organizational Computing and Electronic Commerce*, *15*(1), 35–60. doi:10.1207/s15327744joce1501_3

Janssen, M. (2009). The rebirth of intermediaries: New intermediaries for managing and orchestrating organizational networks using e-services. *International Journal of E-Services and Mobile Applications*, *1*(1). doi:10.4018/jesma.2009010104

Jarvenpaa, S. L., Tractinsky, N., & Vitale, M. (2000). Consumer trust in an Internet store. *Information Technology Management*, *1*(1), 45–71. doi:10.1023/A:1019104520776

Jarvis, C. B., Mackenzie, S. B., & Podsakoff, P. M. (2003). A Critical Review of Construct Indicators and Measurement Model Misspecification in Marketing and Consumer Research. *The Journal of Consumer Research*, *30*, 199–218..doi:10.1086/376806

Javalgi, R. G., Martin, C. L., & Todd, P. R. (2004). The export of e-services in the age of technology transformation: Challenges and implications for international service providers. *Journal of Services Marketing*, *18*(7), 560–573. doi:10.1108/08876040410561884

Jayanti, R. K., & Burns, A. C. (1998). The antecedents of preventative healthcare behavior: An empirical study. *Journal of the Academy of Marketing Science*, *26*(1), 6–15. doi:10.1177/0092070398261002

Jelassi, T., & Enders, A. (2005). *Strategies for e-business: Creating value though electronic and mobile commerce*. UK: Pearson Education.

Jiang, Z., & Benbasat, I. (2007). Investigating the influence of the functional mechanisms of online product presentations. *Information Systems Research, 18*(4), 454–470. doi:10.1287/isre.1070.0124

Jingjun Xu, D. (2006-2007). The Influence of Personalization in Affecting Consumer Attitudes Toward Mobile Advertising in China. *Journal of Computer Information Systems*, •••, 9–19.

Jones, M. C., & Beatty, R. C. (1998). Towards the development of measures of perceived benefits and compatibility of EDI: A comparative assessment of competing first order factor models. *European Journal of Information Systems, 7*(3), 210–220. doi:10.1057/palgrave.ejis.3000299

Jun, M., & Cai, S. (2001). The key determinants of Internet banking service quality: A content analysis. *International Journal of Bank Marketing, 19*(7), 276–291. doi:10.1108/02652320110409825

Jun, M., & Cai, S. (2001). The key determinants of Internet banking service quality: A content analysis. *International Journal of Bank Marketing, 19*(7), 276–291. doi:10.1108/02652320110409825

Jun, J. W., & Lee, S. (2007). Mobile Media use and its Impact on Consumer Attitudes Toward Mobile Advertising. *International Journal of Mobile Marketing, 2*(1), 50–58.

Jungk, R., & Müllert, N. (1996). *Future Workshops - How to Create Desirable Futures*. London: Institute for Social Inventions.

Junglas, I. A., & Spitzmuller, C. (2005). A Research Model for Studying Privacy Concerns Pertaining to Location-Based Services. *Proceedings of the 38th Annual Hawaii International Conference* (pp. 180a-180b).

Kaas, K. P. (1990). Nutzen und Kosten der Werbung. *ZfbF, 42*(6), 492–504.

Kaasinen, E. (2005). User acceptance of location-aware mobile guides based on seven field studies. *Behaviour & Information Technology, 24*, 37–49. doi:10.1080/01449290512331319049

Kahn, J. (2004, May 30). The most populous nation paces a population crisis. *New York Times*.

Kalakota, R., & Whinston, A. B. (1996). *Frontiers of electronic commerce*. Reading, MA: Addison-Wesley Publishing Company, Inc.

Kalakota, R., & Robinson, M. (2001). *M-Business: The Race to Mobility*. McGraw-Hill Companies.

Kallio, T., Kankainen, A., Kaikkonen, A., Kekalainen, A., & Cankar, M. (2005). Usability Testing of Mobile Applications: A Comparison between Laboratory and Field Testing. *Journal of Usability Studies, 1*, 4–16.

Kandampully, J. (2003). B2B relationships and networks in the Internet age. *Management Decision, 41*(5), 443–451. doi:10.1108/00251740310479296

Kangas, E., & Kinnunen, T. (2005). Applying user-centered design to mobile application development. *Communications of the ACM, 48*(7), 55–59. doi:10.1145/1070838.1070866

Karahanna, E., Straub, W. D., & Norman, L. C. (1999). Information technology adoption across time: a cross-sectional comparison of pre-adoption and post-adoption beliefs. *Management Information Systems Quarterly, 23*(2), 183–213. doi:10.2307/249751

Karjaluoto, H., Mallita, M., & Pento, T. (2002). Factors underlying attitude formation towards online banking in Finland. *International Journal of Bank Marketing, 20*(6), 261–272. doi:10.1108/02652320210446724

Karjaluoto, H., & Alatalo, T. (2007). Consumers' attitudes towards and intention to participate in mobile marketing. *International Journal of Services Technology and Management, 8*(2/3), 155–173..doi:10.1504/IJSTM.2007.012866

Karjaluoto, H., Mattila, M., & Pento, T. (2002). Factors underlying attitude formation towards online banking in Finland. *International Journal of Bank Marketing, 20*(6), 261–272. doi:10.1108/02652320210446724

Kay, R. H. (1993). An exploration of theoretical and practical foundations for assessing attitudes toward computers: the computer attitude measure (CAM). *Computers in Human Behavior*, (9): 371–386. doi:10.1016/0747-5632(93)90029-R

Kayworth, T. R., Chatterjee, D., & Sambamurthy, V. (2001). Theoretical justification for IT infrastructure investments. *Information Resources Management Journal, 14*(3), 5. doi:10.4018/irmj.2001070101

Keim, T., & Weitzel, T. (2006). *Strategies for hiring IT professionals: An empirical analysis of employer and job seeker behavior on the IT labor market.* 12th Americas Conference on Information Systems, Acapulco, Mexico.

Keim, T., Malinowski, J., & Weitzel, T. (2005). Bridging the assimilation gap: A user-centered approach to IT adoption in corporate HR processes. In *Proceedings of the 2005 Americas Conference on Information Systems*, Omaha.

Kensing, F., & Halskov Madsen, K. (1991). "enerating Visions: Future Workshops and Metaphorical Design, in J. Greenbaum and M. Kyng (eds.), *Design at Work*, 155-168. Hillsdale: Lawrence Erlbaum Associates, Inc.

Khalifa, M., & Cheng, S. K. N. (2002). Adoption of Mobile Commerce: Role of Exposure. In *Proceedings of the 35th Hawaii International Conference on System Sciences, Hawaii*. (CD-ROM), Computer Society Press (10 pages).

Khalifa, M., & Shen, N. (2005). Effects of electronic customer relationship management on customer satisfaction: A temporal model. In *Proceeding of the 38th Annual Hawaii International Conference on System Sciences* (HICSS'05). IEEE Society Press.

Kim, E. B., & Eom, S. B. (2002). Designing Effective Cyber Store User Interface. *Industrial Management & Data Systems*, *102*(5), 241–251. doi:10.1108/02635570210428276

Kim, S., & Stoel, L. (2004). Apparel Retailers: Website Quality Dimensions and Satisfaction. *Journal of Retailing and Consumer Services*, *11*(2), 109–117. doi:10.1016/S0969-6989(03)00010-9

Kim, W. C., & Mauborgne, R. A. (2005). Blue ocean strategy: From theory to practice. *California Management Review*, *47*(3), 105–121.

Kim, S., Kim, H., & Lee, K. (2005). Unified Optimal Power Allocation Strategy for MIMO Candidates in 3GPP HSDPA. *ETRI Journal*, *27*(6), 769–776.

Kimberly, J. R., & Evanisko, M. J. (1981). Organization innovation: the influence of individual, organizational, and contextual factors on hospital adoption of technological and administrative innovations. *Academy of Management Journal*, *24*(4), 689–713. doi:10.2307/256170

King, J., & Applegate, L. (1997). The crisis in the case of research crisis. In DeGross, J., Lee, A., & Lieberman, J. (Eds.), *Information Systems and qualitative research*. London, UK: Chapman and Hill.

Kirsner, D., & Balbi, D. (1997). Les besoins des banques. *Banque*, *586*, 38–39.

Klecun-Dabrowska, E., & Cornford, T. (2002). *The organizing vision of telehealth*. Gdansk, Poland: ECIS.

Klefsjo, B., Wiklund, H., & Edgeman, R. L. (2001). Six Sigma seen as a methodology for total quality management. *Measuring Business Excellence*, *5*(1), 31-35. (Document ID: 82396437).

Kleijnen, M., Ruyter, K., & Wetzels, M. (2007). An assessment of value creation in mobile service delivery and the moderating role of time consciousness. *Journal of Retailing*, *83*(1), 33–46. doi:10.1016/j.jretai.2006.10.004

Kleijnen, M. H. P., De Ruyter, K., & Wetzels, M. G. M. (2007). An assessment of value creation in mobile service delivery and the moderating role of time consciousness. *Journal of Retailing*, *83*(1), 33–46. doi:10.1016/j.jretai.2006.10.004

Klemettinen, M. (Ed.). (2007). *Enabling Technologies for Mobile Services: The MobiLife Book*. Chichester, England: John Wiley & Sons Ltd.

Kling, R., & Iacono, C. (1989). The institutional character of computerized Information System. *Office Technology & People*, *1*(1), 24–43.

Kling, R., & Scacchi, W. (1982). The Web of computing: Computer technology as social organization. *Advances in Computers*, *21*, 1–90. doi:10.1016/S0065-2458(08)60567-7

Kling, R., & Dutton, W. H. (1982). The computer package, dynamic complexity. In Danziger, J. N., Dutton, W., Kling, R., & Kraemer, K. L. e. (Eds.), *Computers and politics: High technology in American local governments* (pp. 22–50). New York, NY: Columbia University Press.

Knight, L., & Pye, A. (2007). In the search for network learning: Some practical and theoretical challenges in the process research. In Gossling, T., Oerlemans, L. A. G., & Jansen, R. (Eds.), *Inside networks: A process view on multi-organisational partnerships*. UK: Edward Elgar Publishing Limited.

Knutsen, A. (2005). M-Service Expectancies and Attitudes: Linkages and Effects of First Impressions. In *Proceedings of the 38th Hawaii International Conference on System Sciences, Hawaii.* (CD-ROM), Computer Society Press (10 pages).

Kolodinsky, J. M., Hogarth, J. M., & Hilgert, M. A. (2004). The adoption of electronic banking technologies by US consumers. *International Journal of Bank Marketing*, *22*(4), 238–259. doi:10.1108/02652320410542536

Koolwaaij, J., Tarlano, A., Luther, M., Morhs, B., Battestini, A., & Vaidya, R. (2006, July). *ContextWatcher – Sharing context information in everyday life. International Conference on Web Technologies, Applications, and Services*, Calgary, Canada.

Kotler, P., Hayes, T., & Bloom, P. (2002). *Marketing professional services—Forward thinking strategies for boosting your business, your image and your profits* (2nd ed.). Paramius, NJ: Prentice Hall Press.

Kotler, P., Armstrong, G., Saunders, J., & Wong, V. (2002). *Introduction to marketing* (2nd ed.). London, UK: Pearsons Education.

Kotler, P. (1991). *Marketing Management: Analysis, Planning, Implementation and Control* (7th ed.). Englewood Cliffs, New Jersey: Prentice-Hall International Inc.

Kotler, P. (1999). *Principles of Marketing*. London: Prentice Hall Europe.

Kotler, P. Dimaulo, McDougall G.G. & Armstrong G. (1991), *Le Marketing de la théorie à la pratique* [Marketing : From theory to Practice]. Boucherville, Québec: G. Morin, 1991.

Kozlowski, S. W. J., & Hults, B. M. (1987). An exploration of climates for technical updating and performance. *Personnel Psychology*, *40*, 539–563. doi:10.1111/j.1744-6570.1987.tb00614.x

Kozlowski, S. W. J., & Klein, J. K. (2000). A multilevel approach to theory and research in organizations: Contextual, temporal, and emergent process. In Klein, J. K., & Kozlowski, S. W. J. (Eds.), *Multilevel theory, research, and methods in organizations* (pp. 3–90). San Francisco, CA: Jossey-Bass.

KPMG. (2006). *Consumers and Convergence, challenges and opportunities in meeting next generation customer needs*. KPMG International.

Kranenburg, H, & van,., Bargh, M.S., Iacob, S., & Peddemors, A. (2006). A Context Management Framework for Supporting Context Aware Distributed Applications. *IEEE Communications Magazine*, *44*(8), 67–74. doi:10.1109/MCOM.2006.1678112

Krishnan, R., Smith, M. D., & Telang, R. (2003). The economics of peer-to-peer networks. *Journal of Information Technology Theory*, *5*(3), 31–44.

Kroeber-Riel, W., & Weinberg, P. (2003). *Konsumentenverhalten* (8th ed.). München: Vahlen.

Krupinski, E. A., Engstrom, M., Baker, G., Levine, N., & Weinstein, R. S. (2004). The challenges of following patients and assessing outcomes in dermatology. *Journal of Telemedicine and Telecare*, *10*, 21–24. doi:10.1258/1357633043227764149

Kuisma, T., Laukkanen, T., & Hiltunen, M. (2007). Mapping the reasons for resistance to Internet banking: a means-end approach. *International Journal of Information Management*, *27*(2), 75–85. doi:10.1016/j.ijinfomgt.2006.08.006

Kumar, K., & van Dissel, H. G. (1996). Sustainable collaboration: Managing conflict and cooperation in interorganisational systems. *Management Information Systems Quarterly*, *20*(3), 279–300. doi:10.2307/249657

Kurnia, S., & Johnston, R. (2002). A review of approaches to EC-enabled IOS adoption studies. In *Proceedings of the 35th Annual Hawaii Conference on System Sciences* (HICSS'02), IEEE Society Press.

Kuusisto, A. (2008). *Customer roles in business service production - implications for involving the customer in service innovation. Research report 195*. Lappeenranta: Lappeenranta University of Technology.

Kwon, O., Choi, K., & Kim, M. (2007). User acceptance of context-aware services: self-efficacy, user innovativeness and perceived sensitivity on contextual pressure. *Behaviour & Information Technology*, *26*(6), 483–498. doi:10.1080/01449290600709111

Kwon, T. H., & Zmud, R. W. (1987). Unifying the fragmented models of Information Systems implementation. In Boland, R. J., & Hirscheim, R. A. (Eds.), *Critical issues in Information Systems*. New York, NY: John Wiley.

Kwon, T. H. (1987). *A study of the influence of communication network on MIS institutionalization.* Unpublished dissertation, The University of North Carolina at Chapel Hill, United States -- North Carolina.

Labay, D. G., & Kinnear, T. C. (1981). Exploring the consumer decision process in the adoption of solar energy systems. *The Journal of Consumer Research, 8*(3), 271–278. doi:10.1086/208865

LaBelle, C. D., Shaw, K., & Hellenack, L. J. (1980). Solving the turnover problem. *Datamation*, April, 144-152.

Laffie, L. (2005). From the tax adviser: Expanded e-services access. *Journal of Accountancy, 199*(6), 98–98.

Laforet, S., & Xiaoyan, L. (2005). Consumers' attitudes towards online and mobile banking in China. *International Journal of Bank Marketing, 23*(5), 362–380. doi:10.1108/02652320510629250

Langley, A., & Truax, J. (1994). A Process Study of new Technology Adoption. In *Smaller Manufacturing firms*. Press of University of Quebec in Montreal.

Larsson, R., & Bowen, D. E. (1989). Organization and customer: Managing design and coordination of services. *Academy of Management Review, 14*(2), 213–233.

Latour, B. (1992). The sociology of a few mundane artifacts. B. In Bijker, and J. Law (eds.), *Shaping Technology/ Building Society Studies in Sociotechnological Change* (135-150). Cambridge, Massachussets: MIT Press.

Laukkanen, T. (2007a). Customer preferred channel attributes in multi-channel electronic banking. *International Journal of Retail & Distribution Management, 35*(5), 393–412. doi:10.1108/09590550710743744

Laukkanen, T. (2007b). Internet vs. mobile banking: comparing customer value perceptions. *Business Process Management Journal, 13*(6), 788–797. doi:10.1108/14637150710834550

Laukkanen, T., & Lauronen, J. (2005). Consumer value creation in mobile banking services. *International Journal of Mobile Communications, 3*(4), 325–338. doi:10.1504/IJMC.2005.007021

Laukkanen, T., & Pasanen, M. (2008). Mobile banking innovators and early adopters: How they differ from other online users? *Journal of Financial Services Marketing, 13*(2), 86–94. doi:10.1057/palgrave.fsm.4760077

Laukkanen, T., Sinkkonen, S., Kivijärvi, M., & Laukkanen, P. (2007). Innovation resistance among mature consumers. *Journal of Consumer Marketing, 24*(7), 419–427. doi:10.1108/07363760710834834

Laukkanen, T., & Lauronen, J. (2005). Consumer value creation in mobile banking services. *International Journal of Mobile Communications, 3*(4), 325–338. doi:10.1504/IJMC.2005.007021

Laumer, S. (2009). Non-monetary solutions for retaining the IT workforce. In *Proceedings of the 15th Americas Conference on Information Systems* (AMCIS 2009). San Francisco, USA.

Laumer, S., Eckhardt, A., & Weitzel, T. (2008). Recruiting IT professionals in a virtual world. In *Proceedings of the 12th Pacific Asia Conference on Information Systems* (PACIS 2008), Suzhou, China.

Laumer, S., von Stetten, A., & Eckhardt, A. (2009). Online gaming to apply for jobs –The impact of self-assessment and e-assessment on recruiting. In *Proceedings of the 41th Hawaii International Conference on System Sciences* (HICSS 2009), Big Island, USA.

Lauriol, J. (1998). Les représentations sociales dans la décision. In Laroche, H., & Nioche, J.-P. (Eds.), *Repenser la stratégie*. France: Vuibert.

Lee, H. G., & Clark, T. H. (1997). Market process reengineering through electronic market systems: Opportunities and challenges. *Journal of Management Information Systems, 13*(3), 113–136.

Lee, J., & Allaway, A. (2002). Effects of personal control on adoption of self-service technology innovations. *Journal of Services Marketing, 16*(6), 553–572. doi:10.1108/08876040210443418

Lee, W., & Benbasat, I. (2003). Designing an Electronic Commerce Interface: Attention and Product Memory as Elicited by Web Design. *Electronic Commerce Research and Applications*, *2*, 240–253. doi:10.1016/S1567-4223(03)00026-7

Lee, S., Tsai, Y., & Jih, W. (2006). An Empirical Examination of Customer Perceptions of Mobile Advertising. *Information Resources Management Journal*, *19*(4), 39. doi:10.4018/irmj.2006100103

Lee, E.-J., Lee, J., & Eastwood, D. (2003). A two-step estimation of consumer adoption of technology-based service innovations. *The Journal of Consumer Affairs*, *37*(2), 256–282.

Lee, D., Zhu, M., & Hu, H. (2005). When Location-Based Services Meet Databases. *Mobile Information Systems*, *1*(2), 81–90.

Lee, I. (2005a). An integrated economic decision and simulation methodology for e-recruiting process redesign. *International Journal of Simulation and Process Modelling*, *1*(3/4), 179–188. doi:10.1504/IJSPM.2005.007648

Lee, I. (2005b). An analytical model of e-recruiting investment decision: An economic employment approach. *IEEE Transactions on Engineering Management*, *52*(4), 486. doi:10.1109/TEM.2005.857569

Lee, I. (2007). The architecture for a next-generation holistic e-recruiting system. *Communications of the ACM*, *50*(7), 81–85. doi:10.1145/1272516.1272518

Lehner, F., & Watson, R. T. (2001). *From E-Commerce to M-Commerce: Research Directions*. Working paper. University of Regensburg: Chair of Business Informatics.

Lei, D. T. (1997). Competence-building, technology fusion and competitive advantage: The key roles of organizational learning and strategic alliances. *International Journal of Technology Management*, *14*(2/3/4), 208-37.

Lejeune, A., Préfontaine, L., & Ricard, L. (2001). Les chemins vers la performance : L'approche relationnelle et la transformation des entreprises [The path of performance : Relational approach and organisational transformation]. *Gestion*, *26*(3), 45–51. doi:10.3917/riges.263.0045

Lemaître, P. (1997). Les enjeux de la banque à distance [The issues of home banking]. *Banque*, *587*, 63–65.

Leonard-Barton, D. (1988). Implementation as mutual adaptation of technology and organization. *Research Policy*, *17*(5), 251. doi:10.1016/0048-7333(88)90006-6

Leppäniemi, M., & Karjaluoto, H. (2005). Factors influencing consumers' willingness to accept mobile advertising; a conceptual model. *International Journal of Mobile Communications*, *3*(3), 197–213..doi:10.1504/IJMC.2005.006580

Leppäniemi, M., & Karjaluoto, H. (2008). Exploring the effects of gender, age, income and employment status on consumer response to mobile advertising campaigns. *Journal of Systems and Information Technology*, *10*(3), 251–265..doi:10.1108/13287260810916943

Lettl, C., Herstatt, C., & Gemuenden, H. G. (2006). Users' contributions to radical innovation: evidence from four cases in the field of medical equipment technology. *R & D Management*, *36*(3), 251–272. doi:10.1111/j.1467-9310.2006.00431.x

Li, S., Glass, R., & Records, H. (2008). The Influence of Gender on New Technology Adoption and Use–Mobile Commerce. *Journal of Internet Commerce*, *7*(2), 270–289. doi:10.1080/15332860802067748

Lian, J. W., & Lin, T. M. (2008). Effects of Consumer Characteristics on their Acceptance of Online Shopping: Comparisons among Different Product Types. *Computers in Human Behavior*, *24*, 48–65. doi:10.1016/j.chb.2007.01.002

Lin, C. (2003). An interactive communication technology adoption model. *Communication Theory*, *13*(4), 345–365. doi:10.1111/j.1468-2885.2003.tb00296.x

Loane, M. A., Oakley, A., Rademaker, M., Bradford, N., Fleischl, P., & Kerr, P. (2001). A cost-minimization analysis of the societal cost of real-time tele-dermatology compared with conventional care. *Journal of Telemedicine and Telecare*, *7*, 233–238. doi:10.1258/1357633011936453

Lockett, A., & Littler, D. (1997). The adoption of direct banking services. *Journal of Marketing Management*, *13*, 791–811. doi:10.1080/0267257X.1997.9964512

Lohmöller, J.-B. (1989). *Latent Variable Path Modeling with Partial Least Squares*. Heidelberg: Physica-Verlag.

Lohse, G., Bellman, S., & Johnson, E. (2000). Consumer buying behaviour on the Internet: findings from panel data. *Journal of Interactive Marketing, 14*(1), 15–29. doi:10.1002/(SICI)1520-6653(200024)14:1<15::AID-DIR2>3.0.CO;2-C

Loudon, D. L., & Della Bitta, A. J. (1993). *Consumer behaviour: Concepts and applications* (4th ed.). New York, NY: McGraw-Hill.

Lu, J., Yao, J. E., & Yu, C. S. (2005). Personal Innovativeness, Social Influences and Adoption of Wireless Internet Services via Mobile Technology. *The Journal of Strategic Information Systems, 14*(3), 245–268. doi:10.1016/j.jsis.2005.07.003

Luarn, P., & Lin, H.-H. (2005). Toward an understanding of the behavioral intention to use mobile banking. *Computers in Human Behavior, 21*(6), 873–891. doi:10.1016/j.chb.2004.03.003

Lucas, H. C. (1981). An experimental investigation of the use of computerized–based graphics in decision making. *Management Science (Pre-1986), 27*(7), 757.

Luftman, J., & Kempaiah, R. (2008). Key issues for IT executives 2007. *MIS Quarterly Executive, 7*(2), 99–112.

Luftman, J., Kempaiah, R., & Nash, E. (2006). Key issues for IT executives 2005. *MIS Quarterly Executive, 5*(2), 81–99.

Lymer, A., & Debreceny, R. (2003). The auditor and corporate reporting on the Internet: Challenges and institutional responses. *International Journal of Auditing, 7*(2), 103–120. doi:10.1111/1099-1123.00063

Lymer, A., Debreceny, R., Gray, G., & Rahman, A. (1999). *Business reporting on the Internet*. London, UK: IASC.

Lymer, A. (1997). The use of the Internet in company reporting: A survey and commentary on the use of the WWW in corporate reporting in the UK. *Journal of Financial Information Systems.*

Lyytinen, K., & Yoo, Y. (2002). Research Commentary: The Next Wave of Nomadic Computing. *Information Systems Research, 13*(4), 377–388. doi:10.1287/isre.13.4.377.75

MacKenzie, S. B., & Lutz, R. J. (1989). An Empirical Examination of the Structural Antecedents of Attitude Toward the Ad in an Advertising Pretesting Context. *Journal of Marketing, 53*(2), 48–65..doi:10.2307/1251413

Malhotra, A., & Segars, A. H. (2005). Investigating Wireless Web Adoption Patterns in the U.S. *Communications of the ACM, 48*(10), 105–110. doi:10.1145/1089107.1089113

Malhotra, Naresh K., Sung S. K., & Agarwal, J. (2004). Internet Users' Information Privacy Concerns (IUIPC). *The Construct, the Scale, and a Causal Model. Information Systems Research, 15*(4), 336–355. doi:10.1287/isre.1040.0032

Malinowski, J., Keim, T., & Weitzel, T. (2005). Analyzing the impact of IS support on recruitment processes: An e-recruitment phase model. In *Proceedings of The Ninth Pacific Asia Conference on Information Systems*, Bangkok, Thailand

Mallat, N. (2006). *Exploring Consumer Adoption of Mobile Payments - A Qualitative Study*. Helsinki, Finland: Presented at Helsinki Global Mobility Roundtable.

Mallenius, S., Rossi, M. & Tuunainen, V. K. Factors affecting the adoption and use of mobile devices and services by elderly people – results from a pilot study. *Presented at LA Global Mobility Round Table, Los Angeles.*

Malone, T. W., & Crowston, K. (1994). The interdisciplinary study of coordination. *ACM Computing Surveys, 26*(2), 87–119. doi:10.1145/174666.174668

Malone, T. W., Yates, J., & Benjamin, R. I. (1987). Electronic markets and electronic hierarchies. *Communications of the ACM, 30*(6), 484–497. doi:10.1145/214762.214766

March, S. T., & Smith, G. F. (1995). Design and natural science research on Information Technology. *Decision Support Systems, 15*, 251–266. doi:10.1016/0167-9236(94)00041-2

Markus, M. L., & Christiaanse, E. (2003). Adoption and impact of electronic marketplaces. *Information Systems and E-business Management, 1*(2), 139–155. doi:10.1007/s10257-003-0001-7

Markus, H., & Kitayama, S. (1991, April). Culture and the self: implications for cognition, emotion and motivation. *Psychological Review, 98*, 224–252. doi:10.1037/0033-295X.98.2.224

Marr, N., & Prendergast, G. (1993). Consumer adoption of self-service technologies in retail banking: is expert opinion supported by consumer research. *International Journal of Bank Marketing, 11*(1), 3–10. doi:10.1108/02652329310023381

Marsh, H. W., & Hocevar, D. (1985). Application of confirmatory factor analysis to the study of self-concept: First- and higher-order factor models and their invariance across groups. *Psychological Bulletin, 97*, 562–582. doi:10.1037/0033-2909.97.3.562

Mayring, P. (2000). Qualitative content analysis. *Forum Quality Social Research, 1*(2). Retrieved on November 30, 2010, from http://www.qualitative-research.net/fqs-texte/2-00/2-00mayring-e.htm

McCarthy, J., & Matzke, P. (2010). *Coming upheaval in tech services*. Forrester. Retrieved on November 27, 2010, from http://www.forrester.com/rb/Research/coming_upheaval_in_tech_services/q/id/56279/t/2

McCausland, R. (2004, 7 September). Home, sweet home page. *Accounting Technology*.

McKee, J. J., Evans, N. E., & Owens, F. J. (1996). Digital transmission of 12-lead electrocardiograms and Duplex speech in the telephone bandwidth. *Journal of Telemedicine and Telecare, 2*(1). doi:10.1258/1357633961929150

McMillan, S. J. (2000). The microscope and the moving target: The challenge of applying content analysis to the World Wide Web. *Journalism & Mass Communication Quarterly, 77*(1), 80–98.

Meuter, M., Ostrom, A., Roundtree, R., & Bitner, M. (2000). Self-service technologies: Understanding customer satisfaction with technology-based service encounters. *Journal of Marketing, 64*(3), 50–64. doi:10.1509/jmkg.64.3.50.18024

Meuter, M. L., Bitner, M. J., Ostrom, A. L., & Brown, S. W. (2005). Choosing among alternative service delivery modes: An investigation of customer trial of self-service technologies. *Journal of Marketing, 69*(2), 61–83. doi:10.1509/jmkg.69.2.61.60759

Meuter, M. L., Ostrom, A. L., Bitner, M. J., & Roundtree, R. I. (2003). The influence of technology anxiety on consumer use and experience with self-service technologies. *Journal of Business Research, 56*(11), 899–906. doi:10.1016/S0148-2963(01)00276-4

Meyer, A. D., & Goes, J. B. (1988). Organizational assimilation of innovations: A multilevel contextual analysis. *Academy of Management Journal, 31*(4), 897–923. doi:10.2307/256344

Mikkelsen, A., & Gronhaug, K. (1999). Measuring organizational learning climate. *Review of Public Personnel Administration, 19*(4), 31. doi:10.1177/0734371X9901900404

Miles, M. B., & Huberman, A. M. (1994). *Qualitative Data Analysis*. California: Sage Publications, Inc.

Miller, A. (1985). Technological thinking: Its impact on environmental management. *Environmental Management, 9*, 179–190. doi:10.1007/BF01867074

Mingers, J. (2001). Combining IS Research Methods: Towards a Pluralist Methodology. *Information Systems Research, 12*(3), 240–259. doi:10.1287/isre.12.3.240.9709

Mingers, J., & Gill, A. (1997). *Multimethodology: Theory and Practice of Combining Management Science Methodologies*. Chichester, U.K.: Wiley.

Mitchell, A. A., & Olson, J. C. (1981). Are product attribute beliefs the only mediator of advertising effects on brand attitude? *JMR, Journal of Marketing Research, 18*(3), 318–332..doi:10.2307/3150973

Mitchell, B. R., Mitchell, J. G., & Disney, A. P. (1996). User adoption issues in renal telemedicine. *Journal of Telemedicine and Telecare, 2*(2). doi:10.1258/1357633961929835

mITF (mobile IT Forum) (2004). *Flying carpet: Towards the 4th Generation Mobile Communications System, version 2.0*. mITF retrieved from: www.mitf.org/public_e/archives/Flying_Carpet_Ver200.pdf

Modahl, M. (2000). *Now or never*. Harper Collins: New York.

Mohamed, E. K., Oyelere, P., & Al-Busaidi, M. (2009). A survey of Internet financial reporting in Oman. *International Journal of Emerging Markets, 4*(1), 56–71. doi:10.1108/17468800910931670

Mohr, L. (1982). *Explaining Organizational Behaviour*. San Francisco: Jossey-Bass.

Mols, N. P. (1998). The behavioural consequences of PC banking. *International Journal of Bank Marketing, 16*(5), 195–201. doi:10.1108/02652329810228190

Mols, N. P. (2001). Organising for the effective introduction of new distribution channels in retail banking. *European Journal of Marketing, 35*(5/6), 661–686. doi:10.1108/03090560110388150

Morgeson, F. P., & Hofmann, D. A. (1999). The structure and function of collective constructs: Implications for multilevel research and theory development. *Academy of Management Review, 24*(2), 249.

Muenstermann, B., Eckhardt, A., & Weitzel, T. (2010a). The performance impact of business process standardization - An empirical evaluation of the recruitment process. *Business Process Management Journal, 16*(1), 29–56. doi:10.1108/14637151011017930

Muk, A. (2007). Consumer's intentions to opt in to SMS advertising. *International Journal of Advertising, 26*(2), 177–198.

Munos, A. (1999). Technologies & Métier de service [Technologies and service professions]. *Décisions Marketing, 17*(August), 55–65.

Muñoz, P., & Cruz, P. (2004). Flow in eBank Satisfaction Measure. *33rd European Marketing Academy Conference Proceedings*, Book of Abstracts, Murcia, Spain.

Muylle, S., Moenaert, R., & Despotin, M. (2004). The Conceptualization and Empirical Validation of Web Site User Satisfaction. *Information & Management, 41*, 543–560. doi:10.1016/S0378-7206(03)00089-2

Myers, R. (1986). *Classical and modern regression with applications*. Boston, Duxbury Press.

Nambisan, S., & Baron, R. A. (2010). Different roles, different strokes: Organizing virtual customer environments to promote two types of customer contributions. *Organization Science, 21*(2), 554–572. doi:10.1287/orsc.1090.0460

Nantel, J., & Sekhavat, Y. (2008). The impact of SMS advertising on members of a virtual community. *Journal of Advertising Research, 48*(3), 363–374..doi:10.2501/S002184990808032X

Narinder, C. (2007). Fast growth of Mobile Communications in India: Lessons for emerging markets. *6th Annual Global Mobility Roundtable*, Los Angeles, USA.

Necker, M. (2006). A comparison of scheduling mechanisms for service class differentiation in HSDPA networks. *International Journal of Electronics and Communications, 60*, 136–141. doi:10.1016/j.aeue.2005.11.014

Negash, S., Ryan, T., & Igbaria, M. (2003). Quality and effectiveness in Web-based customer Support systems. *Information & Management, 40*(8), 757–768. doi:10.1016/S0378-7206(02)00101-5

Neuman, W. L. (2006). *Social research methods: Qualitative and quantitative approaches* (6th ed.). Boston, MA: Allwyn & Bacon.

Newell, S., Robertson, M., Scarbrough, H., & Swan, J. (2002). *Managing knowledge work*. New York, NY: Palgrave.

Nielsen, J. (1994). *Usability Engineering*. San Francisco: Morgan Kaufmann.

Nieminen, H. (2005). Organizational receptivity – Understanding the inter-organizational learning ability. *Electronic Journal of Knowledge Management, 3*(2), 107–118.

Nobel, S. M., Coast, J., & Benger, J. R. (2005). A cost-consequences of minor injuries telemedicine. *Journal of Telemedicine and Telecare, 11*, 15–19. doi:10.1258/1357633053430421

Nonaka, I. (1994). A dynamic theory of organizational knowledge creation. *Organization Science, 5*(1), 14–37. doi:10.1287/orsc.5.1.14

Nordal, E. J., Moseng, D., Kvammen, B., & Lochen, M. (2001). A comparative study of teleconsultation versus face-to-face consultations. *Journal of Telemedicine and Telecare, 7*(5). doi:10.1258/1357633011936507

Novak, T., Hoffman, D., & Duhachek, A. (2003). The influence of Goal-Directed and experiential activities on online flow experiences. *Journal of Consumer Psychology, 13*(1-2), 3–17. doi:10.1207/S15327663JCP13-1&2_01

Nunnally, J. C. (1978). *Psychometric Theory*. New York: McGraw-Hill.

Nysveen, H., Pedersen, P., & Thorbjørnsen, H. (2005). Explaining intention to use mobile chat services: Moderating effects of gender. *Journal of Consumer Marketing*, *22*(5), 247–256. doi:10.1108/07363760510611671

Nysveen, H., Pedersen, P., Thorbjørnsen, H., & Berthon, P. (2005). Mobilizing the Brand: The Effects of Mobile Services on Brand Relationships and Main Channel Use. *Journal of Service Research*, *7*(3), 257–276. doi:10.1177/1094670504271151

Nysveen, H., Pedersen, P. E., & Thorbjørnsen, H. (2005). Intentions to Use Mobile Services: Antecedents and Cross-Service Comparisons. *Journal of the Academy of Marketing Science*, *33*(3), 330–346. doi:10.1177/0092070305276149

O'Reilly, C., & Tushman, M. (2004, April). The ambidextrous organization. *Harvard Business Review*, *82*(4), 74–81.

OfCom Statistics. (2008). *Communications - The next decade*. Retrieved May 13, 2008, from http://www.ofcom.org.uk /research/

O'Hara, B., & Petrick, A. (2004). *IEEE 802.11 handbook: a designer's companion*. IEEE Standards Association. ISBN 0738144495, 9780738144498

Ojala, T., Korhonen, J., Aittola, M., Ollila, M., Koivumäki, T., Tähtinen, J., & Karjaluoto, H. (2003). SmartRotuaari - Context-aware mobile multimedia services. *2nd International Conference on Mobile and Ubiquitous Multimedia* (pp. 9-18), Norrköping, Sweden.

Okazaki, S., Katsukura, A., & Nishiyama, M. (2007). How Mobile Advertising Works: The Role of Trust in Improving Attitudes and Recall. *Journal of Advertising Research*, *47*(2), 165–178..doi:10.2501/S0021849907070195

Olllo, www.olllo.nl. Accessed 12 November 2007.

Organisation Mondiale de la Santé. (2006). *Rapport sur la santé dans le monde: Travailler ensemble pour la santé*. Genève, Switzerland: OMS.

Orlikowski, W. J. (1992). The duality of technology: Rethinking the concept of technology in organizations. *Organization Science*, *3*(3), 398–427. doi:10.1287/orsc.3.3.398

Orlikowski, W. J. (1993). Learning from notes: Organizational issues in groupware implementation. *The Information Society*, *9*(3), 237–250. doi:10.1080/01972243.1993.9960143

Orlikowski, W. J., & Gash, D. C. (1994). Technological frames: Making sense of Information Technology in organizations. *ACM Transactions on Information Systems*, *12*(2), 174–207. doi:10.1145/196734.196745

Orlikowski, W. J., & Iacono, C. S. (2001). Research commentary: Desperately seeking "it" in IT research - A call to theorizing the IT artifact. *Information Systems Research*, *12*(2), 121. doi:10.1287/isre.12.2.121.9700

Österle, H., Fleisch, E., & Alt, R. (2000). *Business networking. Shaping enterprise relationships on the Internet*. Springer Verlag.

Ostlund, L. E. (1974). Perceived innovation attributes as predictors of innovativeness. *The Journal of Consumer Research*, *1*(2), 23–29. doi:10.1086/208587

Ostlund, L. E. (1969). The role of Product Perceptions in innovative behaviour, in ed. P.R. Mac Donald, *Marketing Involvement in Society and the Economy*, Chicago: American Marketing Association, 259-266.

Oumlil, A., & Williams, A. (2000). Consumer Education Programs for Mature Consumers. *Journal of Services Marketing*, *14*(3), 232–243. doi:10.1108/08876040010327239

Pan, G. S. C. (2005). Information Systems project abandonment: A stakeholder analysis. *International Journal of Information Management*, *25*(2), 173–184. doi:10.1016/j.ijinfomgt.2004.12.003

Pande, P., & Holp, L. (2002). *What is Six Sigma?* New York, NY: McGraw-Hill.

Parasuraman, A., Berry, L. L., & Zeithaml, V. A. (1988). Servqual: A multiple-item scales for measuring consumer perceptions of service quality. *Journal of Retailing*, *64*(1), 12–40.

Parasuraman, A., Zeithaml, V. A., & Berry, L. L. (1985). A conceptual model of service quality and its implications for future research. *Journal of Marketing*, *49*(4), 41–50. doi:10.2307/1251430

Parasuraman, A. (2000). Technology Readiness Index (TRI): a multiple-item scale to measure readiness to embrace new technologies. *Journal of Service Research, 2*(4), 307–320. doi:10.1177/109467050024001

Paré, G., & Sicotte, C. (2004). Les technologies de l'information et la transformation de l'offre de soins. *Cahier Du GReSI, 4*(4).

Park, J., Lennon, S. J., & Stoel, L. (2005). On-Line Product Presentation: Effects on Mood, Perceived Risk, and Purchase Intention. *Psychology and Marketing, 22*(9), 695–719. doi:10.1002/mar.20080

Patel, D. (2002). Managing talent. *HR Magazine*, March.

Patton, M. Q. (2002). *Qualitative Research & Evaluation Methods* (3 ed.). Sage Publications, Inc. California.

Pavlou, P. A., Liang, H., & Xue, Y. (2007). Understanding and mitigating uncertainty in online exchange relationships: A principal-agent perspective. *Management Information Systems Quarterly, 31*(1), 105–136.

Peffers, K., Tuunanen, T., Rothenberger, M., & Chatterjee, S. (2008). A design science research methodology for Information Systems research. *Journal of Management Information Systems, 24*(3), 45–77. doi:10.2753/MIS0742-1222240302

Pejic-Bach, M., & Pejic-Bach, M. (2008). Profightstore.com – Developing an online store for the niche market. In Scupola, A. (Ed.), *Cases on managing e-services* (pp. 76–88). Hershey, PA: Idea Group Inc. doi:10.4018/9781605660646.ch006

Penttinen, E., Saarinen, T., & Sinervo, P. (2008). The role of e-services in the transition from the product focus to the service focus in the printing business. In Scupola, A. (Ed.), *Cases on managing e-services* (pp. 156–165). Hershey, PA: Idea Group Inc.doi:10.4018/9781605660646.ch011

Peppard, J. (2003). Managing IT as a portfolio of services. *European Management Journal, 21*(4), 467–483. doi:10.1016/S0263-2373(03)00074-4

Peppers, D., & Rogers, M. (2001). *One to one B2B: Customer development strategies for the business-to-business world.* New York, NY: Currency Doubleday.

Perrien, J., Filiatrault, P., & Ricard, L. (1993). The implementation of relationship marketing in commercial banking. *Industrial Marketing Management, 22*, 141–148. doi:10.1016/0019-8501(93)90040-E

Peterson, J., & Tarallo, J. (2007). Wireless technology issue overview. Bell Labs Technical Journal (Special Issue: Wireless Network Technology), *11*(4), 1-4

Phillips, J. M. (1998). Effects of realistic job previews on multiple organizational outcomes: A meta-analysis. *Academy of Management Journal, 41*(6), 673–690. doi:10.2307/256964

Piccinelli, G., & Stammers, E. (2001). *From e-processes to e-networks: An e-service-oriented approach.* Retrieved from http://www.research.ibm.com/people/b/bth/OOWS2001/piccinelli.pdf

Pickering, A. (1995). *The mangle of practice.* Chicago, IL: The University of Chicago Press.

Picolo, D., & Smolle, D. (2000). Tele-dermoscopy: Results of a multicentre study on 43 pigmented skin lesions. *Journal of Telemedicine and Telecare, 6*(3), 132–137. doi:10.1258/1357633001935202

Pikkarainen, T., Pikkarainen, K., Karjaluoto, H., & Pahnila, S. (2004). Consumer acceptance of online banking: an extension of the technology acceptance model. *Internet Research, 14*(3), 224–235. doi:10.1108/10662240410542652

Pinch, T., & Bijker, W. (1987). The social connstruction of facts and artifacts. In Bijker, W., Hughes, T., & Pinch, T. (Eds.), *The social construction of technological systems* (pp. 159–187). Cambridge, MA: MIT Press.

Pine, B. J., & Gilmore, J. H. (1999). *The experience economy.* Boston, MA: Harvard Business School Press.

Pinsonneault, A., & Kraemer, K. L. (1993). Survey Research Methodology in Management Information Systems: An Assessment. *Journal of Management Information Systems, 10*(2), 75–105.

Plastic Logic. retrieved February 8th, 2007. Available at:http://www.plasticlogic.com/news.php

Podsakoff, P. M., & Organ, D. W. (1986). Self-reports in organizational research: Problems and prospects. *Journal of Management, 12*, 531–544. doi:10.1177/014920638601200408

Polanyi, M. (1958). *The tacit dimension*. New York, NY: Doubleday.

Polatoglu, V., & Ekin, S. (2001). An Empirical Investigation of the Turkish Consumers Acceptance of Internet Banking Services. *International Journal of Bank Marketing*, *19*(4), 156–165. doi:10.1108/02652320110392527

Politzer, R. M., Harris, D. L., Gaston, M. H., & Mullan, F. (1991). Primary care physician supply and the medically underserved. A status report and recommendations. *Journal of the American Medical Association*, *266*(1), 104–109. doi:10.1001/jama.266.1.104

Polymer Vision. retrieved August 14th, 2008. Available at: http://www.polymervision.com/

Porac, J. F., Thomas, H., & Baden-Fuller, C. (1989). Competitive groups as cognitive communities: The case of Scottish knitwear manufacturers. *Journal of Management Studies*, *26*(4), 397–416. doi:10.1111/j.1467-6486.1989.tb00736.x

Prahalad, C. K., & Ramaswamy, V. (2004). *The future of competition: Co-creating unique value with customers*. Boston, MA: Harvard Business School Press.

Provan, K. G., & Milward, H. B. (1995). A preliminary theory of network effectiveness: A comparative study of four community mental health systems. *Administrative Science Quarterly*, *40*(1), 1–33. doi:10.2307/2393698

Pujari, D. (2004). Self-service with a smile? Self-service technology (SST) encounters among Canadian business-to-business. *International Journal of Service Industry Management*, *15*(2), 200–219. doi:10.1108/09564230410532510

Puri & Radhika. (1996). Measuring and Modifying Consumer Impulsiveness: A Cost-Benefit Accessibility Framework. *Journal of Consumer Psychology*, *5*(2), 87–113. doi:10.1207/s15327663jcp0502_01

Purvis, R. L., Sambamurthy, V., & Zmud, R. (2001). The assimilation of knowledge platforms in organizations: An empirical investigation. *Organization Science*, *12*(2), 117. doi:10.1287/orsc.12.2.117.10115

Putnam, J. M. (2007). Combining telehealth and e-learning: A case study in smoking cessation programming. *American Journal of Health Studies*, *22*(3), 130.

QUALCOMM. (2008). *HSDPA for Improved Downlink DataTransfer.* Retreived on April 2008, from http://www.cdmatech.com/ download_library/pdf/ hsdpa_downlink_wp_12-04.pdf

Quelch, J. A., & Klein, L. R. (1996). The Internet and International Marketing. *Sloan Management Review*, (Spring): 60–75.

Rae, J. (2007, February 27). Debate: Six Sigma vs. innovation. *BusinessWeek*. Retrieved November 23, 2007, from http://www.businessweek.com/innovate/content/feb2007/id20070227_766365.htm

Ragsdale, J. (2010). Service and support technology adoption and spending: The 2010-11 outlook. Retrieved on November 27, 2010, from http://searchcrm.techtarget.com/news/2240024630/Service-and-support-technology-adoption-and-spending-The-2010-11-outlook

Rai, A., & Sambamurthy, V. (2006). Growth and interest in service management. *Information Systems Research*, *17*(4), 327–331. doi:10.1287/isre.1060.0108

Rai, A., & Sambamurthy, V. (2006). Editorial notes – The growth of interest in services management: Opportunities for Information Systems scholars. *Information Systems Research*, *17*(4), 327–331. doi:10.1287/isre.1060.0108

Rainio, A. (2001). Location-Based Services and Personal Navigation in Mobile Information Society. *International Conference FIG Working Week*.

Raju, P. S. (1980). Optimum stimulus level: its relationship to personality, demographics, and exploratory behaviour. *The Journal of Consumer Research*, *7*, 272–282. doi:10.1086/208815

Ram, S. (1989). Successful innovation using strategies to reduce consumer resistance: an empirical test. *Journal of Product Innovation Management*, *6*(1), 20–34. doi:10.1016/0737-6782(89)90011-8

Ram, S., & Jung, H. (1991). Forced Adoption of Innovations in Organizations: Consequences and Implications. *Journal of Product Innovation Management*, *8*(2), 117–126. doi:10.1016/0737-6782(91)90005-J

Ram, S., & Sheth, J. (1989). Consumer resistance to innovations: the marketing problem and its solutions. *Journal of Consumer Marketing*, *6*(2), 5–14. doi:10.1108/EUM0000000002542

Ranganathan, C., & Ganapathy, S. (2002). Key Dimensions of B2C Web Sites. *Information & Management*, *39*, 457–465. doi:10.1016/S0378-7206(01)00112-4

Rao, A., Weber, A., Gollamudi, S., & Soni, R. (2009). LTE and HSPA+: Revolutionary and evolutionary solutions for global mobile broadband. *Bell Labs Technical Journal (Special Issue: 4G Wireless Technologies, 13*(4), 7–34

Raper, J., Gartner, G., Karimi, H., & Rizos, C. (2007). A critical evaluation of location based services and their potential. *Journal of Location Based Services*, *1*(1), 5–45. doi:10.1080/17489720701584069

Raper, J. (2006). Design constraints on operational LBS. In G. Gartner, W. Cartwright, & M. Peterson (Eds), *Location Based Services and TeleCartography. Lecture Notes in Geoinformation and Cartography*. Berlin Heidelberg: Springer.

Rappaport, S. D. (2007). Lessons from Online Practice: New Advertising Models. *Journal of Advertising Research*, *47*(2), 135–141..doi:10.2501/S0021849907070158

Reimann, M., Oliver, S., & Thomas, J. S. (2010). Toward an understanding of industry commoditization: Its nature and role in evolving market competition. *International Journal of Research in Marketing*, *27*, 188–197. doi:10.1016/j.ijresmar.2009.10.001

Rettie, R., Grandcolas, U., & Deakins, B. (2005). Text Message Advertising: Response Rates and Branding Effects. *Journal of Targeting. Measurement and Analysis for Marketing*, *13*(4), 304–312..doi:10.1057/palgrave.jt.5740158

Ricard, L., & Perrien, J. (1999). Explaining and evaluating the implementation of organizational relationship marketing in the banking industry: Client's perceptions. *Journal of Business Research*, *45*, 199–209. doi:10.1016/S0148-2963(97)00226-9

Riddle, D. I. (2001). Cadre d'analyse de rentabilisation pour l'adoption des affaires électroniques dans les petites entreprises, available at : www.strategis.ic.gc.ca/epic/site/sbrp-rppe.nsf/vwapj/framework_f.pdf/$FILE/framework_f.pdf - Consulted in octobre 2003.

Riggins, F. J., & Rhee, H.-S. S. (1998). Toward a Unified View of Electronic Commerce. *Communications of the ACM*, *41*(10), 88–95. doi:10.1145/286238.286252

Rindskopf, D., & Rose, T. (1988). Some theory and applications of confirmatory second-order factor analysis. *Multivariate Behavioral Research*, *23*, 51–67. doi:10.1207/s15327906mbr2301_3

Ringle, C. M., Wende, S., & Will, S. (2005). *SmartPLS 2.0 (M3) Beta*. Hamburg, http://www.smartpls.de

Ritti, R., & Silver, J. (1986). Early processes of institutionalization: The dramaturgy of exchange in interoganizational relations. *Administrative Science Quarterly*, *31*, 25–42. doi:10.2307/2392764

RIZOS, C. (2007). The Future of Global Navigation Satellite Systems. *Technical report of The Institute of Electronics. Information and Communication Engineers*, *107*(2), 25–29.

Robertson, T. S. (1971). *Innovative behaviour and communication*. New York: Holt, Rinehart and Winston.

Rockart, J. F. (1979). Chief executives define their own data needs. *Harvard Business Review*, *57*(2), 81–93.

Rogers, E. M. (1983). *Diffusion of innovations*. New York: The free press.

Rogers, E. M., & Shoemaker, F. F. (1971). *Communications of Innovations*. New York: Free Press.

Roloff, J. (2008). A life cycle model of multi-stakeholder networks. *Business Ethics (Oxford, England)*, *17*(3), 311–325. doi:10.1111/j.1467-8608.2008.00537.x

Roos, N. P., Fransoo, R., Carriere, K. C., & Frohlic, N. (1997). Needs-based planning: The case of Manitoba. *Canadian Medical Association Journal*, *157*(9), 15–1221.

Rosemberg, N. (1994). *Exploring the black box: Technology, economics and history*. Cambridge, UK: Cambridge University Press. doi:10.1017/CBO9780511582554

Rosted, J. (2005). *User-driven innovation–Results and Recommendations*. FORA Report.

Rotchanakitumnuai, S., & Speece, M. (2003). Barriers to Internet banking adoption: A qualitative stud among corporate customers in Thailand. *International Journal of Bank Marketing*, *21*(6/7), 312–323. doi:10.1108/02652320310498465

Rowley, J. (2006). An analysis of the e-service literature: Towards a research agenda. *Internet Research, 16*(3), 339–359. doi:10.1108/10662240610673736

Roy, M., Dewit, O., & Aubert, B. (2001). The impact of interface usability on trust in web retailers. *Internet Research: Electronic Networking Applications and Policy, 11*(5), 388–398. doi:10.1108/10662240110410165

Rubeck, R., & Miller, G. A. (2008). vGOV! Remote video access to government services. In A. Scupola (Ed.), *Cases on managing e-services* (pp. 253-268). Hershey, PA: Idea Group Inc.

RUIS de Sherbrooke. (2007). La téléassistance en soins de plaies, orientée vers une amélioration continue de la qualité des soins. *Manuel D'Organisation De Projet,* 1-63.

Rumberger, J. S. (2005). *Is there a strategic and financial business case for telehealth in home health agencies?* Unpublished doctoral dissertation, The Pennsylvania State University, United States -- Pennsylvania.

Rust, R. T. (2001). Editorial: The rise of e-services. *Journal of Service Research, 3*(1), 283–284. doi:10.1177/109467050134001

Rust, R. T., & Kannan, P. K. (2003). E-service: A new paradigm for business in the electronic environment. *Communications of the ACM, 46*(6), 37–42. doi:10.1145/777313.777336

Rust, R. T., & Kannan, P. K. (2002). *E-Service: New Directions in Theory and Practice*. Armonk, New York, NY: ME Sharpe.

Sadin, M. (2003, October 6). Computers that care. *Newsweek*, p. 6.

Saga, V., & Zmud, R. (1994). The nature and determinants of Information Technology acceptance, routinization and infusion. In Levine, L. (Ed.), *Diffusion, transfer, and implementation of Information Technology* (pp. 67–68). Amsterdam, The Netherlands: North-Holland.

Salle, M. (2004). *IT service management and IT governance: Review, comparative analysis and their impact on utility computing*. Retrieved December 15, 2007 from http://www.hpl.hp.com/techreports/2004/HPL-2004-98.pdf

Sambamurthy, V., & Zmud, R. W. (2000). Research commentary: The organizing logic for an enterprise's IT activities in the digital era - A prognosis of practice and a call for research. *Information Systems Research, 11*(2), 105. doi:10.1287/isre.11.2.105.11780

Sarkar, M. B., Butler, B., & Steinfield, C. (1995). Intermediaries and cybermediaries: A continuing role for mediating players in the electronic marketplace. *Journal of Computer-Mediated Communication, 1*(3).

Sarker, S., & Wells, J. D. (2003). Understanding mobile handheld device use and adoption. *Communications of the ACM, 46*(12), 35–40. doi:10.1145/953460.953484

Sathye, M. (1999). Adoption of Internet banking by Australian consumers: an empirical investigation. *International Journal of Bank Marketing, 17*(7), 324–334. doi:10.1108/02652329910305689

Sawhney, M., Verona, G., & Prandelli, E. (2005). Collaborating to create: The Internet as a platform for customer engagement in product innovation. *Journal of Interactive Marketing, 19*(4), 4–17. doi:10.1002/dir.20046

Sayar, C., & Wolfe, S. (2007). Internet banking market performance: Turkey versus the UK. *International Journal of Bank Marketing, 25*(3), 122–141. doi:10.1108/02652320710739841

Schein, E. (1985). *Organizational culture and leadership.* San Francisco, CA: Jossey-Bass.

Schlageter, M. (2005). Leveraging technology. *Journal of Accountancy, 199*(6), 14.

Schlosser, A. E., Shavitt, S., & Kanfer, A. (1999). Survey of Internet users' attitudes toward Internet advertising. *Journal of Interactive Marketing, 13*(3), 34–54.. doi:10.1002/(SICI)1520-6653(199922)13:3<34::AID-DIR3>3.0.CO;2-R

Schneider, B., & Bowen, D. E. (1995). *Winning the service game*. Boston, MA: Harvard Business School Press.

Schneider, B. (1985). Organizational behavior. *Annual Review of Psychology, 36*, 573–611. doi:10.1146/annurev.ps.36.020185.003041

Schneider, B., & Reichers, A. E. (1983). On the etiology of climates. *Personnel Psychology, 36*(1), 19–39. doi:10.1111/j.1744-6570.1983.tb00500.x

Schneider, B. (1983). Work climates: An interactionist perspective. In Freimer, N. W., & Geller, E. S. (Eds.), *Environmental psychology: Directions and perspectives* (pp. 153–178). New York, NY: Praeger.

Schneider, B. (1990). The climate for service: An application of the climate construct. In Schneider, B. (Ed.), *Organizational climate and culture*. San Francisco, CA: Jossey-Bass.

Schneider, B. (1995). *Personalbeschaffung*. Peter Lang Europäischer Verlag der Wissenschaften: Frankfurt a.M.

Schofield, R. S., Kline, S. E., Schmalfuss, C. M., Carver, H. M., & Aranda, J. M. (2005). Early outcomes of a care coordination-enhanced telehome care program for elderly veterans with chronic heart failure. *Telemedicine and E-Health*, *11*, 20–27. doi:10.1089/tmj.2005.11.20

Scholz, C. (2000). *Personalmanagement: Informationsorientierte und verhaltenstheoretische Grundlagen*, 5th ed. München.

Schultze, U., & Bhappu, A. D. P. (2005). Incorporating self-serve technology into co-production designs. *International Journal of e-Collaboration*, *1*(4), 1–23. doi:10.4018/jec.2005100101

Scupola, A. (2008a). Conceptualizing competences in e-services adoption and assimilation in SMEs. *Journal of Electronic Commerce in Organizations*, *6*(2). doi:10.4018/jeco.2008040105

Scupola, A. (2008b). E-services: Definition, characteristics and taxonomy: Guest editorial preface, Special Issue on E-Services. *Journal of Electronic Commerce in Organizations*, *6*(2), i–iii.

Scupola, A. (Ed.). (2008c). *Cases on managing e-services*. Hershey, PA: Idea Group Inc.

Scupola, A., & Nicolajsen, H. W. (2009). The role of customer involvement in library e-services. In D'Atri, A., & Saccà, D. (Eds.), *Information Systems: People, organizations, institutions, and technologies*. Springer. doi:10.1007/978-3-7908-2148-2_3

Scupola, A. (2008). Special issue on e-services. *Journal of Electronic Commerce in Organizations (JECO)*.

Shackleton, P., Fisher, J., & Dawson, L. (2003). Town hall e-government: A study of local government electronic service delivery. *Proceedings of the 14th Australasian Conference on Information Systems*, Perth, Australia November.

Shandh, J. N. (1968). A factor analytic model of brand loyalty. *JMR, Journal of Marketing Research*, *5*, 395–404. doi:10.2307/3150264

Shea, C. (2009). Kindling controversy. Retrieved February 15th, 2009. Available at: http://www.boston.com/bostonglobe/ideas/brainiac/2009/02/kindling_controversy.html

Sheth, J. (1981). Psychology of innovation resistance: the less developed concept (LDC) in diffusion research. *Research in Marketing*, *4*(3), 273–282.

Shore, J. H., & Brooks, E. (2007). An economic evaluation of telehealth data collection with rural populations. *Psychiatric Services (Washington, D.C.)*, *58*(6), 830–835. doi:10.1176/appi.ps.58.6.830

Shukla, A. (2010). IT managed services market in India to hit $US3.8 billion in 2013. Retrieved on November 27, 2010, from http://www.techworld.com.au/article/345074/it_managed_services_market_india_hit_us3_8_billion_2013/

Siau, K., & Shen, Z. (2003). Mobile communications and mobile services. *International Journal of Mobile Communications*, *1*(1/2), 3–14. doi:10.1504/IJMC.2003.002457

Sigala, M. (2007). Investigating the Internet's impact on interfirm relations: Evidence from the business travel management distribution chain. *Journal of Enterprise Information Management*, *20*(3), 335–355. doi:10.1108/17410390710740772

Simon, K. (2000, December 11). DMAIC versus DMADV. *iSixSigma.com*. Retrieved September 28, 2007, from http://europe.isixsigma.com/library/content/c001211a.asp

Simons, L. P. A., van Loon, J., de Koning, N. M., Kruse, J., & Bouwman, H. (2009). Increasing the loyalty effects of e-CRM across the service delivery cycle. In *Proceedings of Bled eConference – eEnablement: Facilitating an Open, Effective and Representative eSociety*, June 14 – 17, Bled, Slovenia.

Sims, F., Williams, M.-A., & Elliot, S. (2007). Understanding the Mobile Experience Economy: A key to richer more effective M-Business Technologies, Models and Strategies. In *Proceedings of Sixth International Conference on the Management of Mobile Business* (ICMB 2007).

Singh, M., & Byrne, J. (2005). Performance evaluation of e-business in Australia. *Electronic Journal of Information Systems Evaluation, 8*(2).

SiRF Technology, Inc. (2008). Retrieved from http://www.sirf.com/products /gps_chip.html.

Skanning, C. (2005). The cost and benefits of customer self-service. *The Wise Marketer*. Retrieved on November 28, 2010, from http://www.thewisemarketer.com/features/read.asp?id=83

Skhiri, S. (2000). Adoption des innovations en technologie de l'information par les entreprises internationales [The adoption of IT innovations by international companies]. Unpublished Master's thesis in Business Administration, University of Quebec in Montreal, Canada.

Smith, J., & Barclay, D. (1997). The effects of organizational differences and trust on the effectiveness of selling partner relationships. *Journal of Marketing, 61*, 3–21. doi:10.2307/1252186

Solitander, M. (2006). Balancing the flows: Managing the intellectual capital flows in inter-organizational projects. *Electronic Journal of Knowledge Management, 4*(2), 197–206.

Spulber, D. F. (1996). Market microstructure and intermediation. *The Journal of Economic Perspectives, 10*(3), 135–152.

SSPA. (2004). *2005 support demand research series. Tech strategy partners and SSPA*. Retrieved on August 7, 2007, from http://www.thesspa.com/publications/pr/041012b.asp

Stafford, T. (2003). E-services. *Communications of the ACM, 46*(6), 26–28. doi:10.1145/777313.777333

Steinfield, C. (2004). The development of Location Based Services in Mobile Commerce. In B. Preissl, H. Bouwman, & C. Steinfield (Eds), *Elife after the Dot.com bust*. Berlin; Springer Verlag.

Steiniger, S., Neun, M., & Edwardes, A. (2007). Foundations of Location Based Services. *Lecture Notes on LBS*. Retrieved in October 2008, from http://www.geo.unizh.ch /publications/cartouche/ lbs_lecturenotes_steinigeretal2006.pdf

Sternthal, B., Dholakia, R., & Leavitt, C. (1978). The persuasive effect of source credibility: Tests of cognitive response. *The Journal of Consumer Research, 4*(4), 252–260..doi:10.1086/208704

Stokes, S. L. (2000). Attracting and keeping IT talent. *Information Systems Management, 17*, 8–16. doi:10.1201/1078/43192.17.3.20000601/31235.2

Strauss, A. (1978). A social world perspective. In Denzin, N. (Ed.), *Studies in symbolic interaction* (*Vol. 1*). Greenwich, CT: JAI Press.

Stremersch, S., & Tellis, G. (2002). Strategic bundling of Products and Prices: A new synthesis for marketing. *Journal of Marketing, 66*(1), 55–72. doi:10.1509/jmkg.66.1.55.18455

Strohmeier, S. (2007). Research in e-HRM: Review and implications. *Human Resource Management Review, 17*, 19–37. doi:10.1016/j.hrmr.2006.11.002

Subramaniam, A. (2009). *Web self-service will make you great*. Retrieved on November 28, 2010, from http://www.destinationcrm.com/Articles/Web-Exclusives/Viewpoints/Web-Self-Service-Will-Make-You-Great-54931.aspx

Sundbo, J. (1997). Management of innovation in services. *The Service Industries Journal, 17*(3), 432–455. doi:10.1080/02642069700000028

Suoranta, M., & Mattila, M. (2004). Mobile Banking and Consumer Behaviour – New Insights into the Diffusion Pattern. *Journal of Financial Services Marketing, 8*(4), 354–366. doi:10.1057/palgrave.fsm.4770132

Svensson, J., & Hakeröd, J. (2006). Design of navigational support for task oriented reading in e-newspapers. In *Proceedings of the 29th Information Systems Research Seminar in Scandinavia, Denmark.*

Swanson, B. E., & Ramiller, N. C. (1997). The organizing vision in Information Systems innovation. *Organization Science, 8*(5), 458. doi:10.1287/orsc.8.5.458

Sweeney, J. C., & Lapp, W. (2004). Critical service quality encounters on the Web: An exploratory study. *Journal of Services Marketing, 18*(4), 276–289. doi:10.1108/08876040410542272

Swinyard, W. R. (1993). The effects of mood, involvement, and quality of store experience on shopping intentions. *The Journal of Consumer Research, 20*, 271–280. doi:10.1086/209348

Sybase (2008). *Research Reveals Global Appetite for Mobile Banking Services.* Retrieved May 21, 2008, from http://www.sybase.com /detail?id=1054292

Szulanski, G. (2000). The process of knowledge transfer: A diachronic analysis of stickiness. *Organizational Behavior and Human Decision Processes, 82*(1), 9–27. doi:10.1006/obhd.2000.2884

Tabachnik, B. G., & Fidell, L. S. (1996). *Using Multivariate Statistics.* New York: Harper Collins.

Tan, M., & Teo, T. S. H. (2000). Factors Influencing the Adoption of Internet Banking. *Journal of the Association for Information Systems, 1*(5), 1–44.

Tan, G. W., & Wei, K. K. (2006). An Empirical Study of Web Browsing Behaviour: Towards an Effective Website Design. *Electronic Commerce Research and Applications, 5*, 261–271. doi:10.1016/j.elerap.2006.04.007

Tao, G., Pung, H. K., & Zhang, D. Q. (2005). A service-oriented middleware for building context-aware services. *Journal of Network and Computer Applications, 28*(1), 1–18. doi:10.1016/j.jnca.2004.06.002

Taub, E. A. (2008). New E-Newspaper Reader Echoes Look of the Paper. Retreived February 2nd, 2009. Available at:http://www.nytimes.com/2008/09/08/technology/08ink.html

Taylor, S. (1998). *Websites – A missed opportunity?* London: Business Briefing. Institute of Chartered Accountants in England and Wales.

Teerling, M., Haaker, T., & Vos, H. de (2007, October 18-21). *Understanding Consumer Value of Mobile Service Bundles, NAEC*, Lake Garda, Italy.

Tenenhaus, M., Vinzi, V. E., Chatelin, Y.-M., & Lauro, C. (2005). PLS path modeling. *Computational Statistics & Data Analysis, 48*(1), 159–205..doi:10.1016/j.csda.2004.03.005

Teo, T., & Pok, S. (2003). Adoption of WAP-enabled mobile phones among Internet users. *Omega, 31*, 483–498. doi:10.1016/j.omega.2003.08.005

Ter Hofte, G. H., Mulder, I., & Verwijs, C. (2006). Close Encounters of the Virtual Kind: A Study on Place-based Presence. AI & Society. *The Journal of Human-Centred Systems, 20*(2).

The Economist. (2006). Who killed the newspaper? Retrieved September 29th, 2008. Available at: http://www.economist.com/opinion/displaystory.cfm?story_id=7830218.

The Swedish Newspaper Publishers' Association. (2008), retrieved August 14th, 2008. Available at: http://www.tu.se/index_target.asp?c1=1214

Theiss, A., David, C., & Yuan, C. (2005). Global Positioning Systems: an analysis of applications, current development and future implementations. *Computer Standards & Interfaces, 27*(2), 89–100. doi:10.1016/j.csi.2004.06.003

Then, N. K., & Delong, M. R. (1999). Apparel shopping on the web. *Journal of Family and Consumer Sciences, 91*, 65–68.

Thomas, S., & Ray, K. (2000). Recruiting and the Web: High-tech hiring. *Business Horizons, 43*(5/6), 43–52. doi:10.1016/S0007-6813(00)89200-9

Thompson, V. (2007). Talent shortage? How to win with what you've got. *Super Vision, 68*(2), 15–17.

Tilson, D., Lyytinen, K., & Baxter, R. (2004). A Framework for selecting a Location Based Service (LBS) Strategy and Service Portfolio. In *Proceedings of the 37th Hawaii International Conference on System Sciences, Hawaii.* (CD-ROM), Computer Society Press (10 pages).

Tiwana, A., & Balasubramaniam, R. (2001). *E-services, problems, opportunities, and digital platforms.* 34th Hawaii International Conference on System Sciences. March 2001.

Tornatzky, L. G., & Klein, K. J. (1982). Innovation Characteristics and Adoption-Implementation. *IEEE Transactions on Engineering Management*, *29*(1), 28–45.

Totaltelecom (2007). *Location based services.* Retrieved from: www.totaltele.com/View.aspx?ID=91162&t=4, last accessed 12 November 2007.

Toufaily, E., Daghfous, N., & Toffoli, R. (2009). The adoption of e-banking by Lebanese banks: Success and critical factors. *International Journal of E-Services and Mobile Applications*, *1*(1). doi:10.4018/jesma.2009010105

Tracey, J. B., Tannenbaum, S. I., & Kavanagh, M. J. (1995). Applying trained skills on the job: The importance of the work environment. *The Journal of Applied Psychology*, *80*(2), 239–252. doi:10.1037/0021-9010.80.2.239

Traub, R. (1994). *Reliability for the social sciences: theory and applications.* London: Sage Publications.

Trauth, E. M., Quesenberry, J. L., & Yeo, B. (2008). Environmental influences on gender in the IT workforce. *The Data Base for Advances in Information Systems*, *8*(39).

Tsang, M. M., Ho, S.-C., & Liang, T.-P. (2004). Consumer Attitudes Toward Mobile Advertising: An Empirical Study. *International Journal of Electronic Commerce*, *8*(3), 65–78.

Tscherning, H., & Damsgaard, J. (2008). Understanding the Diffusion and Adoption of Telecommunication Innovations: What We Know and What We Don't Know. In IFIP, Volume 287, *Open IT-Based Innovation: Moving Towards Cooperative IT Transfer and Knowledge Diffusion*, eds. León, G., Bernardos, A., Casar, J., Kautz, K., and DeGross, J. (Boston: Springer), 41-62.

Twitter (2009). *Twitter.com*. Accessed 31 March 2009.

U. S. Census Bureau. (2008). Quarterly retail e-commerce sales 4th quarter 2007. Retrieved February, 2008, from: http://www.census.gov/mrts/www/data/html/07Q4.html.

U.S. Department of Labor. (2004). *Bureau of Labor Statistics*.

Udo, G. J., Bagchi, K. K., & Kris, P. J. (2010). An assessment of customers' e-service quality perception, satisfaction and intention. *Information & Management*, *30*(6), 481–492. doi:10.1016/j.ijinfomgt.2010.03.005

Valentin, F., & Hansen, P. V. (2004). *Udvikling af videnservice (Development of knowledge service)*. Copenhagen, Denmark: Nyt Teknisk Forlag.

Van der Heijden, H., & Verhagen, T. (2003). Online Store Image: Conceptual Foundations and Empirical Measurement. *Information & Management*, *41*, 609–617. doi:10.1016/S0378-7206(03)00100-9

Van Maanen, J., & Schein, E. (1979). Toward a theory of organizational socialization. *Research in Organizational Behavior*, *1*, 209–264.

van Sinderen, M. J., Halteren, A. T., Wegdam, M., Meeuwissen, B., & Eertink, H. E. (2006, September). Supporting Context-Aware Mobile Applications: An Infrastructure Approach. *IEEE Communications Magazine*, 96–104. doi:10.1109/MCOM.2006.1705985

Venkatesh, V., Morris, M. G., Davis, G. B., & Davis, F. D. (2003). User Acceptance Of Information Technology: Toward A Unified View. *Management Information Systems Quarterly*, *27*(3), 425–478.

Venkatesh, V., Ramesh, V., & Massey, A. (2003). Understanding usability: In mobile commerce. *Communications of the ACM*, *46*(12), 53–56. doi:10.1145/953460.953488

Venkatraman, N. (1995). Reconfigurations d'entreprises provoquées par les technologies de l'information. In *L'entreprise compétitive au futur* (pp. 151–195). Les Éditions d'Organisation.

Veugelers, P. J., Yip, A. M., & Burge, F. (2004). Les inégalités en santé et dans la prestation des services de santé. *Maladies Chroniques Du Canada, 25*(3/4).

Villates, D. (1997). Demain, la banque à distance [The Distant bank of tomorrow]. *Banque*, *585*, 68–70.

Vlachopoulou, M., & Manthou, V. (2003). Partnership alliances in virtual markets. *International Journal of Physical Distribution & Logistics Management*, *33*(3), 254–267. doi:10.1108/09600030310471998

Vogel, R. L., & Ackerman, R. J. (1998). Is primary care physician supply correlated with health outcomes? *International Journal of Health Services*, 28.

von Hippel, E. (2005). Democratizing Innovation: The evolving phenomenon of user innovation. *Journal für Betriebswirtschaft, 55*(1), 63–78. doi:10.1007/s11301-004-0002-8

Von Stetten, A., Laumer, S., Eckhardt, A., Weitzel, T., & von Westarp, F. (2011). (in press). *Recruiting trends 2011*. Eine empirische Untersuchung mit den Top-1.000-Unternehmen aus Deutschland sowie den [-Unternehmen aus den Branchen Finanzdienstleistung, IT und Oeffentlicher Dienst, Research Report, Bamberg and Frankfurt a. Main.]. *Top (Madrid)*, 300.

Vos,H. de, Haaker, T., Teerling, M., & Kleijnen, M. (2008, June). Consumer Value of Context Aware and Location Based Mobile Services. *21st Bled eConference*, Bled, Slovenia.

Voss, K. E., Spangenberg, E. R., & Grohmann, B. (2003, August). Measuring the Hedonic and Utilitarian Dimensions of Consumer Attitude. *JMR, Journal of Marketing Research, 40*, 310–320. doi:10.1509/jmkr.40.3.310.19238

Vragov, R. (2009). Sponsored search as a strategic e-service. *International Journal of E-Services and Mobile Applications, 1*(1). doi:10.4018/jesma.2009010102

Vrechopoulos, A., Constantiou, I., & Sideris, G. (2003). The critical role of consumer behaviour research in mobile commerce. *International Journal of Mobile Communications, 1*(3), 329–340. doi:10.1504/IJMC.2003.003498

Vrechopoulos, A. P., & Siomkos, G. J. (2002). Store atmosphere in non-store retailing. *Journal of Internet Marketing, 3*(1), Retrieved May, 2008, from: http://www.arraydev.com/commerce/JIM/0207-03.htm.

Wah, L. (1999). Banking on the Internet. *American Management Association, 88*(11), 44–48.

Walker, R. H., Craig-Lees, M., Hecker, R., & Francis, H. (2002). Technology-enabled service delivery: An investigation of reasons affecting customer adoption and rejection. *International Journal of Service Industry Management, 13*(1), 91–106. doi:10.1108/09564230210421173

Walls, J. G., Widmeyer, G. R., & El Sawy, O. A. (1992). Building an Information System design theory for vigilant EIS. *Information Systems Research, 3*(1), 36–59. doi:10.1287/isre.3.1.36

Walsh, S. T., Kirchhoff, B. A., & Newbert, S. (2002). Differentiating Market Strategies for Disruptive Technologies. *IEEE Transactions on Engineering Management, 49*, 341–351. doi:10.1109/TEM.2002.806718

Walsham, G. (1995). The emergence of interpretivism in IS research. *Information Systems Research, 6*(4), 376–394. doi:10.1287/isre.6.4.376

WAN. (2006). *World Association of Newspapers*. World Press Trends.

Wang, S., & Archer, N. (2007). Business-to-business collaboration through electronic marketplaces: An exploratory study. *Journal of Purchasing and Supply Chain Management, 13*(2), 113–126. doi:10.1016/j.pursup.2007.05.004

Wang, J., Yao, J., Yan, R., & Hsu, J. (2008). E-bay: An e-titan success story. In Scupola, A. (Ed.), *Cases on managing e-services* (pp. 126–140). Hershey, PA: IGI Global. doi:10.4018/9781605660646.ch009

Ward, M., Gruppen, L., & Regehr, G. (2002). Measuring self-assessment: Current state of the art. *Advances in Health Sciences Education : Theory and Practice, 7*, 63–80. doi:10.1023/A:1014585522084

Ward, T. (2004a). *Is there a light at the end of the tunnel - Can we resolve the physician distribution challenge in Canada?* Washington, DC: International Health Workforce Conference October 7-9.

Warner, M., & Zhu, Y. (2002). Human resource management with Chinese characteristics: A comparative study of the People's Republic of China and Taiwan. *Asia Pacific Business Review, 9*(2), 21–43.

Webster, F. E. J. (1969). New Product Adoption in Industrial Markets: A Framework for Analysis. *JMR, Journal of Marketing Research, 33*(3), 35–39.

Weick, K. (1990). The vulnerable system: An analysis of the Tenerife Air disaster. *Journal of Management, 16*(3), 571–593. doi:10.1177/014920639001600304

Weill, P., Subramani, M., & Broadbent, M. (2002). Building IT infrastructure for strategic agility. *MIT Sloan Management Review, 44*(1), 57.

Weinberg, A. (2007). *Focusing on the customer: An interview with the head of Merrill Lynch's operations and IT*. Retrieved July 21, 2007 from http://www.mckinseyquarterly.com/article_abstract.aspx?ar=2021&L2=13

Weitzel, T., Eckhardt, A., & Laumer, S. (2009). A framework for recruiting IT talent: Lessons from Siemens. [MISQE]. *MIS Quarterly Executive*, *8*(4), 175–189.

Westwick, R. (2008). Mobile WiMAX versus LTE: A comparison of next generation mobile broadband technologies. *Journal of Telecommunications Management*, *1*(1), 79–85.

WhatIs.com. (2008). *IT service management definition*. Retrieved July 5, 2008, from http://www.bitpipe.com/tlist/IT-Service-Management.html

WhatIs.com. (n.d.). *E-services*. Retrieved from http://searchcio.techtarget.com/sDefinition/0,sid19_gci535452,00.html

Winters, J. M., & Winters, J. M. (2004). A telehomecare model for optimizing rehabilitation outcomes. *Telemedicine Journal and e-Health*, *10*, 200–212. doi:10.1089/tmj.2004.10.200

Wise, R., & Morrison, D. (2000). Beyond the exchange: The future of B2B. *Harvard Business Review*, *78*(6), 86–89.

Wittea, T., & Wilson, A. (2004). Accuracy of non-differential GPS for the determination of speed over ground. *Journal of Biomechanics*, *37*(12), 1891–1898. doi:10.1016/j.jbiomech.2004.02.031

Wold, H. (1982). Soft Modeling: the basic design and some extensions. In Jöreskog, K. G., & Wold, H. (Eds.), *Systems under indirect observation: causality, structure and prediction. Part II*. Netherlands: North-Holland Publishing company.

Wooton, R., Patil, N. G., Scott, R. E., & Ho, K. (2009). *Telehealth in the developing world*. London, UK: Royal Society of Medicine Press LTD.

Worthen, B. (2005). ITIL and Sarbanes-Oxley (Sarbox). *CIO Magazine*. Retrieved on November 15, 2007, from http://www.cio.com/article/10519/_ITIL_and_Sarbanes_Oxley_Sarbox_

Wu, J.-H., & Wang, S.-C. (2005). What drives mobile commerce? An empirical evaluation of the revised technology acceptance model. *Information & Management*, *42*(5), 719–729. doi:10.1016/j.im.2004.07.001

Xiao, Z., Jones, M., & Lymer, A. (2002). Immediate trends in Internet reporting. *European Accounting Review*, *11*(2), 245–275. doi:10.1080/09638180020017087a

Yang, K. (2005). Exploring factors affecting the adoption of mobile commerce in Singapore. *Telematics and Informatics*, *22*(3), 257–277. doi:10.1016/j.tele.2004.11.003

Yant, B. (2005). *Impact of telehealth on job satisfaction of nurses and organizational performance of home health agencies*. Unpublished doctoral dissertation, The Pennsylvania State University, United States -- Pennsylvania.

Yin, R. K. (1989). *Case study research: Design and methods*. Newbury Park, CA: Sage Publications.

Yu, S., & Rijia, D. (2009). *Comparison of Internet financial reporting between China and United States*. 2009 International Conference on Information Management, Innovation Management and Industrial Engineering, (p. 211). Xian, China, IEEE.

Zanot, E. J. (1984). Public attitudes toward advertising: The American experience. *International Journal of Advertising*, *13*, 3–15.

Zeithaml, V., Parasuraman, A., & Malhotra, A. (2003). Service quality delivery through Web sites: A critical review of extant knowledge. *Journal of the Academy of Marketing Science*, *30*(4), 362–375. doi:10.1177/009207002236911

Zeithaml, V. A., Parasuraman, A., & Malhotra, A. (2001). *A conceptual framework for understanding e-service quality: Implications for future research and managerial practice*, (pp. 1-49). (MSI Working Paper Series, No 00-115), Cambridge, MA.

Zhang, X., & Myers, M. (2005). Web design and e-commerce. In Gao, Y. (Ed.), *Web System Design and Online Consumer Behavior* (pp. 205–221). Hershey: Idea Group Publishing.

Zhou, K. Z., & Nakamoto, K. (2007). How do Enhanced and Unique Features Affect New Product Preference? The Moderating Role of Product Familiarity. *Journal of the Academy of Marketing Science*, *35*, 53–62. doi:10.1007/s11747-006-0011-3

Zinkhan, G. M., & Locander, W. B. (1988). ESSCA: A Multi-dimensional Analysis Tool for Marketing Research. *Journal of the Academy of Marketing Science*, *16*, 36–46. doi:10.1007/BF02723324

Zipf, A., & Joest, M. (2004). User Expectations and Preferences regarding Location-based Services - results of a survey. *2nd Symposium on Location Based Services and TeleCartography*. Vienna, Austria.

Zmud, W. R., & Apple, L. E. (1992). Measuring technology incorporation/infusion. *Journal of Product Innovation Management*, *9*, 148–155. doi:10.1016/0737-6782(92)90006-X

Zucker, L. (1977). The role of institutionalization in cultural persistence. *American Sociological Review*, *42*(5), 726–743. doi:10.2307/2094862

About the Contributors

Ada Scupola is an Associate Professor at the Department of Communication, Business and Information Technologies, Roskilde University, Denmark. She holds a Ph.D in business administration from Roskilde University, an MBA from the University of Maryland at College Park, USA, and a M.Sc. from the University of Bari, Italy. She is the editor-in-chief of *The International Journal of E-Services and Mobile Applications*. In the last years she has participated to a number of national and international projects often requiring close collaboration with industrial partners mostly focusing on user driven innovation in service industries, innovation and outsourcing in software services, the role of ICTs supply chain in facility services, and digitalization of services. Her research, counting circa 90 publications, has been widely published in several international journals, book chapters, and conference proceedings.

* * *

Mohammad Al Nabhan was born in Jordan, in 1983. He finished his Bachelor's degree in Computer Science in 2004, from Mu'tah University, Jordan. In 2006, he received his Master's degree in computer science where he focused his research in computer multimedia applications at Anglia Ruskin University (ARU), UK. Al Nabhan is currently perusing his PhD in computer science at Brunel University, UK, where he is focusing on designing and developing novel software algorithms for navigation oriented applications. He is a member of IET, IEEE, and the Royal Institute of Navigation (RIN). AL Nabhan's research interest includes mobile computing, Information Systems, multimedia applications QoS, and Location Based Services (LBS).

Suleiman Almasri received his Diploma in telecommunications engineering from The Telecom College- Jordan in 1992, BSc in computer science from Philadelphia University- Jordan in 1998, MSc in Computer Science from Amman Arab University- Jordan in 2005 and PhD degree in Computer Science from Anglia Ruskin University, UK, in 2009. Almasri worked for Orange Telecom- Jordan as a Systems Administrator between 1998 and 2006. He subsequently joined the Department of Computing and Technology at ARU where he commenced his doctoral research. His study was focusing on overcoming the issues related to Mobile Computing such as location-based services. The outcome of his research study has been published in 12 refereed papers.

John Breen is the Director of the Centre for Tourism and Services Research at Victoria University, Australia. He has conducted research in collaboration with industry partners and government agencies across a variety of industries including specific projects investigating enterprise education, tourism, accounting practices, child care, community banking, and agriculture. More generally he has undertaken research in the small business sector considering communication channels, learning and training, business growth, home based businesses, taxation, ICT, export, and entrepreneurship. John has served on industry committees and provided advice to governments on small business policy.

Stephen Burgess has research and teaching interests that include the use of ICTs in small businesses (particularly in the tourism field), the strategic use of ICTs, and B2C electronic commerce. He has received a number of competitive research grants in these areas. He has completed several studies related to website features in small businesses and how well websites function over time, including his PhD from Monash University, Australia (completed in 2002). He has authored/ edited three books and special editions of journals in topics related to the use of ICTs in small business and been track chair at the international ISOneWorld, IRMA, Conf-IRM, and ACIS conferences in related areas. More recently, Stephen has extended his research interests to include the use of websites by community based organisations. He has published in journals such as the *Journal of Information Science, Information Systems Frontiers,* the *International Journal of Tourism Research* and the *Journal of Hospitality, Marketing and Management.*

Peter Chan, PMP, is a business planning manager at Hewlett Packard (HP). He runs a program management office within HP.com division. Before his dotcom era, Peter spent over eleven years on the road as an architect and consultant providing High Availability and IT Service Management services to companies from Fortune 500 to small and medium sized businesses across the world. Peter also worked at Bell Northern Research, a subsidiary of Nortel, for five years where he was a Member of Scientific Staff. Peter received his Bachelor's degree in Computer Information Systems from Georgia State University. He holds a Master's degree and a e-Commerce certification both from University of Maryland University College. Peter's current research interests include operation management efficiency, project and program management excellence, and how to apply all these disciplines into the virtual world.

Vanessa Cooper is a Senior Lecturer in the School of Business Information Technology, RMIT University, Melbourne, Australia. She has also lectured at the University of Tasmania and Deakin University in Computing, Information Systems, and Accounting. Vanessa has worked as a Systems Engineer in the IT services industry in both Australia and the UK. Her research has investigated IT services, knowledge management, relationship management, and electronic business.

Pedro Cruz is currently a Professor of marketing and head of research unit at the ISG-Business School (Lisbon). He held visiting positions in University of Salamanca (Spain) and Polytechnic Institute of Leiria (Portugal). He received his PhD (business administration) from University of Salamanca (Spain). Previously, he had worked at the hospitality sector. He is interested in the management of IT adoption, CEM and consumer behaviour in electronic bank contexts. He has published several book chapters, seven referred journals (Journal of Euromarketing, among others), and in international conferences proceedings (EMAC, ANZMAC, ICE-B, IACIS).

Naoufel Daghfous, PhD, is a Marketing Professor at the "Université du Québec à Montréal" (UQÀM) since 1994. He published many papers in Canadian, American, and French marketing reviews. He is the author of many books in consumer behavior, marketing management, and quantitative methods. His research domains are basically in international marketing, innovations management, and consumer behavior. For the last two decades, he collaborates actively in research project with the public and private organisations in international business and e-commerce.

Shauntell Durant is a consultant with 10+ years of significant IT and Project Management experience and over 15+ years working in the public sector healthcare industry which include expertise in supporting all phases of software development life cycle and project management life cycle. Shauntell has a Master's of Science in Information Technology and an associate certificate in Project Management. Shauntell is a member of Project Management Institute (PMI) organization and Silver Springs Comment Chapter. She has been selected as an "approved" volunteer contributor to participate in upcoming Exam Development Meetings for PMI to review, edit, and write questions/items for use in future PMI credential exams. In Shauntell's spare time, she enjoys working out, watching and playing sports, listening to music, and studying for the PMP exam. Shauntell believes that she has an obligation to the community so her goal is to tutor kids in math.

Andreas Eckhardt is an Assistant Professor at the Institute of Information Systems of Goethe University Frankfurt and a postdoctoral researcher at the "Centre of Human Resources Information Systems (CHRIS)", a research cooperation project of two German Universities. Prior to this, he received his PhD from Goethe University Frankfurt and worked as an HR project manager at Daimler AG in Taiwan. As part of his work at CHRIS he is co-author of the book "Recruiting 2010" and the study series "Recruiting Trends" as well as "Bewerbungspraxis." Additionally he consults companies in HRIS related projects and serves as a lecturer at congresses and seminars. His research interests include social influence in IT adoption, IT resistance, user personality, and computer personnel-related issues like recruiting, retaining, and skilling the IT workforce. He has published numerous articles in journals including Journal of Information Technology, MIS Quarterly Executive, Information Systems Frontiers, Business Process Management Journal, and Business & Information Systems Engineering, as well as in the proceedings of various conferences (including ICIS and ECIS). He is also the author of one book and several book chapters.

Carlos Flavián holds a PhD in business administration and is Professor of marketing in the faculty of Economics and Business Studies at the University of Zaragoza (Spain). His research has been published in several academic journals, specialized in marketing (European Journal of Marketing, Journal of Consumer Marketing, Journal of Strategic Marketing, International Journal of Market Research, etc.) and new technologies (information & management, industrial management and data systems, internet research, etc.). He is a member of the editorial board of the Industrial Marketing Management, the Journal of Retailing and Consumer Services, the International Journal of Services and Standards, and the Journal of Marketing Communications.

Verna Gall, PMP, has 13 years of experience in Information Technology. For the last 8 years, she has practiced the discipline of Web analytics for a financial services firm. She also has e-commerce experience as an EDI Analyst within the health care industry. Verna holds a Graduate Project Management Certificate and is actively pursuing a Master's of Science in Information Technology from the University of Maryland University College.

Călin Gurău is an Associate Professor of Marketing at GSCM - Montpellier Business School, France, since September 2004. His present research interests are focused on marketing strategies for high-technology firms, Internet marketing, and sustainable development. He has published more than 50 papers in internationally refereed journals, such as International Marketing Review, Journal of Consumer Marketing, Journal of Marketing Communications, et cetera.

Raquel Gurrea holds a PhD in business administration and is Assistant Professor in the faculty of Economics and Business Studies at the University of Zaragoza (Spain). Her main research lines are online consumer behavior, multichannel distribution, and the analysis of the advantages and limitations of the Internet in the development of the economic activity. Her work has been published in several journals, such as International Journal of Market Research, Information & Management, Journal of Retailing and Consumer Services, Internet Research, Online Information Review, and Journal of Targeting, Measurement and Analysis for Marketing.

Timber Haaker received his PhD in mathematical analysis. He is a senior researcher at Novay, working in the Networked Business group since 2003. He focuses on methods and tools for supporting ICT enabled business innovation. He has written several papers and book chapters on business models, service innovation, and evaluation studies. His research interests include networked innovation in financial services and healthcare.

Anders Henten is Professor at Center for Communication, Media and Information Technologies (CMI) at the Department of Development and Planning at Aalborg University in Copenhagen. He is a graduate in communications and international development studies from Roskilde University in Denmark (1989) and holds a Ph.D. from the Technical University of Denmark (1995). Anders Henten has worked professionally in the area of communications economy and policy for more than 20 years. He holds academic positions as member of the scientific committee of the European Communications Policy Research Conference, member of the board of directors of the International Telecommunication Society, editor of the Nordic and Baltic Journal of Information and Communication Technologies, associate editor of Telematics and Informatics, member of the editorial board of Telecommunications Policy, and member of the editorial secretariat of Communications & Strategies.

Ziad Hunaiti was born in Amman, Jordan, in 1976. He received Diploma in Aeronautical Communications Engineering in 1996 from Queen Noor Civil Aviation College, Jordan, the BSc degree in Electrical and Electronics Engineering in 2001 form Near East University, Cyprus, and PhD degree in Systems Engineering from Brunel University, UK, in 2005. He also worked as Instructor between 1996 and 2002 in both Queen Noor Civil Aviation College, Jordan and Near East University, Cyprus. Between 2002 and 2008 he worked as researcher at Brunel University and Anglia Ruskin University, UK. He currently works as Lecturer at Anglia Ruskin University, UK. He has published over 50 Journal papers, book chapters, and conference papers. Dr. Hunaiti is a Fellow of the Higher Education Academy (HEA), UK. His main research interests include networking, satellite navigation systems, wireless networks, mobile Information Systems, m-health, m-learning, location based services, and communication systems applications.

Carina Ihlström Eriksson (PhD) is Assistant Professor in informatics at Halmstad University and leads the Media IT research group within the EIS (embedded and intelligent systems) research environment. Her research interest lies within the fields of digital publishing, and she focus on new innovative media and end-user involvement in the innovation process. She is the leader of Halmstad Living Lab and several research projects. Her research is published in international journals such as IT & People, Electronic Markets, Journal of Organizational Computing and Electronic Commerce, and International Journal of Services Technology and Management.

Marijn Janssen is Director of the interdisciplinary SEPAM Master programme and an Associate Professor within the Information and Communication Technology section of the Technology, Policy and Management Faculty of Delft University of Technology. He has been a consultant for the Ministry of Justice and received a Ph.D. in Information Systems (2001). He serves on several editorial boards (including International Journal of E-Government Research, Government Information Quarterly, and Information Systems Frontiers), has many research projects, and is involved in the organization of a number of conferences. His research interests are in the field of e-government, design, orchestration, and shared services. He was ranked as one of the leading e-government researchers in 2009 and published over 180 refereed publications. More information: www.tbm.tudelft.nl/marijnj.

Joachim Jean-Jules is a GRECUS junior Fellow and an Instructor in the Business school of the Université de Sherbrooke. His research areas are strategic and performance management and their applications through business process analysis and improvement, balance scorecard development, and management of technology and innovation. His works have been published in journal and proceedings of several international conferences including IJESMA, IRMA, Conf-IRM, AIM, and AIMS. He has been the recipient of numerous prestigious scholarships and fellowships, including the Fonds Québécois de Recherche sur la Société et la Culture (2007-2009).

Mirella Kleijnen currently is an Assistant Professor as the Vrije Universiteit Amsterdam. She obtained her PhD from the Faculty of Economics and Business Administration, Maastricht University, the Netherlands. Her PhD research focused on consumer evaluations of mobile service innovations. Moreover, she is interested in adoption versus resistance of innovations, interactive media, and services marketing, thereby focusing on consumer perspectives. Her work has been published in Journal of Retailing, Journal of Services Research, Journal of Interactive Marketing, Journal of Financial Services Marketing, Mobile Commerce: Technology, Theory, and Applications, International Journal of Service Industry Management, and Total Quality Management. She has served as an "ad hoc" reviewer for several journals, among which Journal of Marketing, Journal of the Academy of Marketing Science, Journal of Interactive Marketing, and Journal of Electronic Commerce Research.

Tommi Laukkanen is currently a Professor of marketing at the Department of Economics and Business Administration, University of Joensuu (Finland). He received his PhD from Lappeenranta University of Technology. His research interests include innovation adoption and resistance, consumer behavior, and tourism marketing. He has published 14 refereed international journal articles in journals such as Journal of Consumer Marketing, International Journal of Information Management, and International Journal of Bank Marketing. He serves in the editorial boards of the Journal of Euromarketing and International Journal of Electronic Finance.

Sven Laumer is graduated research assistant and doctoral student at the department for Information Systems and Services. He is part of the research project "Centre of Human Resources Information Systems" and is interested in the support of Information Systems for HR tasks in organizations. As part of his work at CHRIS, he is co-author of the study series "Recruiting Trends" and " Bewerbungspraxis," serves as consultant, and gives talks at congresses and seminars about the usage of IT in recruiting. His research results about IT adoption and usage, IT resistance, social influence and IT, and the IT workforce have been published among others in Journal of Information Technology, Information Systems Frontiers, Business & Information Systems Engineering, and MIS Quarterly Executive, as well as in the proceedings of various conferences (including ICIS and ECIS).

Sharman Lichtenstein is an Associate Professor in the School of Information Systems, Deakin University, Melbourne, Australia. She has spent more than thirty years in industry, academia, and consulting, in Information Technology and Systems. Sharman has been a member of faculty at RMIT, Monash University, and Deakin University. Her research has focused on information security management, Internet use in the workplace, knowledge management, and electronic business. Sharman has published widely and is currently a member of Editorial boards for leading international journals.

Christian Maier is a doctoral student at the department for Information Systems and Services and a PhD fellow of the German state of Bavaria. Prior to this, he graduated in Information Systems from Bamberg University. He works in the research project "Centre of Human Resources Information Systems" and is particularly interested in the integration of psychological aspects of IT users' personality into Information Systems research. His research about individual social networking, IT adoption, IT resistance, and user personality has been published in various proceedings such as ECIS, DIGIT, WCSB, and HRIS.

Pablo Muñoz is Professor of marketing and head of at business administration department at the University of Salamanca (Spain). He holds a PhD from the University of Oviedo (Spain). His research interests include retailing, electronic commerce, and consumer behaviour. He has published over 20 articles in refereed national and international journals, including in the Journal of Retailing and Journal of Retailing and Consumer Services.

Hanne Westh Nicolajsen holds a position as Associate Professor at Department for Communication and Psychology at Aalborg University in Copenhagen. Her research interests include organizational implementation and use of ICT especially with focus on knowledge management and computer-mediated communication. Innovation in services is another research area of interest; here, involvement of users through the use of new ICT is of particular interest. HWN holds a Ph.D. from the Technical University of Denmark.

Carlos Orús is predoctoral grant holder in the faculty of Economics and Business Studies at the University of Zaragoza (Spain). His main research lines are online consumer behavior and the relevance of website design for developing of the economic activity on the Internet. His work has been published in several journals such as Electronic Government, an International Journal and International Journal of Services and Standards, and has been presented in national and international conferences with high impact such as International Conference on Recent Advances in Retailing and Services Science, International Conference on Corporate and Marketing Communications, and Management International Conference.

Mahesh S. Raisinghani is an Associate Professor in the Executive MBA program at the TWU School of Management. He is a Certified E-Commerce Consultant (CEC) and a Project Management Professional (PMP). Dr. Raisinghani was awarded the 2008 Excellence in Research & Scholarship award and the 2007 G. Ann Uhlir Endowed Fellowship in Higher Education Administration. He was also the recipient of TWU School of Management's 2005 Best Professor Award for the Most Innovative Teaching Methods; 2002 research award; 2001 King/Haggar Award for excellence in teaching, research, and service; and a 1999 UD-GSM Presidential Award. His research has been published in several academic journals such as IEEE Transactions on Engineering Management, Information & Management, Information Resources Management Journal, International Journal of Innovation and Learning, Journal of IT Review, Journal of Global IT Management among others and international/national conferences. Dr. Raisinghani is included in the millennium edition of Who's Who in the World, Who's Who among Professionals, Who's Who among America's Teachers, and Who's Who in Information Technology.

Risto Rajala researches and lectures at the Helsinki School of Economics. He is also Research Manager of the centre of expertise in service research and education at the Aalto University. His research interests are focused on service systems, innovation strategy, and business models. His work has been published in a number of journals including California Management Review, Industrial Marketing Management and Journal of Systems and Information Technology, and Journal of Theoretical and Applied Electronic Commerce Research.

Jari Salo is currently a Professor of Information and Communication Business at the University of Oulu. He received his DSc (Econ. & Bus. Adm.) from the University of Oulu. Before joining the faculty, he worked in the clothing and retailing industry. Salo has published in Industrial Marketing Management, Journal of Business and Industrial Marketing, Business Process Management Journal, Online Information Review, and other journals. Besides journal articles, Salo has also published several book chapters and participated in the leading conferences in the marketing and Information Technology area. Salo is the editor-in-chief of Journal of Digital Marketing. His current research interests include business relationship digitization, electronic commerce, mobile marketing, and new product development.

Eliamani Sedoyeka is a PhD Candidate at Anglia Ruskin University researching low cost ways to provide connectivity to rural and remote parts developing countries. He is researching new generation wireless technologies like WiMAX and the way they can be adopted by communities facing digital isolation and financial difficulties. He also has research interests in QoS, SLA and data mining, and data warehousing techniques.

Ross Smith is a Professor in the School of Business Information Technology, RMIT University, Melbourne, Australia. Since completing his doctoral studies in computational physics, in 1977, he has been a member of faculty at the Australian National University, Swinburne University of Technology, Deakin University and, since January 2007, at RMIT University. Ross also worked for 4 years as a Senior Research Scientist with the Defence Science and Technology Organisation (DSTO). Ross has researched and published widely in systems methodologies, requirements engineering, and knowledge management. Research in the area of knowledge management, together with colleagues at Deakin University and RMIT, has focussed on knowledge transfer via Web-based self-service systems.

Jesper Svensson is Lecturer in informatics at Halmstad University and is also doing PhD work within the Media IT research group. His research interests are related to open user centered innovation processes with a specific interest in living labs. He has been doing research work in several projects such as "DigiNews," "Designing the future e-newspaper," and "Designing Ubiquitous Media Services by Action Research." Currently he is co-leading several living lab research activates at Halmstad Living Labs.

Marije Teerling is a currently working as a researcher at Novay in Enschede, Netherlands. She works on the intersection of marketing and ICT, for instance on research relating to (re)directing customers to specific channels, multichannel management, the adoption of mobile services, and ICT innovations in the way we shop. She holds a PhD in Marketing from the Faculty of Economics of the University of Groningen, the Netherlands. She has presented at numerous conferences – both academic and business - related to e-commerce, marketing, and e-government, such as the International Conference on Electronic Business and Marketing Science. She has published several papers – both in peer reviewed outlets and in popular magazines – and book chapters.

Roy Toffoli, PhD, is currently Professor of Marketing at the "Université du Québec à Montréal" (UQAM), having joined the university in 1997. He has over twenty years of experience in the teaching of Marketing, having taught numerous courses in the field such as: Marketing management, international marketing, consumer behaviour, marketing of high technology products and innovations, and advertising. Professor Toffoli has also been a member of the faculty at McGill University. He is presently also the coordinator for the executive MBA program in the Management of Technology, and director of the Center for research in marketing communications at UQAM. Prior to his academic career, he served as the administrative director of the Industrial Innovation Center at the École Polytechnique de Montréal. He is the author or co-author of numerous papers, conference proceedings, and cases, some of which have appeared in the Journal of Business Research, Advances in Consumer Research, International Journal of Advertising, Journal of Applied Social Psychology, and the Canadian Journal of Administrative Sciences. He has also presented at numerous scientific conferences such as: the Association for Consumer Research, the Academy of Marketing Science, and the Administrative Science Association of Canada.

Elissar Toufaily has an MBA from the "Université du Québec à Montréal" (UQÀM) (Canada) and "École Supérieure et Internationale de Gestion" (Lebanon), and is currently a PhD candidate in marketing at the School of Business at the "Université du Québec à Montréal". Before graduating, she was a lecturer at the School of Business at the University of UQAM, and researcher at the chair of management of financial services at UQÀM. She has participated in various international conferences in marketing and management of Information Systems such as International Business Information Management Association IBIMA, Administrative Sciences Association of Canada ASAC, Association Française de Marketing, Association Internationale du Management Stratégique AIMS, Association Tunisienne en Marketing ATM, International Conference on Information System ICIS, and others, and has published several research papers and conference proceedings.

Tuure Tuunanen is a Senior Lecturer in the Department of Information Systems and Operations Management at the University of Auckland. He holds a D.Sc. (Econ) from the Helsinki School of Economics. His current research interests lie in the areas of IS development methods and processes, requirements engineering, risk management, and convergence of IS and marketing disciplines, specifically in design of interactive consumer services and products. His research has been published in Information & Management, Journal of Database Management, Journal of Information Technology Theory and Application, Journal of Management Information Systems, and Technology Analysis and Strategic Management. Tuunanen was recently appointed as editor-in-chief for Journal of Information Technology Theory and Application – A Publication of The Association for Information Systems.

Alain O. Villeneuve holds a DBA degree in Information Systems from Boston University. As a practitioner from 1975 to 1990, Dr. Villeneuve has been involved in all stages of systems development and has actively participated in hundreds of systems implementation. Many IS projects in which he participated were in the area of healthcare management and networking. He has joined Université de Sherbrooke as a faculty in 1996. Dr. Villeneuve's current research interests are in the area of knowledge management, knowledge creation and transfer, knowledge management supporting technologies, research methods, human expertise, and decision-making. His works have been published in Decision Support Systems and Information and Management among others in addition to being presented at several international conferences. Dr. Villeneuve was also an active member of KTE (Knowledge Transfer and Exchange) committee of the CIHR (Canadian Institutes for Health Research) for several years.

Henny de Vos is a researcher at Novay, working together with companies on creating viable and feasible ICT-based services. To be able to create such services, she focuses on their business models from four perspectives, e.g. the service and its value proposition, technology, organizations arranged in a value network, and costs and benefits. She has written several books and papers on a method to achieve viable services.

Roumen Vragov received his PhD in economic systems design with minor in management of Information Systems at the University of Arizona. He spent a year as an IFREE Post-Doctoral Research Fellow in the area of combinatorial auctions at George Mason University. He is currently an Assistant Professor of Computer Information Systems and Director of the Information Systems Experimental Laboratory at the Zicklin School of Business at Baruch College, CUNY. His current research interests include design of Internet auctions, e-government applications, and other decentralized exchange systems.

Mika Westerlund researches and lectures at the Helsinki School of Economics. His research interests are focused on service innovation, business strategy, and interorganizational networks. His work has been published in a number of journals including California Management Review and Industrial Marketing Management, the International Journal of Entrepreneurship and Innovation, and Journal of Business and Industrial Marketing.

Index

A

B

C

D

E

F

G

H

I

K

L

M

T

U

V

W

Y

CPSIA information can be obtained at www.ICGtesting.com
264416BV00003B/1-76/P

9 781609 606077